Hormonally Induced Changes in Mind and Brain

Curt Richter's "multiple activity cage" of 1922. At The Johns Hopkins University, Richter pioneered the study of the hormonal regulation of behavior. Richter is legendary for his simple methodological innovations, from how to catch wild rats in Baltimore, to how to remove the parathyroid gland. He understood that methodological innovations exist at many levels of analysis. He also understood that synthesis requires working at endocrine, anatomical, and behavioral levels of analysis.

Hormonally Induced Changes

in Mind and Brain

Edited by

Jay Schulkin

Behavioral Neuroscience Unit
Neuroendocrinology Branch
National Institute of Mental Health
Bethesda, Maryland

ACADEMIC PRESS, INC.

A Division of Harcourt Brace & Company

San Diego   New York   Boston   London
Sydney   Tokyo   Toronto

Academic Press, Inc.
1250 Sixth Avenue, San Diego, California 92101-4311

United Kingdom Edition published by
Academic Press Limited
24–28 Oval Road, London NW1 7DX

Library of Congress Cataloging-in-Publication Data

Hormonally induced changes in mind and brain / edited by Jay Schulkin.
    p.   cm.
    Includes bibliographical references and index.
    ISBN 0-12-631330-X
    1. Neuroendocrinology. 2. Psychoneuroendocrinology.
  I. Schulkin, Jay.
    [DNLM: 1. Behavior-physiology. 2. Hormones-physiology.
  3. Neurotic Disorders--etiology. 4. Endocrine Glands-
-physiopathology. WK 102 H81157 1993]
  QP356.4.H685 1993
  612.8'22--dc20
  DNLM/DLC
  for Library of Congress   93-9947
                                         CIP

PRINTED IN THE UNITED STATES OF AMERICA
93 94 95 96 97 98   QW   9 8 7 6 5 4 3 2 1

This book is dedicated to Curt Richter. He is no household name, but he is perhaps America's greatest psychobiologist. He did a great deal of work on the hormonal regulation of the brain and behavior during his long career at Johns Hopkins University.

The book is also now dedicated to Alan Epstein of the University of Pennsylvania. He died in a tragic accident as this book was near completion. Though not a contributor to this book he is well represented by friends, students, colleagues, and teachers. Alan was a great fan of Curt Richter and spent a lifetime inquiring into relationships between hormones and their effects on behavior and the brain. He will be missed very much. He was a loyal friend and dedicated researcher who loved science. His office was next to mine for over 12 years. We shared common interests and collaborated together on many experiments. He was generous and provided laboratory space for me during this period, and was a valued senior colleague.

Jay Schulkin

Contents

Neuroendocrinology of Sodium Hunger:
Angiotensin, Corticosteroids, and Atrial
Natriuretic Hormone   13

Cholecystokinin: A Neuroendocrine Key to
Feeding Behavior   51

CHAPTER 3

Sexual Behavior: Endocrine Function and Therapy 71

Kim Wallen and Jennifer Lovejoy

CHAPTER 4

Hormones and Aggression 99

Allan Siegel and Melissa K. Demetrikopoulos

CHAPTER 5

Anabolic Steroids: Misuse or Abuse? 129

Lynn H. O'Connor and Theodore J. Cicero

CHAPTER 6
_____

Adrenal Steroid Effects on the Brain: Versatile Hormones with Good and Bad Effects   157

Bruce S. McEwen, Randall R. Sakai, and Robert L. Spencer

CHAPTER 7
_____

Psychoimmunology: The Missing Links   191

David Saphier

CHAPTER 8

Oxytocin and the Neuroendocrine Basis of Affiliation   225

Thomas R. Insel

CHAPTER 9

Neurohormones in Depression and Anxiety   253

Margaret Altemus and Philip W. Gold

CHAPTER 10

Hormones, Rhythms, and the Blues   287

Donald L. McEachron and Jonathan Schull

CHAPTER 11

Bright Light Therapy and Circadian Neuroendocrine
Function in Seasonal Affective Disorder   357

David Avery and Kitty Dahl

Contributors

Numbers in parentheses indicate the pages on which the authors' contributions begin.

Margaret Altemus (253), Laboratory of Clinical Science, Division of Intramural Research Programs, National Institute of Mental Health, Bethesda, Maryland 20892

David Avery (357), Harborview Medical Center, and Department of Psychiatry and Behavioral Sciences, University of Washington School of Medicine, Seattle, Washington 98104

Theodore J. Cicero (129), Department of Psychiatry, Washington University School of Medicine, St. Louis, Missouri 63110

Kitty Dahl (357), Harborview Medical Center, and Department of Psychiatry and Behavioral Sciences, University of Washington School of Medicine, Seattle, Washington 98104

Melissa K. Demetrikopoulos (99), Department of Neuroscience, New Jersey Medical School and the Graduate School of Biomedical Sciences, University of Medicine and Dentistry of New Jersey, Newark, New Jersey 07103

Steven J. Fluharty (13), Departments of Animal Biology and Pharmacology, School of Medicine and Veterinary Medicine, Institute of Neurological Sciences, University of Pennsylvania, Philadelphia, Pennsylvania 19141

James Gibbs (51), Department of Psychiatry, Cornell University Medical College, and Edward W. Bourne Behavioral Research Laboratory, The New York Hospital—Cornell Medical Center, White Plains, New York 10605

Danielle Greenberg (51), Department of Psychiatry, Edward W. Bourne Behavioral Research Laboratory, The New York Hospital—Cornell Medical Center, White Plains, New York 10605

Thomas R. Insel (225), Laboratory of Neurophysiology, National Institute of Mental Health, Poolesville, Maryland 20837

Jennifer Lovejoy (71), Department of Psychology and, Yerkes Regional Primate Research Center, Emory University, Atlanta, Georgia 30322

Donald L. McEachron (287), Department of Psychiatry, University of Pennsylvania School of Medicine, and Biomedical Engineering and Science Institute, Drexel University, Philadelphia, Pennsylvania 19104

Bruce S. McEwen (157), Laboratory of Neuroendocrinology, The Rockefeller University, New York, New York 10021

Lynn H. O'Connor (129), Department of Psychiatry, Washington University School of Medicine, St. Louis, Missouri 63110

Randall R. Sakai (157), Laboratory of Neuroendocrinology, The Rockefeller University, New York, New York 10021

David Saphier (191), Departments of Pharmacology and Therapeutics, and Psychiatry, Louisiana State University Medical Center, Shreveport, Louisiana 71130

Jay Schulkin (1, 13), Behavioral Neuroscience Unit, Neuroendocrinology Branch, National Institute of Mental Health, Bethesda, Maryland 20892

Jonathan Schull (287), Department of Psychology, Haverford College, Haverford, Pennsylvania 19041

Allan Siegel (99), Department of Neuroscience, New Jersey Medical School, and the Graduate School of Biomedical Sciences, University of Medicine and Dentistry of New Jersey, Newark, New Jersey 07103

Gerard P. Smith (51), Department of Psychiatry, Edward W. Bourne Behavioral Research Laboratory, The New York Hospital—Cornell Medical Center, White Plains, New York 10605

Robert L. Spencer (157), Laboratory of Neuroendocrinology, The Rockefeller University, New York, New York 10021

Kim Wallen (71), Department of Psychology and, Yerkes Regional Primate Research Center, Emory University, Atlanta, Georgia 30329

Preface

What we have in this volume, appropriately dedicated to Curt Richter and Alan Epstein, is an exciting progress report that takes us from Claude Bernard's concept of the "milieu interieur," through Cannon's homeostasis and Richter's self-regulatory behavior, to a fascinating review of our current understanding of the psychobiology of stress, adaptation, and the clinical relevance of stress in human disorder. The brain, especially the hypothalamus and the neuroendocrine system, is key to the operation and regulation of all complex behaviors. At the physiological level, there is regulation of the internal environment. At the behavioral level, there is the development of the integrated appetitive and consummatory motivated behavior, so crucial to the survival of the individual and the species. One consequence of these mechanisms is that the consummatory behavior produces a rewarding state in the brain that both reinforces and shapes appetitive behaviors, producing new instrumental or operant learning, thus expanding the range of behaviors that can become adaptive, greatly increasing the possibilities of survival. Viewed in this paradigm, reward and reinforcement are hedonic states that reveal themselves in animals through the appetitive instrumental and operant responses, whether we are dealing with the reward value of certain tastes, of water for the thirsty animal, of sex, of maternal behavior, or even of aggression (Stellar, 1980). In humans, the same mechanisms are at work, only now we have the possibility of subjective reports of the hedonic states of pleasure, displeasure, and pain that parallel the behavioral states of approach and avoidance. Thus the hormonal pathways discussed in this volume lead not only to brain and behavior, but also to the subjective awareness of mind, to its hedonic processes, and to their disturbances in "dis-ease."

In historical perspective, we've come a long way since Richter introduced the concept of self-regulatory behavior to refer to behaviors essential to the preservation of the internal environment (Richter, 1942–43). He was the first to study these behaviors systematically by applying the methods of endocrine gland extirpation and hormone replacement therapy, first to impair the homeostatic mechanism and then to restore it. Nowhere is his concept more clearly illustrated than in the study of salt appetite (Richter, 1936). In this case, adrenalectomy makes it impossible for an animal like the rat to retain salt at the kidney, so it ingests large quantities of 3% NaCl solutions and keeps itself alive. Replacement therapy, using the adrenal hormone aldosterone, restores sodium retention and eliminates the need to ingest large amounts of salt. To make the example more dramatic, Wilkins and Richter (1940) reported the case of a three-year-old boy with tumors that destroyed his adrenal glands who kept himself alive by avidly ingesting salt, even by the spoonful, until he was placed on a strict hospital diet and unfortunately died.

Richter, of course, had his predecessors in Darwin (1873), Claude Bernard (1865), and Cannon (1932), all of whom recognized the importance of adaptive behavior in the preservation of the internal environment and in the survival of the organism. As a matter of fact, at the turn of the century, Andre Mayer (1900) demonstrated the basic principles of regulatory behavior in his study of thirst and drinking behavior. But it was Richter who generalized and who investigated a wide range of self-regulatory behaviors: specific hungers for salt, calcium, carbo-hydrate, fat, and protein, dramatically illustrated in his cafeteria feeding experi-ments in which rats selected a balanced diet from an array of macronutrients, vitamins, and minerals, supplied in separate cups and solutions (Richter et al., 1937). Again, the phenomenon was demonstrated in humans, this time by Davis (1927), using infants who selected nutritionally balanced diets from an array of foods. To top it all off, Richter showed that these behaviors had a circadian rhythm, driven by an internal clock. Furthermore, he wrote that the circadian rhythm contributed to the control of mood and psychiatric disorder (Richter, 1960).

Others picked up the thread of Richter's work and extended it to other behaviors in which hormones were important, such as sexual behavior, maternal behavior, and aggression. It was Lashley (1938), however, who conceived that the hormones had to act through the brain to arouse behavior; he conceptualized a central excitatory mechanism that the hormones aroused. From then on, the field took off. Morgan (1943) thought that there had to be a central inhibitory state as well as a central excitatory state. Stellar (1954) tried to relate these states to hypothalamic structures and built on Lashley's idea that the effects of hormones and sensory stimulation were additive in the central nervous system, producing motivational states, whether in service of homeostasis or not.

A singular exemplar of these concepts was Alan Epstein, who gave us such clear conceptions of the nature of ingestive behavior and, in the case of salt appetite, showed that the avidity for salt depended on the synergistic action of the hormones angiotensin and aldosterone in the brain (Epstein, 1984). His work brought us to the next generation of research and the fruits of much of what he exemplified are reported in this volume.

The field of neuroendocrinology has expanded greatly, and its relationship to behavior is becoming more and more explicit and, at the same time, more and more diverse. The explicit mechanisms revealed are demonstrated everywhere in this volume, for now we know a great deal about the specific structures in the hypothalamus and the broader limbic system that are involved and we know some of the circuits extending between the brainstem and the forebrain that make up the neural systems. Furthermore, within these systems there are specific neurotransmitters and neuromodulators and dedicated receptors on these neurons where the action of hormones is generated, permitting the hormone to have more than one effect, depending on which receptors are involved and their locations.

Interested readers will also find explored within these pages a diverse range of behaviors affected by neuroendocrine mechanisms, including behaviors that are not directly involved in homeostasis, but that are important in successful adaptation and the survival of the species (e.g., reproductive behaviors). Sexual behavior, maternal behavior, filial behavior, and aggression are all topics that fascinated Richter. In addition, he was fascinated by stress and how the process of domestication reduced the stress reaction in the rat, particularly in the adrenal glands. Now we know not only that stress affects the endocrine system, but also that the consequence of these persistent endocrine effectors can lead to changes in other organ systems, as Selye (1946) has pointed out, leading to disease states. Many organ systems are affected by stress, including the respiratory system, the cardiovascular system, the immune system, the adipose tissue system, the joints, and the brain itself, including normal cell death in the hippocampus. As explicit neuroendocrine pathways become elucidated we learn, step by step, about the mechanisms whereby stress leads to disease and the important role that adaptive control in the psychological sense can play in the final common path of dysfunction.

In those early human studies, stimulated by Selye's observations and the experience of the Second World War, urinary steroids were measured in many subjects, from Harvard oarsmen to polar explorers, and indeed there were many surprises. In humans, apparently, it is not easy to objectively assess the stressor; personal meaning and novelty appear to be extremely important in opening the pathway to physiological arousal. Learning and contingent (i.e., hedonic) reinforcement play a major role in shaping the stress response, and in the young such learning ideally takes place within the context of maternal affiliation and social bonding.

Now, many years later, with the enormous advances in our molecular and cellular understanding of these mechanisms we begin to sense an integrative science of mind and brain. Furthermore, neuroendocrinology and our expanding understanding of brain, environmental challenge, and the immune system have become essential windows through which we can increasingly focus a clinical science of behavior. This volume makes an extraordinary contribution in this regard; each chapter can stand alone but when taken together the whole enhances our understanding of both fundamental processes and clinical dysfunction. Thus we can draw upon the insights of a molecular understanding of the effects of steroids on the brain as we approach the clinical challenges of how chronic stress can lead to depression and anxiety, and how the fundamental challenge of a periodic planet and the consequent circadian and seasonal variation must be understood if we are to grapple with the episodic nature of human mood and its disorder.

This volume provides just what it sets out to explicate and celebrate in the lives of Curt Richter and Alan Epstein—a fascination with the dynamics of living systems carefully researched and eruditely set forth so that serious students may continue to build for themselves and in turn educate us all.

References

Bernard, C. (1957). "An introduction to the study of experimental medicine." New York: Dover Press, Reprint (original published in 1865).

Cannon, W. B. (1932). "The wisdom of the body." New York: Norton.

Darwin, C. (1873). "The expression of the emotions in man and animals." New York: Appleton.

Davis, C. M. (1927). Self selection of diet by new weaned infants. *American Journal of Diseases of Children 36*, 651–679.

Epstein, A. N. (1984). Mineralocorticoids and cerebral angiotensin may act together to produce sodium appetite. *Peptides, 3*, 493–494.

Lashley, K. S. (1938). An experimental analysis of instinctive behavior. *Psychological Review, 45*, 445–471.

Mayer, A. (1900). Variation de la tension asmotique du song chez les animous prives de liquedes. *Compte Rendu des Seances de la Societe le Biologie, 52*, 153–155.

Morgan, C. T. (1943). *Physiological Psychology*, New York: McGraw Hill.

Richter, C. P. (1936). Increased salt appetite in adrenalectomized rats. *American Journal of Physiology. 115*, 155–161.

Richter, C. P. (1942-43). Total self-regulatory functions in animals and human beings. *Harvey Lecture Series, 38*, 63–103.

Richter, C. P. (1960). Biological clocks in medicine and psychiatry: Shock-phrase hypothesis. *Proceedings of the National Academy of Sciences, 46*, 1506–1530.

Richter, C. P., Holt, L. E., & Barelore, B. (1937). The effect of self-selection of diet-food (protein, lacbohydrates, and fats), minerals and vitamins, on growth, activity, and reproduction in rats. *American Journal of Physiology, 119*, 388–389.

Selye, H. (1946). The general adaptation syndome and the diseases of adaptation. *Journal of Clinical Endocrinology. 6*, 117–196.

Stellar, E. (1954). The physiology of motivation. *Psychological Review, 61*, 5–22.

Stellar, E. (1980). Brain mechanisms and hedonic processes. *Acta Neurobiologiae Experimentalis, 40,* 313–324.

Wilkins, L. & Richter, C. P. (1940). A great craving for salt by a child with cortico-adrenal insufficiency. *Journal of the American Medical Association, 114,* 866–686.

Eliot Stellar
Peter Whybrow
University of Pennsylvania

Acknowledgments

I thank my wife, April, and our little girl, Danielle, for my joy; and Bob and my friends for their encouragement of my scientific pursuits. I thank Hal Pashler, Paul Rozin, John Sabini, and David Sarokin for their comments on the book.

I also thank Bruce McEwen. Over a 5-year period he and I were Co-Principal Investigators of a section of a Program Project Grant from the National Institute of Mental Health. It was during this time, and through my association with his laboratory at Rockefeller University, that my broader appreciation of the role of hormones in the expression of behavior was kindled. He in fact encouraged me to put this book together and is a much valued colleague and friend.

Introduction

Jay Schulkin

Behavioral Neuroscience Unit, Neuroendocrinology Branch
National Institute of Mental Health

In this introductory chapter, I have three aims. The first is to convey some facts about hormones, brain, and behavior. The second is to inform the reader somewhat about the contents of the book. The third is make some comments about Curt Richter and Alan Epstein, to whom the book is dedicated.

I. Historical Context

Hormonal messages are phylogenetically ancient. Consider that various hormones in insects produce molting and metamorphoses, or evoke odors that attract other conspecifics. Hormones are, therefore, found in invertebrates and vertebrates. Hormones are distributed diversely in the body. They are found in skin, heart, and brain, as well as elsewhere.

Hormones generally are separated into two main classes. Peptide hormones typically act on membranes of cells to produce their effects; steroid hormones act on the nucleus of cells to promote protein synthesis. The first effect is usually fast and the second slow. Although this distinction is beginning to blur somewhat (Wehling et al., 1992), it is still to some extent true. Both peptide and steroid hormones have profound effects on behavior.

Hormones such as insulin, cholecystokinin, or bombesin, all essential for food regulation, are produced in gastrointestinal organs. Hormones such as aldosterone, angiotensin, and atrial natriuretic factor, essential for hydromineral and water balance, are produced in the adrenal glands, kidney, and heart. Hormones such as vitamin D, parathyroid, thyroid, and melatonin, essential for calcium balance, circadian rhythmicity, and mental health, are produced in the skin, parathyroid, thyroid, and pineal glands. Hormones such as testosterone and estrogen, produced in the testes and ovaries, are fundamental to gender and to expressions of aggression. Finally, adrenalcorticotropic hormone (ACTH) and corticosterone, essential in the adaptation to challenging situations and glucose transport, are produced in the pituitary and adrenal gland.

Hormones once were defined as chemical messengers that are blood-borne to distal organs where they affect development and function. Now we know that hormones can act proximally and distally. We also know that many peptide hormones produced in the periphery also are produced in the central nervous system. For example, peptide hormones such as insulin, oxytocin, angiotensin, cholecystokinin, natriuretic factor, parathyroid, and calcitonin are produced in endocrine cells in the periphery and in the brain. In the periphery, we call them hormones. In the brain, we call them neurotransmitters or neuromodulators. Moreover, a third class of compounds can be construed as hormones. The catecholamines, for example, are produced in the adrenal medulla and in cells in the brain; they are (among other things) involved in the regulation of behavioral arousal.

Consider for a moment a very brief historical introduction to the functional segmentation of hormones and their specific effects on behavior. For many centuries, castration has been known to result in the loss of aggression and libido, and prevent the normal adolescent development of the male voice. In young boys who had the bad fortune to have a good voice, castration was used to maintain their position in the choir. Of course, the procedure was also commonly performed on individuals thought to have committed a sex crime. One of the most famous examples of this is the castration imposed on Abelarde, the great medieval philosopher, who ran off with one of his female students; her uncles brought the vengeance of castration on him. He lamented this fact and documented it in his work "My Great Misfortune" (Abelarde, 1972).

Experimentally, over 100 years ago, the removal of the testes was discovered to result in the loss of male sexual behavior in roosters; subsequent reattachment resulted in the onset of the behavior. At the time, why this occurred was not known since hormones had not been discovered.

The word "hormone" was used first by Bayliss and Starling in 1905. By the 1920s, adrenal medullary catecholamine hormones were known to contribute to the regulation of emotional states and the adaptation to regulatory duress (Cannon, 1929).

The discovery of a number of hormonal substances (e.g., insulin) in the early part of this century prompted a search for an understanding of their role in the physiological control of the body. What followed was the discernment of their role in behavior. That endocrine functions were contributors in the expression and regulation of behavior became apparent.

II. Hormonal Control of Ingestive Behavior

Since the time of Pavlov, we have known that a "cephalic phase" is orchestrated by the brain in the body's anticipatory adaptations (see Powley, 1977). "Cephalic phase" means the brain's anticipatory regulation of bodily responses. For example, insulin, which is so fundamental to the regulation and utilization of nutrients, is secreted in response to gustatory activation in anticipation of its use in the body; the function of this anticipation is to facilitate transport of macronutrients. Insulin secretion also is subject to conditioning; its secretion can be associated with the time of day of an occurrence of a meal (Woods et al., 1977). In other words, insulin is secreted in anticipation of the needs of the body.

Hormones also decline rapidly after the onset of a behavior. For example, angiotensin and aldosterone, elevated during body fluid and sodium deprivation, are reduced immediately after ingestion of sodium, even if sodium is not absorbed in the body (Denton, 1982). This response also occurs for vasopressin (ADH) secretion. Vasopressin regulates water excretion by its action on the kidney. When extracellular fluid volume is depleted, or when osmolarity is increased, ADH is secreted. However, when water is ingested ADH levels decrease at least in part by the actions of water in the mouth (Nicolaidis, 1977).

The loss of hormones that regulate behavior also results in compensatory behavioral responses. Curt Richter's contribution in this domain is extraordinary. For example, when deprived of the adrenal gland, mammals lose sodium, but with access to sodium salts rats ingest them and, therefore, live (Richter, 1936; Epstein and Stellar, 1955). When the sodium-retaining mineralocorticoid hormones secreted by the adrenal gland are returned to the animal, its salt intake returns to normal, as does its sodium excretion. The animals now can retain sodium. However, when larger doses of the hormone are given, the appetite is reinstated (see review by Richter, 1956). Why does this occur? The way we understand the hunger for sodium, at least in the rat, is that the hormone serves two functions: one is to retain sodium when sources of it are low in the body; the second role is to signal the brain to search for and then ingest salt (see Schulkin, 1991; Chapter 1). Alan Epstein, in the context of a community in which others had similar ideas, had an important insight into the hormonal mechanisms of sodium hunger. When the rat loses sodium because the adrenal gland has been removed, angiotensin (another natriorexigenic hormone that regulates body sodium balance through behavioral means) generates this salt-seeking behavior by its actions in the brain (Epstein, 1984); that is, the brain's response to the hormone is to generate both appetitive and consummatory behavior (Chapter 1). The hormones of sodium homeostasis—angiotensin and aldosterone—play fundamental roles in the behavioral regulation of sodium by their actions on the brain as well as in physiological regulation of sodium by their actions on the kidney (Epstein, 1984).

Lack of relevant hormones also drives behavioral compensatory responses, particularly in the case of alterations to the internal milieu. One of the most important examples of such a response is the regulation of calcium. Depriving animals (rats and monkeys) of calcium via parathyroidectomy results in calcium loss, but with access to calcium solutions the animals ingest greater amounts of the calcium to compensate for the loss of calcium that results from the lack of parathyroid hormone. Once again, when the hormone is given back, the appetite for calcium subsides (Richter, 1943, 1956).

In both the calcium and the sodium examples, the level of the calcium and sodium hormones affects the ingestion of minerals. The presence or absence of particular hormones affects other ingestive behaviors. Several chapters demonstrate that the line of inquiry into the hormonal regulation of ingestive behavior is advancing. For example, the reduction of food intake, the feelings of satiety, and the phenomenon of obesity may be related to alterations in gastrointestinal peptide hormones such as cholecystokinin (Chapter 2). Angiotensin generates both water and salt intake (Chapter 1). Aldosterone generates a salt craving (Chapters 1 and 6). Glucocorticoids potentiate aldosterone-induced cravings for salt, but also generate a craving for carbohydrates (Chapters 1 and 6). Atrial natriuretic peptide signals water and salt satiety (Chapter 1).

Note an important fact about the research on the hormonal regulation of behavior: the same hormone can play a number of different roles in the regulation of behavior. For example, nest building is generated by prolactin, but prolactin also generates thirst (reviewed by Schulkin, 1991). Oxytocin plays a role in milk let-down, uterine contraction, osmotic regulation, salt appetite, or affiliation (Chapter 8). Vasopressin, fundamental in the regulation and maintenance of water balance, also may play a small role in anorexia nervosa (Gold et al., 1983). What two events could be more disssimilar? However, nature economizes on its resources, using the same material in a variety of ways.

III. Endocrine Control of Sexually Dimorphic Behaviors and Cycles

Sexual differences are paradigms of the hormonal regulation of behavior. The hormonal regulation of ingestive behavior is pronounced particularly during the metabolic and mineral demands that females face during reproduction; many minerals and nutrients are ingested in greater amounts during pregnancy and lactation (Richter, 1943, 1956). In fact, female hormones of reproduction are involved importantly in this change in ingestive behavior. For example, prolactin, estrogen, and progesterone are known to increase the ingestion of minerals (Denton, 1982). Of course, these same hormones are well known for their involvement in the regulation of female sexual and maternal behavior (Beach, 1948; Pfaff, 1980), in addition to the regulation of water balance during lactation in mammals

(Denton, 1982). Again, nature has used hormonal signals to generate a number of behavioral as well as physiological adaptations.

Behavior is altered following changes in hormonal levels during the estrous cycle. For example, the running behavior of female rats is increased at this time (Richter, 1943). Removal of the ovaries disrupts this behavior; their replacement or hormonal replacement reinstates it. Subsequently, ingestive behaviors were discovered also to be altered during the ovarian cycle (e.g., Danielson and Buggy, 1980). Typically, food, water, and salt intake decreases during this period.

The hormonal regulation of sexual responsiveness also has been studied extensively (Beach, 1948). Estrogen and progesterone are well known to generate female sexual behavior; testosterone generates male sexuality (reviewed by Goy and McEwen, 1980). Chapter 3 addresses the role of these hormones in sexual behavior of animals and humans. In other animals, sexual behavior is almost totally dependent on the presence of the hormones; in humans, this behavior is more complicated. Some evidence also suggests that sexual preferences in humans may reflect anatomical differences in known sexually dimorphic nuclei in the hypothalamus (LeVay, 1991). The gonadal steroid hormones not only generate consummatory behavior but also appetitive behavior. From rats to monkeys, animals treated with the gonadal steroid hormones will perform operants to gain access to animals of the other sex (Goy and McEwen, 1980).

Whereas Chapter 3 addresses the role of testosterone and estrogen on sexual behavior, Chapter 4 reviews testosterone's contribution to aggression and dominance. Male dominance in several species has been correlated with levels of testosterone, as has rough play. Aggressive behaviors that are related to testosterone can be seen within days of birth (Frank, Glickman, & Light, 1991). Aggressive behaviors that result from testosterone also seem to occur in humans (Hines, 1982).

Anabolic steroids (used to promote muscle mass and strength in athletes) also may affect aggression. Abuse of these substances may reflect the addictive euphoria of feeling powerful, but also is known to depress the immune system (Chapters 5 and 7) and, consequently, increase possibility of diseases. A major health question is what the effects of steroid abuse in athletes will be especially over the long term. What are the psychological states generated by abusing such steroids (Chapter 5)?

IV. Critical Stages in Development and the Expression of Sexually Dimorphic Behaviors: Organizational and Activational Effects of Hormones

Steroid hormones have two distinct effects on the brain. One is to induce organizational changes in the structure of the brain; typically, this occurs during critical stages in development, although some evidence suggests that it can occur in adulthood as well (Arnold and Breedlove, 1985). The second effect is activational, that is, the arousal of the neural circuit for the onset of a behavior. Estrogen acts on a circuit in the brain, that was organized or established during the critical stage in development.

In critical developmental stages, the gonadal steroid hormones act on the brain to induce long-term changes (Goy and McEwen, 1980). The actions of the gonadal steroid hormones at critical stages produce long-term changes in brain morphology, connectivity, and neurotransmitter synthesis. The activation of the behavior is elicited later when the hormones act on hormone-sensitive neural circuits in the brain. In other words, in some psychobiological stages, hormonal activation has long-lasting effects on the brain and behavior later in life.

The experiments of Goy and his colleagues demonstrated these ideas about organization and activation effects of steroid hormones. First, these investigators deprived a variety of animals (reviewed by Goy and McEwen, 1980) of estrogen or testosterone during a critical window of their development (first week following birth). Then, when mature, the animals were tested for normal sexual responsiveness to these gonadal steroid hormones. The animals did not show the normal sexual behavior; that is, by depriving the male of testosterone during a critical stage in development the male did not respond to testosterone in the elicitation of male sexual behavior. Similar results were achieved for female sexual behavior elicited by estrogen. Depriving the male or female after this critical window by gonadectomy did not result in the abolition of sexual behavior when these animals were tested with the gonadal steroid hormones in maturity. These hormones and others (McEwen and Pfaff, 1985) have profound effects on brain morphology and neurotransmitter synthesis.

Sex differences in the behaviors of many animals constitute one consequence of hormonal actions on the brain. The brain in either sex is feminine until converted to a masculine brain by the actions of testosterone release and its conversion in the brain to estradiol (Goy and McEwen, 1980). The result is the masculinization of not only the brain but also behavior. The masculinization, or defeminization, of the brain not only affects sexual behavior, ingestive behavior, play, and aggressive behavior, but also affects physiological processes (as do the estrous cycle, noted earlier, LSH, and other hormonal secretions; see Chapters 3, and 4). Although this point is controversial, hormonal actions on the brains of humans has been suggested to affect sexual identity, preference, and expression (Hines, 1982; LeVay, 1991). In other animals, the manipulation of the gonadal steroid hormones during critical stages of development clearly affects everything from running activity, sweet and salt preference, and extracellular fluid balance (Schulkin, 1991) to aggression (Chapter 4).

Several chapters make reference to critical stages in development in which hormones have consequences for brain and behavior. The idea of critical stages in which both the brain and behavior are altered for a lifetime has permeated research not only in the field of behavioral endocrinology, but in other fields of neurobehavior. We know that visual, auditory, and even gustatory input at a critical window in development is essential to the normal acquisition of sensory acuity and neural connectivity and morphology. Experience during critical windows of development play an important role in more functions, such as those of bird song (Nottebohm, 1980) or perhaps human language (Lenneberg, 1967; Chomsky, 1975). In this context of a model system for understanding hormones and behavior, consider song production in the male bird. First, the experience of hearing song from conspecifics at a critical stage in development is essential for the normal expression of song. Second, in the case of bird song, if the animal is deprived of testosterone during a critical window of development of the brain, the song is known to be not expressed or muted when mature (Nottebohm, 1980).

V. Adaptation to Stress: Hormonal Contribution in Coping with the Psychological Challenges of Everyday Life

Perhaps critical stages exist during which experiences and hormonal activation of the brain have long-lasting consequences for normal human behavior and for aberrations in its expression. Thus, various psychopathologies conceivably are the consequences of various childhood experiences and possible endocrine reactions to these events. One speculative hypothesis is that the exaggerated activation of endocrine functions that results in the psychopathology of affective disorders, and the attempt to cope with the affliction, may cause further deterioration over time. In fact, several pathologies may result from prolonged stress on the body. An endocrine-based adaptation to stress has long been known though the works of Hans Selye (1956,1976). The pituitary–adrenal axis is particularly active in the adaptation to stress, in addition to hypothalamic corticotropin releasing hormone (e.g. Gold, Goodwin, & Chrousos, 1988). The activation of ACTH and corticosterone, in addition to corticotropin releasing hormone, is discussed in Chapters 6, 7, 8, and 10. We know, for example, that the exaggerated and prolonged release of corticosterone during stress can in some, but not all, conditions result in neuronal cell death (Chapter 6); we know that this hormone can be elevated during depression (Gold et al., 1988a,b). The impact of corticosterone may be greater during critical stages in development in promoting neuronal cell death (Meaney et al., 1991). Perhaps the first event is an attempted adaptation to the stress brought on by the illness; over time, the body and the brain and the mind are broken down. This idea is only speculation. However, various kinds of hormonal activation have been correlated with various emotional disorders. Inquirers should consider whether critical times exist during which the activation of these hormones can have long-lasting consequences for brain and behavior.

Consider a set of interesting findings from Sapolsky's laboratory (1992) about corticosterone and adaptations to social related stress. Sapolsky and Ray (1989) showed that social factors influence corticosterone circulation; the lower the social rank of a male baboon, the greater the levels of corticosterone typically circulating in his blood. Why should this be? The alpha male, in common terms, is secure in his position; the lower ranking males are not. The greater levels of corticosterone in the lower ranking males are part of their endocrine adaptation to their insecure position. Of course this insecure position also further affects endocrine function; these males have lower levels of circulating testosterone, which decreases their chances for successful reproduction. In any event, one wonders whether elevated corticosterone levels at critical times in development have greater effects than at others. As indicated, prolonged levels of corticosterone result in neuronal cell death in hippocampal tissue, which perhaps compounds the status of these males. Evidence suggests that, whereas newborn male and female humans have the same amount of circulating levels of corticosterone, the male secretes greater amounts than the female in response to stimulation. What are the long-term effects of this behavior? Although this book does not address these questions, no doubt future research will.

Several chapters address the role of various hormones in coping with everyday life (e.g., Chapter 9), including the effects of prolonged stress on the immune system. Psychoimmunology is one important interface between the psychology and endocrinology of adaptation (Chapter 7). The endocrinology of arousal in humans indicates the important role of corticotropin releasing hormone in addition to other neurohormonal signals (Chapter 9). A fuzzy continuum exists along which some stress activates the body, but prolonged or intense stress breaks it down. Richter long ago noted that wild rats had larger adrenal glands than domesticated rats. He assumed that these glands were more responsive to demanding situations because the everyday experiences of wild rats demanded that; the enlarged adrenal glands were suggestive of this greater endocrine capability (see Richter, 1976). Of course, the other end of the continuum is the breakdown of bodily and mental functions with too much stress.

These results support the common notion that a sense of hopelessness can lead to physical disease. Curt Richter thought that feelings of what he called "hopelessness" (1957) and what others have called "helplessness" (Seligman, 1975) were at the root of a number of physical and mental impairments. The adaptive endocrine response to stress can lead to feelings of hopelessness, when solutions seem distant or nonexistent. With an increased sense of futility, perhaps the hyperaroused state becomes hypoaroused (Chapter 9) with an increased susceptibility for disease because of the suppression of the immune system (Chapter 7).

VI. Hormonal Control of Endogenous Biological Clocks and Their Importance to Mental Health

Conceptually, hopelessness and some forms of depression are close. Alterations of endogenous biological rhythmicity, which is under neural and endocrine control, are important factors in the regulation of behavior and certain kinds of depression (Richter, 1965). The regulation of calcium (perhaps by calcium-retaining hormones) has, for some time, been thought to be tied to the expression of manic–depressive illness (see Chapter 10). Fluctuations in calcium might underlie some of the effects of manic–depressive illness. This illness might occur by the alteration of biological rhythmicity (e.g., circadian rhythmicity). Richter gave calcium to people suffering from manic–depressive experiences at the Phipps Clinic at Johns Hopkins; he claimed that the mineral ameliorated the discrepancies in their wake–sleep cycle and their affective disorders (1967, 1976).

We know now that circadian and other forms of biological rhythmicity are involved in normal mental health (see Chapters 10 and 11). We know that brain regions such as the suprachiasmatic nucleus of the hypothalamus are involved critically in circadian rhythmicity, and that damage to this brain region disrupts endogenously generated circadian rhythmicity (Rosenwasser and Adler, 1986). Considering that sunlight is used to help ameliorate depression (Rosenthal et al., 1984) and that seasonal depression is correlated with decreased exposure to the sun (Chapter 11), Richter may have been prescient in postulating a role for calcium, and perhaps the calcium regulating hormones in parathyroid and thyroid glands, in biological rhythmicity (see Chapter 10). Vitamin D, a steroid hormone synthesized by ultraviolet light, also may have some role in manic–depressive illness and the regulation of biological rhythmicity (Stumpf and Privette, 1989). This involvement is not known, and other hormones such as parathyroid might be involved (Chapter 10). Hormones or neurotransmitters such as melatonin that are involved in bodily rhythmicity and sensitivity to light might be involved in biologically based affective disorders (Chapters 10 and 11). In fact, hormonal factors are influenced by seasonal or environmental events, which in turn influence the state of the animal or person.

Durkheim (1897, 1951), the great French sociologist of the 19th century, suggested that suicide was more prominent in Protestant cultures; his examples were those of Northern Europe. He thought these individuals experienced greater isolation, less of a community sense, and therefore a susceptibility to suicide. Of course, although this hypothesis may be somewhat true, Northern Europe is colder, has less sun and exposure to ultraviolet light, and therefore fewer hormones that are responsive to sunlight (e.g., Vitamin D and other hormonal signals such as melatonin). These levels may lead a person to this state. The psychology meshes with the endocrinology in explaining the pathology.

VII. Curt Richter and Alan Epstein

This book is dedicated to two behavioral endocrinologists: Curt Richter and Alan Epstein. As the reader has probably realized, I am an admirer of the late Curt Richter. He was one of the leaders in the study of the hormonal regulation of behavior (Schulkin, Rozin, & Stellar, 1994). From 1918 until his death in 1989, Richter was inquiring into the role of endocrine function and the expression of behavior in many areas of psychobiology (see Richter, 1976). The range was from adrenal hormones, gonadal steroids and biological rhythms, and hormonal adaptation to stress to hormonal fluctuations during mental pathology. His inquiry was performed with imagination, great scope, and experimental insight. His experiments were diverse. The legacy of his work is continued by the expression of that rich role that hormones play in behavior as displayed by these chapters. Perhaps it is fitting that three of the authors have won the prize for outstanding work in psychoendocrinology (Phil Gold, Tom Insel, and David Saphier); the prize is the Curt Richter Award.

Alan Epstein, my close senior colleague for 15 years, lived in the tradition of Richter. The laboratory infused him with enthusiasm and tenacity. His field of inquiry was the study of ingestive behavior. Thus, his principle interests were feeding, thirst, and mineral hungers and the hormonal control of these behaviors. This research is represented in several chapters. Following Richter, he had a life-long interest in the study of sodium hunger and he studied the hormonal control of ingestive behavior.

VIII. Limitations and Scope of the Book

Although this book covers many of the relationships between hormones and the brain and behavior, other topics could not be covered (e.g., field work in behavioral endocrinology). I hope the reader will find this book a useful guide to this rapidly expanding and critical area of studies linking behavior to neuroendocrinology.

This book describes the behavioral and neural consequences of endocrine activity. It is intended to integrate laboratory experiments in animals and humans. The relevance to mental health is balanced with a basic understanding of what hormones do to behavior and to the brain. Scientific research in this field is growing at a rapid rate, with implications that are profound in terms of both basic knowledge and therapeutic relief.

I have assembled a group of scholars who are among the leading experts in the world on these matters. The 11 chapters discuss hormonal involvement in ingestive behaviors, sexual behavior, aggression, pharmacological abuse, adaptation to stress, attachment, depression, anxiety, and biological cyclicity. The goal of the book is to reveal the major advances in the field and, where possible, to integrate psychological with biological and therapeutic approaches. The endocrinology is rooted in psychology as well as in the functions of the brain.

References

Abelarde, P. (1972). *The story of my misfortune.* New York: MacMillan Press.

Arnold, A. P., & Breedlove, S. M. (1985). Organizational and activational effects of sex and steroids on brain and behavior: A reanalysis. *Hormones and Behavior, 19,* 469–498.

Beach, F. A. (1948). *Hormones and behavior.* New York: Paul B. Hoeber.

Cannon, W. B. (1929). *Bodily changes in pain, hunger, fear, and rage.* New York: Appleton–Century.

Chomsky, N. (1975). *Reflections on language.* New York: Pantheon Press.

Danielson, J., & Buggy, J. (1980). Depression of ad lib and angiotensin-induced sodium intakes at oestrus. *Brain Research Bulletin, 5,* 501–504.

Denton, D. A. (1982). *The hunger for sodium.* New York: Springer-Verlag.

Durkheim, E. (1897,1951). *Suicide* (translated by J. A. Spaulding and G. Simpson). New York: Free Press.

Epstein, A. N. (1984). The dependence of the salt appetite of the rat on the hormonal consequences of sodium deficiency. *Journal of Physiology, 79,* 494–498.

Epstein, A. N., & Stellar, E. (1955). The control of salt preference in the adrenalectomized rat. *Journal of Comparative and Physiological Psychology, 48,* 176–172.

Frank, L. G., Glickman, S. E., & Light, P. (1991). Fatal sibling aggression, precocial development, and androgens in neonatal spotted hyenas. *Science, 252,* 702–704.

Gold, P. W., Kaye, W. H., Robertson, G. L., and Ebert, M. (1983). Abnormalities in plasma cerebrospinal fluid arginine vasopressin in patients with anorexia nervosa. *New England Journal of Medicine, 308,* 1117–1123.

Gold, P. W., Goodwin, F. K., & Chrousos, G. P. (1988a). Clinical and biochemical manifestations of depression, Part 1. *New England Journal of Medicine, 319,* 348–353.

Gold, P. W., Goodwin, F. K., & Chrousos, G. P. (1988b). Clinical and biochemical manifestations of depression, Part 2. *New England Journal of Medicine, 319,* 413–420.

Goy, R. W., & McEwen, B. S. (1980). *Sexual differentiation of the brain.* Cambridge: MIT Press.

Hines, M. (1982). Prenatal gonadal hormones and sex differences in human behavior. *Psychological Bulletin, 92,* 56–80.

Lenneberg, E. H. (1967). *Biological foundations of language.* New York: Wiley.

LeVay, S. (1991). A difference in hypothalamic structure between heterosexual and homosexual men. *Science, 253,* 1034–1037.

McEwen, B. S., & Pfaff, D. (1985). Hormonal effects on hypothalamic neurons: Analyzing gene expression and neuromodulator action. *Trends in Neurological Science Reviews, March,* 105–110.

Meaney, M. J., Mitchell, J. B., Aitken, D. H., Bhatnagar, S., Bondoff, S. R., Iny, L. J. and Sarriegy, A. (1991). The effects of neonatal handling on the development of the adrenalcortical response to stress: Implications for neuropathology and cognitive deficits in later life. *Psychoneuroendocrinology, 16,* 85–103.

Nicolaidis, S. (1977). Sensory-neuroendocrine reflexes and their anticipatory and optimizing role on metabolism. *In The chemical senses and nutrition.* New York: Academic Press.

Nottebohm, F., & Arnold, A. P. (1976). Sexual dimorphism in vocal control areas of the songbird brain. *Science, 194,* 211–213.

Nottebohm, F. (1980). Brain pathways for vocal learning in birds: A review of the first 10 years. *In*

A. N. Epstein and J. M. Sprague (eds.), *Progress in psychobiology and physiological psychology.* New York: Academic Press.

Pfaff, D. (1980). *Estrogens and the brain.* New York: Springer-Verlag.

Powley, T. L. (1977). The ventralmedial hypothalamic syndrome, satiety, and a cephalic phase hypothesis. *Psychological Review, 84,* 89–126.

Richter, C. P. (1936). Increased salt appetite in adrenalectomized rats. *American Journal of Physiology, 111,* 155–161.

Richter, C. P. (1943). Total self regulatory functions in animals and human beings. *Harvey Lecture Series, 38,* 63–103.

Richter, C. P. (1956). Salt appetite of mammals: Its dependence on instinct and metabolism. *In* "L'instinct dans le comportement des animaux et de l'homme," pp. 577–629. Paris: Masson.

Richter, C. P. (1957). On the phenomenon of sudden death in animals and man. *Psychosomatic Medicine, 19,* 191–198.

Richter, C. P. (1965). *Biological clocks in medicine and psychiatry.* Springfield, Illinois: Charles T. Thomas.

Richter, C. P. (1976). *In* E. Blass (Ed.), *The psychobiology of Curt Richter.* Baltimore: York Press.

Rosenthal, N. E., Sack, D. A., Gillin, J. C., Lewy, A. J., Goodwin, F. K., Davenport, Y., Mueller, P. S., Newsome, D. A., & Wehr, T. A. (1984). Seasonal affective disorder. *Archives of General Psychiatry, 41,* 72–80.

Rosenwasser, A. M., & Adler, N. T. (1986). Structure and function in circadian timing systems: Evidence for multiple coupled circadian oscillators. *Neuroscience and Biobehavioral Reviews, 10,* 431–448.

Sapolsky, R. M. (1992). Stress, The Aging Brain and the Mechanisms of Neuron Death. Cambridge: MIT Press.

Sapolsky, R., & Ray, J. C. (1989). Styles of dominance and their endocrine correlates among wild baboons *(Papio anubis). American Journal of Primatology, 18,* 1–13.

Schulkin, J. (1991). *Sodium hunger.* Cambridge: Cambridge University Press.

Schulkin, J., Rozin, P., & Stellar, E. (1994). *Curt Richter: A great American inquirer.* Washington: National Academy of Sciences. (In press).

Seligman, M. E. P. (1975). *Helplessness.* San Francisco: Freeman.

Selye, H. (1956,1976). *The stress of life.* New York: McGraw Hill.

Stumpf, W. E., & Privette, T. (1989). Light, vitamin D, and psychiatry. *Psychopharmacology, 97,* 285–294.

Wehling, M., Eisen, C., and Christ, M. (1992). Aldosterone-specific membrane receptors and rapid non-genomic actions of mineralocorticoids. *Molecular and Cellular Biology, 90,* 5–9.

Woods, S. C., Vasseli, J. R., Kaestner, E., Szakmary, G. A., Milburn, P., & Vitello, M. V. (1977). Conditioned insulin secretion and meal feeding in rats. *Journal of Comparative and Physiological Psychology, 91,* 128–133.

Neuroendocrinology of Sodium Hunger: Angiotensin, Corticosteroids, and Atrial Natriuretic Hormone

Jay Schulkin

Behavioral Neuroscience Unit
Neuroendocrinology Branch
National Institute of Mental Health
and

Steven J. Fluharty

Departments of Animal Biology and Pharmacology
School of Medicine and Veterinary Medicine
Institute of Neurological Sciences
University of Pennsylvania, Philadelphia

I. Introduction

The two inquirers to whom this book is dedicated spent a great part of their career first, as did Curt Richter, discovering the phenomenon of sodium appetite in the laboratory and then trying to understand it. Alan Epstein's first project with his mentor Eliot Stellar was the phenomenon of adrenalectomy-induced sodium appetite (Epstein & Stellar, 1955), which first had been documented by Richter (1936).

Years later, Epstein remarked that he finally understood the project that launched his scientific career. He and his colleagues hypothesized that, under natural conditions, mineralocorticoid hormones in conjunction with angiotensin

13

aroused the appetite for sodium in the brain of the sodium depleted animal (Fluharty & Epstein, 1983; Sakai, 1986). The so-called "synergy hypothesis" recognized that either hormone alone could arouse the appetite but that both are elevated during sodium depletion. From this hypothesis came the prediction that the effect of the two hormones when they are elevated concurrently is larger than the sum of their individual effects, which is why Epstein thought these hormones were synergistic with one another in their induction of a sodium appetite.

According to Epstein, the avidity for sodium of adrenalectomized rats was an example of the singular action of cerebral angiotensin in the brain. Without mineralocorticoid hormone, as in the adrenalectomized rat, the appetite now depends entirely on angiotensin. In support of this hypothesis, Epstein and his colleagues demonstrated that central and not peripheral angiotensin receptor blockade abolished salt appetite in the adrenalectomized rat, suggesting that the appetite is dependent upon cerebral angiotensin (Sakai and Epstein, 1990a, b).

Richter was never as successful explaining the phenomenon of sodium appetite. However, his early insights into the likely cerebral mechanisms underlying his behavior were remarkably prophetic. He believed that the hunger for sodium resulted in changes in the brain and that the brain orchestrated behavior in response to the hormones of sodium homeostasis. His primary focus was on putative changes in the gustatory neural axis (Richter, 1939,1956). Indeed, salty tastes are perceived differently in the sodium hungry rat (cf. Contreras, 1977; Jacobs, Mark, & Scott, 1988; Nakamura & Norgren, 1992) and in humans placed on sodium deficient diets or treated with diuretics (Beauchamp et al., 1990). At present, the role of these gustatory changes and their relationship to the hormones of sodium homeostasis is not clear. We are still far from understanding this phenomenon in complete detail. (For alternative views, see Denton, 1982; Stricker & Verbalis, 1990.)

Behaviorally, the hormones that generate an appetite for sodium do so by eliciting a craving for a substance, in this case a salty substance (Schulkin, 1982). The sodium hungry animal searches for salty substances; recalls where they were when the animal did not need them (Krieckhaus & Wolf, 1968), what they were associated with (Berridge & Schulkin, 1989), or at what time of day they would appear (Rosenwasser, Schulkin, & Adler, 1988); and utilizes that knowledge to satisfy its hunger for sodium (reviewed by Wolf, 1969; Schulkin, 1991). Sodium appetite is a paradigmatic example of a hormonally induced motivated behavior. Sodium hungry animals express both appetitive and consummatory phases of motivated behavior. For example, mineralocorticoid-treated rats will go down a runway for sodium-taste rewards (Schulkin, Arnell, & Stellar, 1985; Figure 1). This is the appetitive phase of motivated behavior.

The consummatory phase includes the ingestion of the desired substance, in this case salty commodities. For example, when hypertonic NaCl is infused into the oral cavity, sodium hungry rats emit species-specific stereotyped oral–facial

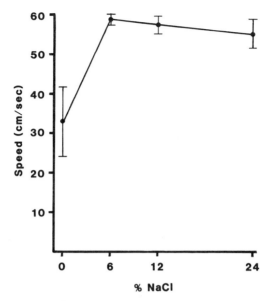

Figure 1.   The running speed for sodium in mineralocorticoid-treated rats. Note that the highest concentration is roughly 8 times the concentration of sea water, which the rats will not ingest and yet still express appetitive behavior toward (Schulkin, Arnell, & Stellar, 1985).

responses to the salt (3% NaCl) as though they were ingesting something good such as sucrose; when not sodium hungry, they behave as though the salt tastes bad like quinine water (Berridge, Flynn, Schulkin, & Grill, 1984; Berridge & Schulkin, 1989; Figure 2). Thus, both appetitive and consummatory phases of motivated behavior are altered when the hormones of sodium homeostasis are elevated and act on the brain.

The conditions for sodium hunger generally emerge from perturbations of extracellular fluid homeostasis, including hypovolemia and extreme sodium loss. In all these situations plasma renin, aldosterone, and also corticosterone levels are elevated (see Stricker, Vagnucci, McDonald, & Leenen, 1979; Denton, 1982; Figure 3).

Sodium hunger, or sodium craving, is a phenomenon of nature. Several species are known to travel great distances to ingest salt at mineral licks and other sources of salt when the sodium content of their diet is reduced (see Denton, 1982). The range of this phenomenon is expressed in a variety of species of omnivorous and herbivorous birds and mammals. This behavior is less clear or perhaps non-existent in carnivores, who ingest sodium by the consumption of their prey.

Several natural observations have found correlations between sodium consumption and the search for and ingestion of environmental sodium (reviewed in

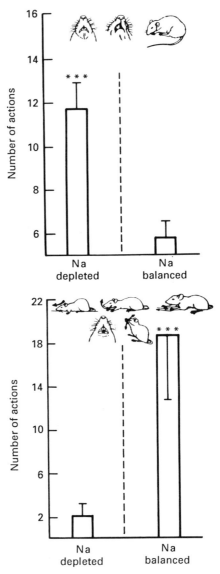

Figure 2. The taste reactivity profile to intraoral infusions of hypertonic NaCl when the hormones of sodium homeostasis were elevated and when they were not. (Top) Combined mean (SEM) number of ingestive actions (rhythmic tongue protrusions, nonrhythmic lateral tongue protrusions, and paw licks). (Bottom) Combined mean (SEM) number of aversive actions (chin rubs, head shakes, paw treads, gapes, face washes, and forelimb flails) (Berridge et al., 1984; Berridge & Schulkin, 1989).

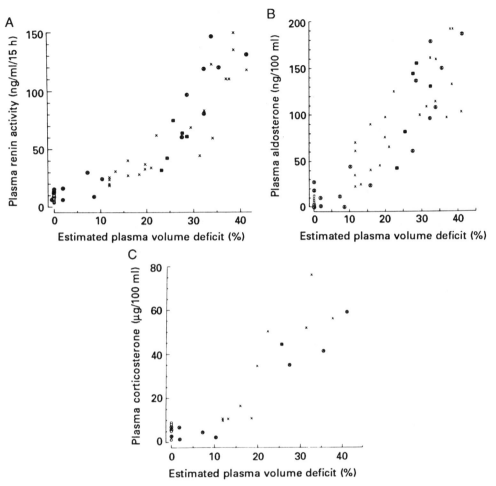

Figure 3. Renin activity (A), aldosterone concentration (B), and corticosterone concentration (C) as a function of estimated plasma volume deficit. Reprinted with permission from Stricker et al. (1979).

Denton, 1982; Schulkin, 1991). Elegant studies during the 1960s by Denton's group demonstrated relationships between activity levels of the renin–angiotensin–aldosterone system (and corticosterone levels), the reduced sodium content of the diet being consumed, and the ingestion of sodium (Figure 4).

Our focus in this chapter is on the role of angiotensin and the corticosteroids in arousing the appetite for sodium, and that of atrial natriuretic peptide in reducing the appetite for sodium. In the following discussion, we first address

Figure 4. A kangaroo at a salt peg during a time when sources of sodium are diluted and the hormones of sodium homeostasis are elevated. Reprinted with permission from Denton (1982).

angiotensin; then we discuss adrenal steroid hormones and the combination of the two corticosteroids in generating a sodium hunger. We then discuss the inhibition of the renin–angiotensin–aldosterone system and sodium appetite by atrial natriuretic peptide hormone. We end with a discussion of a putative neural system that underlies the behavioral regulation of sodium and its activation when the hormones of sodium homeostasis are elevated.

II. Angiotensin-Induced Sodium Appetite

Angiotensin is one of the primary hormones for the regulation of extracellular fluid balance (Fitzsimons, 1979). Behaviorally angiotensin is known to arouse the ingestion of water and sodium (Buggy & Fisher, 1974; Chiaraviglio, 1976; Fluharty & Manaker, 1983). Both behaviors are essential to maintaining extracellular fluid volume (Stricker & Wolf, 1969).

Witnessing animals drink abruptly when angiotensin is injected in the brain is remarkable. They drink water, and then exhibit a more delayed ingestion of sodium. The phenomena are clear, although natriuresis does occur over time and is related to the amount of sodium ingested (e.g., Buggy & Fisher, 1974; Fluharty & Manaker, 1983). Both ingestive responses are expressed in neonates well before they are able to ingest water or sodium independently (e.g., Leshem & Epstein, 1989), suggesting that the appetite is innate and that the hormone activates a neural circuit designed to orchestrate these behaviors. Moreover, the appetite that is expressed after administration of central angiotensin results in persistent increases in sodium consumption, even after the hormone is removed (Bryant, Epstein, Fitzsimons, & Fluharty, 1980). Angiotensin-induced sodium appetite is expressed in different species before the onset of natriuresis (e.g., mice Denton et al., 1990; see also Rowland & Fregly, 1988). The ingestion of sodium is augmented further when angiotensin is combined with sodium depletion in sheep (Weisinger et al., 1986). Finally, central injections of angiotensin increase the appetitive phase of this motivated behavior, for example, running down a runway for water or sodium (Zhang, Stellar, & Epstein, 1984).

In addition to the behavioral actions of angiotensin, this hormone also participates in the physiological regulation of body fluid homeostasis. Hypovolemia is one of several stimuli that activates the renin–angiotensin system (Peach & Chiu, 1974; Denton, 1982). This condition can occur naturally during blood loss or water or sodium deprivation. The rate-limiting step in the synthesis of angiotensin is the release of renin from the juxtaglomerular cells of the kidney. Renin converts the plasma α-globulin protein angiotensinogen to angiotensin I, which subsequently is converted to angiotensin II by a carboxyl dipeptidase known as angiotensin converting enzyme. Circulatory angiotensin has numerous peripheral actions including vasoconstriction, aldosterone release, augmentation of sympathetic nervous system function, and renal conservation of sodium and water (Philips, 1978). Circulating angiotensin also has important central nervous system effects although, like most other peptide hormones, has restricted effects on cerebral structures because of the blood–brain barrier (Simpson & Routtenberg, 1973; Philips, 1978). By acting on forebrain circumventricular organs (e.g., subfornical organ), blood-borne angiotensin can regulate pituitary function, stimulate central pressor responses, and elicit thirst. In addition to the well-established peripheral system, angiotensin acts in the central nervous system (Ganten, Hutchinson, & Schelling, 1975; Denton, 1982) in vasopressin release and in the regulation of thirst and extracellular fluid balance.

In the brain, the two major types of angiotensin receptors are referred to as AT1 and AT2 (Rowe, Saylor, & Speth, 1992). Whereas the dipsogenic response to angiotensin may be mediated solely by AT1 receptors (Rowland, Rozelle, Riley, & Fregly, 1992), the expression of the appetite for sodium appears to depend on both subtypes of angiotensin receptors (AT1 and perhaps AT2; Rowland et al., 1992).

Many of the cellular actions of the angiotensin in its peripheral target organs as well as in the brain involve the intracellular mobilization of calcium. This response is mediated by AT1 receptors and involves membrane-associated phosphoinositide hydrolysis and the subsequent production of inositol trisphosphate (IP₃). In neuron-like cultures, evidence suggests that a β-phosphoinositide-specific phospholipase C, is coupled to angiotensin receptors (Mah et al., 1992) via a G protein. A second cellular action of angiotensin in neurons involves the modulation of cGMP levels (Sumners, Myers, Kalberg, & Raizada, 1990; Sumners & Myers, 1991) that may involve both increases and decreases in the level of this cyclic nucleotide. When angiotensin stimulates the production of cGMP, it appears to do so through the intermediate formation of nitric oxide; a response that involves both AT1 and AT2 receptor subtypes (Zaharn, Ye, Ades, Regan, & Fluharty, 1992). Moreover, the interaction of both receptor subtypes, nitric oxide formation, and second messenger production may require interactions between angiotensin receptors expressed on both neurons and glial cells. Decreases in cGMP levels appear to involve AT2 receptors and activation of a cGMP-specific phosphodiesterase exclusively. This arrangement may exist only in neurons (Sumners & Myers, 1991). In both cases of cGMP regulation, the end result of the second messenger action appears to involve modulation of ionic conductances that determine membrane excitability (Sumners & Myers, 1991). These cellular changes ultimately underlie the behavioral and physiological changes elicited by the action of this peptide in the brain.

III. Site of Action for Angiotensin-Induced Sodium Appetite

Systemic angiotensin is well known to act on the subfornical organ (a circumventricular region of the brain) to elicit the ingestion of water (Simpson & Routtenberg, 1973). Through connection of the subfornical organ to the anterior third ventricular region [AV3V, including the median preoptic nucleus and the organum vasculosum lamina terminalis (OVLT)], the two sites regulate the ingestion of water in response to angiotensinergic signals (e.g., Buggy & Johnson, 1977; Philips, 1978; Miselis, 1981; Lind, 1988). However, the subfornical organ and its efferent projections (Schulkin, Eng, & Miselis, 1983; Thunhorst, Fitts, & Simpson, 1987; Masson & Fitts, 1989) are not essential for a number of forms of hormonally induced sodium appetite (but see Thunhorst, Ehrlich, & Simpson, 1990; Weisinger et al., 1990). Therefore, the anatomical sites that regulate water and perhaps sodium ingestion in response to an angiotensinergic signal appear to be different.

Angiotensin receptors are localized in circumventricular organs (e.g., subfornical organ, OVLT) and the median preoptic nucleus, in addition to magnocellular cells of the paraventricular hypothalamus, supraoptic nucleus, zona incerta, lateral hypothalamus, parabrachial nucleus, and solitary nucleus (see Lind, 1988). Angiotensin-containing neurons and terminal fields also have been found in many of these same regions, including the amygdala and the bed nucleus of the stria terminalis (Lind, Swanson, & Ganten, 1985; Lind, 1988).

One site that is immediately obvious as potentially important for the angio-tensinergic arousal of a sodium appetite is the AV3V region because of its involvement in body fluid and cardiovascular regulation (Buggy & Johnson, 1977; Johnson, 1985). Moreover, evidence suggests that lesions of this region reduce the salt appetite of rats on a sodium-deficient diet (Bealer & Johnson, 1979) or after sodium depletion (Chiaraviglio & Perez Guaita, 1984). Our own lesion studies of this region of the brain resulted in the abolition of renin–angiotensin-induced sodium appetite (Figure 5), whereas deoxycorticosterone (DOCA)-induced so-dium appetite remained intact (DeLuca, Galaverna, Schulkin, Yao, & Epstein, 1992). Other investigators independently found similar results (Fitts & Mason, 1990; Fitts, Tjepkes, & Bright, 1990).

Importantly, infusions of angiotensin within this region increase sodium ingestion, whereas the same infusions near the subfornical organ produce no such effect (Fitts & Mason, 1990; Figure 6). In conjunction with the facts that peri-pherally administered angiotensin results in natriuresis-induced sodium appetite (e.g., Findlay & Epstein, 1980; Yang & Epstein, 1991) and that peripheral block-age of angiotensin does not reduce the sodium appetite depletion (Sakai & Epstein, 1990a), these results argue that this region is importantly involved in the arousal of sodium appetite derived from central angiotensin.

IV. Adrenal Steroid-Induced Sodium Appetite

Richter and later others (e.g., Braun-Menedez, 1952) noted that deoxycorticoster-one, a mineralocorticoid hormone, would raise the ingestion of sodium in adre-nally intact rats (Richter, 1941; Rice & Richter, 1943). In adrenalectomized rats, this compound first reduced and then raised the appetite, depending on the dosage. Years later, two inquirers simultaneously discovered that aldosterone, the natu-rally occurring mineralocorticoid hormone, could either increase or decrease the appetite for sodium in the adrenalectomized rat (Fregly & Waters, 1966; Wolf, 1966) or could elicit an appetite for sodium in adrenally intact rats (Wolf, 1964; Wolf & Handel, 1966).

The first time aldosterone is elevated, it results in sodium ingestion, as measured by blood content of sodium and by monitoring aldosterone levels of rats raised on a sodium-rich diet (therefore, the rats have never been in a sodium-deficient state and without elevated levels of aldosterone). After the first time they were given aldosterone, rats ingested sodium. On the basis of this and other

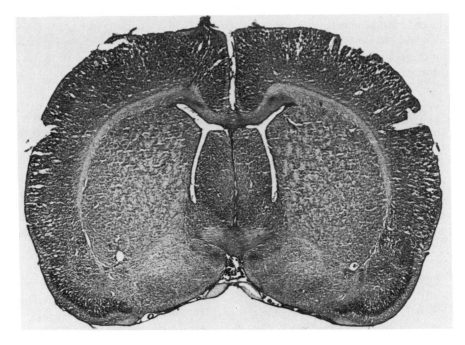

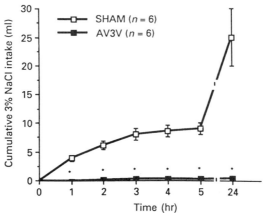

Figure 5. A lesion of the anterior third ventricular region (AV3V) and its effect on renin–angiotensin-induced sodium ingestion. It abolished the ingestion of sodium. Reprinted with permission from DeLuca Jr. et al. (1992). *Brain Research Bulletin, 8,* 73–87, with permission of Pergamon Press Ltd., Headington Hill Hall, Oxford OX3 OBW, United Kingdom.

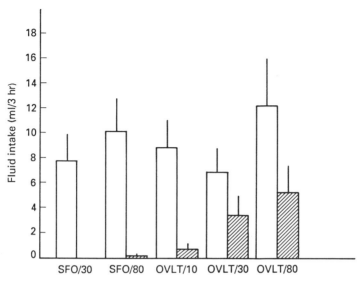

Figure 6. The water (open bars) and salt (hatched bars) intake of rats injected with angiotensin either within the subfornical organ (SFO) or within the organum vasculosm of the lamina terminalis (OVLT). It is clear from the figure that only water intake increased following angiotensin injected in the SFO, but water and sodium ingestion increased following OVLT injections. Reprinted with permission from Fitts and Masson (1990).

evidence, we argue that the appetite that arises from this hormone is innate (Schulkin, 1978). The expression of sodium ingestion in response to mineralocorticoid hormone can be demonstrated experimentally in neonates, before adult ingestive behavior actually is expressed (Thompson & Epstein, 1991). Mineralocorticoid-induced sodium appetite is expressed in several mammalian species (hamsters, Fitts, Yang, Corp., & Simpson, 1983; sheep, Denton, 1982; pigeons, Epstein & Massi, 1987). Mineralocorticoids also promote motivated behavior; rats will run down a runway for small rewards of sodium (Schulkin et al., 1985, as depicted in Figure 1) or will bar-press in an operant chamber to obtain sodium (Quartermain & Wolf, 1967).

However, mineralocorticoid activation obviously is not necessary for the expression of the appetite, since adrenalectomized rats express a sodium appetite (Richter, 1936). Moreover, hypovolemia induces a sodium appetite in therapeutically mineralocorticoid-treated adrenalectomized rats (given enough mineralocorticoid hormone to hold sodium, but unable to increase mineralocorticoid output in response to extracellular sodium loss; Stricker & Wolf, 1969). Perhaps in this case, central angiotensin activates the behavioral function, or removal of putative inhibitory signals (Strickler and Verbalis, 1987, 1988).

Considering the issue of how the same hormone can both decrease and increase the appetite for sodium in the adrenalectomized rat may provide valuable insights into the role of the hormone. Aldosterone has several important physiological effects. It reduces sodium excretion at the level of the kidney, mobilizes sodium transport out of sodium reserves such as bone, and redistributes sodium from salivary and alimentary tracts, ultimately to maintain extracellular fluid volume (Blair-West, Coghlan, Denton, Godling, & Wright, 1963; Denton, 1965,1982). In the adrenalectomized rat, returning the mineralocorticoid hormone reverses the chronic sodium loss (Denton, 1965). The behavioral adaptation of ingesting sodium to compensate for the loss of sodium then can be terminated. However, higher levels of circulating mineralocorticoid activates the rat brain to promote the search and then the ingestion of sodium as though it needs sodium. In other words, regions of the brain that regulate the behavioral end of body sodium homeostasis are activated and the animal acts as though it is in need of sodium. The hormone, therefore, maintains sodium under normal circumstances by promoting physiological responses to conserve and redistribute sodium, and then acts on the brain to elicit the behavior of sodium ingestion (Wolf, 1965; Schulkin, 1991). Both the physiological and behavioral responses have the same end—to maintain extracellular sodium balance.

We already have indicated that mineralocorticoids and glucocorticoids, in addition to angiotensin, play a role in the regulation of body sodium homeostasis and that, during sodium depletion or extracellular fluid loss, mineralocorticoid and glucocorticoid hormones are elevated (Stricker et al., 1979; Denton, 1982). In rats, glucocorticoid activation typically does not result in elevated sodium ingestion by itself (e.g., Braun-Menendez, 1952; Wolf, 1965). However, when glucocorticoids are combined with mineralocorticoids, the ingestion of sodium is greater than when mineralocorticoids are given alone (Wolf, 1965; Coirini, Schulkin, & McEwen, 1988; Ma, McEwen, Sakai, & Schulkin, 1993; Zhang, Epstein, & Schulkin, 1993). This phenomenon is noted when corticosterone is combined with aldosterone (Coirini et al., 1988; Ma et al., 1993; Zhang et al., 1993); when aldosterone is combined with dexamethasone (Ma et al., 1993), when deoxycorticosterone is combined with dexamethasone or with corticosterone (Wolf, 1965; Coirini et al., 1988); or when aldosterone is combined with RU28362, a glucocorticoid agonist (Devenport & Stith, 1992; Ma et al., 1993). The enhanced appetite seems to be independent of sodium loss and is, therefore, primary to the corticosteroid activation in brain (Ma et al., 1993).

Why should glucocorticoids potentiate mineralocorticoid-induced sodium ingestion? First, both hormones evolved from a common molecular ancestor; the distinction between the two hormones is less clear in lower organisms (Arriza, Simerly, Swanson, & Evans, 1988). Second, and importantly, glucocorticoids can increase the aldosterone-preferring corticosteroid Type I sites in brain (Brinton & McEwen, 1988; Coirini et al., 1988). This increase in binding could increase the natriorexegenic efficacy of the mineralocorticoid hormones. Indeed, since fewer aldosterone-preferring corticosteroid Type I sites than glucocorticoid-preferring corticosteroid Type II sites exist in brain (e.g., Chapter 6), this mechanism would insure the uptake of the mineralocorticoid, an event that occurs through an enzyme (11-β-hydroxysteroid dehydrogenase) that degrades and thereby reduced glucocorticoid binding to Type I sites (Funder, Pearce, Smith, & Smith, 1988). Therefore, glucocorticoid hormones may have an important role in the expression of sodium appetite.

Perhaps one can now understand one of the puzzling facts about mineralocorticoid-induced sodium appetite, namely, that deoxycorticosterone produces greater ingestion of sodium than does aldosterone, the naturally occurring mineralocorticoid hormone (Figure 7). Why? Aldosterone is a pure mineralocorticoid; deoxycorticosterone is not and has glucocorticoid and mineralocorticoid properties. Hence, deoxycorticosterone may be more natriorexegenic because administration of this hormone results in giving the animal both glucocorticoid and mineralocorticoid hormone.

V. Sites of Action for Adrenal Steroid-Induced Sodium Appetite

Evidence suggests that the phenomenon of adrenal steroid-induced sodium appetite is centrally induced. Mineralocorticoid antagonists are known to block min-

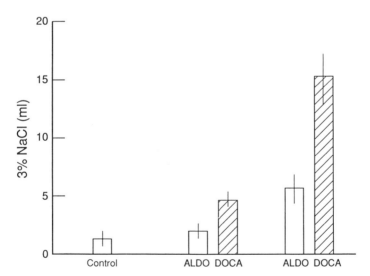

Figure 7. The sodium intake of aldosterone- or deoxycorticosterone-treated rats, or during control condition. Each rat was its own control and injected SC with either hormone once weekly with 2.5 or 5.0 mg (Goldman and Schulkin, unpublished observations).

eralocorticoid-induced sodium ingestion (Wolf, 1969; Ma et al., 1992) while leaving other forms of sodium ingestion intact (e.g., pregnancy-induced sodium appetite). Peripheral blockade of mineralocorticoid receptors does not affect mineralocorticoid-induced sodium appetite, whereas central blockade does (Ma et al., 1992). Which sites in the brain would be likely to be affected?

Several receptor binding studies in the brain have indicated the amygdala as a site for adrenal steroid binding (e.g., Coirini et al., 1988). Although the hippocampus contains aldosterone-preferring sites (e.g., Ermisch & Ruhe, 1978; Birmingham, Sar, & Stumpf, 1984), this area is not importantly involved in regulatory homeostatic behaviors such as sodium appetite. The medial amygdala is involved in other steroid-mediated behaviors (e.g., Lehman & Winans, 1980). Importantly, the amygdala is known to be involved in sodium appetite (e.g., Nachman & Ashe, 1974; Zolovick, Avrith, & Jalowiec, 1977). In addition, one site in the brain that contains both Type I and Type II corticosteroid-preferring sites and mRNA for these receptors (Arriza et al., 1988; Coirini et al., 1988) is the amygdala, specifically the medial nucleus which caudally becomes the amygdala–hippocampal transition zone (Canteras, Simerly, & Swanson, 1992). We therefore looked at the role of this nucleus in corticosteroid-induced behavior.

Electrolytic lesions of the medial nucleus of the amygdala reduced or abolished aldosterone or deoxycorticosterone-induced sodium appetite; the effect was specific for the adrenal steroid-induced appetite (Nitabach, Schulkin, & Epstein, 1989; Schulkin, Marini, & Epstein, 1989). In more recent work, we extended these observations by showing that infusions of ibotenic acid—which destroy cell bodies leaving fibers intact within the medial nucleus of the amygdala (Figure 8)—abolishes corticosteroid-induced sodium appetite (Figure 9)—whereas angiotensin-induced sodium appetite is unaffected (Zhang et al., 1993).

In addition to lesion studies, other evidence suggests that central administration of mineralocorticoid hormone to the medial amygdala elicits sodium ingestion, whereas angiotensin infusions and implants of aldosterone placed elsewhere in the brain do not elicit this response (Figure 10; Reilly, Mamadi, Schulkin, McEwen, & Sakai, 1993; M. Nitabach & J. Schulkin, unpublished observations). Moreover, mineralocorticoid antagonists (RU28318) applied to this region abolish the sodium appetite induced by intravenous aldosterone (Ma et al., 1992). This evidence, coupled with the receptor localization studies and lesion studies, argues that the medial amygdala is one site importantly involved in the arousal of sodium appetite by corticosteroid hormones.

Figure 8. Reconstruction of the minimum (fine strips) and maximum (coarse strips) extent of medial amygdala damage, in addition to ingestion of sodium in adrenal steroid-treated rats. A, aldosterone; C, corticosterone; N-F, need-free sodium ingestion; IBOL, ibotenic acid lesion. (Reprinted with permission from Zhang et al. (1993).

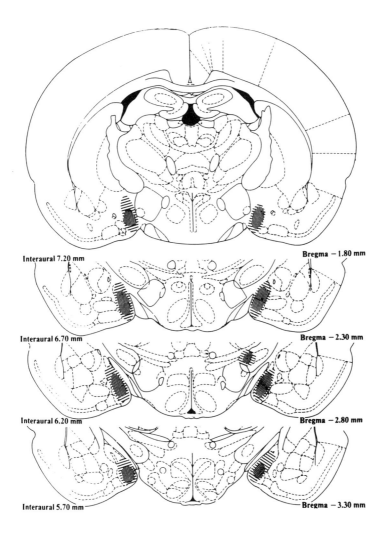

Interaural 7.20 mm — Bregma −1.80 mm

Interaural 6.70 mm — Bregma −2.30 mm

Interaural 6.20 mm — Bregma −2.80 mm

Interaural 5.70 mm — Bregma −3.30 mm

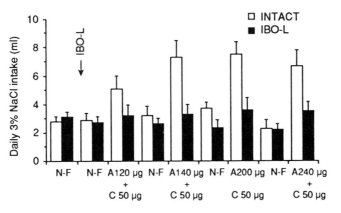

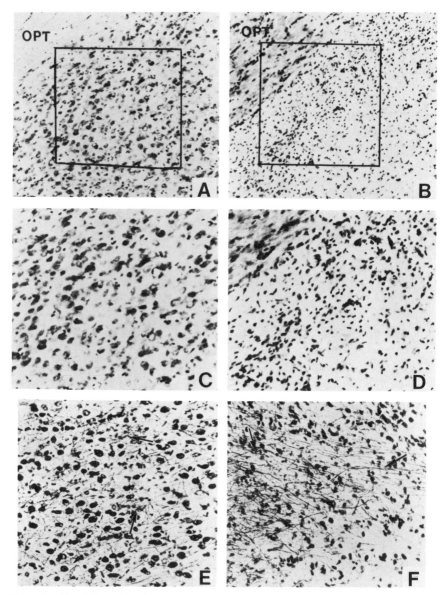

Figure 9. Photomicrograph of the medial amygdala. (A,C) Normal cells. (B,D) The same region after injections of ibotenic acid. Note that cell body loss and extensive glial accumulation can be seen throughout the ibotenic acid-lesioned area. C and D represent higher magnification (× 35.2) of the enclosed areas of A and B (×22). OPT, Optic tract; (E, F) (× 35.2) Fibers in both E (normal) and F (ibotenic acid) are apparent. Reprinted with permission from Zhang et al. (1993).

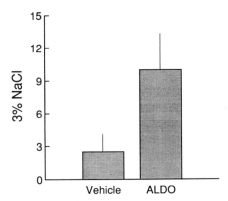

Figure 10. Sodium intake of aldosterone implanted in the medial amygdala and vehicle treated rats over a 23-hr period. The sodium intake increased with the aldosterone treatment (Nitabach and Schulkin, unpublished observations).

Our evidence also suggests that, under some conditions, relatively rapid natriorexigenic effects (i.e., within minutes) when the hormone is applied to the medial amygdala (Reilly et al., 1993), suggesting that some of these effects may not reflect genomic actions via classical intracellular changes, but instead be mediated by the actions of neuroactive steroids, generated by metabolism of deoxycorticotsterone in brain (Paul & Purdy, 1992). Nonetheless, corticosteroid hormones bind to receptors and promote protein synthesis by DNA-related RNA synthesis (Edelman, 1978; Szerlip et al., 1989). The mineralocorticoid activates Na^+ cellular electrogenic sodium pumps via protein-dependent changes (Edelman, 1978; Will et al., 1985). Interestingly, sodium transport changes occur in the amygdala after mineralocorticoid treatment (Grillo, Coirini, McEwen, & DeNicola, 1989), and sodium transport is known to influence sodium ingestion (Michell, 1979; Vivas & Chiaraviglio, 1987; Denton, 1982). Both events depend on cellular protein synthesis mechanisms (Edelman, 1978), suggesting that one mechanism by which mineralocorticoids induce sodium ingestion may be changes in sodium transport-dependent protein synthesis in the medial amygdala.

VI. Angiotensin and Adrenal Steroid-Induced Sodium Appetite

Earlier studies suggested that the hormones of sodium homeostasis might be working together to arouse a sodium appetite (see review by Fregly & Rowland, 1985). However, Epstein and his students (Fluharty and Sakai) demonstrated convincingly that the hormones of sodium homeostasis and extracellular fluid balance, when given exogenously in rats with normal sodium balance, would arouse the appetite for sodium.

The "synergy hypothesis" was demonstrated first using DOCA in combination with central angiotensin (Fluharty & Epstein, 1983; Zhang et al., 1984; Fuller & Fitzsimons, 1986; Figure 11). When doses of angiotensin and mineralocorticoid hormone that alone would not raise the appetite for sodium were combined, the sodium appetite was expressed. Moreover, at higher doses of angiotensin that would raise the appetite, the appetite was even greater when angiotensin was combined with mineralocorticoid hormone. The same phenomenon was demonstrated later using aldosterone in concert with central angiotensin (Sakai, 1986). Stricker (1983) examined the sodium ingestion of rats on a sodium-deficient diet

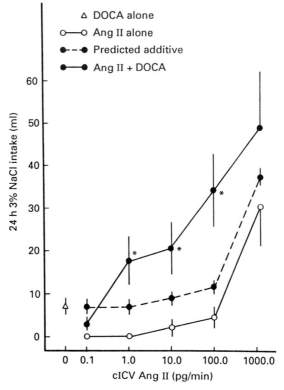

Figure 11. Cumulative mean intake of 3% NaCl. Infusions of angiotensin alone (ANG), deoxycorticosterone (DOCA), both, or predicted values. Reprinted with permission from Fluharty and Epstein (1983).

that were treated with an agent that depleted extracellular fluid. This investigator concluded that the exaggerated appetite that develops resulted from a synergistic interaction of exogenous aldosterone and angiotensin. Moreover, the two hormones combined produce motivated behavior (running down a runway; Zhang et al., 1984) that reflects the level of hormones; that is, the greater the level of hormone administered the greater the level of running speed and sodium ingestion. Finally, the expression of the synergy-induced sodium appetite appears early in the life of the neonate before the independent ingestion of sodium (Thompson & Epstein, 1991) and is demonstrated in other species (pigeons, Massi & Epstein, 1990).

The hormonal treatment effects are central. Peripheral angiotensin combined with aldosterone does not produce such effects on sodium appetite (Sakai & Epstein, 1990a,b). Moreover, when both hormones are blocked in the brain, the arousal of sodium ingestion that results from sodium depletion is abolished (Sakai, Nicolaidis, & Epstein, 1989); when the hormones are blocked peripherally, this does not occur (Sakai & Epstein, 1990a); when either hormone is blocked alone the appetite is expressed, but in muted form (Buggy & Jonklaas, 1984; Weiss, Moe, & Epstein, 1986; Sakai et al., 1989). Thus, the infusion studies of the hormones in conjunction with the blocker studies provide evidence that these hormones of sodium homeostasis raise the appetite by their actions in the brain; when these hormones are blocked, no sodium appetite is expressed in response to sodium deprivation.

Other studies demonstrate functional relationships between adrenal steroids and the central regulation of angiotensin. For instance, mineralocorticoids are known to increase angiotensin receptors in brain (Wilson, Sumners, Hathaway, & Fregly, 1986; King, Harding, & Moe, 1988). The same treatment results in up-regulation of angiotensin receptors in primary culture of neonatal rat brain and is mediated by Type I corticosteroid-preferring receptors (Sumners, Tang, Zelezna, & Raizadan, 1991b). The same dosages of mineralocorticoids, when combined with central angiotensin, result in greater sodium ingestion than when angiotensin is given alone (Fluharty & Epstein, 1983; King et al., 1988). In these studies, the mineralocorticoid hormone that was used was DOCA, which contains both adrenal steroids (e.g., aldosterone and corticosterone). Interestingly, when one compares the ingestive behavior of rats treated with either aldosterone or deoxycorticosterone in combination with intraventricular infusions of angiotensin, the intake of the deoxycorticosterone group is greater (c.f., Fluharty & Epstein, 1983; Sakai, 1986). In addition, subsequent studies of mRNA for angiotensinogen have shown that corticosterone, not aldosterone, results in increased activation of angiotensinogen mRNA in brain (Angulo, Schulkin, & McEwen, 1988; Angulo, Rifting, LeDaux, & McEwen, 1991). In addition, corticosterone increases angiotensin-induced water drinking (Sumners et al., 1991b) as well as mineralocorticoid-induced sodium appetite.

Additional insights into the interaction of both adrenal steroid hormones on central angiotensin have emerged from work on cultured neuronal cell lines. Specifically, in several neuroblastoma cell lines, corticosterone and the glucocorticoid agonist RU23862 up-regulate both AT1 and AT2 receptors; this effect also was demonstrated in brain tissue (Maki, He, Zhang, Williamson, & Fluharty, 1992). Aldosterone, in contrast, only seemed to up-regulate the AT2 angiotensin sites and amplify the stimulatory cGMP response (Figure 12). Type 2 corticosteroid agonists such as RU23986 increase both receptor subtypes and enhance angiotensin-stimulated calcium mobilization and cGMP production. The endogenous steroid corticosterone produces the same pattern of receptor–effector changes. In this regard, PLC-α which appears to be coupled to angiotensin receptors in neuronal tissue, also is up-regulated by steroid action (Maki et al., 1992). This up-regulation of angiotensin appears to depend on genomic effects, since it is blocked with actinomycin, a protein synthesis blocker. Injections of RNA isolated from steroid-treated cells into oocytes result in 2- to 3-fold increases in angiotensin receptor expression (S. J. Fluharty and M. M. White, unpublished observations). Collectively, these results suggest that the gene(s) encoding the multiple subtypes of angiotensin receptors possesses a steroid-responsive element that may underlie adrenal steroid facilitation of angiotensin in brain and, subsequently, its effects on sodium or water ingestion.

The elevation of corticosteroid activation along with angiotensin mimics the endocrine profile during body sodium and extracellular fluid depletion (Stricker et

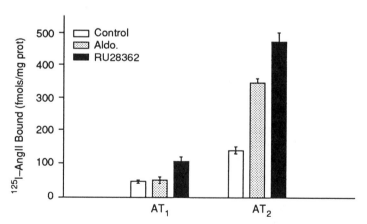

Figure 12. Effects of corticosteroids on angiotensin receptor subtypes in cultured neuronal cell lines. Cells were treated with steroids for 20 hrs and AngII-Rs was measured in membrane preparations (Fluharty, unpublished observations).

al., 1979; Denton, 1982). The result is the ingestion of sodium. Adrenal steroid hormones have genomic actions that prime the brain for angiotensin by up-regulating angiotensin receptors and increasing gene transcription. Thus, both adrenal steroid hormones in concert with angiotensin arouse the hunger for sodium.

We should alter the framework originally proposed by Epstein and his colleagues to include corticosterone, in addition to aldosterone and central angiotensin, in the control of the behavioral expression of sodium appetite that results from extracellular and sodium depletion (Schulkin, 1991). Corticosterone appears to have two important roles in the arousal of a sodium appetite. The first is inducing greater uptake of aldosterone-preferring Type I corticosteroid receptors in brain (see Brinton & McEwen, 1988; Coirini et al., 1988). The second, is increasing the levels for the angiotensin precursor angiotensinogen (Angulo et al., 1988,1991). This precursor is found only in glial cells, suggesting that this interaction ultimately would involve neuron–glial interactions, as in the second messenger system. In other words, multiple cellular mechanisms appear to underlie the amplification of angiotensin action induced by adrenal steroids in the brain. On the one hand, Type I occupancy increases angiotensin production, thus providing

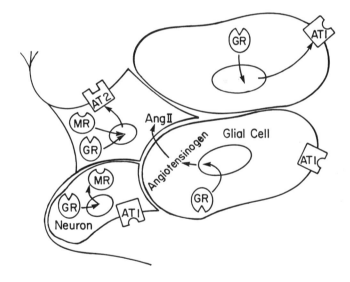

Figure 13. Characterization of the possible cellular mechanisms underlying the synergistic interactions between angiotensin II and corticosteroids in the central nervous system. The model emphasizes that corticosteroid can up-regulate AngII-Rs as well as increase the cerebral production of AngII. The model also proposes important interactions between neurons and glial cells (S. J. Fluharty, unpublished observations).

more ligand to the receptors. In addition, both Type I and Type II genomic actions result in an increased density of receptors to bind the ligand and ultimate applification of the cellular signaling mechanisms that transduce the binding of angiotensin to cellular actions.

VII. Site of Action of Angiotensin and Adrenal Steroid-Induced Sodium Appetite

Although functional studies described implicate the AV3V region for angiotensin-induced sodium appetite and the medial amygdala for adrenal steroid-induced sodium appetite, which sites are responsive for both adrenal steroid and angiotensin hormonal signals? The central nucleus of the amygdala appears to be a plausible candidate.

The central nucleus of the amygdala region receives the densest projections from brainstem gustatory sites as well as more general visceral information (Herrick, 1905; Norgren, 1976,1984). This area concentrates both hormones and is rich in angiotensin-responsive neuron terminals (e.g., Lind, 1988). The central nucleus of the amygdala is involved in cardiovascular regulation (Galeno, Hoeson, Maixner, Johnson, & Brody, 1982). Moreover, recent investigators, using c-*fos* as an anatomical marker, found that angiotensin infusions increase c-*fos* activation in the central nucleus of the amygdala during sodium depletion (McKinley, personal communication).

Since body sodium homeostasis is tied intimately to cardiovascular regulation and gustation is the specific sensory channel in identifying sodium in the environment (Schulkin, 1982; Norgren, 1984), we investigated the effects of lesioning this nucleus on sodium ingestion. We found that both central renin–angiotensin-induced sodium appetite and mineralocorticoid-induced sodium appetite are abolished by central nucleus lesions; water ingestion in response to angiotensin remained relatively intact (Figure 14; Galaverna et al., 1992).

The medial amygdala and central amygdala are tied anatomically to the bed nucleus of the stria terminalis and, in fact, are continuous with it (Johnston, 1923; Alheid & Heimer, 1988). Lesions of either the medial or the central amygdala affect sodium ingestion. Perhaps sites within the bed nucleus of the stria terminalis also are important in these behaviors. Lesions of the bed nucleus are known to reduce sodium depletion-induced sodium appetite (Zardetto-Smith, Beltz, & Johnson, 1991) and mineralocorticoid-induced sodium appetite (R. Maki and J. Schulkin, unpublished observations). Studies must be performed with both hormones directly applied to this region to determine whether the appetite for sodium is aroused.

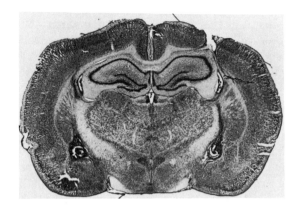

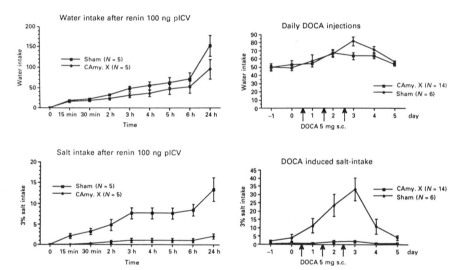

Figure 14. Representative central nucleus lesion. Graphs show that salt intake response to either renin or DOCA is abolished whereas the water intake is not different in lesion and control groups. Reprinted with permission from Galaverna et al. (1992), *Brain Research Bulletin, 28,* 89–98, with permission from Pergamon Press Ltd., Headington Hill Hall, Oxford OX3 OBW, United Kingdom.

VIII. Atrial Natriuretic Peptide and the Inhibition of Sodium Ingestion

Atrial natriuretic peptide (ANP) hormone has behavioral effects on water and sodium ingestion that are logical consequences of known involvement in body sodium homeostasis and its regulation of the renin–angiotensin–aldosterone system (see subsequent discussion). Centrally administered ANP decreases the sodium appetite of sodium-depleted rats (Fitts, Thunhorst, & Simpson, 1985) or rabbits (Tarjian, Denton, & Weisinger, 1988). Peripherally administered ANP is without such effects (Fitts et al., 1985; Tarjian et al., 1988). ANP also reduces the motivated behavior of the sodium-hungry rat. Sodium-depleted and adrenal steroid-treated rats have reduced running speed for sodium salts after central administration of ANP (P. Arnell, E. Stellar, et al., unpublished observations). Moreover, centrally administered angiotensin-induced water and sodium appetite are reduced by central administration of ANP (Masotto & Negro-Vilar, 1985; Figure 15; see also MacCann et al, 1989). Other dipsogenic responses are not reduced by ANP treatment (e.g., carbachol or intracellular induced thirst; Fitts et al., 1985). Moreover, ANP does not seem to reduce other forms of ingestive behavior (sucrose consumption; P. Arnell, E. Stellar, et al., unpublished observations). Collectively, these results suggest that ANP selectively acts on the hormones of sodium-appetite elicitation in reducing sodium ingestion.

The physiological background for the behavioral role of ANP is described briefly. For years a natriuretic hormone was thought to exist, based on studies on sodium excretion. Morphological studies then demonstrated that cells in the atria resembled endocrine cells. These cells were increased by sodium loading and were decreased by sodium restriction. Crude atrial extract was demonstrated to exhibit some of the most potent natriuretic–diuretic properties (deBold, Brownstein, Veress, & Sonnenberg, 1981). Cardiac volume receptors are coupled to the endocrine function of the atria; thus, increased plasma volume stimulates atrial receptors by

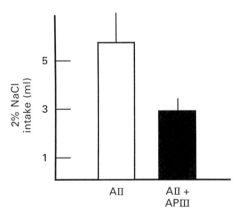

Figure 15. Effects of atriopeptin III (AP III) on angiotensin (AII)-induced sodium intake. Reprinted from Masotto and Negro-Vilar (1985). Inhibition of spontaneous or angiotensin II-stimulated water intake by atrial natriuretic factor. *Brain Research Bulletin, 15,* 523–526, with permission from Pergamon Press, Ltd., Headington Hill Hall, Oxford OX3 0BW, United Kingdom.

distension, in addition to the ANP released from isolated atria in response to increased sodium concentration (Cantin & Genest, 1985). Once in the circulation, ANP activates a variety of target organs—including the kidney and adrenal cortex, vasculature, and brain—to reduce plasma volume.

One of the ways in which ANP participates in body fluid homeostasis is by interacting with other hormonal regulators such as the renin–angiotensin–aldosterone system. ANP is an antagonist of the renin–angiotensin–aldosterone system. For example, synthetic ANP has been shown to inhibit *in vitro* and *in vivo* renin release (Brands & Freeman, 1988). The effect appears to reflect direct action of ANP on the juxtaglomerular cells, which may be mediated in part by changes in cyclic nucleotide production, particularly a decrease in cAMP and an increase in cGMP (Siedman & Bloch, 1987). ANP appears to have a tonic inhibition of renin release, in addition to limiting the production of angiotensin. ANP also opposes many of the physiological actions of angiotensin as well as the release of aldosterone (Brands & Freeman, 1988). The observation that ANP inhibits aldosterone production as elicited by dibutyryl-cAMP suggests that the site of action for ANP is distal to the hormone receptors and may involve inhibition of adenylyl cyclase or stimulation of guanylyl cyclase activity in adrenal cortical membranes and in brain (Israel, Garrido, Barbella, & Becemberg, 1988; Seidman & Bloch, 1988).

IX. Sites of Action for Atrial Natriuretic Peptide Inhibition of Sodium Appetite

Although its name implies that ANP is confined to the heart, clearly one of the most interesting extracardiac sites of action is the brain. In general, ANP-immunoreactive cell bodies with varicose fibers have been shown to be widely distributed throughout the brain (Skofitsch, Jacobowitz, Eskay, & Zamir, 1985). In the telencephalon, immunoreactivity for ANP has been localized in the bed nucleus of the stria terminalis and in both the central and the medial amygdala; the diencephalon sites include very dense immunoreactivity in the AV3V region, in addition to hypothalamic nuclei. Immunoreactivity has been observed in brain stem sites, including the parabrachial and solitary nuclei (Zamir, Skofitsch, Eskay, & Jacobowitz, 1986). In addition to the apparent presence of ANP-containing neurons, autoradiographic techniques have demonstrated the existence of receptors for this peptide in a variety of brain regions, including circumventricular organs and brainstem sites such as the solitary nucleus (Gibson, Wildey, Manaker, & Glembotski, 1986). These receptors are specific for ANP, are saturable, and exhibit high affinity.

The presence of ANP binding sites in circumventricular organs provides a window through which circulating ANP might regulate some aspects of the central nervous systems control of body sodium homeostasis. ANP infused directly into the subfornical organ reduces angiotensin-induced water drinking; also, ANP directly implanted within this region has these effects (Ehrlich & Fitts, 1990). Interestingly, ANP injected into the AV3V region (OVLT) reduces sodium ingestion in response to angiotensin-related captopril treatment (D. A. Fitts, unpublished observations). These data suggest that ANP may be acting on one site in the brain to decrease the signal for thirst to angiotensin and at another to affect hunger for sodium. The possible functions of ANP in other brain sites are not known but, like other peptides, this hormone may be acting as a neurotransmitter or neuromodulator within these circuits.

X. Anatomical Circuit for Sodium Appetite in Response to the Hormones That Regulate Extracellular Fluid and Sodium Balance

The appetite for sodium depends on forebrain structures for its expression. Decerebrated rats do not increase their ingestion of sodium when sodium is infused directly into the oral cavity while the hormones of sodium homeostasis are elevated (Grill, Schulkin, & Flynn, 1986). However, rats will increase their food intake when metabolically deprived (Miller & Sherrington, 1915; Grill, 1980). Moreover, infusions of angiotensin into the fourth ventricle do not elicit sodium ingestion (Fitts & Mason, 1989). Perhaps this difference in neural level of control (Jackson, 1884,1958) reflects the fact that the behavioral regulation for mineral and body fluid homeostasis was selected for when animals emerged from the sea, which is why the appetites for both water and sodium are more encephalized within the nervous system than hunger is (Grill et al, 1986; Schulkin, 1991).

Sites in the forebrain that are responsive to the hormones of sodium homeostasis include the medial and central nuclei of the amygdala, regions of the bed nucleus or extended amygdala, and the AV3V region. These sites are depicted in Figure 16. Also depicted are other sites known to be involved in body fluid and sodium homeostasis (e.g., lateral hypothalamus, zona incerta; Wolf & Schulkin, 1980; Grossman, 1990). These regions include gustatory sites in the solitary and parabrachial nuclei in the brainstem (Flynn, Grill, Schulkin, & Norgren, 1992). These brainstem sites send visceral afferents and receive efferents from the amygdala, the bed nucleus, and the AV3V region, and are known to be involved in sodium appetite and body fluid balance. This ventral pathway from the parabrachial nucleus into the forebrain may be the afferent limb of viscerally generated motivated behavior (Pfaffman, Norgren, & Grill, 1977). Moreover, although the stria terminalis is not essential for this hormonally induced behavior (Black, Weingarten, Epstein, Maki, & Schulkin, 1992), the amygdalafugal pathway is;

transection of this pathway reduces both adrenal steroid- and angiotensin-induced sodium ingestion, and is known to contain angiotensin fibers (Alheid, Schulkin, & Epstein, 1992). This pathway may be an efferent limb from the amygdala to midline sites in generating the appetite for sodium in response to the hormones of sodium homeostasis.

Thus, during sodium and extracellular fluid depletion, the renin–angiotensin–aldosterone system and corticosterone are activated. Peripherally, these hormones act to conserve and redistribute sodium to maintain sodium balance. Centrally they act to generate a hunger for sodium. Both behavior and physiology have the same end point: the maintenance of sodium and extracellular balance. According to the

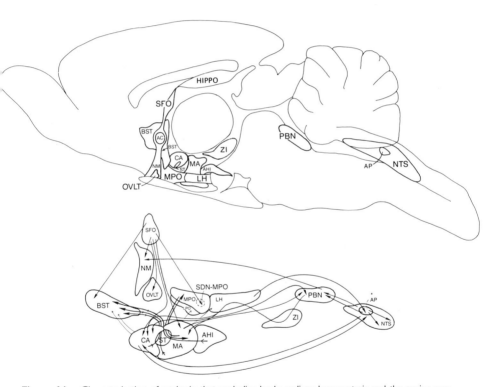

Figure 16. Characterization of rat brain that underlies body sodium homeostasis and the major connections. AHI, amygdala–hippocampal regions; AC, anterior commissure; AP, area postrema; BST, bed nucleus of the stria terminalis; CA, central nucleus of the amygdala; Hippo, hippocampus; LH, lateral hypothalamus; LPO, lateral preoptic nucleus; MA, medial nucleus of the amygdala; MPO, medial preoptic nucleus; NM, nucleus medianus; NTS, nucleus of the solitary tract; OVLT, organum vasculosum of the lamina terminalis; PBN, parabrachial nucleus; SDN-MPO, sexually dimorphic nucleus of the medial preoptic area; SFO, subfornical organ; ST, stria terminalis; Zi zona incerta. Reprinted with permission from Schulkin (1991).

current working hypothesis, the hormones of sodium homeostasis act in the brain as follows: (1) Angiotensin acts within midline structures surrounding the third ventricle (AV3V), where atrial peptide has its inhibition on the appetite. (2) The medial amygdala mediates the corticosteroid activation of the appetite. (3) The central nucleus of the amygdala, and perhaps the extended amygdala (bed nucleus), is responsive to both natriorexegenic signals.

XI. Conclusion

For some investigators, the study of the hormonal regulation of sodium appetite is like a first love. We began our careers investigating it and, although we have gone on to study other subjects as well, we always return to study this problem. This is no less true of Curt Richter and Alan Epstein. The hormonal basis of sodium appetite is interesting and important because it is amenable to study and because the behaviors generated by the search for sodium and its ingestion, although not discussed at length, are interesting, varied, and biologically important (see Denton, 1982; Rozin & Schulkin, 1990). Moreover, the research is also of medical import since the hormones of sodium homeostasis are tied to both extracellular fluid and cardiovascular regulation.

We did not, in this chapter, address all aspects of the hormonal control of sodium hunger. We have not discussed the ingestion of sodium during pregnancy and lactation (see Richter, 1956; Denton, 1982; Schulkin, 1991), nor have we covered release of inhibitory control on sodium hunger via oxytocin (Stricker & Verbalis, 1988), effects of tachykinins on sodium appetite (Massi & Epstein, 1989; Massi, Gentili, Perfumi, deCaro, & Schulkin, 1990), or the role of brain sodium composition on sodium consumption (Denton, 1982; Weisinger et al., 1987). We did not discuss how any of these behaviors reflect species differences and even strain differences (Denton, 1982; Schulkin, Liebman, Ehrman, Norton, & Ternes, 1984; Rowland & Fregly, 1988), nor that stress (restraint, crowding, or treatment with corticotropin-releasing factor, or ACTH) also elicit sodium ingestion (Denton, 1982; Tarjian & Denton, 1991). We have concentrated on the hormones that are elevated during sodium depletion and that arouse the hunger for sodium by their actions on the brain. An additional reason for this focus is that both Curt Richter and Alan Epstein, to whom this book is dedicated, tried to understand these hormonal factors in the regulation of sodium homeostasis.

The tradition that both inquirers followed had its roots in Claude Bernard (1865,1957) and his study of the regulation of the internal milieu, as well as Walter Cannon (1932) and his notion of homeostatic regulation. Curt Richter and Alan Epstein encouraged study of the behavioral end point; both behavior and physiology serve to maintain the internal milieu. This tradition is supported by study of the central nervous system, of what Lashley called "central excitatory states"

(1938) or what Morgan called "central motive states" (1943; see also Stellar, 1954). The hormones of sodium homeostasis induce a central motive state, the state of craving sodium, then searching for it, and finally ingesting it.

We have shown that the hormones of sodium homeostasis arouse the appetite for sodium by their actions in critical regions of the brain and that the analysis of this phenomenon is programmatic and speaks to a number of levels of analysis: behavioral, anatomical, endocrinological, physiological, cellular, and molecular. The phenomenon of sodium hunger remains a model system in which to inquire how hormones affect the brain in the production of motivated behavior.

References

Alheid, G. F., and Heimer, L. (1988). New perspectives in basal forebrain organization and special relevance for neuropsychiatric disorders: The striatopallidal, amygdaloid and corticopetal components of substantia innominata. *Neuroscience, 22*, 1–39.

Alheid, G. F., Schulkin, J., and Epstein, A. N. (1992). Amygdala pathways involved in sodium appetite in the rat, anatomical considerations. *Society for Neuroscience Abstracts.*

Angulo, J. A., Schulkin, J., and McEwen, B. S. (1988). Effect of sodium depletion and aldosterone treatment on angiotensinogen mRNA in the brain of the rat. *Society for Neuroscience Abstracts.*

Angulo, J. A., Riftina, F., LeDoux, M., and McEwen, B. S. (1991). Adrenal glucocorticoid regulation of angiotensinogen expression in the rat brain. *Society for Neuroscience Abstracts.*

Arriza, J. L., Simerly, R., Swanson, L. W., and Evans, R. M. (1988). The neuronal mineralocorticoid receptor as a mediator of glucocorticoid response. *Neuron, 1*, 887–900.

Arriza, J. L., Weinberger, C., Cerelli, G., Glaser, T. M., Handelin, B. L., Housman, D. E., and Evans, R. M. (1987). Cloning of human mineralocorticoid receptor complementary DNA: structural and functional kinship with the glucocorticoid receptor. *Science, 237*, 268–275.

Atunes-Rodrigues, J., McCann, S. M., and Samson, W. K. (1986). Central administration of atrial natriuretic factor inhibits saline preference in rats. *Endocrinology, 118*, 1726–1728.

Bealer, S. L., and Johnson, A. K. (1979). Sodium consumption following lesions surrounding the anteroventral third ventricle. *Brain Research Bulletin, 4*, 287–290.

Bern, H. A. (1967). Hormones and endocrine glands of fishes. *Science, 158*, 455–462.

Bernard, C. A. (1865,1957). "An introduction to the study of experimental medicine." New York: Dover Press.

Berridge, K. C., Flynn, F. W., Schulkin, J., and Grill, H. J. (1984). Sodium depletion enhances salt palatability in rats. *Behavioral Neuroscience, 98*, 652–660.

Berridge, K. C., Grill, H. J., and Norgren, R. (1981). Relation of consummatory responses and preabsorptive insulin release to palability and taste aversions. *Journal of Comparative and Physiological Psychology, 95*, 363–382.

Berridge, K. C., and Schulkin, J. (1989). Palatability shift of a salt-associated incentive drive during sodium depletion. *Quarterly Journal of Experimental Psychology, 41B*, 121–138.

Birmingham, M. K., Sar, M., and Stumpf, W. E. (1984). Localization of aldosterone and corticosterone in the central nervous system, assessed by quantitative autoradiography. *Neurochemical Research, 9*, 333–350.

Black, R. M., Weingarten, H. P., Epstein, A. N., Maki, R., and Schulkin, J. (1992). Transection of the stria terminalis without damage to the medial amygdala does not alter sodium regulation in rats. *Acta Neurobiologiae Experimentalis, 52*, 9–15.

Blaine, E. H., Covelli, M. D., Denton, D. A., Nelson, J. F., and Schulkes, A. A. (1975). The role of ACTH and adrenal glucocorticoids in the salt appetite of wild rabbits. *Coryctolagus cuniculus. Endocrinology, 97,* 793–801.

Blair-West, J. R., Coghlan, J. P., Denton, D. A., Goding, J. R., and Wright, R. D. (1963). The effect of aldosterone, cortisol, and corticosterone upon the sodium and potassium content of sheep's parotid saliva. *Journal of Clinical Investigation, 4,* 484–492.

Brands, M. W., and Freeman, R. H. (1988). Aldosterone and renin inhibition by physiological levels of atrial natriuretic factor. *American Journal of Physiology, 23,* R1011–R1016.

Braun-Menendez, E. (1953). Modificadores del apetito especifico para la sal en ratas blancas. *Revista de la Sociodad Argentina de Biologia, 11,* 92–102.

Braun-Menendez, E., and Brandt, P. (1952). Aumento del apetito especifico para lasal provocada por la desoxicorticosterona. 1. Caracteristicas. *Revista de la Sociedad Argentina de Biologia, 28,* 15–23.

Brinton, R. E., and McEwen, B. S. (1988). Regional distinctions in the regulation of type 1 and type 11 adrenal steroid receptors in the central nervous system. *Neuroscience Research Communications, 2,* 37–45.

Bryant, R. W., Epstein, A. N., Fitzsimons, J. T., and Fluharty, S. J. (1980). Arousal of a specific and persistent sodium in the rat with continuous intracerebroventricular infusion of angiotensin II. *Journal of Physiology, 301,* 365–382.

Buggy, J., and Fisher, A. E. (1974). Evidence for a dual central role for angiotensin in water and sodium intake. *Nature, 250,* 735–735.

Buggy, J., Huot, S., Pamnani, M., and Haddy, F. (1984). Periventricular forebrain mechanisms for blood pressure regulation. *Federation Proceedings, 43,* 25–31.

Buggy, J., and Johnson, A. K. (1977). Preoptic-hypothalamic periventricular lesions: Thirst deficits and hypernatremia. *American Journal of Physiology, 5,* R44–R52.

Buggy, J., and Jonklaas, J. (1984). Sodium appetite decreased by central angiotensin blockade. *Physiology and Behavior, 33,* 749–753.

Cannon, W. B. (1932). "The wisdom of the body." New York: Norton.

Canteras, N. S., Simerly, R. B., and Swanson, L. W. (1992). Connections of the posterior nucleus of the amygdala. *Journal of Comparative Neurology, 324,* 143–179.

Cantin, M., and Genest, J. (1985). The heart and the atrial natrriuretic factor. *Endocrinological Reviews, 6,* 107–127.

Chao, H. M., Choo, P. H., and McEwen, B. S. (1989). Glucocorticoid and mineralocorticoid receptor mRNA expression in rat brain. *Neuroendocrinology, 50,* 365–370.

Chiaraviglio, E. (1971). Amygdaloid modulation of sodium chloride and water intake in the rat. *Journal of Comparative and Physiological Psychology, 76,* 407–407.

Chiaraviglio, E. (1976). Effects of renin-angiotensin system on sodium intake. *Journal of Physiology, 255,* 57–66.

Chiaraviglio, E., and Perez Guaita, M. F. (1984). Anterior third ventricle (A3v) lesions and homeostasis regulation. *Journal of Physiology, 79,* 446–452.

Coirini, H., Marusic, E. T., DeNicola, A. F., Rainbow, T. C., and McEwen, B. S. (1983). Identification of mineralocorticoid binding sites in rat brain by competition studies and density gradient centrifugation. *Neuroendocrinology, 37,* 354–360.

Coirini, H., Schulkin, J., and McEwen, B. S. (1988). Behavioral and neuroendocrine regulation of mineralocorticoid and glucocorticoid action. *Society for Neuroscience Abstracts.*

Contreras, R. J. (1977). Changes in gustatory nerve discharges with sodium deficiency: A single unit analysis. *Brain Research, 121,* 373–378.

Crabbe, J. (1961). Stimulation of active sodium transport by the isolated toad bladder with aldosterone *in vitro. Journal of Clinical Investigation, 40,* 2103–2110.

deBold, A. J., Brownstein, H. B., Veress, A. T., and Sonnenberg, H. (1981). A rapid potent natriuetic response to intravenous injection of atrial myocardial extracts in rats. *Life Science, 28,* 89–94.

De Kloet, E. R. (1991). Brain corticosteroid receptor balance and homeostatic control. *Frontiers in Neuroendocrinology, 12*, 95–165.

DeLuca, L. A., Galaverna, O., Schulkin, J., Yao, S. Z., and Epstein, A. N. (1992). The anteroventral wall of the third ventricle and the angiotensinergic component of need-induced sodium intake in the rat. *Brain Research Bulletin, 8*, 73–87.

Denton, D. A. (1982). "The hunger for salt." Springer-Verlag, New York.

Denton, D. A., Blair-West, J. R., McBurnie, M., Osborne, P. G., Tarjan, E., Williams, R. M., and Weisinger, R. S. (1990). Angiotensin and salt appetite of BALB/c mice. *American Journal of Physiology, 28*, R729–R735.

Devenport, L., and Stith, R. (1992). Mimicking corticosterone's daily rhythm with specific receptor agonists: Effects on food, water and sodium intake. *Physiology and Behavior, 51*, 1247–1255.

Edelman, I. S. (1978). Candidate mediators in the action of aldosterone on Na^+ transport. In (J. F. Hoffman, Ed.) "Membrane transport processes," Vol. 1, pp. 125–140. New York: Raven Press.

Ehrlich, K. J., and Fitts, D. A. (1990). Atrial natriuretic peptide in the subfonical organ reduces drinking induced by angiotensin or in response to water deprivation. *Behavioral Neuroscience, 2*, 366–372.

Epstein, A. N. (1982). Mineralocorticoids and cerebral angiotensin may act to produce sodium appetite. *Peptides, 3*, 493–494.

Epstein, A. N. (1984). The dependence of the salt appetite in the rat on the hormonal consequences of sodium deficiency. *Journal of Physiology, 79*, 494–498.

Epstein, A. N., and Massi, M. (1987). Salt appetite in the pigeon to pharmacological treatments. *Journal of Physiology, 393*, 555–568.

Epstein, A. N., and Stellar, E. (1955). The control of salt preference in the adrenalectomized rat. *Journal of Comparative Physiology and Psychology, 48*, 167–172.

Epstein, A. N., Fitzsimons, J. T., and Rolls, B. J. (1970). Drinking induced by injection of angiotensin into the brain of the rat. *Journal of Physiology, 210*, 457–475.

Ermisch, R., and Ruhe, H-J. (1978). Autoradiographic demonstration of aldosterone-concentrating neuron populations in rat brain. *Brain Research, 147*, 154–158.

Findlay, A. L. R., and Epstein, A. N. (1980). Increased sodium intake is somehow induced in rats by intravenous angiotensin. *Hormones and Behavior, 14*, 86–92.

Fitts, D. A., and Mason, D. B. (1989a). Forebrain sites of action for drinking and salt appetite to angiotensin or captopril. *Behavioral Neuroscience, 103*, 865–872.

Fitts, D. A., and Mason, D. B. (1989b). Subfornical organ connectivity and drinking to captopril or carbachol in rats. *Behavioral Neuroscience, 103*, 873–880.

Fitts, D. A., and Mason, D. B. (1990). Preoptic angiotensin and salt appetite. *Behavioral Neuroscience, 104*, 643–650.

Fitts, D. A., Thunhorst, R. L., and Simpson, J. B. (1985). Diuresis and reduction of salt appetite by lateral ventricular infusions of atriopeptin II. *Brain Research, 348*, 118–124.

Fitts, D. A., Tjepkes, D. S., and Bright, R. O. (1990). Salt appetite and lesions of the ventral part of the ventral medial preoptic nucleus. *Behavioral Neuroscience, 103*, 818–827.

Fitts, D. A., Yang, O. O., Corp, E. S., and Simpson, J. B. (1983). Sodium retention and salt appetite following deoxycorticosterone in hamsters. *American Journal of Physiology*, R78–83.

Fitzsimons, J. T. (1979). "The Physiology of thirst and sodium appetite." Cambridge, England: Cambridge Press.

Fluharty, S. J., and Epstein, A. N. (1983). Sodium appetite elicited by intracerebroventricular infusion of angiotensin 11 in the rat: 11. Synergistic interaction with systemic mineralocorticoids. *Behavioral Neuroscience, 97(5)*, 746–758.

Fluharty, S. J., and Manaker, S. (1983). Sodium appetite elicited by intracerebroventricular infusion of angiotensin II in the rat: 1. Relation to urinary sodium excretion. *Behavioral Neuroscience, 97*, 738–745.

Flynn, F. W., Grill, H. J., Schulkin, J., and Norgren, R. (1992). Central gustatory lesion II: Effects on sodium appetite, taste aversion learning, and feeding behaviors. *Behavioral Neuroscience, 105,* 944–954.

Fregly, M. J., and Rowland, N. E. (1985). Role of renin-angiotensin-aldosterone system in NaCl appetite of rats. *American Journal of Physiology,* R1–R11.

Fregly, M. J., and Waters, W. (1966). Effect of mineralocorticoids on spontaneous sodium chloride appetite of adrenalectomized rats. *Physiology of Behavior, 1,* 65–74.

Fuller, L. M., and Fitzsimons, J. T. (1986). Influence of sodium load on angiotensin-induced sodium appetite. In (G. DeCaro, A. N. Epstein, and M. Massi, Eds.), "The physiology of thirst and sodium appetite" New York: Plenum Press.

Funder, J. W., Feldman, D., and Edelman, I. S. (1972). Specific aldosterone binding in rat kidney and parotid. *Journal of Steroid Biochemistry, 3,* 209–218.

Funder, J. W., Pearce, P. T., Smith, R., and Smith, A, I. (1988). Mineralocorticoid action: Target tussue specificity is enzyme, not receptor, mediated. *Science, 242,* 583–585.

Galaverna, O., DeLuca, L. A., Schulkin, J., Yal, S. Z., and Epstein, A. N. (1992). Deficits in NaCl ingestion after damage to the central nucleus of the amygdala in the rat. *Brain Research Bulletin, 28,* 89–98.

Galeno, T. M., Hoesen, G. W. V., Maixner, W., Johnson, A. K., and Brody, M. J. (1982). Contribution of the amygdala to the development of spontaneous hypertension. *Brain Research, 246,* 1–6.

Ganong, W. F. (1984). The brain renin-angiotensin system. *Annual Review of Physiology, 46,* 71–31.

Gansean, R., and Summers, C. (1989). Glucocorticoids potentiate the dipsogenic action of angiotensin II. *Brain Research, 499,* 121–130.

Ganten, D., Hutchinson, S., and Schelling, P. (1975). The intrinsic brain iso-renin angiotensin system: Its possible role in central mechanisms of blood pressure regulation. *Clinical Science and Molecular Medicine, 48,* 265–268.

Gibson, T. R., Wildey, G. M., Manaker, S., and Glembotski, C. C. (1986). Autoradiograph localization and characterization of atrial natriuretic peptide binding sites in the rat central nervous system and adrenal gland. *The Journal of Neuroscience, 7,* 2004–2011.

Giguere, V., Yang, N., Segui, P., and Evens, R. M. (1988). Identification of a new class of steroid hormones receptors. *Nature, 331,* 91–95.

Grill, H. J. (1980). Production and regulation of ingetive consummatory behavior in the chronic decrebrate rat. *Brain Research Bulletin, 5,* 79–87.

Grill, H. J., Schulkin, J., and Flynn, F. W. (1986). Sodium homeostasis in chronic decerebrate rats. *Behavioral Neuroscience, 100,* 536–543.

Grillo, C., Coirini, H., McEwen, B. S., and DeNicola, A. F. (1989). Changes of salt intake of (Na K)-ATPase activity in brain after high dose treatment of dexoycorticosterone. *Brain Research, 499,* 225–233.

Grossman, S. P. (1990). "Thirst and sodium appetite." San Diego: Academic Press.

Herrick, C. J. (1905). The central gustatory pathway in the brain of bony fishes. *Journal of Comparative Neurology, 15,* 375–456.

Israel, A., Garrido, M. R., Barbella, Y., and Becemberg, I. (1988). Rat atrial natriuretic peptide (99–126) stimulates guanylate cyclase activity in rat subfornical organ and choroid plesus. *Brain Research Bulletin, 20,* 253–256.

Jackson, H. (1884,1958). Evolution and dissolution of the nervous system. In (J. Taylor, Ed.), "Selected writings of John Hughlings Jackson," Vol. II. London: Staples Press.

Jacobs, K. M., Mark, G. P., and Scott, T. R. (1988). Taste responses in the nucleus tractus solitarius of sodium-deprived rats. *Journal of Physiology, 406,* 393–410.

Johnson, A. K. (1985). The periventricular anteroventral third ventrical (AV3V): Its relationship with the subfornical organ and neural systems involved in maintaining body fluid homeostasis. *Brain Research Bulletin, 15,* 595–601.

Johnston, J. B. (1923). Further contributions to the study of the evolution of the forebrain. *Journal of Comparative Neurology, 5*, 337–381.

King, S. J., Harding, J. W., and Moe, K. E. (1988). Elevated sodium appetite and brain binding of angiotensin II in mineralocorticoid-treated rats. *Brain Research, 448*, 140–149.

Krieckhaus, E. E., and Wolf, G. (1968). Acquisition of sodium by rats: interaction of innate mechanisms and latent learning. *Journal of Comparative and Physiological Psychology, 2*, 197–201.

Lashley, K. S. (1938). An experimental analysis of instinctive behavior. *Psychological Review, 45*, 445–471.

Lehman, M. N., and Winans, S. S. (1980). Medial nucleus of the amygdala mediates chemosensory control of male hamster sexual behavior. *Science, 210*, 5576–5600.

Lind, R. W. (1988). Sites of action of angiotensin in the brain. In (J. Harding, H. Wright, R. C. Speth, and N. Barnes, Eds.), "Angiotensin and blood pressure regulation." New York: Academic Press.

Lind, R. W., Swanson, L. W., and Ganten, D. (1985). Organization of angiotensin II immunoreactive cells and fibers in the rat central nervous system. *Neuroendocrinology, 40*, 2–24.

Lynch, K. R., Hawelu-Johnson, C. L., and Guyenet, P. G. (1987). Localization of brain angiotensinogen mRNA by hybridization histochemistry. *Molecular Brain Research, 2*, 149–158.

Ma, L. Y., McEwen, B. S., Sakai, R. R., and Schulkin, J. (1993). Glucocorticoids facilitate mineralocorticoid-induced sodium intake in the rat. *Hormones and Behavior, 27*, 240–250.

Ma, L. Y., Polidori, C., Schulkin, J., Epstein, A. N., Stellar, E., McEwen, B. S., and Sakai, R. R. (1992). Effect of centrally administered mineralocorticoid or glucocorticoid receptor antagonist on aldosterone-induced sodium intake in rats. *The Society of Neuroscience Abstracts.*

MacCann, S. M., Franci, C. R., and Antunes-Rodrigues, J. A. (1989). Hormonal control of water and electrolyte intake and output. *Acta Physiologica Scandinavica, 583*, 97–104.

McEwen, B. S., Jones, K. J., and Pfaff, D. W. (1987). Hormonal control of sexual behavior in the female rat: Molecular, cellular and neurochemical studies. *Biology of Reproduction, 36*, 37–45.

McEwen, B. S., and Pfaff, D. W. (1985). Hormone effects on hypothalamic neurons: Analysing gene expression and neuromodulator action. *Trends in Neurological Science Reviews*, 105–110.

Mah, S. J., Ades, A. M., Mir, R., Siemens, I. R., Williamson, J. R., and Fluharty, S. J. (1992). Association of solubilized angiotensin II-receptores with phospholipase c-α in murine neuroblastoma N1E-115 cells. *Molecular Pharmacology, 42*, 428–437.

Maki, R., He, P., Zhang, D. M., Williamson, J. R., and Fluharty, S. J. (1992). Corticosteroid regulation of PLC-a in rat brain and cultured neuronal cells. *The Society of Neuroscience Abstracts.*

Massi, M., and Epstein, A. N. (1989). Suppression of salt intake in the rat by neurokinin A: comparison with the effect of kassinin. *Regulatory Peptides, 24*, 233–244.

Massi, M., and Epstein, A. N. (1990). Angiotensin/aldosterone synergy governs the salt appetite of the pigeon. *Appetite, 14*, 181–192.

Massi, M., Gentili, L., Perfumi, M., deCaro, G., and Schulkin, J. (1990). Inhibition of salt appetite in the rat following injection of tachykinins into the medial amygdala. *Brain Research, 513*, 513–517.

Masotto, C., and Negro-Vilar, A. (1985). Inhibition of spontaneous or angiotensin II-stimulated water intake by atrial natriuretic factor. *Brain Research Bulletin, 15*, 523–526.

Mendelsonhn, F. A. O., Allen, A. M., Clevers, J., Denton, D. A., Tarjan, E., and McKinley, M. J. (1988). Localization of angiotensin II receptor binding in rabbit brain by in vitro autoradiography. *Journal of Comparative Neurology, 270*, 372–384.

Miller, F. R., and Sherrington, C. S. (1915). Some observations on the bucco-pharyngeal reflex deglutition in the cat. *Quarterly Journal of Experimental Physiology, 9*, 147–186.

Morgan, C. T. (1943). "Physiological psychology." New York: McGraw Hill.

Nachman, M., and Ashe, J. H. (1974). Effects of basolateral amygdala lesions on neuphobia learned taste aversions, and sodium appetite in rats. *Journal of Comparative and Physiological Psychology, 87*, 622–643.

Nakamura, K., and Norgren, R. (1992). Salt deprivation alters taste responses in the nucleus of the solitary tract of behaving rats. *The Society of Neuroscience Abstracts.*

Nitabach, M., Schulkin, J., and Epstein, A. N. (1989). The medial amygdala is part of a mineralocorticoid-sensitive circuit controlling NaCl intake in the rat. *Behavioral Brain Research, 35,* 127–134.

Norgren, R. (1976). Taste pathways to hypothalamus and amygdala. *Journal of Comparative Neurology, 166,* 17–30.

Norgren, R. (1984). Central neural mechanisms of taste. In (J. M. Brookhart and V. B. Mountcastle, Eds.), "Handbook of physiology. The nervous system," pp. 1087–1128. American Physiological Society, Bethesda, Maryland.

Paul, S. M., and Purdy, R. H. (1992). Neuroactive steroids. *FASEB J., 6,* 2311–2322.

Peach, M. J., and Chiu, A. T. (1974). Stimulation and inhibition of alosterone biosynthesis in vintro by angiotensin II and analogs. *Circulatory Research, 34,* 1–14.

Pfaff, D. W. (1980). "Estrogens and brain function: Neural analysis of a hormone-controlled mammalian reproductive behavior." New York: Springer-Verlag.

Pfaffmann, C., Norgren, R., and Grill, H. J. (1977). Sensory affect and motivation. In (B. M. Wenzel and H. P. Ziegler, Eds.), "Tonic function of sensory systems." New York: New York Academy of Sciences.

Philips, M. I. (1978). Angiotensin in the brain. *Neuroendocrinology, 25,* 354–377.

Plunkett, L. M., Shigematsu, K., Kurihara, M., and Saavedra, J. M. (1987). Localization of angiotensin II receptors along the anteroventral third ventricle area of the rat brain. *Brain Research, 405,* 205–212.

Pressley, K., and Funder, J. W. (1975). Glucocorticoid and mineralocorticoid receptor in gut mucosa. *Endocrinology, 97,* 588–596.

Quartermain, D., and Wolf, G. (1967). Drive properties of mineralocorticoid-induced sodium appetite. *Physiology and Behavior, 2,* 261–263.

Reagan, L. P., Ye, X., Maretzki, C. H., and Fluharty, S. J. (1993). Down regulation of angiotensin II receptors and desensitization of cGMP production in differentiated murine NIE-115 neuroblastoma cells. *Journal of Neurochemistry.*

Reilly, J. J., Mamadi, D. B., Schulkin, J., McEwen, B. S., and Sakai, R. R. (1993). The effect of amygdala adrenal steroid implants in the rat brain on sodium intake. 11th International Conference on the Physiology of Food and Fluid Intake, Oxford, England.

Rice, K. K., and Richter, C. P. (1943). Increased sodium chloride and water intake of normal rats treated with desoxycorticosterone acetate. *Endocrinology, 33,* 106–115.

Richter, C. P. (1936). Increased salt appetite in adrenalectomized rats. *American Journal of Physiology, 115,* 155–161.

Richter, C. P. (1941). Sodium chloride and dextrose appetite of untreated and treated adrenalectomized rats. *Endocrinology, 29,* 115–125.

Richter, C. P. (1956). Salt appetite of mammals: Its dependence on instinct and metabolism. L'instinct dans le comportement des animaux et de l'homme. pp. 577–629. Paris: Masson.

Rowe, B. P., Saylor, D. L., and Speth, R. C. (1992). Analysis of angiotensin II receptor subtypes in individual rat brain. *Neuroendocrinology, 55,* 563–573.

Rowenwasser, A. R., Schulkin, J., and Adler, A. N. (1988). The behavior of salt hungry rats to limited periods of salt. *Animal Behavior, 16,* 324–329.

Rowland, N. E., and Fregly, M. J. (1988). Sodium appetite: Species and strain differences and role of renin-angiotensin-aldosterone system. *Appetite, 11,* 143–178.

Rowland, N. E., Rozelle, A., Riley, P. J., and Fregly, M. J. (1992). Effect of nonpeptide angiotensin receptor antagonists in water intake and salt appetite in rats. *Brain Research Bulletin, 29,* 389–393.

Rozin, P. N., and Schulkin, J. (1990). Food selection. In (E. M. Stricker, Ed.), "Handbook of behavioral neurobiology," pp. 297–328. New York: Plenum Press.

Sakai, R. R. (1986). The hormones of renal sodium conservation act synergistically to arouse a sodium appetite in the rat. In (G. deCaro, A. N. Epstein, and M. Massi, Eds.), "The physiology of thirst and sodium appetite" New York: Plenum Press.

Sakai, R. R., and Epstein, A. N. (1990a). Peripheral angiotensin II is not the cause of sodium appetite in the rat. *Appetite, 15,* 161–170.

Sakai, R. R., and Epstein, A. N. (1990b). The dependence of adrenalectomy-induced sodium appetite on the action of angiotensin II in the brain of the rat. *Behavioral Neuroscience, 104,* 167–176.

Sakai, R. R., Nicolaidis, S., and Epstein, A. N. (1989). Salt appetite is suppressed by interference with angiotensin II and aldosterone. *American Journal of Physiology, 251,* R762–R768.

Sapher, C. B., and Levisohn, D. (1983). Afferent connections of the median preoptic nucleus in the rat: Anatomical evidence for a cardiovascular integrative mechanism in the anteroventral third ventricular region. *Brain Research, 288,* 21–31.

Schulkin, J. (1978). Mineralocorticoids, dietary conditions and sodium appetite. *Behavioral Biology, 23,* 197–205.

Schulkin, J. (1982). Behavior of sodium deficient rats: The search for a salty taste. *Journal of Comparative and Physiological Psychology, 96,* 628–634.

Schulkin, J. (1991). "Sodium hunger." Cambridge: Cambridge University Press.

Schulkin, J., Arnell, P., and Stellar, E. (1985). Running to the taste of salt in mineralocorticoid-treated rats. *Hormones and Behavior, 19,* 413–425.

Schulkin, J., Eng, R., and Miselis, R. R. (1983). The effects of disconnecting the subfornical organ on behavioral and physiological responses to alterations of body sodium. *Brain Research, 263,* 351–355.

Schulkin, J., Liebman, D., Ehrman, R. N., Norton, D. L., and Ternes, J. (1984). Sodium hunger in the rhesus monkey. *Behavioral Neuroscience, 4,* 753–756.

Schulkin, J., Marini, J., and Epstein, A. N. (1989). A role for the medial region or the amygdala in mineralocorticoid-induced salt hunger. *Behavioral Neuroscience, 103,* 178–185.

Semple, P. F., Nichols, M. G., Tree, M., and Fraser, R. (1978). Angiotensin II in the dog: Blood levels and effect on aldosterone. *Endocrinology, 4,* 1476–1482.

Simpson, J. B., and Routtenberg, A. (1973). Subfornical organ: site of drinking elicitation by angiotensin II. *Science, 181,* 1172–1174.

Skofitsch, G., Jacobowitz, D. M., Eskay, R. L., and Zamir, N. (1985). Distribution of atrial natriuretic factor-like immunoreactive neurons in the rat brain. *Neuroscience, 4,* 917–948.

Stellar, E. (1954). The physiology of motivation. *Psychological Review, 61,* 5–22.

Stricker, E. M. (1966). Extracellular fluid volume and thirst. *American Journal of Physiology, 211,* 232–238.

Stricker, E. M. (1983). Thirst and sodium appetite after colloid treatment in rats: Role of the renin–angiotensin–aldosterone system in rats. *Behavioral Neuroscience, 97,* 725–737.

Stricker, E. M., and Jalowiec, J. E. (1970). Restoration of intravascular fluid volume following acute hypovolemia in rats. *American Journal of Physiology, 218,* 191–196.

Stricker, E. M., Vagnucci, A. H., McDonald, R. H., Jr., and Leenen, F. H. (1979). Renin and aldosterone secretions during hypovolemia in rats: relation to NaCl intake. *American Journal of Physiology,* R45–R51.

Stricker, E. M., and Verbalis, J. G. (1987). Central inhibitory control of sodium appetite in rats: Correlation with pituitary oxytocin secretion. *Behavioral Neuroscience, 4,* 560–567.

Stricker, E. M., and Verbalis, J. G. (1988). Hormones and Behavior: the biology of thirst and sodium appetite. *American Scientist, 76,* 261–267.

Stricker, E. M., and Verbalis, J. G. (1990). Sodium Appetite. In "Handbook of behavioral neurobiology," Vol. 10. New York: Plenum Press.

Stricker, E. M., and Wolf, G. (1969). Behavioral control on intravascular fluid volume: Thirst and sodium appetite. *Annals of New York Academy of Sciences, 157,* 553–568.

Sumners, C., Gault, T. R., and Fregly, M. J. (1991a). Potentiation of angiotensin II-induced drinking by glucocorticoids is a specific glucicorticoid Type II receptor (GR)-mediated event. *Brain Research, 552*, 283–290.

Sumners, C., and Myers, L. M. (1991). Angiotensin II decrease cGMP levels in neural cultures from rat brain. *American Journal of Physiology, 226*, C79–C87.

Sumners, C., Myers, L. M., Kalberg, C. J., and Raizada, M. K. (1990). Physiological and pharmacological comparison of angiotensin II receptors in neuronal and astrocyte glial cultures. *Progress in Neurobiology, 34*, 355–385.

Sumners, C., Tang, W., Zelezna, B., and Raizada, M. K. (1991b). Angiotensin II receptor substypes are coupled with distinct signal transduction mechanisms in neurons and astroycytes from rat brain. *Proceedings of the National Academy of Sciences U.S.A., 88*, 7567–7571.

Tarjian, E., and Denton, D. A. (1991). Sodium/water intake of rabbits following administration of hormones of stress. *Brain Research Bulletin, 26*, 133–136.

Tarjian, E., Denton, D. A., and Weisinger, R. S. (1988). Atrial natriuretic peptide inhibits water and sodium intake in rabbits. *Regulatory Peptides, 23*, 63–75.

Thunhorst, R. L., Ehrlich, K. J., and Simpson, J. B. (1990). Subfornical organ participates in salt appetite. *Behavioral Neuroscience, 4*, 637–642.

Thunhorst, R. L., Fitts, D. A., and Simpson, J. B. (1987). Separation of captopril effects on salt and water intake by subfornical organ lesions. *American Journal of Physiology, 252*, R409–R418.

Vinson, G. P., Whitehouse, B. J., Goddard, C., and Sibley, C. P. (1979). Comparative and evolutionary aspects of aldosterone secretion and zona glomerulosa function. *Journal of Endocrinology, 81*, 5P–24P.

Weisinger, R. S., Denton, D. A., Nicolantonio, R., McKinley, M. J., Muller, A. F., and Tarjan, E. (1987). Role of angiotensin in sodium appetite of sodium-depleted sheep. *The American Journal of Physiology, 11*, R-51–R63.

Weisinger, R. S., Denton, D. A., Di Nicolantonio, R., Hards, D. K., McKinley, M. J., Oldfield, B., and Osborne, P. G. (1990). Subfornical organ lesion decreases sodium appetite in the sodium-depleted rat. *Brain Research, 420*, 135–143.

Weiss, M. L., Moe, K. E., and Epstein, A. N. (1986). Interference with central actions of angiotensis II. *American Journal of Physiology*, R250–R259.

Wilson, K. M., Sumners, C., Hathaway, S., and Fregly, M. J. (1986). Mineralocorticoids modulate central angiotensin II receptors in rats. *Brain Research, 382*, 87–96.

Wolf, G. (1965). Effect of deoxycorticosterone on sodium appetite of intact and adrenalectomized rats. *American Journal of Physiology, 208*, 1281–1285.

Wolf, G. (1969a). Effects of a mineralocorticoid antagonist on sodium appetite. VII International Congress of Nutrition, Prague.

Wolf, G. (1969b). Innate mechanisms for regulation of sodium appetite. In (C. Pfaffman, Ed.), "Olfaction and taste." New York: Rockefeller University Press.

Wolf, G., and Handel, P. J. (1966). Aldosterone induced sodium appetite: dose response and specificity. *Endocrinology, 78*, 1120–1124.

Wolf, G., and Schulkin, J. (1980). Brain lesions and sodium appetite: An approach to the neurological analysis of homeostatic behavior. In (M. Kare, Ed.), "Biological and behavioral aspects of salt intake." New York: Academic Press.

Yang, Z.-H., and Epstein, A. N. (1991). Blood-borne and cerebral angiotensin and the genesis of salt intake. *Hormones and Behavior, 25*, 461–476.

Yongue, B. G., and Roy, E. J. (1987). Endogenous aldosterone and corticosterone in brain cell nuclei of adrenal-intact rats: Regional distribution and effects of physiological variations in serum steroids. *Brain Research, 436*, 49–61.

Zamir, N., Skofitsch, G., Eskay, R. L., and Jacobowitz, D. M. (1984). Distribution of immunoreactive atrial natriuetic peptides in the central nervous system of the rat. *Brain Research, 365*, 105–111.

Zarahn, E. D., Ye, X., Ades, A. M., Regan, L. P., and Fluharty, S. J. (1992). Angiotensin-induced cGMP production is mediated by multiple receptor subtypes and nitric oxide in N1E-115 neuroblastoma cells. *Journal of Neurochemistry, 58,* 1960–1963.

Zardetto-Smith, A., Beltz, T. S., and Johnson, A. K. (1991). Bed nucleus of the stria terminalis lesion reduces yohimbine and furosemide-induced salt ingestion.

Zhang, D. M., Epstein, A. N., and Schulkin, J. (1993). Medial region of the amygdala involvement in adrenal steroid-induced salt appetite. *Brain Research, 600,* 20–26.

Zhang, D. M., Stellar, E., and Epstein, A. N. (1984). Together intracranial angiotensin and systemic mineralocorticoid produce avidity for salt in the rat. *Physiology and Behavior, 32,* 677–681.

Zolovick, A. J., Avrith, D., and Jalowiec, J. E. (1980). Reversible colchicine-induced disruption of amygdaloid function in sodium appetite. *Brain Research, 5,* 35–39.

CHAPTER

2

Cholecystokinin: A Neuroendocrine Key to Feeding Behavior

James Gibbs, Gerard P. Smith, and Danielle Greenberg

Department of Psychiatry
Cornell University Medical College
and
Edward W. Bourne Behavioral Research Laboratory
The New York Hospital–Cornell Medical Center
White Plains, New York 10605

Why do behaviors occur at some times and not at others? This deceptively simple question—the central issue in the study of motivation—fascinated Curt Richter throughout his life in science. Through it he recognized a fundamental (and still unsolved) problem in feeding behavior: why do animals and humans eat meals?

In 1921, Richter initiated a classic effort to approach this question. Characteristically, he began by recasting the question in an accessible form: *How* do rats eat? To answer it, he created a simple, direct, ingenious device by mounting the corners of two connected cages on rubber tambours that transmitted activity-induced pressure changes from each of the cages to two levers that would mark a smoked drum. Since the larger cage served as the rat's living quarters and the smaller cage was only large enough to accommodate a food cup, the arrangement produced a continuous record of how the rat ordered its eating and how general activity related to eating activity (Richter, 1922). The device neatly illustrates one description of Richter as " . . . a master of the art of measurement, of finding ways to reliably monitor interesting behavior" (Rozin, 1976).

Even in the strikingly impoverished environment that Richter provided, rats *did not nibble continuously.* Rather, they chose to arrange their feeding in an unmistakable sequence of discrete bouts (Figure 1), just as humans do. The figure captures, in miniature, one of the great themes of Curt Richter's life-work—the physiological mysteries so forcefully presented by the rhythmic nature of many behavioral phenomena. In this miniature, sharply intensified activity always precedes and always follows each meal; the very brief meals are always separated by intermeal intervals of approximately 3 hr.

How does the rat arrange meals in this way? Richter's record suggests one perspective: the pattern results from the *inhibitory* power of food to dominate the behavior. Thus, within a few minutes of meal initiation, ingested food stops feeding behavior for hours, generating a fast, total and enduring abolition of the behavior. If the phenomenon were not so familiar to each of us, it would be amazing. The clarity of Richter's demonstration demands renewed respect for this everyday behavioral drama. This inhibitory perspective suggested to us that the more accessible goal for a biological inquiry into mechanism was to identify the physiological events that *stop* feeding when food is eaten, not those that drive feeding. We chose to study satiety, not hunger.

This chapter reviews the experimental consequences of that choice. First, a series of anatomically oriented studies revealed the importance of the small intestine as the site of origin of an unidentified preabsorptive satiety signal. Next, the characteristics of the behavioral response to food in the intestine led to the identification of a classic intestinal hormone, cholecystokinin, as a physiological satiety signal. Studies now in progress focus on determining the mechanism of action of endogenous cholecystokinin, particularly its mode of release, its peripheral site of action, and its route of transmission to the brain.

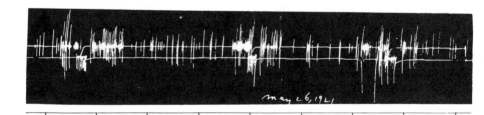

Figure 1. Smoked-drum record made in 1921 by Curt Richter, demonstrating the relationship between general activity and meal-taking behavior in one rat. Excursions from the upper line indicate the rat's movement in a large living cage, whereas excursions from the lower line indicate movement in a smaller, connected food box which the rat entered only to eat. Hatched line below the record represents time in hours. Note the discrete meals separated by intermeal intervals of 2 to 3 hr, and the periods of markedly increased activity that immediately precede and follow meals; the postmeal activity probably reflects grooming and exploration, facets of the behavioral sequence of satiety (Antin, Gibbs, Holt, Young, & Smith, 1975). For details, refer to Richter (1922).

I. Where Does Food Act to Produce Satiety?

As Figure 1 illustrates, the inhibitory effect of ingested food on feeding is rapid—meals end within minutes of their initiation. The fact that the food-activated mechanisms responsible for satiety must be equally rapid biased our attention toward the earliest actions of ingested food on the gastrointestinal tract, thus promoting the possibility of peripheral mechanisms as the most immediate and important. Because meals end before large amounts of food can be absorbed, ingested food might produce signals that are sufficient for satiety at some site (or sites) along the surface of the gastrointestinal tract. To determine whether these signals originate from oral, gastric, or intestinal sites, various surgical techniques have been employed over the past 50 years. When cannulas or catheters are implanted chronically in the esophagus or stomach of laboratory animals and opened temporarily during test meals to allow diversion and removal of the ingested food, all species tested overeat (Janowitz & Grossman, 1949; James, 1963; Davis & Campbell, 1973). This behavior is the "sham feeding" response first noted experimentally by Pavlov and Schumova-Simanovskaia (1895).

The results of this type of behavioral testing can be more striking and more interesting. When total recovery of an ingested liquid food is achieved in sham feeding, satiety is *absent* (Young, Gibbs, Antin, Holt, & Smith, 1974). In Figure 2, note that sham feeding in rats (produced by temporarily opening chronically implanted gastric cannulas) is practically uninterrupted over a period of hours, even in the first sham feeding experience after extensive real feeding experience with the same food. Nonhuman primates, under similar test conditions, display a similar absence of satiety (Gibbs, Maddison, & Rolls, 1981). Thus, the normal taste and swallowing of a familiar food, as well as the past experience and any learning associated with the taste and texture of the familiar food, are not sufficient for satiety under these conditions; the accumulation of food in the stomach and/or the entry of food into the small intestine are necessary. An early instance of human sham feeding (the result of a chronic exteriorizing upper intestinal fistula formed following a penetrating wound of the abdomen) reinforces these results and interpretations. In this case, the patient's physician noted, "It is not easy to imagine the intense hunger and greed with which the patient consumed colossal amounts of food . . . without reaching the feeling of satiation" (Busch, 1858).

To determine the gross anatomical source of the satiety signals that have been short-circuited by gastric sham feeding, we determined whether food infused directly into the small intestine would reinstate satiety in animals in the midst of continuous sham feeding. Again, the results in rats (Liebling, Eisner, Gibbs, & Smith, 1975) and rhesus monkeys (Gibbs et al., 1981) were practically identical: duodenal infusions of small amounts of the same type of liquid food that animals were sham feeding produced highly significant dose-related suppressions of on-

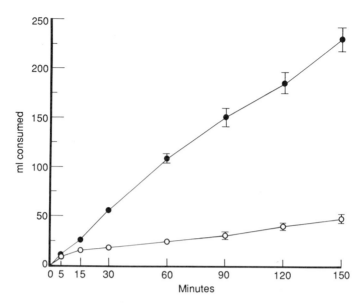

Figure 2. Cumulative intakes of liquid food (mean in ml ± SEM) by three rats, after overnight food deprivations, on a day when the gastric fistulas were closed (○) and all ingested food remained within the gastrointestinal tract (real feeding) and on a day when the gastric fistulas were open (●) and all ingested food drained immediately to the outside (sham feeding). Note the absence of satiety—almost continuous sham feeding—when the gastric fistulas were open. For details, refer to Young et al. (1974).

going sham feeding (see Figure 3). This result was interesting on several counts. First, the important role of the small intestine in satiety was clear. Second, the intestinal signal *did not require the presence of gastric distention* (a textbook explanation for satiety), since gastric distention cannot occur when animals sham feed when gastric cannulas are open. Third, food in the intestine not only produced satiety, but produced the other behaviors normally observed in animals when a meal ends–grooming, exploration, and sleep—a cascade we have referred to as the "behavioral sequence of satiety" (Antin, Gibbs, Holt, Young, & Smith, 1975). Traces of this sequence appear after each meal on Richter's smoked drum (Figure 1). Finally, this "intestinal satiety" effect was fast, occurring within minutes, and transient (see Figure 3).

Thus, the satiating action of food in the small intestine, in the absence of gastric distention, can produce several of the characteristics of satiety that Richter demonstrated in Figure 1—the speed of onset, the resulting brief period of activation (grooming and exploration), and the prolonged period of quiescence (sleep). The results indicated that the small intestine was the source of an important satiety signal that could account for most, and perhaps all, of the behavioral consequences

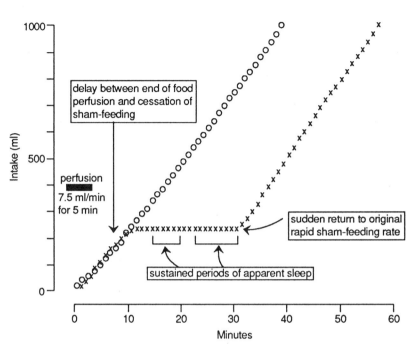

Figure 3. Cumulative sham intakes of liquid food by one rhesus monkey at 1-min intervals on two different test days. The animal was sham feeding (gastric cannula open) on both days. On the control day (O),physiological saline (0.15 M) was delivered to the duodenum during a 5-min period beginning at 0 min; on the test day (x), the same volume of liquid food was delivered to the duodenum at the same rate. After a 7-min delay, food in the intestine stopped sham feeding and elicited the behaviors characteristic of satiety. For details, refer to Gibbs et al. (1981).

of food ingestion. This result focused our attention on an exploration of possible intestinal mechanisms.

II. How Is Intestinal Satiety Produced?

The stimulus of food could act through one or more of several pathways: directly, following the absorption of nutrients into the portal or systemic circulations; by the release of substances from the wall of the small intestine, that would in turn act as satiety signals (through a circulating endocrine or local paracrine mechanism); or by the activation of extrinsic or intrinsic intestinal nerves.

For one nutrient, fat, powerful evidence suggests that the small intestine itself is the source of a potent satiety signal, and that absorbed fats play little or no role in the short-term control of meal size. In rats, we have shown that a powerful dose-related suppression of sham feeding is produced by delivery of a lipid emulsion to the lumen of the duodenum (Greenberg, Smith, & Gibbs, 1990; see Figure 4). The same emulsion delivered to the hepatic portal vein (Greenberg, Becker, Gibbs, & Smith, 1986) or inferior vena cava (Greenberg, Gibbs, & Smith, 1989) was without effect. In addition, after delivery to the duodenal lumen, radio-labeled fats appear in the circulation over a time course that is *clearly inconsistent* with a role in satiety (Greenberg, Kava, Wojnar, & Greenwood, 1989). Similarly, in humans, a significant inhibition of food intake occurs at a test meal after jejunal and ileal infusions of a lipid emulsion, but equivalent intravenous infusions fail to have any effect on food intake under the same conditions (Welch, Saunders, & Read, 1985). Thus, these results limit the possible mechanisms of satiety, at least for fats, to the stimulation of intestinal nerves and/or the release of intestinal hormones.

Richter's record showed that the inhibitory action of food is transient—feeding begins again after an interval of hours (see Figure 1). This characteristic suggested to us that gastrointestinal hormones, which are known to be released for a limited time into the circulation in response to ingested food, might serve not

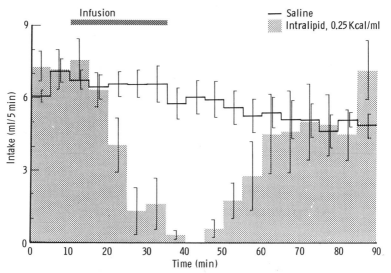

Figure 4. Sham feeding of liquid food (mean intake in ml ± SEM, during sequential 5-min intervals) throughout 90-min test periods on two different days. Seven rats were sham feeding (gastric cannulas open) on both days. On the control day (continuous line), physiological saline was infused intraduodenally during the period indicated; on the test day (shaded area), an emulsified mixture of long-chain triglycerides (Intralipid, 0.25 kcal/ml) was delivered at the same time, volume, and rate. Fat in the intestine produced a rapid, marked, and transient inhibition of sham feeding. In other tests, larger caloric loads produced similarly rapid and marked inhibitions, but for longer periods. For details, refer to Greenberg et al. (1990).

only as signals to curtail a meal, but as signals to control the duration of the intermeal interval. The behavioral results noted in Figure 3 also directed attention to a possible hormonal mechanism. Examination of the results of every test on each rhesus monkey involved in this study revealed a delay of 5–8 min between the end of the intestinal nutrient infusion and the cessation of sham feeding, a latency that suggested that intestinal satiety was not mediated solely by neural elements and that some humoral (and, we believed, hormonal) mechanism was involved.

III. Cholecystokinin and Intestinal Satiety

Some early work had noted that impure intestinal extracts could decrease food intake (Maclagan, 1937; Ugolev, 1960; Schally, Redding, Lucien, & Meyer, 1967). In the early 1970s, based on these leads, we began testing the classic gastrointestinal hormones for their satiating potencies. We found that peripheral administration of an impure preparation of cholecystokinin (CCK, a peptide released into circulation from cells of the small intestine by the stimulus of ingested food arriving in the intestinal lumen) produced a dose-related inhibition of food intake in rats (Gibbs, Young, & Smith, 1973a). Fulfilling the requirements demanded by Richter's record, the inhibitory action of CCK was rapid in onset and limited in duration.

No obvious evidence of disability or malaise was detected when rats ate less after CCK injections. They began eating eagerly, but stopped sooner; also, doses of CCK that reduced liquid food intake after food deprivation failed to reduce water intake after water deprivation. This latter dissociation provided simple clear evidence for the behavioral specificity of the action, since the peptide discriminated between two different motivated behaviors (eating and drinking) that employed the identical motor act—licking a liquid from a spout. Chemical specificity was demonstrated by the finding that the synthetic sulfated C-terminal octapeptide of CCK (a biologically active form in visceral systems, referred to as CCK-8) exerted a potent satiating effect whereas the desulfated form was ineffective (Gibbs, Young, & Smith, 1973b). This difference is important because it indicates that exogenous CCK acts at the Type A CCK receptor (which has a much higher affinity for sulfated than for desulfated CCK and predominates in the periphery) and not at the Type B receptor (which has approximately equivalent affinities for the two forms and predominates in the brain). Further, both gastrin, with a very similar but not identical C-terminal structure, and secretin, a structurally unrelated intestinal peptide, showed little or no activity (Gibbs et al., 1973a; Lorenz, Kreielsheimer, & Smith, 1979).

Peripheral injections of CCK produced inhibitions of *sham feeding* in rats (Gibbs et al., 1973b) and in rhesus monkeys (Falasco, Smith, & Gibbs, 1979).

Figure 5 illustrates an early example of this action. Note that, like the effect of ingested food, the effect of CCK is potent, rapid, and transient. In addition, the use of the sham feeding preparation revealed strong parallels between the satiating action of food delivered to the small intestine, the satiating action of fat (a potent CCK releaser) delivered to the small intestine, and the peripheral administration of CCK (compare Figures 3, 4, and 5). Each manipulation inhibited sham feeding in a dose–response fashion; each elicited the characteristic behavioral sequence of satiety—grooming, exploration, and apparent sleep; and none required gastric distention for its action. In a final parallel, we were able to show that the satiety action of CCK (Antin, Gibbs, & Smith, 1978), like that of food delivered to the small intestine (Antin, Gibbs, & Smith, 1977), requires the synergistic input of oropharyngeal food stimulation in close temporal association. Thus, CCK, a peptide released by nutrients in the intestinal lumen, mimics intestinal satiety.

The satiety action of CCK has been observed in many animal species, including rhesus monkeys (Gibbs, Falasco, & McHugh, 1976) and baboons (Stein, Woods, Figlewicz, & Porte, 1986), across a wide variety of test conditions (Smith

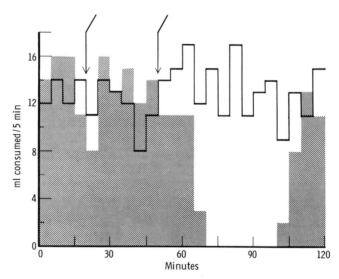

Figure 5. Sham feeding of liquid food by one rat (in ml, during sequential 5-min intervals) throughout 2-hr test periods on two different days. On the control day (continuous line), physiological saline (0.15 M) was injected intraperitoneally at the two points indicated by arrows; on the test day (shaded area), physiological saline (0.15 M) was injected at the first arrow and a partially purified preparation of CCK (approximately 2 µg/kg bioequivalents) was injected at the second arrow. CCK produced a rapid, marked, and transient inhibition of sham feeding. In addition, this solution produced the characteristic sequence of behaviors that accompanies normal satiety—grooming, exploration, and apparent sleep. For details, refer to Gibbs et al. (1973b) and Antin et al. (1975).

et al., 1981). In addition, CCK reduces food intake in normal-weight human volunteers after intravenous infusion (Kissileff, Pi-Sunyer, Thornton, & Smith, 1981; Stacher, Steinringer, Schmierer, Schneider, & Winklehner, 1982; Schick et al., 1991); as in animals, the action in humans is dose-related. The subjective reports of human volunteers receiving CCK have been of particular interest because of controversies concerning whether CCK might inhibit food intake in animals by producing illness or discomfort (Deutsch & Hardy, 1977; Stricker & Verbalis, 1990). However, during the slow intravenous infusions employed to mimic physiological release in the double-blind studies of food intake at test meals in humans, reports of mild discomfort occurred in less than 15% of subjects (Smith & Gibbs, 1987). Further, volunteers routinely were unable to identify days on which the peptide or placebo was administered (Kissileff et al., 1981). Thus, a notable feature of this work is the concordance of studies in animals and humans. Careful thorough tests of efficacy and behavioral specificity in rodents not only have forecast repeatedly the results found in many other species, including nonhuman primates, but have predicted efficacy and the lack of significant side effects in humans with striking accuracy.

Cholecystokinin also produces satiety at test meals in humans with idiopathic obesity (Pi-Sunyer, Kissileff, Thornton, & Smith, 1982). In view of the potential for the therapeutic use of CCK in obesity, it is encouraging to note that the repeated administration of CCK to animals at sequential meals does not provoke evidence of tolerance (West, Fey, & Woods, 1984); nevertheless, whether repeated administration of CCK to obese humans over extended periods of time can reduce body weight has not been determined. In a final parallel between findings in animals and humans, peripheral administration of CCK exerts its full satiating effect in obese rats after electrolytic lesions of the ventromedial hypothalamus (Kulkosky, Breckenridge, Krinsky, & Woods, 1976; Smith, Jerome, Cushin, Eterno, & Simansky, 1981) and in obese patients suffering from hypothalamic tumors or injury (Boosalis et al., 1992).

In summary, an extensive and convergent body of evidence has accumulated to establish that the administration of exogenous CCK can produce the objective signs and the subjective sensations of normal satiety. What is the physiological relevance of this conclusion? This concern—the importance of *endogenous* CCK for the natural termination of a meal—has been a point of frequent dispute. Findings from a rapidly growing list of experiments that employed highly potent and highly selective second- and third-generation CCK receptor antagonists have provided clarification of this crucial question. The results illustrated in Figure 6 provide powerful evidence supporting the specific hypothesis developed in this chapter—that endogenous CCK, released by the preabsorptive action of fat in the small intestine, inhibits food intake. This figure shows *reversal* of the inhibitory effect of intraintestinal fat on sham feeding by peripheral administration of the selective Type A CCK receptor antagonist lorglumide (Greenberg, Torres, Smith,

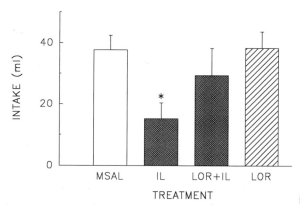

Figure 6. Reversal of the inhibitory action of intestinal fat on sham feeding by a Type A CCK receptor antagonist. Bars represent the total amount of liquid food sham-fed (mean in ml ± SEM) on different days under the following conditions: during an intraduodenal infusion of physiological saline following an intraperitoneal injection of physiological saline (MSAL); during an intraduodenal infusion of Intralipid following a saline injection (IL); during Intralipid infusion following an injection of lorglumide, the selective antagonist (LOR + IL); and during saline infusion following lorglumide injection (LOR). The CCK antagonist reversed fat-induced inhibition of food intake and had no effect when administered alone. *marks statistically significant difference from MSAL, LOR + IL, and LOR results ($p<0.05$); LOR + IL results are not significantly different from MSAL or LOR results. For details, refer to Greenberg et al. (1989).

& Gibbs, 1989; see also Yox, Stokesberry, & Ritter, 1989). The evidence for a physiological role of endogenous CCK is not limited to such specific conditions. Type A receptor blockade has been shown to increase food intake in rats (Watson et al., 1988; Dourish, Rycroft, & Iversen, 1989; Reidelberger & O'Rourke, 1989; Garlicki, Konturek, Majka, Kwiecien, & Konturek, 1990; Miesner, Smith, Gibbs, & Tyrka, 1992), mice (Silver, Flood, Song, & Morley, 1989), hamsters (Adrian, Bilchik, Zucker, & Modlin, 1988), and pigs (Ebenezer, de la Riva, & Baldwin, 1990). Although the initial findings using antagonists in humans still conflict (Wolkowitz et al., 1990; Drewe, Gadient, Rovati, & Beglinger, 1991), most evidence has established firmly that CCK plays a necessary role in limiting meal size in a variety of species under a variety of conditions. Further work should establish the limits and conditions that define this role, and should determine its relevance to humans.

IV. Ontogeny of the Cholecystokinin Satiety System

Peripherally administered CCK can inhibit food intake as early as the first week of life, but the response depends on the experimental conditions. In the neonate,

Houpt and Houpt (1979) and Anika (1983) demonstrated a significant inhibition of natural suckling from the dam as early as 1–3 days of age. However, in pups in which milk was delivered through a tongue cannula during suckling from the dry nipple of an anesthetized dam, CCK failed to inhibit intake at 5 and 10 days; at 15 and 20 days, the satiety response appeared (Blass, Beardsley, & Hall, 1979).

In a different model (independent ingestion of milk from a surface), CCK inhibited intake in a dose–response fashion at 1, 3, 6, and 10 days of age (Robinson, Moran, & McHugh, 1988). This response depends on Type A, not Type B, CCK receptors (Smith, Tyrka, & Gibbs, 1991). Finally, in 9- to 12-day-old pups, we demonstrated that Type A CCK receptor blockade reverses the inhibition of independent ingestion produced by intubation with trypsin inhibitor, a procedure known to increase endogenous peripheral CCK (Weller, Smith, & Gibbs, 1990; see Figure 7). Thus, the endogenous CCK satiety system is intact in preweanling rats. Further work is required to determine the conditions under which the system operates physiologically to control suckling and independent ingestion.

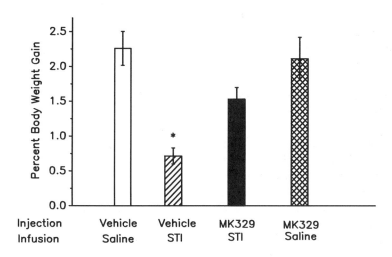

Figure 7. The effect of treatment with MK-329, a selective antagonist of Type A CCK receptors, on suppression of feeding induced by soybean trypsin inhibitor (STI), a releaser of endogenous CCK. Rat pups (9- to 12-day-old) were treated with intraperitoneal injections of MK-329, intragastric intubations of STI, or the combination; independent ingestion of milk from a surface was measured as percentage gain in body weight. The CCK antagonist reversed STI-induced inhibition of food intake and had no effect when administered alone. *marks statistically significant difference from the vehicle/saline-treated group ($p<0.05$) and the MK-329/saline-treated group $p<0.05$); MK-329/STI-treated and MK-329/saline-treated groups were not significantly different. For details, refer to Weller et al. (1990).

V. How Does Cholecystokinin
Act to Elicit Satiety?

A guiding and heuristic assumption of much of the work reviewed here has been that endogenous CCK, once released by ingestion of food, functions through its classic *endocrine* mode of action to produce satiety. Recent experiments have challenged this assumption. Since intestinal endocrine CCK is released into the hepatic portal system, exogenous CCK delivered to the hepatic portal vein should inhibit food intake at least as well as CCK delivered by intraperitoneal injection. When we carried out this experiment, however, hepatic portal injections of CCK-8 failed to inhibit feeding (Greenberg, Smith, & Gibbs, 1987; see also Strubbe, Wolsink, Schutte, & Prins, 1989), suggesting that this small form of the peptide was being degraded in the liver. Although the larger known intestinal forms of CCK, such as CCK-22, may not be degraded, the result suggests that the assumption of an endocrine mode of action should be reconsidered. Additional strong evidence comes from the demonstration that, although the inhibition of sham feeding produced by intestinal infusions of maltose can be reversed by Type A CCK receptor antagonists, these infusions fail to raise levels of circulating CCK bioactivity (Yox, Brenner, & Ritter, 1993). The findings suggest that intestinal CCK released by endocrine cells might act locally, through a paracrine mechanism, to initiate satiety.

Further, an understanding of the mechanism of action of CCK subsequent to its release is far from complete. The intestinal CCK that is released by the presence of ingested food appears to act at a peripheral site when it elicits satiety. This conclusion follows from surgical and chemical lesion studies that have demonstrated that afferent (Smith, Jerome, & Norgren, 1985), capsaicin-sensitive (South & Ritter, 1988) neurons of the abdominal vagus nerve are required for the satiety action of CCK when it is injected peripherally. In addition, surgical removal of the pyloric sphincter of the stomach, an area rich in Type A CCK receptors (Moran, Robinson, & McHugh, 1985) can attenuate the effects of high doses of CCK (Moran, Shnayder, Hostetler, & McHugh, 1988), suggesting that action of the sphincter is necessary for at least part of the satiety action. Whether a corresponding gastric function (such as emptying rate) plays any role in the action is unclear (Moran & McHugh, 1988).

How does peripheral CCK-initiated satiety information enter the brain? Afferent neurons of the vagus have their first synaptic relay in the nucleus tractus solitarius of the dorsal hindbrain. Bilateral electrolytic lesions of this region and the neighboring area postrema abolished the satiating effect of low doses of peripherally injected CCK (Edwards, Ladenheim, & Ritter, 1986). Note that surgical removal of the pyloric sphincter decreases the effect of high doses of CCK and that lesions of the dorsal hindbrain block the effect of low doses. This lack of concordance is one indication that other peripheral sites containing Type A CCK receptors—for example, pancreas and afferent fibers of the vagus nerve itself (Schwartz, McHugh, & Moran, 1991)—must be explored for possible roles in mediating satiety produced by peripheral administration of CCK. The attenuation of peripheral CCK-induced satiety after dorsal hindbrain lesions indicates that this region is a relay for CCK-evoked signals. Central projection sites of dorsal hindbrain neurons therefore are candidates for further processing of such signals, but lesions of these projection sites and their ascending pathways have yielded conflicting results. Some investigators have reported large reductions in the satiating potency of peripherally injected CCK after lesions of paraventricular nucleus (Crawley & Kiss, 1985) or after midbrain transections (Crawley, Kiss, & Mezey, 1984), whereas others have reported an intact response (Smith et al., 1981; Grill & Smith, 1988). Much additional work is needed to identify the brain sites and brain pathways involved in processing the satiety action of peripherally administered CCK.

Peripherally administered CCK does not appear to penetrate the blood–brain barrier (Oldendorf, 1981). The brain itself, however, contains widely and heterogeneously distributed neuronal stores of CCK and CCK receptors. This *central* system can be activated by direct administration of exogenous CCK. By the intracerebroventricular route, the peptide has been found to inhibit food intake in a wide variety of animal species (Maddison, 1977; Della-Fera & Baile, 1979; Schick, Yaksh, & Go, 1986a; Figlewicz, Sipols, Green, Porte, & Woods, 1989). As for peripheral injections, the satiety effect of intracerebroventricular injections is behaviorally specific, and seems to require action of CCK at Type A receptors (Zhang, Bula, & Stellar, 1986), which occur in a limited number of locations in brain tissue (Moran, Robinson, Goldrich, & McHugh, 1986). Only three sites— dorsal hindbrain, parabrachial nucleus, and medial hypothalamus—have been tested for responsiveness to direct injections of CCK. Whereas injections into dorsal hindbrain (nucleus tractus solitarius) were without effect in one study (Crawley, 1985), injections of low doses of CCK-8 into the parabrachial nucleus (Hofbauer et al., 1991) or the paraventricular nucleus of hypothalamus (Faris, Scallet, Olney, Della-Fera, & Baile, 1983) reduced feeding. Further, administration of two relatively nonspecific first-generation CCK antagonists (benzotript and proglumide) into the paraventricular nucleus reliably increased feeding (Schwartz, Dorfman, Hernandez, & Hoebel, 1988), suggesting that endogenous CCK at that site may play an important role in satiety. How might neuronal stores of CCK be activated relative to meals? The answer is unknown, but it is of interest that the release of endogenous CCK-8 at hypothalamic sites has been demonstrated after intragastric administration of nutrients in cats (Schick, Yaksh, & Go, 1986b) and in monkeys (Schick, Reilly, Roddy, Yaksh, & Go, 1987), but not after the intravenous administration of high doses of CCK-8 (Schick, Yaksh, & Go, 1986b).

VI. Therapeutic Implications

What is the promise of this research for an understanding of obesity, bulimia nervosa, and anorexia nervosa? Although the potential is obviously great, much additional work is required. Alterations in CCK levels or function have not been shown to play a role in the development or the maintenance of obesity, although this is a topic of current work (Strohmayer, Greenberg, von Heyn, Dornstein, & Balkman, 1988; Woolf et al., 1988; Strohmayer, von Heyn, Dornstein, & Greenberg, 1989; Boosalis et al., 1992). In an interesting series of studies, circulating levels of CCK in response to test meals have been found to be very low in bulimic patients and markedly elevated in anorectic patients—changes that are consistent with a role for CCK in these disorders (Geracioti & Liddle, 1988; Harty, Pearson, Solomon, & McGuigan, 1991; Philipp, Pirke, Kellner, & Krieg, 1991). Nevertheless, whether these changes are related causally to the psychopathology or simply the result of chronically abnormal patterns of eating behavior is not clear.

Even if CCK is not involved in the pathophysiology of obesity or bulimia, its use in treatment bears consideration. For example, developing releasing agents for the endogenous peptide may be possible (Smith & Gibbs, 1976). Such agents, if calorically trivial, might be used in obese or bulimic patients with the aim of limiting meal size and/or extending the length of the intermeal interval. The exogenous use of CCK itself presents several problems, including the lack of an orally effective form and ignorance about its effectiveness in reducing the intake of highly palatable or preferred foods, those usually chosen for binge eating. Since almost all studies to date have examined the effects of CCK only in single tests of individual meal size, a major unresolved issue is whether repeated administration over time would promote weight loss or simply provoke regulatory countermeasures that would neutralize their short-term actions (West, Fey, & Woods, 1984). Finally, of course, the safety of this biologically potent agent when given chronically is unknown and must be established unequivocally.

VII. Conclusion

Curt Richter's artful clear demonstrations taught students of feeding behavior that the individual meal—brief and infrequent—is the functional unit of analysis. This fundamental lesson prompted questions about where ingested food acts so quickly to stop feeding, and led us to search for gastrointestinal sites at which food rapidly and unambiguously elicits satiety, that is, sites of tight stimulus–response coupling. The behavioral characteristics of the satiety response to food delivered to the upper small intestine suggested neuroendocrine mechanisms, leading in turn to a sustained analysis of the gut–brain peptide CCK as a putative satiety signal. A large, reliable satiety action of peripherally administered exogenous CCK has been demonstrated in many species, including humans. Numerous reports using antagonists of CCK receptors prove that endogenous CCK exerts a physiological role as a signal for satiety—the first peptide signal to be so recognized. The breadth and importance of this role has not yet been explored systematically. Much work must be done before an understanding is reached regarding the peripheral mechanism of the satiety action of CCK and the subsequent processing of this signal, carried by vagal afferent fibers, after its entry into brain tissue. In addition, how the satiety effect of centrally administered CCK, which models the release of CCK in brain, relates to the satiety effect of peripherally administered CCK, must be unraveled. Finally, how successfully these studies will spur the recognition of other peptidergic satiety signals, and whether this area of research can contribute to the understanding or the treatment of eating disorders in humans, remains to be determined. Clearly, these evolving questions provide a rich source of problems for contemporary behavioral neuroscience and for clinical psychiatry.

Acknowledgments

We wish to acknowledge the late Alan Epstein's constructive criticism and constant encouragement of our work with cholecystokinin. His analysis of the role of the peptide angiotensin in thirst and salt appetite remains a rich resource for us.

The work reviewed here was supported, in part, by a grant from the United States Public Health Service (National Institutes of Health DK33248) and a James A. Shannon Directors Award to James Gibbs; from the National Institutes for Mental Health (MH40010 and RSA MK00149) to Gerard P. Smith; and from the National Institutes of Health (DK38757), the Weight Watchers Foundation, and the International Life Sciences Institute to Danielle Greenberg. We thank Jane Magnetti for expert processing of the manuscript.

References

Adrian, T. E., Bilchik, A. J., Zucker, K. A., & Modlin, I. M. (1988). CCK receptor blockade increases hamster body weight and food intake. *FASEB Journal, 2,* A737.

Anika, S. M. (1983). Ontogeny of cholecystokinin satiety in rats. *European Journal of Pharmacology, 89,* 211–215.

Antin, J., Gibbs, J., Holt, J., Young, R. C., & Smith, G. P. (1975). Cholecystokinin elicits the complete sequence of satiety in rats. *Journal of Comparative and Physiological Psychology, 89,* 784–790.

Antin, J., Gibbs, J., & Smith, G. P. (1977). Intestinal satiety requires pregastric food stimulation. *Physiology and Behavior, 18,* 421–425.

Antin, J., Gibbs, J., & Smith, G. P. (1978). Cholecystokinin interacts with pregastric food stimulation to elicit satiety in the rat. *Physiology and Behavior, 20,* 67–70.

Blass, E. M., Beardsley, W., & Hall, W. G. (1979). Age-dependent inhibition of suckling by cholecystokinin. *American Journal of Physiology, 236,* E567–E570.

Boosalis, M. G., Gemayel, N., Lee, A., Bray, G. A., Laine, L., & Cohen, H. (1992). Cholecystokinin and satiety: Effect of hypothalamic obesity and gastric bubble insertion. *American Journal of Physiology, 262,* R241–R244.

Busch, W. (1858). Contribution to the physiology of the digestive organs. *Archiv für Pathologische Anatomie und Physiologie und für Klinische Medizin, 14,* 140–186.

Crawley, J. N. (1985). Neurochemical investigation of the afferent pathway from the vagus nerve to the nucleus tractus solitarius in mediating the "satiety syndrome" induced by systemic cholecystokinin. *Peptides, 6* (Suppl. 1), 133–137.

Crawley, J. N., & Kiss, J. Z. (1985). Paraventricular nucleus lesions abolish the inhibition of feeding induced by systemic cholecystokinin. *Peptides, 6,* 927–935.

Crawley, J. N., Kiss, J. Z., & Mezey, E. (1984). Bilateral midbrain transections block the behavioral effects of cholecystokinin on feeding and exploration in rats. *Brain Research, 322,* 316–321.

Davis, J. D., & Campbell, C. S. (1973). Peripheral control of meal size in the rat: Effect of sham feeding on meal size and drinking rate. *Journal of Comparative and Physiological Psychology, 83,* 379–387.

Della-Fera, M., & Baile, C. A. (1979). Cholecystokinin octapeptide: Continuous picomole injections into the cerebral ventricles of sheep suppress feeding. *Science, 206,* 471–473.

Deutsch, J. A., & Hardy, W. T. (1977). Cholecystokinin produces bait shyness in rats. *Nature (London), 266,* 196.

Dourish, C. T., Rycroft, W., & Iversen, S. D. (1989). Postponement of satiety by blockade of brain cholecystokinin (CCK-8) receptors. *Science, 245,* 1509–1511.

Drewe, J., Gadient, A., Rovati, L. C., & Beglinger, C. (1991). Role of cholecystokinin (CCK) in the control of food intake in man. *Gastroenterology, 100,* A438.

Ebenezer, I. S., de la Riva, C., & Baldwin, B. A. (1990). Effects of the CCK receptor antagonist MK-329 on food intake in pigs. *Physiology and Behavior, 47,* 145–148.

Edwards, G. L., Ladenheim, E. E., & Ritter, R. C. (1986). Dorsal hindbrain participation in cholecystokinin-induced satiety. *American Journal of Physiology, 251,* R971–R977.

Falasco, J. D., Smith, G. P., & Gibbs, J. (1979). Cholecystokinin suppresses sham feeding in the rhesus monkey. *Physiology and Behavior, 23,* 887–890.

Faris, P. L., Scallet, A. C., Olney, J. W., Della-Fera, M. A., & Baile, C. A. (1983). Behavioral and immunohistochemical analysis of the function of cholecystokinin in the hypothalamic paraventricular nucleus. *Society for Neuroscience Abstracts, 9,* 184.

Figlewicz, D. P., Sipols, A. J., Green, P., Porte, D., Jr., & Woods, S. C. (1989). IVT CCK-8 is more effective than IV CCK-8 at decreasing meal size in the baboon. *Brain Research Bulletin, 22,* 849–852.

Garlicki, J., Konturek, P. K., Majka, J., Kwiecien, N., & Konturek, S. J. (1990). Cholecystokinin receptors and vagal nerves in control of food intake in rats. *American Journal of Physiology, 258,* E40–E45.

Geracioti, T. D., Jr., & Liddle, R. A. (1988). Impaired cholecystokinin secretion in bulimia nervosa. *New England Journal of Medicine, 319,* 683–688.

Gibbs, J., Falasco, J. D., & McHugh, P. R. (1976). Cholecystokinin decreased food intake in rhesus monkeys. *American Journal of Physiology, 230,* 15–18.

Gibbs, J., Maddison, S. P., & Rolls, E. T. (1981). Satiety role of the small intestine examined in sham-feeding rhesus monkeys. *Journal of Comparative and Physiological Psychology, 95,* 1003–1015.

Gibbs, J., Young, R. C., & Smith, G. P. (1973a). Cholecystokinin decreases food intake in rats. *Journal of Comparative and Physiological Psychology, 84,* 488–495.

Gibbs, J., Young, R. C., & Smith, G. P. (1973b). Cholecystokinin elicits satiety in rats with open gastric fistulas. *Nature (London), 245,* 323–325.

Greenberg, D., Becker, D. C., Gibbs, J., & Smith, G. P. (1986). Intraportal administration of fats fails to elicit satiety. *Society for Neuroscience Abstracts, 12,* 795.

Greenberg, D., Gibbs, J., & Smith, G. P. (1989). Infusions of lipid into the duodenum elicit satiety in rats while similar infusions into the vena cava do not. *Appetite, 12,* 213.

Greenberg, D., Kava, R., Wojnar, Z., & Greenwood, M. R. C. (1989). Satiation following intraduodenal infusion of intralipid occurs prior to appearance of [^{14}C]-intralipid in plasma. *Society for Neuroscience Abstracts, 15,* 1280.

Greenberg, D., Smith, G. P., & Gibbs, J. (1987). Infusion of CCK-8 into the hepatic-portal vein fails to reduce food intake in rats. *American Journal of Physiology, 252,* 1015–1018.

Greenberg, D., Smith, G. P., & Gibbs, J. (1990). Intraduodenal infusions of fats elicit satiety in the sham feeding rat. *American Journal of Physiology, 259,* R110–R118.

Greenberg, D., Torres, N. I., Smith, G. P., & Gibbs, J. (1989). The satiating effects of fats is attenuated by the cholecystokinin antagonist lorglumide. *Annals of the New York Academy of Science, 575,* 517–520.

Grill, H. J., & Smith, G. P. (1988). Cholecystokinin decreases sucrose intake in chronic decerebrate rats. *American Journal of Physiology, 254,* R853–R856.

Harty, R. F., Pearson, P. H., Solomon, T. E., & McGuigan, J. E. (1991). Cholecystokinin, vasoactive intestinal peptide and peptide histidine methionine responses to feeding in anorexia nervosa. *Regulatory Peptides, 36,* 141–150.

Hofbauer, R. D., Semrad, K. A., Wheat, H. S., Howard, J. J., Cozzari, C., Hartman, B. K., & Faris, P. L. (1991). Involvement of cholecystokinin in the parabrachial nucleus (PBN) in the control of food intake: Anatomical and behavioral studies. *Society for Neuroscience Abstracts, 17,* 542.

Houpt, K. A., & Houpt, T. R. (1979). Gastric emptying and cholecystokinin in the control of food intake in suckling rats. *Physiology and Behavior, 23,* 925–929.

James, W. T. (1963). An analysis of esophageal feeding as a form of operant reinforcement in the dog. *Psychological Reports, 12,* 31–39.

Janowitz, H. D., & Grossman, M. I. (1949). Some factors affecting the food intake of normal dogs and dogs with esophagostomy and gastric fistula. *American Journal of Physiology, 159,* 143–148.

Kissileff, H. R., Pi-Sunyer, F. X., Thornton, J., & Smith, G. P. (1981). Cholecystokinin-octapeptide (CCK-8) decreases food intake in man. *American Journal of Clinical Nutrition, 34,* 154–160.

Kulkosky, P. J., Breckenridge, C., Krinsky, R., & Woods, S. C. (1976). Satiety elicited by the C-terminal octapeptide of cholecystokinin in normal and VMH-lesioned rats. *Behavioral Biology, 18,* 227–234.

Liebling, D. S., Eisner, J. D., Gibbs, J., & Smith, G. P. (1975). Intestinal satiety in rats. *Journal of Comparative and Physiological Psychology, 89,* 955–965.

Lorenz, D. N., Kreielsheimer, G., & Smith, G. P. (1979). Effect of cholecystokinin, gastrin, secretin and GIP on sham feeding in the rat. *Physiology and Behavior, 23,* 1065–1072.

Maclagan, N. F. (1937). The role of appetite in the control of body weight. *Journal of Physiology, 90,* 385–394.

Maddison, S. (1977). Intraperitoneal and intracranial cholecystokinin depress operant responding for food. *Physiology and Behavior, 19,* 819–824.

Miesner, J., Smith, G. P., Gibbs, J., & Tyrka, A. (1992). Intravenous infusion of CCKA-receptor antagonist increases food intake in rats. *American Journal of Physiology, 262,* R216–R219.

Moran, T. H., & McHugh, P. R. (1988). Gastric and nongastric mechanisms for satiety action of cholecystokinin. *American Journal of Physiology, 254,* R628–R632.

Moran, T. H., Robinson, P. H., Goldrich, M. S., & McHugh, P. R. (1986). Two brain cholecystokinin receptors: Implications for behavioral actions. *Brain Research, 362,* 175–179.

Moran, T. H., Robinson, P. H., & McHugh, P. R. (1985). Pyloric CCK receptors: Site of mediation for satiety? *Annals of the New York Academy of Science, 448,* 621–623.

Moran, T. H., Schnayder, L., Hostetler, A. M., & McHugh, P. R. (1988). Pylorectomy reduces the satiety action of cholecystokinin. *American Journal of Physiology, 255,* R1059–R1063.

Oldendorf, W. H. (1981). Blood-brain barrier permeability to peptides: Pitfalls in measurement. *Peptides, 2 (Suppl. 2)*, 109–111.

Pavlov, I. P., & Schumova-Simanovskaia, E. O. (1895). Die innervation der magendrüsen beim hunde. *Archiv für Anatomie und Physiologie* [Cited in I. P. Pavlov, (1910) *The work of the digestive glands* (W. H. Thompson, translator), 2^d English Ed. London: Charles Griffin & Company.

Philipp, E., Pirke, K.-M., Kellner, M. B., & Krieg, J.-C. (1991). Disturbed cholecystokinin secretion in patients with eating disorders. *Life Science, 48*, 2442–2450.

Pi-Sunyer, X., Kissileff, H. R., Thornton, J., & Smith, G. P. (1982). C-Terminal octapeptide of cholecystokinin decreases food intake in obese men. *Physiology and Behavior, 29*, 627–630.

Reidelberger, R. D., & O'Rourke, M. F. (1989). Potent cholecystokinin antagonist L-364,718 stimulates food intake in rats. *American Journal of Physiology, 257*, R1512–R1518.

Richter, C. (1922). A behavioristic study of the activity of the rat. *Comparative Psychology Monograph, 1*, 1–55.

Robinson, P. H., Moran, T. H., & McHugh, P. R. (1988). Cholecystokinin inhibits independent ingestion in neonatal rats. *American Journal of Physiology, 255*, R14–R20.

Rozin, P. (1976). Curt Richter: The compleat psychobiologist. In E. M. Blass (Ed.), *The Psychobiology of Curt Richter* (pp. xv–xxviii). Baltimore: York Press.

Schally, A. V., Redding, T. W., Lucien, H. W., & Meyer, J. (1967). Enterogastrone inhibits eating by fasted mice. *Science, 157*, 210–211.

Schick, R. R., Reilly, W. M., Roddy, D. R., Yaksh, T. L., & Go, V. L. W. (1987). Neuronal cholecystokinin-like immunoreactivity is postprandially released from primate hypothalamus. *Brain Research, 418*, 20–26.

Schick, R. R., Yaksh, T. L., & Go, V. L. W. (1986a). Intracerebroventricular injections of cholecystokinin octapeptide suppress feeding in rats—Pharmacological characterization of this action. *Regulatory Peptides, 14*, 277–291.

Schick, R. R., Yaksh, T. L., & Go, V. L. W. (1986b). An intragastric meal releases the putative satiety factor cholecystokinin from hypothalamic neurons in cats. *Brain Research, 370*, 349–353.

Schick, R. R., Schuszdziarra, V., Mössner, J., Neuberger, J., Schröder, B., Segmüller, R., Maier, V., & Classen, M. (1991). Effect of CCK on food intake in man: physiological or pharmacological effect? *Zeitschrift für Gastroenterologie 29*, 53–58.

Schwartz, D. H., Dorfman, D. B., Hernandez, L., & Hoebel, B. G. (1988). Cholecystokinin: 1. CCK antagonists in the PVN induce feeding 2. Effects of CCK in the nucleus accumbens on extracellular dopamine turnover. In R. Y. Wang & R. Schoenfeld (eds.), *Cholecystokinin antagonists* (pp. 285–305). New York: Alan R. Liss.

Schwartz, G. J., McHugh, P. R., & Moran, T. H. (1991). Integration of vagal afferent response to gastric loads and cholecystokinin in rats. *American Journal of Physiology, 261*, R64–R69.

Silver, A. J., Flood, J. F., Song, A. M., & Morley, J. E. (1989). Evidence for a physiological role for CCK in the regulation of food intake in mice. *American Journal of Physiology, 256*, R646–R652.

Smith, G. P., & Gibbs, J. (1976). Cholecystokinin and satiety: theoretic and therapeutic implications. In D. Novin, W. Wyrwicka, & G. Bray (Eds.), *Hunger: Basic mechanisms and clinical implications* (pp. 349–355). New York: Raven Press.

Smith, G. P., & Gibbs, J. (1987). The effect of gut peptides on hunger, satiety, and food intake in humans. *Annals of the New York Academy of Science, 499*, 132–136.

Smith, G. P., Gibbs, J., Jerome, C., Pi-Sunyer, F. X., Kissileff, H. R., & Thornton, J. (1981). The satiety effect of cholecystokinin: A progress report. *Peptides, 2*, 57–59.

Smith, G. P., Jerome, C., Cushin, B. J., Eterno, T., & Simansky, K. J. (1981). Abdominal vagotomy blocks the satiety effect of cholecystokinin in the rat. *Science, 213*, 1036–1037.

Smith, G. P., Jerome, C., & Norgren, R. (1985). Afferent axons in abdominal vagus mediate satiety effect of cholecystokinin in rats. *American Journal of Physiology, 249*, R638–R641.

Smith, G. P., Tyrka, A., & Gibbs, J. (1991). Type-A CCK receptors mediate the inhibition of food intake and activity by CCK-8 in 9- to 12-day-old rat pups. *Pharmacology, Biochemistry and Behavior, 38,* 207–210.

South, E. H., & Ritter, R. C. (1988). Capsaicin application to central or peripheral vagal fibers attenuates CCK satiety. *Peptides, 9,* 601–612.

Stacher, G., Steinringer, H., Schmierer, G., Schneider, C., & Winklehner, S. (1982). Cholecystokinin octapeptide decreases intake of solid food in man. *Peptides, 3,* 133–136.

Stein, L., Woods, S. C., Figlewicz, D. P., & Porte, D., Jr. (1986). The effect of fasting interval on CCK-8 suppression of food intake in the baboon. *American Journal of Physiology, 250,* R851–R855.

Stricker, E. M., & Verbalis, J. G. (1990). Control of appetite and satiety: Insights from biologic and behavioral studies. *Nutrition Review, 48,* 49–56.

Strohmayer, A. J., Greenberg, D., von Heyn, R., Dornstein, L., & Balkman, C. (1988). Blockade of cholecystokinin (CCK) satiety in genetically obese Zucker rats. *Society for Neuroscience Abstracts, 14,* 1196.

Strohmayer, A., von Heyn, R., Dornstein, L., & Greenberg, D. (1989). CCK receptor blockade by L-364,718 increases food intake and meal taking behavior in lean but not obese Zucker rats. *Appetite, 12,* 240.

Strubbe, J. H., Wolsink, J. G., Schutte, A. M., & Prins, A. J. A. (1989). Hepatic-portal and cardiac infusion of CCK-8 and glucagon induce different effects on feeding. *Physiology and Behavior, 46,* 643–646.

Ugolev, A. M. (1960). The influence of duodenal extracts on general appetite. *Doklady Akademii Nauk SSSR, 133,* 1251–1254.

Watson, C. A., Schneider, L. H., Corp, E. S., Weatherford, S. C., Shindledecker, R., Murphy, R. B., Smith, G. P., & Gibbs, J. (1988). The effects of chronic and acute treatment with the potent peripheral cholecystokinin antagonist L-364,718 on food and water intake in the rat. *Society for Neuroscience Abstracts, 14,* 1196.

Welch, I., Saunders, K., & Read, N. W. (1985). Effect of ileal and intravenous infusions of fat emulsions on feeding and satiety in human volunteers. *Gastroenterology, 89,* 1293–1297.

Weller, A., Smith, G. P., & Gibbs, J. (1990). Endogenous cholecystokinin reduces feeding in young rats. *Science, 247,* 1589–1591.

West, D. B., Fey, D., & Woods, S. C. (1984). Cholecystokinin persistently suppresses meal size but not food intake in free-feeding rats. *American Journal of Physiology, 246,* R776–R787.

Wolkowitz, O. M., Gertz, B., Weingartner, H., Beccaria, L., Thompson, K., & Liddle, R. A. (1990). Hunger in humans induced by MK-329, a specific peripheral-type cholecystokinin receptor antagonist. *Biological Psychiatry, 28,* 169–173.

Woolf, G. M., Howard, J. M., Flower, M. A., McLennan, C. E., Weingarten, H. P., & Collins, S. M. (1988). Postprandial plasma cholecystokinin (CCK) and food intake in lean and obese subjects. *Gastroenterology, 94,* A502.

Young, R. C., Gibbs, J., Antin, J., Holt, J., & Smith, G. P. (1974). Absence of satiety during sham-feeding in the rat. *Journal of Comparative and Physiological Psychology, 87,* 795–800.

Yox, D. P., Brenner, L., & Ritter, R. C. (1992). CCK receptor antagonists attenuate suppression of sham feeding by intestinal nutrients. *American Journal of Physiology, 262,* R554–R561.

Yox, D. P., Stokesberry, H., & Ritter, R. C. (1989). Suppression of sham feeding by intraintestinal oleate: blockade by a CCK antagonist and reversal of blockade by exogenous CCK-8. In J. Hughes, G. Dockray, & G. Woodruff (Eds.), *The neuropeptide cholecystokinin (CCK), anatomy and biochemistry, receptors, pharmacology and physiology* (pp. 218–222). Chichester, England: Ellis Horwood.

Zhang, D.-M., Bula, W., & Stellar, E. (1986). Brain cholecystokinin as a satiety peptide. *Physiology and Behavior, 36,* 1183–1186.

Sexual Behavior: Endocrine Function and Therapy

Kim Wallen and Jennifer Lovejoy[1]

Department of Psychology
and
Yerkes Regional Primate Research Center
Emory University
Atlanta, Georgia 30322

I. Hormonal Regulation of Sexual Behavior: General Issues

A. Ability and Desire

The extent to which ovarian and gonadal hormones influence human sexual behavior is controversial. Although investigators generally agree that male sexual behavior depends on testicular hormones, no consensus has been reached about which aspects of male sexual functioning require testicular hormones. For women, the role of hormones in behavior is even less clear, with no clear agreement that human female sexuality depends on ovarian hormones at all. In a survey of 14 textbooks on human sexuality, we found that 3 of the 14 contended that hormones were irrelevant to female sexual functioning. An additional 7 concluded that hormonal influences on female sexuality existed, but the important hormones were produced by the adrenal cortex and ovarian hormones were of no consequence to

[1]Present address: Pennington Biomedical Research Center, Louisiana State University, 6400 Perkins Road, Baton Rouge, Louisiana 70808-4124.

female sexual functioning. This lack of agreement stems partly from an incomplete reading of the available research findings, as is discussed later in this chapter, but also from our unclear understanding of how human sexual behavior is affected by hormones. In particular, the conceptual framework used to consider hormonal influences on sexual behavior rarely distinguishes hormonal effects on the physical ability to engage in sexual activity from hormonal influences on interest in engaging in sexual activity, particularly when considering human male sexuality in which erection, which physically enables males to engage in sexual intercourse, is considered synonymous with sexual arousal, which is a psychological state that may or may not perfectly match the occurrence of erection. Because of this view, few studies look specifically at males sexual interest and focus instead on control of erection. Similarly, in women, since no obvious physical barrier to engaging in sex exists at any time, the potential of hormones to affect sexual interest routinely has been ignored. Instead, the focus of research on ovarian hormones has been on the occurrence of intercourse and its frequency in relation to the female ovarian cycle.

This perspective began to change in the mid-1970s with the recognition that lack of sexual interest is a clinically important problem in both males and females (Lief, 1977). This recognition resulted in the inclusion of "Inhibited Sexual Desire" in the *Diagnostic and Statistical Manual of Mental Disorders* (DSM-II) [American Psychiatric Association (APA), 1980], which was later changed to "Hypoactive Sexual Desire Disorder" (HSDD) in the revised manual (APA, 1987). This awareness has led to an increasing focus on sexual desire in human sexuality (Lieblum & Rosen, 1988) and possible hormonal regulation of sexual desire (Segraves, 1988). One of the themes to be developed in this chapter is that steroid hormones primarily influence the sexuality of both males and females by affecting sexual desire. A second theme is that hormonal effects on sexual behavior are influenced strongly by social context and are not strict regulators of sexual behavior. Thus, under some social circumstances hormone levels will predict accurately the occurrence of sexual behavior whereas, in a different social context, the same hormonal conditions will appear unrelated to the occurrence of sexual activity (Wallen, 1990).

Current research supports the general conclusion that hormones do not influence greatly the ability of males and females to engage in sexual activity, but modulate sexual interest instead. Ironically, Miller (1931) argued 60 years ago that a unique human adaptation was the ability to engage in sexual activity without hormonal influence, allowing sexual activity to be used in other than reproductive contexts. Subsequent research demonstrated that Miller's view applied to many primate species other than humans as well, and that the ability to mate at any time in the female ovarian cycle may be characteristic of primates (Wallen, 1990). Over the last 15 years, interest in the role hormones play in regulating sexual desire has increased. The separation of hormonal regulation of the physical ability to mate from hormonal regulation of sexual desire allows reconciliation of seemingly contradictory research results in both animals and humans.

Although relating adult responsiveness to gonadal hormones to the effects of hormonal exposure during sexual differentiation is beyond the scope of this chapter (Goy and McEwen, 1980), the characteristics of some of the hormonal effects described here are likely to be related to early androgen exposure. In addition, some differences between males and females, particularly the higher incidence of "deviant hypersexuality" (Cooper, Ishmail, Phanjoo, & Love, 1972) in males, may reflect male exposure to high levels of androgens during sexual differentiation. The exact relationship between behavioral predispositions influenced by prenatal hormonal exposure and adult response to gonadal steroids still must be explored fully.

B. Relevance of Animal Models to Human Sexual Behavior

The conceptual framework for understanding human sexual behavior has come primarily from animal studies. Of these, the overwhelming research emphasis has been on the sexual behavior of laboratory rodents; with a much less concentrated investigation of the sexual behavior of nonhuman primates. The emphasis on rodent sexual behavior has produced great advances in our understanding of how hormones can regulate sexual behavior. Research in rodents first successfully demonstrated that female sexual behavior could be predicted by understanding the underlying ovarian cycle and that sexual behavior could be restored completely by replacing ovarian hormones after ovariectomy (Young, 1937, 1961). Similarly, studies of rodent males demonstrated the critical importance of testicular function on male sexual behavior. Although very successful, this focus on rodent sexual behavior inadvertently delayed the appropriate study of human sexual behavior.

One of the most striking differences between rodent and human sexual behaviors is the stereotyped nature of the former compared with the great flexibility and lability of human sexual behavior. This marked difference led some researchers to argue that the relationship between the behavior of humans and rodents was small and that human behavior would not follow the principles found to regulate rodent sexual behavior. Thus, a tendency evolved to regard results from studies of rodents as irrelevant to an understanding of human behavior. This view stems partly from a failure to appreciate that, in rodents, steroids regulate both the ability to engage in sex and sexual interest, whereas hormones primarily influence sexual desire in humans.

The evidence for hormonal regulation of the ability to mate in rodents is striking, but has been regarded in this context only recently. For example, the vagina of adult female guinea pigs is sealed by a thin membrane for 11–13 days of her 15-day cycle. This membrane ruptures just before ovulation and the onset of sexual receptivity, and returns with the onset of luteal function (Young, 1937). Thus, a female guinea pig cannot mate for most of her cycle, even if she is interested. Similarly, ovarian hormones strictly determine the occurrence of lordosis in female rodents (Young, 1961). However, not until detailed X-ray cinematographic studies revealed that the postural changes during lordosis permitted the male to intromit could lordosis be seen as regulating the ability to mate (Pfaff, Diakow, Montgomery, & Jenkins, 1978). In males, early work demonstrated that penile structure, determining genital feedback, was dependent on testicular hormones (Beach & Levinson, 1950). Without appropriate penile sensitivity, male rats cannot achieve erection, making intromission impossible. In these examples, hormones determine the ability of male and female rodents to mate. In contrast, hormones play a less important role in the regulation of the ability to mate in humans. Thus, castrated male humans achieve penile erections in response to sexual stimulation (Kwan, Greenleaf, Mann, Crapo, & Davidson, 1983) and ovariectomized women can engage in sexual intercourse without hormonal therapy. Physical aspects of sexual functioning, such as vaginal lubrication and the rigidity of the male erection, are affected by hormones. Unlike the situation in rodents, these hormonal effects are a matter of degree rather than determining factors for the ability to mate.

Studies of nonhuman primates suggest strong similarities between hormonal regulation of their sexual behavior and that of sexual behavior in humans. Because nonhuman primate sexual behavior can be studied under more controlled conditions than that of humans, the extent to which gonadal hormones affect the sexual behavior of some primate species was discovered to vary with the social environment of the animals. Thus, the sexual activity of rhesus monkeys appeared to be coupled loosely to the female ovarian cycle when studied in male–female pairs in small laboratory cages, but coupled tightly to the stage of the female ovarian cycle when studied in multifemale groups in large areas (Wallen, 1990). Similarly, the effect of eliminating testicular steroids on male rhesus monkey sexual behavior was less profound when studied in male–female pairs than when studied in a multimale–multifemale group (Phoenix, Slob, & Goy, 1973; Wallen, Eisler, Tannenbaum, Nagell, & Mann, 1991). This lability in the extent to which hormones regulate rhesus monkey sexual behavior was perplexing, for it had not been reported in any other mammalian species. However, when rhesus monkeys, like humans, were realized not to depend on hormones to mate, research interest shifted to how hormones modulate sexual interest in nonhuman primates (Wallen, 1990). A conceptual framework that distinguishes hormonal regulation of ability from regulation of sexual interest allows development of a general framework for understanding hormonal influences on sexuality across species and allows the reconciliation of research in humans with that in other species. Essentially, differences among species in the degree to which hormones appear to control their sexual behavior result from differences in the degree to which their sexual behavior depends on hormonally regulated mechanisms that control the ability to mate.

This view is an elaboration of the "encephalization" of sexual behavior proposed by Beach (1947,1967). The "encephalization of sexual behavior" idea argues that, as the neocortex becomes elaborated in higher animals, hormones have less influence on sexual behavior since cortical mechanisms substitute for the previously hormonally regulated lower mechanisms. Beach did not describe the nature of this change in regulation explicitly, but suggesting that the lower reflexive mechanisms are regulators of the ability to engage in sex is consistent with his proposal. As cortical systems developed, reflexive control of sexual behavior through regulation of sexual capability was replaced by motivational systems regulating choice and sexual interest. This conceptual framework provides a consistent context for understanding the way in which hormones affect the sexual behavior of humans and nonhumans.

II. Female Sexual Behavior

The role of hormones in regulating female sexual behavior has been examined in a variety of species and a variety of contexts. In nonprimate mammals, ovarian hormones have been found to regulate sexual behavior more clearly than in primates, including women. The following sections address the influence of hormones on female sexual behavior in several ways. First, changes in sexual behavior across the estrous or menstrual cycle are examined, followed by analysis of the role of the adrenal gland in female sexual behavior. Finally, hormonal therapies for loss of libido or other sexual dysfunction in women are discussed.

A. Cyclic Changes in Female Sexual Behavior

The influence of the ovarian steroids estrogen and progesterone on sexual behavior was studied first by examining the covariation of hormonal changes and behavior across the female estrous or menstrual cycle. In female rodents (e.g., rats, mice, and guinea pigs) and other nonprimate mammals (e.g., dogs), a tight correlation is seen between levels of estradiol or estradiol and progesterone and behavior (Young, 1961). In rodents under normal conditions, lordosis only occurs in response to sexual stimulation when a preovulatory progesterone surge occurs after a period of elevated follicular estradiol.

In primate females, human and nonhuman, the relationship between ovarian hormones and sexual behavior is less clear. In fact, under typical laboratory conditions, the relationship is so weak that some investigators have considered female primate sexual behavior to be hormonally "emancipated," a view that has not proven to be completely accurate. In the most widely studied nonhuman primate, the rhesus macaque, the importance of ovarian hormones in modulating female sexual behavior depends on the social and environmental conditions under which the animals are studied (Wallen, 1990). Figure 1 illustrates the difference in how tightly associated rhesus female sexual behavior is with her ovarian cycle under these two testing conditions. In caged rhesus monkeys, social factors such as forced proximity to a male and the absence of competing females may be more important than hormonal changes. When caged male–female pairs are tested on each day of the female ovarian cycle, a midcycle increase in sexual activity is seen but, in contrast to rodents, female rhesus monkeys mate throughout their menstrual cycle and may continue to mate even when their ovaries are removed (Michael & Zumpe, 1970; Goy, 1979). In contrast, rhesus monkey females studied under group-living conditions show highly reliable correlations between cycle stage and sexual behavior: behavior peaks at midcycle and ceases completely in the early follicular phase and in the luteal phase (Gordon, 1981; Wallen, Winston, Gaventa, Davis-DaSilva, & Collins, 1984). When correlations between hormone levels and sexual behavior were examined in these studies, female sexual initiation behavior (proceptivity; Beach, 1976) was positively correlated with levels of estradiol and

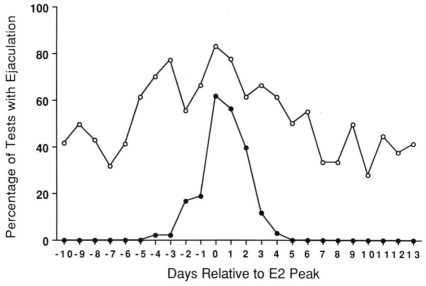

Figure 1.   Comparison of the relationship between the female ovarian cycle and the occurrence of sexual behavior in rhesus monkeys under two different social conditions. Follicular tests are on the left side of the figure and luteal tests are on the right side. Ovulation occurs on approximately day +1. The pair-test [O; Goy (1979)] consisted of a single male and female in a small area, whereas the group test [●; Wallen et al. (1984)] had one male and nine females. Note the marked difference between the two conditions in the occurrence of early follicular and luteal copulation.

inversely correlated with levels of progesterone (Wallen et al., 1984). In contrast, demonstrating significant effects of hormones on female sexual initiation in laboratory pair tests is difficult (Michael, Zumpe, & Bonsall, 1982). Thus, in the monkey, ovarian hormones are clearly important for female sexual initiation; however, the degree of hormonal influence is modified by social and environmental context.

In humans, the data are even more conflicting, in part because of the lack of clearly meaningful measures of female sexual motivation and the challenge, especially in early studies, of obtaining frequent hormone samples to correlate with behavior. Early studies reported peaks in sexual behavior pre- and/or postmenstrually (Schreiner-Engel, Schiavi, Smith, & White, 1981) or peaks at midcycle (Udry and Morris, 1968; Matteo and Rissman, 1984). In these studies, no hormone measurements were available and phase of cycle was estimated by counting backward from menstruation, a technique that only partly aligns the critical periovulatory endocrine events. In studies in which blood samples were taken to provide precise hormonal measurements, the findings also were ambiguous. Although Persky and colleagues (Persky et al., 1978) and Abplanalp and co-workers (Abplanalp, Donnelly, & Pose, 1979) found no changes in female sexual behavior across the menstrual cycle, several studies (Schreiner-Engel et al., 1981; Sanders & Bancroft, 1982; Bancroft, Sanders, Davidson, & Warner, 1983) found behavioral peaks pre- and postmenstrually. However, in these cases, either the studies had few blood samples per cycle or the authors chose to collapse data across cycle phases rather than investigate daily changes in behavior in relation to daily hormonal changes. In the one case in which sexual behavior and hormones were evaluated on a daily basis, Hedricks and colleagues (Hedricks, Piccinino, Udry, & Chimbira, 1987) found that peak coital rate occurred on the day of the luteinizing hormone (LH) surge, which would be the day of peak estradiol secretion. Unfortunately, this study only measured intercourse frequency, making it impossible to determine whether female sexual interest bore the same relationship to hormonal changes. Thus, in women, a relationship exists between sexual behavior and the menstrual cycle, but the exact relationship to sexual desire and the specific hormonal events underlying changes in behavior are unclear.

B. Suppression or Removal of Ovarian Function

The limitations of a strictly correlational approach can be overcome by examining the effects of removing ovarian steroids through surgery, pharmacological suppression, or natural menopause.

In nonprimate mammals, ovariectomy completely abolishes female sexual behavior, which is restored by the administration of exogenous estrogens or a sequential estrogen–progesterone treatment, depending on the species (Young, 1961). In nonhuman primates, although variable results are acquired depending on the social environment, ovariectomy reduces female sexual behavior under all conditions. As in studies of covariation, the effect of removing ovarian steroids is more profound in a complex social environment with multiple females than in laboratory pair tests, in which approximately 20% of ovariectomized females continue to mate for years after ovariectomy (Wallen & Goy, 1977; Wallen et al., 1986). However, under both testing conditions, ovariectomy completely eliminates female sexual initiation, suggesting comparable effects of hormones on motivational systems under diverse social circumstances (Wallen & Goy, 1977; K. Wallen, unpublished observations).

A similar consensus concerning the effects of ovariectomy-hysterectomy on female sexual behavior has been reached for women. Lieblum and co-workers (Lieblum, Bachmann, Kemmann, Colburn, & Schwartzman, 1983) report a precipitous drop in sexual desire after ovariectomy. Similarly, Sherwin and colleagues (Sherwin, 1985; Sherwin, Gelfand, & Brender, 1985; Sherwin & Gelfand, 1987) and Dennerstein and Burrows (1977) report dramatically diminished desire in surgically menopausal women, which is restored with the administration of exogenous hormones. However, this consensus is recent since the general view until the mid-1970s was that ovariectomy had no effect on female sexual behavior. A more careful reading of earlier studies that actually measured sexual desire (e.g., Waxenberg, Drellich, & Sutherland, 1959) reveals evidence of decreased sexual desire after ovariectomy, although desire was not eliminated completely. Thus, some of the confusion in the earlier literature may have stemmed from using intercourse frequency as the sole measure of female sexual interest rather than attempting to assess female desire independently.

Changes in libido and female sexual behavior also have been studied widely in women undergoing natural menopause. Here again, the most consistent finding is that menopause results in a decline in female sexual interest and coital frequency in perimenopausal women (Bachmann et al., 1985; Davidson, 1985; McCoy & Davidson, 1985; Sarrel & Whitehead, 1985; Osborn, Hawton, & Gath, 1988; Morley, 1991). Some of these changes probably are caused by a combination of psychological and motivational factors and physiological factors such as decreased vaginal lubrication, and may provide evidence of the importance of ovarian hormones to the physiological aspects of sexual functioning. Note also that, although sexual behavior diminishes in postmenopausal women, a substantial proportion of older women continue to engage in intercourse well past the age of 80 years (Bretschneider & McCoy, 1988). However, to date no studies have been done that specifically investigate sexual desire in older women rather than the occurrence of intercourse.

Another paradigm for understanding the role of ovarian hormones in female sexuality results from the widespread use of hormonal contraceptives. Estrogen and progestin-based contraceptives, such as the "combined" oral contraceptive pill, have been shown to reduce female sexual initiation and desire in women

(Adams, Gold, & Burt, 1978). In fact, this discovery presented the first clear evidence suggesting that ovarian hormones modulate female sexual initiation. Similarly, Warner and Bancroft (1988) provided additional support for the role of hormones in female sexual behavior when they found that monophasic pill users, who receive a constant hormonal stimulus during the pill cycle, showed less variation in sexual behavior across the cycle than did triphasic pill users, who experience three distinctly different hormonal conditions.

Gonadotropin-releasing hormone (GnRH) agonists have been used in humans to suppress ovarian function, either for contraception or as treatment for gynecological disorders. Unfortunately, none of the human contraceptive trials has evaluated the effects of ovarian suppression on female sexual desire. Results from rhesus monkeys would suggest that women using these compounds would experience reduced sexual interest. GnRH agonist administration in female rhesus monkeys under group-living conditions completely eliminates female sexual initiation (Wallen et al., 1986). Similarly, Lemay and colleagues (Lemay, Maheux, Huot, Blanchet, & Faure, 1988) suggested that women receiving GnRH agonist as treatment of endometriosis often experienced decreased libido. Thus, in all situations studied, artificial or natural suppression of ovarian hormones in female mammals, possibly including humans, reduces sexual behavior, the most consistent effect is on female sexual desire. The effects of these manipulations on intercourse frequency are more variable, as would be expected from a behavior that is not strictly under control of either the male or the female.

The consistency of the relationship between ovarian hormones and human female sexual interest makes it particularly unfortunate that the effects of new hormonal contraceptive therapies in humans (e.g., Norplant) have not included assessment of their effects on female sexual motivation.

C. Adrenal–Cortical Function and Female Sexual Behavior

Although the idea that estrogen and progesterone might influence female sexual behavior seems obvious, another less obvious area of extensive research exists that links female sexual motivation with adrenal androgens. Understanding the development of this idea begins with examination of early methods of treatment for "female hypolibido" in this century.

Research on the effects of androgens in women began with reports from the medical community in the 1930s and 1940s that suggested that large doses of testosterone (frequently inducing clitoral growth and deepening of the voice) increased libido in women (Salmon & Geist, 1943). A study by Waxenberg et al. (1959) on the effects of ovariectomy and adrenalectomy in terminally ill cancer patients had great impact by suggesting that ovariectomy had no effect on female sexual desire but adrenalectomy completely eliminated sexual interest. These researchers concluded that, unlike all previously studied mammals, women depended on adrenocortical androgens for sexual desire. Despite the facts that the study used women who were quite ill, actually showed a comparable decrease in sexual desire from either ovariectomy or adrenalectomy, and lacked the appropriate experimental group in which the ovaries remained intact but the adrenal cortex was removed, this study still is cited routinely as evidence of the critical role of the adrenal cortex in determining female sexuality.

After these early human studies, the role of adrenocortical secretions was investigated in rhesus monkeys using pair tests. These studies suggested that adrenalectomy (or suppression of the adrenals with dexamethasone) diminished or abolished female sexual receptivity (Everitt & Herbert, 1969; Everitt, Herbert, & Hamer, 1972; Johnson & Phoenix, 1976). Testosterone and, in some cases, androstenedione reinstated the behavior in adrenal-suppressed animals. Again, in these studies, an aromatizable androgen (e.g., testosterone) was required to reinstate the behavior whereas nonaromatizable androgens (e.g., 5α-dihydrotestosterone) were ineffective (Wallen & Goy, 1977). In contrast to these early studies in rhesus monkeys, no effect of adrenalectomy or adrenal suppression has been seen in female marmosets (Dixson, 1987) or stumptail macaques (Baum, Slob, de Jong, & Westbroek, 1978; Goldfoot, Weigand, & Scheffler, 1978).

We studied the effects of dexamethasone-induced adrenal suppression in group-living rhesus monkey females (Lovejoy and Wallen, 1990). This treatment reduced total androgens by more than 60%, but did not interfere with ovarian function. Unlike ovarian suppression, which completely eliminated female sexual initiation (Wallen et al., 1986), adrenal suppression had no measurable effect on any aspect of female sexual behavior. Thus, although both adrenal suppression and ovarian suppression reduce circulating androgens by 50% or more, only ovarian suppression, which also eliminates ovarian estradiol, significantly decreases female sexual behavior, strongly suggesting that estrogens but not androgens are crucial for sexual functioning in rhesus monkeys. The evidence in primates does not strongly support a role for adrenocortical function in normal female sexuality. As shown in Table 1, across several different hormonal conditions, only estradiol level predicts the occurrence of female sexual behavior. Testosterone level, in contrast, bears no consistent relationship to female sexual behavior in rhesus monkeys.

In contrast, the musk shrew, an insectivore, appears to rely on adrenal function for the initiation of sexual behavior (Fortman, Dellovade, & Rissman, 1992). Unlike other mammals, female musk shrews do not have a spontaneous ovarian cycle. Instead, ovarian function and sexual behavior are induced within 30 min to 2 hr of exposure to a male (Rissman & Bronson, 1987). Suppression of adrenal function blocks this male-induced onset of sexual behavior and can be reversed by the administration of testosterone (Fortman et al., 1992). Whether or not this behavior represents a unique adaptation of the musk shrew or is relevant to human female sexuality remains unresolved.

TABLE I

Relationship between Estradiol and Testosterone Levels and Female Sexual Initiation in Female Rhesus Monkeys under Various Hormonal Conditions

Hormonal condition	Estradiol (pg/ml±SEM)	Testosterone (pg/ml±SEM)	Behavioral intensity
Periovulatory			
Untreated[a]	324±30	375±20	High
Adrenal-suppressed[a]	361±24	176±18	High
Luteal: Untreated[b]	91±07	375±20	Low
60 days of GnRH-AG[b]	43±01	359±41	Low

[a]Experimental conditions described in Lovejoy and Wallen (1990).
[b]Experimental conditions described in Wallen et al. (1986).

At present, then, no credible evidence exists that adrenal androgens influence sexuality in healthy women. Some support for an influence of androgens on female sexuality comes from studies of supraphysiological doses administered to premenopausal women or more physiological doses to postmenopausal women. In animal studies, although sexual behavior of some species under some conditions appears to be influenced by androgens, the extent of the role of androgens under normal physiological and social conditions is unclear and probably has been overstated.

D. Female Sexual Dysfunction

Probably the most commonly reported sexual problem in women is hypolibido or HSDD (APA, 1987). Several studies have examined the role of hormonal alterations in HSDD with the hypothesis that, if hormones are important in regulating sexuality in women, abnormal hormone levels might be observed in this population.

A thorough study conducted by Schreiner-Engel and colleagues (Schreiner-Engch, Schiavi, White, & Ghizzani, 1989) compared 17 women who met DSM-IIIR criteria for HSDD with 13 normal controls for one cycle. No differences in estrogen, progesterone, LH, testosterone (free or bound), sex hormone binding globulin (SHBG), or prolactin were seen in the women with HSDD compared with the controls. These researchers also found no hormonal differences when comparing women with severe or mild HSDD or with long or short duration of the disorder. They concluded that no evidence is available that reproductive hormones play a role in individual differences in sexual desire. In contrast, Riley (1984) observed elevated prolactin levels in a small percentage (6–18%) of women complaining of loss of libido. Hyperprolactinemia has been associated with decreased libido in several other studies (Muller, Musch, & Wolf, 1979; Buckman

& Kellner, 1985), but not all (Koppelman, Parry, Hamilton, Algna, & Loriaux, 1987).

Despite the lack of clear evidence for a hormonal etiology of female hypo-libido, several types of hormonal therapy have been employed. As mentioned previously, testosterone therapy was popular in earlier studies and seemed to produce some increase in libido (Carney, Bancroft, & Matthews, 1978), although the doses used were frequently supraphysiological. The beneficial effects of testosterone therapy may be less attributable to effects on sexual motivation than to feelings of well-being produced by anabolic steroids. Some investigators have reported that the sense of well-being is the best predictor of the level of female sexual interest (Warner & Bancroft, 1988). Thus, the effects of testosterone may not reflect a direct effect of the hormone on sexual desire.

Perhaps the most research has been done on hormonal replacement therapy for postmenopausal women experiencing decreased sexual desire. In surgically menopausal women, one of the first studies (Dennerstein & Burrows, 1977) suggested that estrogen replacement therapy did not affect overall sexual behavior, although a specific decrease in painful intercourse (dyspareunia) was noted. A subsequent study by these authors, however, suggested that estradiol significantly increased sexual desire and orgasmic frequency, compared with progesterone or placebo in surgically menopausal women (Dennerstein, Burrows, Wood, & Hyman, 1980). More recently, the effects of estrogen therapy alone and estrogen plus androgen or progesterone have been studied by Sherwin and colleagues (Sherwin, 1985, 1991; Sherwin et al., 1985; Sherwin & Gelfand, 1987). In a double-blind, placebo-controlled study by this group on women who were recently surgically menopausal, a combined estrogen/androgen treatment produced higher levels of sexual desire, arousal, and fantasy than estrogen alone or placebo (Sherwin et al., 1985). However, the results of this study are not completely consistent with a role for androgen in replacement therapy, since the surgical control group (who had low androgen levels but elevated estradiol levels) reported sexual desire comparable to that of the group receiving estrogen and androgen replacement, who had much higher androgen titers. Further, the surgical control group during the placebo month showed higher levels of desire and fantasy than any other group, despite low androgen levels (Sherwin et al., 1985).

A subsequent study compared the effectiveness of combined estrogen/androgen treatment, estrogen alone, or placebo in long-term surgically postmenopausal women (Sherwin & Gelfand, 1987). Subjects previously had been receiving hormonal therapy and stopped treatments 8 weeks prior to the experimental phase. In the initial baseline assessment, all three groups of women reported very low levels of sexual desire, fantasy, and orgasm. After 1 week of the hormonal treatments, only women receiving the combined treatment reported increased sexual desire, although both hormonal treatments resulted in increased orgasm frequency. Although these data seem to support most strongly the importance of androgen in female sexual desire, the report also contains data that contradict this view. Hormonal measurements at baseline, when all three groups reported very low sexual desire, revealed that women who had received the combined estrogen/androgen treatment 8 weeks previously still had circulating testosterone levels 4–5 times the peak seen during the ovarian cycle, but had very low estradiol levels, as did the other experimental groups. Thus, as in the previous study, androgen levels did not predict the level of female sexual desire, but estradiol level at baseline did. More recent data (Sherwin, 1991) in naturally menopausal women suggest that estrogen alone *increases* sexual desire and arousal in postmenopausal women relative to a period of no hormonal administration. Progesterone added to estrogen under these conditions did not alter libido, although it increased scores of negative psychological symptoms. A similar finding was reported by Iatrakis and co-workers (Iatrakis, Haronis, Sakellaropoulos, Kourkoubas, & Gallos, 1986).

Other studies in naturally menopausal women suggest that estrogen therapy alone (or in combination with progesterone) has little effect on improving sexual desire or behavior (Utian, 1972; Coope, 1976; Campbell & Whitehead, 1977; Studd et al., 1977; Furuhjelm, Karlgren, & Carstrom, 1984). With respect to testosterone therapy in natural menopause, Lieblum et al. (1983) and McCoy and Davidson (1985) were able to correlate androgen levels but not estrogens to sexual interest. Dow and colleagues (Dow, Hart, & Forest, 1983), however, could show no advantage of estrogen plus testosterone administration over estrogen alone. An interesting report by Myers and co-workers (Myers, Dixen, Morrissette, Carmichael, & Davidson, 1990) demonstrated that estrogen replacement, with or without progesterone or testosterone, had no effect on mood rating, sexual behaviors, or sexual arousal in 40 postmenopausal women. Estrogen plus testosterone therapy, however, increased reported pleasure from masturbation. These authors suggested that androgens may not influence female sexual behavior in general, but could have a specific effect on self-stimulatory behavior. The reason for the discrepancy between these studies on the importance of estrogen is unclear, but may be related to the bioavailability of different steroid preparations and the influence of SHBG on free estradiol levels (Burke & Anderson, 1972; Anderson, 1974).

E. Conclusions

Clear evidence exists for hormonal influence on female sexual behavior in all species studied, although the degree of effect varies with species and environmental and social conditions. In rodents, sexual behavior is linked tightly to ovarian steroid levels. In rhesus monkeys, tight correlations between estradiol and progesterone, but not testosterone, and behavior are seen also, although social situation (e.g., pair tests as opposed to group-living animals) modulate the extent to which hormones influence female sexual behavior. In women, some evidence suggests that sexual desire is linked to cyclical fluctuations in hormone levels. Part of the problem may be in measuring female desire, since human studies rely on self-report rather than observations of sexual initiating behavior, as is possible in animal studies. Although cyclic data are not clear, there can be no doubt the loss or diminution of estrogen levels—as a result of surgical or natural menopause or contraceptive therapy—diminishes libido in women. Whether replacement therapy with either estrogen or testosterone in menopausal women is truly effective in reinstating psychological components of sexual desire (rather than merely ameliorating physical symptoms of steroid withdrawal) requires further study.

Given the large number of studies in this area and the conflicting results, clearly such studies must be planned carefully and be methodologically flawless. Finally, we believe that the role of androgens in influencing libido in nonhuman primates and women has been overstated. Although some evidence exists that supraphysiological doses of testosterone enhance feelings of well-being and sexual desire, to date the evidence that testosterone replacement in healthy women regulates female libido is weak.

III. Male Sexual Behavior

Male sexual behavior has been studied extensively in a variety of nonprimate and primate mammalian species. In all species, testicular function has been found to be necessary for the full male copulatory pattern of mounting, intromission, and ejaculation. However, species differ in the extent to which removal of testicular androgens completely eliminates all aspects of male sexual function. The situation in human males is less well understood, but the data suggest strong similarities to aspects of the nonhuman data.

In females, the effects of hormones were discussed in relation to their effects on female desire to engage in sex and to effects on female physical ability to engage in sex. A similar dichotomy can be applied to males by looking at the effects of testicular hormones on interest in mating with females and at effects on male erectile capacity. As with females, these two potential effects of hormones on male sexual behavior have been investigated only recently. Although the common view is that male erectile capacity is related to testicular function, data from humans and nonhuman primates suggest that testicular hormones do not primarily influence erections. This point is particularly important when discussing the role of testicular hormones in the treatment of sex offenders.

A. Relationship between Testicular Function and Male Sexual Behavior

In nonprimate species, male sexual behavior is completely dependent on the functioning of the male testes. Castration produces a complete disappearance of sexual behavior but, unlike the effect of ovariectomy in nonprimate females, male sexual behavior gradually declines following gonadectomy rather than cease immediately (Young, 1961). Early quantification of postcastration behavioral change in male rats (Beach & Levinson, 1950) noted that sexual behavior declined in an orderly progression: ejaculation disappeared first, followed by intromission, mounting, and finally interest in a sexually receptive female. This early study also noted that the decline in behavior was associated with a change in the surface of the glans penis, which is normally rough due to presence of keratinaceous penile papillae (Phoenix, Copenhaver, & Brenner, 1976). Castration causes these spines to slough off and fail to regenerate, resulting in the eventual appearance of a smooth glans penis (Phoenix et al., 1976). Testosterone therapy for castrated males reinstated sexual behavior; the components returned in the reverse order in which they disappeared and penile papillae were restored to the glans penis. The orderly disappearance of the components of sexual behavior and associated changes of the glans penis, combined with the restoration of both behavior and penile morphology by testosterone, led to the suggestion that the postcastration decline in behavior resulted from reduced sensory feedback from a penis deprived of testosterone stimulation (Beach & Levinson, 1950). This view underscored the importance that penile sensitivity and function were seen to have for male sexual behavior and the lesser role of hormones in influencing male sexual motivation. Although other work in male rats described the importance of arousal mechanisms in male sexual behavior (Beach, 1942), the connection between male testicular function and male sexual interest in females was not made until relatively recently (Everitt, 1990).

The view that male testicular hormones primarily influenced penile sensory feedback was clearly only a partial explanation when androgens other than testosterone were discovered to maintain penile structure without maintaining male sexual behavior. Thus, Whalen and Luttge (1971) found that 5α-dihydrotestosterone (DHT), an androgen that cannot be metabolized to estrogen, restored penile morphology after castration but did not reinstate male sexual behavior. In contrast, 19-hydroxytestosterone (an androgen that can be converted to estrogen but not to DHT) restored male sexual behavior without completely reinstating penile structure. These findings suggest that conversion of testosterone to an estrogen is important to the maintenance of male sexual functioning. This idea, termed the "aromatization hypothesis," has gained support in several species in which aromatization blockers administered to intact males have been shown to eliminate sexual behavior. Thus, although testosterone appears to influence both penile morphology and sexual behavior in male rats, the two are not necessarily coupled, and testicular androgens appear to affect both penile structure and male sexual motivation, possibly through different metabolic pathways. Whereas the effects of androgens on penile structure may not be of primary importance in maintaining male rat sexual behavior, penile feedback is of critical importance to male rats. If penile sensitivity is reduced by anesthetic (Carlsson & Larsson, 1964; Adler & Bermant, 1966), or dorsal nerve transection (Larsson & Sodersten, 1973), male rats continue to mount but fail to achieve erections, resulting in decreased intromissions and ejaculations. Thus, penile feedback is important for the full copulatory sequence but cannot account for the changes in mounting and interest in the female that occur after castration.

In male guinea pigs, the role of penile sensitivity in the maintenance of male sexual behavior is less apparent. Reducing penile sensitivity via topical anesthetic only affected the temporal patterning of male copulatory behavior, but did not eliminate either ejaculation or intromissions (K. W. Slimp, personal communication). As in the rat, castration in this species produced a gradual disappearance of male sexual behavior that was reinstated with testosterone therapy (Grunt & Young, 1953). In contrast to the male rat, treatment of castrated male guinea pigs with either testosterone or DHT reinstated male sexual behavior (Alsum & Goy, 1974). Similarly, treatment of castrated male guinea pigs with testosterone and an aromatase blocker still reinstated sexual behavior (Roy & Goy, 1988; Roy, 1992). Thus, aromatization apparently is not involved in the maintenance or restoration of male guinea pig sexual behavior.

Nonhuman primates show similarities and marked differences to the species just described in the regulation of male sexual behavior. As in rodents, aspects of male primate sexual function are dependent on testicular function. However, the exact relationship continues to elude investigators, since no one has demonstrated convincingly that the level of male sexual activity can be predicted from circulating steroid levels. As in females, the most striking difference between nonprimates and primates has been reported for macaque males in relation to the effect of castration. Although castration reduces overall frequencies of mating in both rhesus (Phoenix et al., 1973; Michael & Wilson, 1974) and stumptail macaques (Schenk & Slob, 1986), castrated rhesus monkey males continue to intromit and show ejaculatory reflexes for at least 6 years postcastration (Phoenix, 1975). Castration does not eliminate the ability of the male to produce erections and, unlike the rat, transection of the dorsal nerves of the penis does not interfere with erections and intromissions, although it virtually eliminates ejaculation (Herbert, 1973). In contrast to many rodents, but like guinea pigs, either DHT or testosterone will restore male sexual behavior after castration in rhesus monkeys (Phoenix et al., 1973; Phoenix, 1974). In addition, the androgen 19-hydroxytestosterone, which can be aromatized to estrogen but cannot be 5α-reduced to DHT, does not restore sexual behavior in castrated male rhesus monkeys, suggesting that aromatization is not important for rhesus monkey male sexual behavior (Phoenix, 1976). Similarly, estradiol does not maintain the sexual behavior of castrated rhesus monkey males and actually interferes with their ability to intromit (Michael, Zumpe, and Bonsall, 1990). Thus, estrogen does not appear to have a role in male rhesus monkey sexual behavior. However, responsiveness to DHT for the restoration of male sexual behavior is not necessarily a primate characteristic, since physiological levels of DHT do not restore the sexual behavior of castrated crab-eating macaques (*Macaca cynomolgous*; Michael, Bonsall, and Zumpe, 1987). Other primate species show similarities to rodent males in the regulation of sexual behavior. For example, the common marmoset (*Callithrix jacchus*) has penile spines similar to those found in male rats; spine removal in intact males eliminates intromission and ejaculation (Dixson, 1991). Unlike male rats, male marmosets with either transection of the penile nerves or spines removed continue to mount and produce erections (Dixson, 1988,1991).

Work using GnRH agonists and antagonists to suppress testicular function in rhesus monkeys has demonstrated that, at least for this species, the extent to which suppressing testicular function eliminates sexual behavior depends on the social context. When male testicular function is suppressed in a multimale group, it produces a significant reduction in male sexual behavior within 1 week, as shown in Figure 2 (Wallen et al., 1991), in contrast to a minimum of 7 weeks to produce a significant decline after castration in male–female pairs (Phoenix et al., 1973). Testicular suppression in a single-male–multiple-female group setting requires a

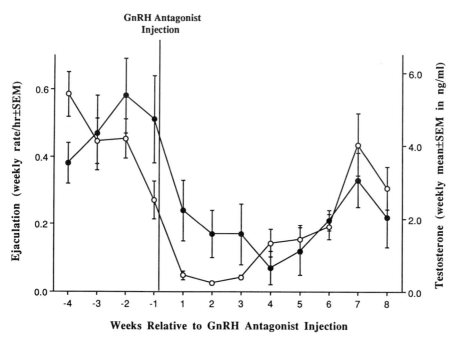

Figure 2. Weekly ejaculation rate (●) in relation to testosterone levels (○) for seven adult rhesus males before and after a single injection of a GnRH antagonist (Antide) (vertical line). Ejaculation rate decreased significantly in the first week after the GnRH antagonist injection.

period of time that is intermediate to that required in pair tests and multimale groups to produce a significant decline in behavior (Davis-DaSilva & Wallen, 1989). This effect of social context is seen most clearly for two males who were used in both testicular suppression studies (Davis-DaSilva & Wallen, 1989; Wallen et al., 1991). In the multifemale social group, these males continued to mate routinely after 4 weeks of testicular suppression. These same males stopped mating 1 week after testicular suppression when tested in a multimale group (Wallen et al., 1991). Thus, like female rhesus monkeys, the opportunity for intrasexual competition influences how tightly coupled sexual behavior is to gonadal function.

The regulation of sexual behavior in human males has many characteristics in common with that in rhesus monkeys. No clear relationship exists between circulating levels of androgen and the frequency of sexual activity, except when androgen titers fall into the castrate range (Davidson, Camargo, & Smith, 1979). However, even castrated males continue to engage in sexual activity, albeit at a reduced level (Heim, 1981). Interestingly, two findings strongly support the idea that testicular hormones primarily influence male sexual interest and not the ability to perform sexually. The first comes from a retrospective study of castrated male sex offenders in Europe (Heim, 1981). These males were asked to estimate their level of sexual intercourse and masturbation before and after castration. Although castration reduced both types of sexual activity significantly, the effect on masturbation was significantly greater than the effect on intercourse. Since masturbation is more dependent on male sexual desire whereas intercourse can be in response to the interest of the partner, this finding suggests that castration primarily affected sexual interest and not ability to engage in sex. This view received further support in a study by Kwan and co-workers (Kwan et al., 1983), who investigated the erectile response of hypogonadal males to sexually explicit films. Surprisingly, males with castrate levels of testosterone developed erections as rapidly as control males and maintained the erection longer than control males. Bancroft and Wu (1983) similarly reported that hypogonadal males produce erections in response to sexually explicit stimuli. Thus the ability to get an erection appears not to be under hormonal control in humans and may be related to exposure to androgens during sexual differentiation. In contrast, sexual desire, and possibly sexual imagery, is strongly dependent on concurrent exposure to testicular hormones.

B. Pharmacological Interference with Testicular Function

The discovery of androgen receptor blockers such as cyproterone acetate (CA) and flutamide ushered in an era of pharmacological methods to control male sexual functioning in the hopes of creating humane treatments for male sex offenders. Although these compounds have dramatic effects when administered to rats and guinea pigs, eliminating male sexual behavior, their effects in primates (nonhuman and human) is less clear. The single study in rhesus monkeys reports that daily administration of CA reduces, but does not eliminate male ejaculatory behavior after several weeks of treatment (Michael, Plant, & Wilson, 1973). In contrast, long-term administration of CA to male stumptail macaques had no detectable effect on their sexual behavior (Slob & Schenk, 1981). Some of this lack of effect may result from the social conditions of testing, or may reflect the incomplete nature of action of these compounds.

The results in humans suggest that these compounds influence male sexual interest but may have little effect on the ability to engage in sex. Bancroft and colleagues (Bancroft, Tennett, Loucas, & Cass, 1974) studied the effect of administering CA to males who had admitted themselves voluntarily for various sexual concerns ranging from pedophilia to excessive sexual demands on their wives. CA treatment markedly decreased sexual desire and interest, but had no effect on erections in response to sexually explicit materials. Similarly, Cooper (1981) found that CA reduced sexual interest as well as spontaneous sexual activity. Thus, treatment of sex offenders with androgen receptor blockers may decrease male interest in seeking out sexual situations, but may have no effect on his responsiveness if he is in the presence of his preferred sexual stimulus. The hope for a simple treatment that will make males uninterested and also incapable of engaging in sex seems unlikely to be developed soon.

C. Male Sexual Dysfunction

Historically, male sexual dysfunction has been synonymous with erectile failure. In fact, male erectile capacity is identified so intimately with male sexual identity that erectile failure is referred to as impotence, although realistically this event should not have that impact on a male's life. Prior to the studies that demonstrated undiminished erectile capacity in hypogonadal males, erectile failure was thought to be related to low levels of androgens. Clearly this is not the case since normal male androgen levels are not required for a male to be capable of producing erections and intromitting. Erectile failure now appears to be unrelated to steroid levels, and may be psychogenic in origin or related to organic causes such as the circulatory consequences of diabetes or other disorders affecting the control of blood flow through the penis. Bancroft and Wu (1983) suggest that erectile failure seen by clinicians in hypogonadal males may be psychogenically induced as a result of performance anxiety in males with low sexual interest. Thus, although erection and male hormones are intimately linked in common parlance, little relationship actually appears to exist between these two male characteristics.

Another male sexual dysfunction is HSDD. Characterizing this condition as a dysfunction of desire has been valuable because it focuses attention on sexual interest and avoids the catch-all term "impotence" that confounds sexual desire and sexual performance, making it difficult to determine the primary dysfunction. HSDD has been reported to occur significantly less frequently in males than in females; males experience HSDD at older ages than females (Segraves & Segraves, 1991). Unlike the condition in females, HSDD in males appears to be related to lower levels of androgen, which may be the result of hypogonadism or age-related declines in testicular function (Schiavi, Schreiner-Engel, White, & Mandeli, 1991). Schavi and colleagues (Schiavi, Schreiner-Engel, White, & Mandeli, 1988) reported significantly lower nocturnal erections and nocturnal levels of testosterone in 17 HSDD males than in 17 matched controls. Further, nocturnal testosterone level in HSDD males, but not in control males, was correlated positively to sexual behavior frequency, suggesting a role for androgen deficiency in this disorder for males. In cases in which HSDD appeared to be related to declines in androgen, hormonal therapy often successfully reversed HSDD (Seagraves, 1988). However, cases of HSDD occur in which hypoandrogenicity as well as other endocrinopathies, such as hyperprolactinemia (Franks, Jacobs, Martin, & Nabarro, 1978), can be ruled out and no clear cause can be determined. Clearly, as in the female, male sexual desire is multiply determined, but a more direct relationship between androgen levels and sexual desire appears to exist, even if the two are not always strictly correlated.

The complementary dysfunction to HSDD might be "hypersexual desire" or hyperlibido. Hypersexuality has been described in the clinical literature typically in reference to deviant or compulsive sexual behavior. Its exact meaning, however, is unclear since the term has been used to refer to the selection of socially inappropriate sexual stimuli, a discordance between the levels of sexual desire in a couple, and even extremely frequent masturbation (Cooper et al., 1972; Cooper, 1981). This lack of precision in the use of these terms makes reaching general conclusions about its cause or treatment difficult. However, if hypersexuality refers to compulsive sexual behavior of high frequency or disturbing intensity, therapies that reduce endogenous testosterone production or action may reduce the compulsive aspects of this sexuality. The androgen receptor blocker CA, as previously described, reduces male sexual desire and has been reported to reduce compulsive sexual behavior effectively (Cooper et al., 1972; Cooper, 1981). Medroxyprogesterone acetate (MPA) therapy, which reduces endogenous testosterone production, also has been reported to reduce the compulsive aspect of sexual behavior in sex offenders (Gagné, 1981). Thus, to the extent that hypersexuality results from excessive or compulsive sexual desire, antiandrogen therapies appear to reduce the occurrence or intensity of this behavior. Note that this aspect of sexuality is almost exclusively a male domain; reports of female compulsive sexuality are exceedingly rare. Male deviant hypersexuality may be a distorted manifestation of a gender difference in sexual desire. For example, a study of Danish women reported that almost one-third never experienced spontaneous sexual desire (Garde & Lunde, 1980). Similarly, 47% of long-term lesbian couples, in which there is no male influence on sexual frequency, reported engaging in sex less than once per month, in contrast to 15% of long-term heterosexual couples. Further, two-thirds of heterosexual couples have weekly sex, as opposed to one-third of lesbian couples (Blumenstein & Schwartz, 1983). Perhaps the fact that males are exposed to relatively constant levels of gonadal steroids predisposes them to more consistent levels of sexual desire, which may become misdirected or compulsive much more easily than in women.

D. Conclusions

Separating male sexual desire from erectile functioning has transformed our understanding of the role of hormones in male sexuality. Although the published information is only slowly becoming generally known, the evidence now overwhelmingly supports the notion that androgens primarily affect male sexual desire and not erectile capacity. Thus, as was the case in women, gonadal hormones in humans act primarily on motivational systems, not on physical mechanisms that allow sexual intercourse to occur. In this regard, human males differ from males of many mammalian species, but may bear some similarities to nonhuman primates in the way hormones regulate sexual behavior. Unlike that of nonhuman primates, human male sexuality also is influenced strongly by social expectations and may be affected adversely by an individual male's expectations and need to function normally. Hormonal therapy still appears to be effective only in those cases with clear evidence of endocrine deficiency; many aspects of male sexuality are unresponsive to hormonal treatment. An understanding of the pivotal role hormones play in male sexual motivation, but not erectile capability, should result in reevaluating expectations for the effectiveness of antiandrogen treatments for male sex offenders. Thinking these treatments will prevent a sexually aggressive male from committing further offenses under the right circumstances is unwarranted. However, thinking these treatments, by reducing the male's level of sexual interest, will reduce the likelihood that the offender will seek out the contexts in which the offenses occurred, thus reducing recidivism, is realistic.

IV. Concluding Remarks

The study of sexual behavior has emerged from its early beginnings, when it was not considered a respectable topic for study, to its current status of being subjected to the same analysis as any other behavioral system. This new objectivity about sexual behavior has allowed investigators to divest the study of sexual behavior of some of its social trappings and to think more clearly about the role of hormones in modulating sexual behavior. Aside from regulating fertility, the most profound affects of gonadal hormones on both males and females increasingly are being found to be psychological. Hormones influence our interest in others sexually and our willingness to engage in sexual activity. We have moved from a mechanistic view of the endocrinology of sexual behavior, in which hormones regulated the expression of stereotyped patterns of sexual behavior, to a view in which hormones set the basic responsivity of individuals to their sexual environment. The final pattern of response then is determined by the current context, the individual's history, and the motivational state. In this view, hormones are not seen as the sole regulators of sexual behavior but as an integral part of the complex of social and physiological systems that coordinate sexual behavior with fertility and social context.

References

Abplanalp, J. M., Donnelly, A. F., & Rose, R. M. (1979). Psychoendocrinology of the menstrual cycle: II. The relationship between enjoyment of activities, moods, and reproductive hormones. *Psychosomatic Medicine, 40*, 523–535.

Adams, D., Gold, A. R., & Burt, A. D. (1978). Rise in female-initiated sexual activity at ovulation and its suppression by oral contraceptives. *New England Journal of Medicine, 299*, 1145–1150.

Adler, N. T., & Bermant, G. (1966). Sexual behavior of male rats: Effects of reduced sensory feedback. *Journal of Comparative Physiology and Psychology, 61*, 240–243.

Alsum, P., & Goy, R. W. (1974). Action of esters of testosterone, dihydrotestosterone, or estradiol on sexual behavior in castrated male guinea pigs. *Hormones and Behavior, 5*, 207–217.

American Psychiatric Association (1980). *Diagnostic and statistical manual of mental disorders*, 3d ed. Washington, D.C.: American Psychiatric Press.

American Psychiatric Association (1987). *Diagnostic and statistical manual of mental disorders* 3d ed., rev. Washington, D.C.: American Psychiatric Press.

Anderson, D. C. (1974). Sex hormone binding globulin. *Clinical Endocrinology, 3*, 60–96.

Bachmann, G. A., Leiblum, S. R., Sandler, B., Ainsley, W., Narcessian, R., Shelden, R., & Hymans, H. N. (1985). Correlates of sexual desire in post-menopausal women. *Maturitas, 7*, 211–216.

Bancroft, J., Tennet, T. G., Loucas, K., & Cass, J. (1974). Control of deviant sexual behavior by drugs: Behavioral effects of oestrogens and antiandrogens. *British Journal of Psychiatry, 125*, 310–318.

Bancroft, J., & Wu, F. C. W. (1983). Changes in erectile responsiveness during androgen replacement therapy. *Archives of Sexual Behavior, 12*, 59–68.

Baum, M. J., Slob, A. K., de Jong, F. H., & Westbroek, D. L. (1978). Persistence of sexual behavior in ovariectomized stumptail macaques following dexamethasone treatment or adrenalectomy. *Hormones and Behavior, 11*, 323–347.

Beach, F. A. (1942). Analysis of the factors involved in the arousal, maintenance and manifestation of sexual excitement in male animals. *Psychosomatic Medicine, 4*, 173–198.

Beach, F. A. (1947). Evolutionary changes in the physiological control of mating behavior in mammals. *Psychology Reviews, 54*, 297–315.

Beach, F. A. (1967). Cerebral and hormonal control of reflexive mechanisms involved in copulatory behavior. *Physiology Reviews, 47*, 289–316.

Beach, F. A. (1976). Sexual attractivity, proceptivity, and receptivity in female mammals. *Hormones and Behavior, 7*, 105–138.

Beach, F. A., & Levinson, G. (1950). Effects of androgen on the glans penis and mating behavior of castrated rats. *Journal of Experimental Zoology (London), 114*, 159–171.

Blumenstein, P., & Schwartz, P. (1983). *American couples: Money, work, and sex.* New York: Morrow.

Bretschneider, J. G., & McCoy, N. L. (1988). Sexual interest and behavior in healthy 80- to 102-year-olds. *Annals of Sexual Behavior, 17,* 109–129.

Buckman, M. T., & Kellner, R. (1985). Reduction of distress in hyperprolactinemia with bromo-criptine. *American Journal of Psychiatry, 142,* 242–244.

Burke, C. W., & Anderson, D. C. (1972). Sex-hormone binding globulin is an oestrogen amplifier. *Nature (London), 240,* 38–40.

Campbell, S., & Whitehead, M. (1977). Oestrogen therapy and the menopausal syndrome. *Clinical Obstetrics and Gynecology, 4,* 31–47.

Carlsson, S. G., & Larsson, K. (1964). Mating in male rats after local anesthetization of the glans penis. *Zeitschrift der Tierpsychologie, 21,* 85–86.

Carney, A., Bancroft, J., & Matthews, A. (1978). Combination of hormonal and psychological treatment for female sexual unresponsiveness: A comparative study. *British Journal of Psychiatry, 132,* 339–346.

Coope, J. (1976). Double-blind crossover study of estrogen replacement. S. Campbell (Ed.) *The management of the menopausal and post-menopausal years* (pp. 159–168). Baltimore: University Park Press.

Cooper, A. J. (1981). A placebo-controlled trial of the antiandrogen cyproterone acetate in deviant hypersexuality. *Comprehensive Psychiatry, 22,* 458–465.

Cooper, A. J., Ishmail, A. A., Phanjoo, A. L., & Love, D. L. (1972). Antiandrogen therapy in deviant hypersexuality. *British Journal of Psychiatry, 120,* 59–63.

Davidson, J. (1985). Sexual behavior and its relationship to ovarian hormones in the menopause. *Maturitas, 7,* 193–201.

Davidson, J., Camargo, C. A., & Smith, E. R. (1979). Effects of androgen on sexual behavior in hypogonadal men. *Journal of Clinical Endocrinology and Metabolism, 48,* 955–958.

Davis-DaSilva, M., & Wallen, K. (1989). Suppression of male rhesus testicular function and sexual behavior by a gonadotropin-releasing-hormone agonist. *Physiology and Behavior, 54,* 263–268.

Dennerstein, L., & Burrows, G. D. (1977). Sexual response following hysterectomy and oophorectomy. *Obstetrics and Gynecology, 49,* 92–96.

Dennerstein, L., Burrows, G. D., Wood, C., & Hyman, G. (1980). Hormones and sexuality: Effect of estrogen and progesterone. *Obstetrics and Gynecology, 56,* 316–322.

Dixson, A. F. (1987). Effects of adrenalectomy upon proceptivity, receptivity and sexual attractiveness in ovariectomized marmosets (*Callithrix jacchus*). *Physiology and Behavior, 39,* 495–499.

Dixson, A. F. (1988). Effects of dorsal penile nerve transection upon the sexual behavior of male marmosets (*Callithrix jacchus*). *Physiology and Behavior, 43,* 235–238.

Dixson, A. F. (1991). Penile spines affect copulatory behavior in a primate (*Callithrix jacchus*). *Physiology and Behavior, 49,* 557–562.

Dow, M. G. T., Hart, D. M., & Forrest, C. S. (1983). Hormonal treatments of sexual unresponsiveness in postmenopausal women. *British Journal of Obstetrics and Gynecology, 90,* 361–366.

Everitt, B. J. (1990). Sexual motivation: A neural and behavioral analysis of the mechanisms underlying appetitive and copulatory responses of male rats. *Neuroscience and Behavioral Reviews, 14,* 217–232.

Everitt, B. J., & Herbert, J. (1969). Adrenal glands and sexual receptivity in female rhesus monkeys. *Nature (London), 222,* 1065–1066.

Everitt, B. J., Herbert, J., & Hamer, J. D. (1972). Sexual receptivity of bilaterally adrenalectomized female rhesus monkeys. *Physiology and Behavior, 8,* 409–415.

Fortman, M., Dellovade, T. L., & Rissman, E. F. (1992). Adrenal contribution to the induction of sexual behavior in the female musk shrew. *Hormones and Behavior, 26,* 76–86.

Franks, S., Jacobs, H. J. S., Martin, N., & Nabarro, J. D. N. (1978). Hyperprolactinemia and impotence. *Clinical Endocrinology, 8*, 277–287.

Furuhjelm, M., Karlgren, E., & Carstrom, K. (1984). The effect of estrogen therapy on somatic and psychical symptoms in postmenopausal women. *Acta Obstetrica Gynecologica Scandinavica, 63*, 655–661.

Gagné, P. (1981). Treatment of sex offenders with medroxyprogesterone acetate. *American Journal of Psychiatry, 138*, 644–646.

Garde, K., & Lunde, I. (1980). Female sexual behavior: A study in a random sample of 40-year-old women. *Maturitas, 2*, 240–255.

Goldfoot, D. A., Weigand, S. J., & Scheffler, G. (1978). Continued copulation in ovariectomized adrenal-suppressed stumptail macaques *(Macaca arctoides)*. *Hormones and Behavior, 11*, 89–99.

Gordon, T. (1981). Reproductive behavior in the rhesus monkey: Social and endocrine variables. *American Zoologist, 21*, 185–195.

Goy, R. W. (1979). Sexual compatibility in rhesus monkeys: Predicting sexual behavior of oppositely sexed pairs of adults. *Ciba Foundation Symposium, vol. 62*, 227–244.

Goy, R. W., & McEwen, B. (1980). *Sexual differentiation of the brain.* Cambridge, Massachusetts: MIT Press.

Grunt, J. A., & Young, W. C. (1953). Consistency of sexual behavior patterns in individual male guinea pigs following castration and androgen therapy. *Journal of Comparative Physiology and Psychology, 6*, 139–144.

Hedricks, C., Piccinino, L. J., Udry, J. R., & Chimbira, T. H. K. (1987). Peak coital rate coincides with onset of luteinizing hormone surge. *Fertility and Sterility, 48*, 234–238.

Heim, N. (1981). Sexual behavior of castrated sex offenders. *Archives of Sexual Behavior, 10*, 11–19.

Herbert, J. (1973). The role of the dorsal nerves of the penis in the sexual behavior of the male rhesus monkey. *Physiology and Behavior, 10*, 293–300.

Iatrakis, G., Haronis, N., Sakellaropoulos, G., Kourkoubas, A., & Gallos, M. (1986). Psychosomatic symptoms in postmenopausal women with or without hormonal treatment. *Psychotherapy and Psychosomatics, 46*, 116–121.

Johnson, D. F., & Phoenix, C. H. (1976). Hormonal control of female sexual attractiveness, proceptivity and receptivity in rhesus monkeys. *Journal of Comparative Physiology and Psychology, 90*, 474–483.

Koppelman, M. C. S., Parry, B. L., Hamilton, J. A., Algna, S. W., & Loriaux, D. L. (1987). Effect of bromocriptine on mood, affect and libido in hyperprolactinemia. *American Journal of Psychiatry, 144*, 1037–1041.

Kwan, M., Greenleaf, W. J., Mann, J., Crapo, L., & Davidson, J. M. (1983). The nature of androgen action on male sexuality: A combined laboratory-self-report study on hypogonadal men. *Journal of Clinical Endocrinologyj and Metabolism, 57*, 557–562.

Larsson, K., & Sodersten, P. (1973). Mating in male rats after section of the dorsal penile nerve. *Physiology and Behavior, 10*, 567–571.

Lemay, A., Maheux, R., Huot, C., Blanchet, J., & Faure, N. (1988). Efficacy of intranasal or subcutaneous luteinizing hormone-releasing hormone agonist inhibition of ovarian function in the treatment of endometriosis. *American Journal of Obstetrics and Gynecology, 158*, 233–236.

Lieblum, S., Bachmann, G., Kemmann, E., Colburn, D., & Schwartzman, L. (1983). Vaginal atrophy in the postmenopausal woman: The importance of sexual activity and hormones. *Journal of the American Medical Association, 249*, 2195–2198.

Lieblum, S., & Rosen, R. C. (1988). *Sexual desire disorders.* New York: Guilford Press.

Lief, H. I. (1977). Inhibited sexual desire. *Medical Aspects of Human Sexuality, 7*, 94–95.

Lovejoy, J., & Wallen, K. (1990). Adrenal suppression and sexual initiation in group-living female rhesus monkeys. *Hormones and Behavior, 24*, 256–269.

McCoy, N. L., & Davidson, J. M. (1985). A longitudinal study of the effects of menopause on sexuality. *Maturitas, 7*, 203–210.

Matteo, S., & Rissman, E. F. (1984). Increased sexual activity during the midcycle portion of the human menstrual cycle. *Hormones and Behavior, 18*, 249–255.

Michael, R. P., Plant, T. M., & Wilson, M. I. (1973). Preliminary studies on the effects of cyproterone acetate on sexual activity and testicular function in adult male rhesus (*Macaca mulatta*). In. *Advances in the Biosciences*, Vol. 10, G. Raspe (Ed.), Pergamon, Press, Oxford.

Michael, R. P., Bonsall, R. W., & Zumpe, D. (1987). Testosterone and its metabolites in male cynomolgous monkeys (*Macaca fascicularis*): Behavior and biochemistry. *Physiology and Behavior, 40*, 527–537.

Michael, R. P., & Wilson, M. (1974). Effects of castration and hormone replacement in fully adult male rhesus monkeys (*Macaca mulatta*). *Endocrinology, 95*, 150–159.

Michael, R. P., & Zumpe, D. (1970). Rhythmic changes in the copulatory frequency of rhesus monkeys (*Macaca mulatta*) in relation to the menstrual cycle and a comparison with the human cycle. *Journal of Reproduction and Fertility, 21*, 199–201.

Michael, R. P., Zumpe, D., & Bonsall, R. W. (1982). Behavior of rhesus during artificial menstrual cycles. *Journal of Comparative Physiology and Psychology, 96*, 875–885.

Michael, R. P., Zumpe, D., & Bonsall, R. W. (1990). Estradiol administration and sexual activity of castrated male rhesus monkeys (*Macaca mulatta*). *Hormones and Behavior, 24*, 71–88.

Miller, G. (1931). The primate basis of human sexual behavior. *Quarterly Review of Biology, 6*, 379–410.

Morley, J. E. (1991). Endocrine factors in geriatric sexuality. *Clinical Geriatric Medicine, 7*, 85–93.

Muller, P., Musch, K., & Wolf, A. S. (1979). Prolactin: Variables of personality and sexual behavior. In L. Zichella & P. Pancheri (Eds.), *Psychoneuroendocrinology in reproduction* (pp. 357–372). Amsterdam: Elsevier.

Myers, L. S., Dixen, J., Morrissette, D., Carmichael, M., & Davidson, J. M. (1990). Effects of estrogen, androgen, and progestin on sexual psychophysiology and behavior in postmenopausal women. *Journal of Clinical Endocrinology and Metabolism, 70*, 1124–1131.

Osborn, M., Hawton, K., & Gath, D. (1988). Sexual dysfunction among middle-aged women in the community. *British Medical Journal, 296*, 959–962.

Persky, H., Charney, N., Lief, H. I., O'Brien, C. P., Miller, W. R., & Strauss, D. (1978). The relationship between plasma estradiol to sexual behavior in young women. *Psychosomatic Medicine, 40*, 523–535.

Pfaff, D. W., Diakow, C., Montgomery, M., & Jenkins, F. A. (1978). X-ray cinematographic analysis of lordosis in female rats. *Journal of Comparative Physiology and Psychology, 92*, 937–941.

Phoenix, C. H. (1974). The effects of dihydrotestosterone propionate on the sexual behavior of castrated male rhesus monkeys. *Physiology and Behavior, 12*, 105–1055.

Phoenix, C. H. (1975). Sexual behavior of castrated male rhesus monkeys treated with 19-hydroxytestosterone. *Physiology and Behavior, 16*, 305–310.

Phoenix, C. H., Copenhaver, K. H., & Brenner, R. M. (1976). Scanning electron microscopy of penile papillae in intact and castrated rats. *Hormones and Behavior, 7*, 217–227.

Phoenix, C. H., Slob, A. K., & Goy, R. W. (1973). Effects of castration and replacement therapy on sexual behavior of adult male rhesus monkeys. *Journal of Comparative Physiology and Psychology, 84*, 472–481.

Riley, A. J. (1984). Prolactin and female sexual function. *British Journal of Sexuality and Medicine, 11*, 14–17.

Rissman, E. F., & Bronson, F. H. (1987). Role of the ovary and adrenal gland in the sexual behavior of the musk shrew, *Suncus murinus*. *Biology of Reproduction, 36*, 664–668.

Roy, M. M. (1992). Effects of prenatal testosterone and ATD on reproductive behavior in guinea pigs. *Physiology and Behavior, 51*, 105–109.

Roy, M. M., & Goy, R. W. (1988). Sex differences in the inhibition by ATD of testosterone-activated mounting behavior in guinea pigs. *Hormones and Behavior, 22*, 315–323.

Salmon, U. J., & Geist, S. (1943). Effect of androgens upon libido in women. *Journal of Clinical Endocrinology and Metabolism, 3*, 235–238.

Sanders, D., & Bancroft, J. (1982). Hormones and the sexuality of women—The menstrual cycle. *Journal of Clinical Endocrinology and Metabolism, 11*, 631–651.

Sarrel, P., & Whitehead, M. I. (1985). Sex and menopause: Defining the issues. *Maturitas, 7*, 217–224.

Schenk, P. E., & Slob, A. K. (1986). Castration, sex steroids, and heterosexual behavior in adult laboratory-housed stumptailed macaques (*Macaca arctoides*). *Hormones and Behavior, 20*, 336–353.

Schiavi, R. C., Schreiner-Engel, P., White, D., & Mandeli, J. (1988). Pituitary-gonadal function during sleep in men with hypoactive sexual desire and in normal controls. *Psychosomatic Medicine, 50*, 304–318.

Schiavi, R. C., Schreiner-Engel, P., White, D., & Mandeli, J. (1991). The relationship between pituitary–gonadal function and sexual behavior in healthy aging men. *Psychosomatic Medicine, 53*, 363–384.

Schreiner-Engel, P., Schiavi, R. C., Smith, H., & White, D. (1981). Sexual arousability and the menstrual cycle. *Psychosomatic Medicine, 43*, 199–214.

Schreiner-Engel, P., Schiavi, R. C., White, D., & Ghizzani, A. (1989). Low sexual desire in women: The role of reproductive hormones. *Hormones and Behavior, 23*, 221–234.

Segraves, K. B., & Segraves, R. T. (1991). Hypoactive sexual desire disorder: Prevalence and co-morbidity in 906 subjects. *Journal of Sex and Marital Therapy, 17*, 55–58.

Segraves, R. T. (1988). Hormones and libido. In R. Lieblum & R. C. Rosen (Eds.), *Sexual Desire Disorders.* (pp. 271–311). New York: Guilford Press.

Sherwin, B. B. (1985). Changes in sexual behavior as a function of plasma sex steroid levels in postmenopausal women. *Maturitas, 7*, 225–233.

Sherwin, B. B. (1991). The impact of different doses of estrogen and progestin on mood and sexual behavior in postmenopausal women. *Journal of Clinical Endocrinology and Metabolism, 72*, 336–343.

Sherwin, B. B., & Gelfand, M. M. (1987). The role of androgen in the maintenance of sexual functioning in oophorectomized women. *Psychosomatic Medicine, 49*, 397–409.

Sherwin, B. B., Gelfand, M. M., & Brender, W. (1985). Androgen enhances sexual motivation in females: A retrospective, crossover study of sex steroid administration in surgical menopause. *Psychosomatic Medicine, 47*, 339–351.

Slob, A. K., & Schenk, P. J. (1981). Chemical castration with cyproterone acetate (Androcur) and sexual behavior in the laboratory-housed male stumptailed macaque (*Macaca arctoides*). *Physiology and Behavior, 27*, 629–636.

Studd, J. W. W., Collins, W. P., Chakravarti, S., Newton, J. R., Oram, D., & Parsons, A. (1977). Oestradiol and testosterone implants in the treatment of psychosexual problems in postmenopausal women. *British Journal of Obstetrics and Gynecology, 84*, 314–315.

Udry, J. R., & Morris, N. M. (1968). Distribution of coitus in the menstrual cycle. *Nature (London), 220*, 593–596.

Utian, W. H. (1972). The true clinical features of postmenopause and oophorectomy and their response to oestrogen therapy. *South African Medical Journal, 46*, 732–737.

Wallen, K. (1990). Desire and ability: Hormones and the regulation of female sexual behavior. *Neuroscience and Biobehavioral Reviews, 14*, 233–241.

Wallen, K., Eisler, J. A., Tannenbaum, P. L., Nagell, K. M., & Mann, D. R. (1991). Antide (NAL-LYS GnRH antagonist) suppression of pituitary–testicular function and sexual behavior in group-living rhesus monkeys. *Physiology and Behavior, 50*, 429–435.

Wallen, K., & Goy, R. W. (1977). Effects of estradiol benzoate, estrone, and propionates of testosterone or dihydrotestosterone on sexual and related behaviors of ovariectomized rhesus monkeys. *Hormones and Behavior, 9,* 228–248.

Wallen, K., Mann, D. R., Davis-DaSilva, M., Gaventa, S., Lovejoy, J., & Collins, D. C. (1986). Chronic gonadotropin-releasing hormone agonist treatment suppresses ovulation and sexual behavior in group-living female rhesus monkeys. *Physiology and Behavior, 36,* 369–375.

Wallen, K., Winston, L., Gaventa, S., Davis-DaSilva, M., & Collins, D. C. (1984). Periovulatory changes in female sexual behavior and patterns of steroid secretion in group-living rhesus monkeys. *Hormones and Behavior, 18,* 431–450.

Warner, P., & Bancroft, J. (1988). Mood, sexuality, oral contraceptives and the menstrual cycle. *Journal of Psychosomatic Medicine, 32,* 417–427.

Waxenberg, S. E., Drellich, M. G., & Sutherland, A. M. (1959). The role of hormones in human behavior. I. Changes in female sexuality after adrenalectomy. *Journal of Clinical Endocrinology, 19,* 193–202.

Whalen, R. E., & Luttge, W. G. (1971). Testosterone, androstenedione, and dihydrotestosterone: Effects on mating behavior of male rats. *Hormones and Behavior, 2,* 117–125.

Young, W. C. (1937). The vaginal smear picture, sexual receptivity, and the time of ovulation in the guinea pig. *Anatomical Records, 67,* 305–325.

Young, W. C. (1961). The hormones and mating behavior. In W. C. Young (Ed.), *Sex and internal secretions* (Vol. II, pp. 1173–1239). Baltimore: Williams and Wilkins.

CHAPTER

4

Hormones and Aggression

Allan Siegel and Melissa K. Demetrikopoulos

Department of Neuroscience
New Jersey Medical School and the Graduate School of Biomedical Sciences
University of Medicine and Dentistry of New Jersey
Newark, New Jersey 07103

I. Introduction

Over the past three decades, an increasing body of literature, discussing studies at both the animal and the human level, has attempted to identify the roles played by various hormonal systems in the regulation of aggressive behavior. Several excellent review chapters have been written on this subject in previous years (Brain, 1977,1978; Bouissou, 1983; Coe and Levine, 1983; Barfield, 1984) as has a text (Svare, 1983). In this chapter, our aim is to review the more recent literature on how hormones may modify aggressive reactions. This review focuses principally on studies that involve androgens and estrogens at different levels over the phylogenetic scale and also includes a brief review of experiments that involve adrenal steroids. The intent is to evaluate the status of these hormones with respect to their effects on aggressive behavior and to suggest possible underlying mechanisms by which these hormones could modulate aggressive responses. Part of the problem associated with our attempts to analyze the experiments in this area of research is related to the fact that different models of aggression have been employed in different laboratories. The concern is that different models of aggression may reflect different behaviors categorized under the rubric of "aggression" rather than a single process. Therefore, we have included a preliminary section outlining different models of aggression, used mainly in rodents, that also have been extended to other species.

Hormonally Induced Changes
in Mind and Brain

II. Animal Models of Aggression

As noted, many experiments aimed at identifying the neural, behavioral, and endocrine substrates of aggressive behavior have used a variety of animal models to examine aggressive reactions. The best elucidation of these models has been offered by Moyer (1976). He describes in detail seven different kinds of aggressive behavior that are associated with different conditions in the environment:

1. fear-induced—a form of aggression that occurs in the presence of an aversive stimulus when escape is not possible
2. maternal—a type of aggression by a lactating female that follows a threat to its recent offspring
3. irritable—a form of aggression that occurs in response to a threat, intimidation, or an annoying environmental condition. This response can be directed against a wide range of environmental objects and is not generally associated with flight or escape behavior
4. sex-related—aggression that occurs in the presence of the same stimuli that elicit sexual behavior and quite often during the act of copulation
5. territorial (resident–intruder)—a response consisting of defensive attack that occurs when an intruder enters an animal's defended area. The territorial or resident–intruder model makes use of the response pattern of the subject following the introduction of an intruder into its environment. Several different intruders typically are used to test this response, including a lactating female, an olfactory bulbectomized male, or an unfamiliar male. Attack on a lactating female intruder is considered independent of testosterone (Whalen & Johnson, 1988) since an intruding lactating female is more likely to be attacked by an intact female than by an intact male. Conversely, olfactory bulbectomized male mice intruders consistently elicit attack from intact male residents and do not typically initiate attack themselves. Therefore, it is evident, that the target stimulus is an important variable in determining the likelihood of aggression elicited by the resident.
6. intermale—spontaneous encounters of fighting between two males that greatly outnumber the frequency of bouts that occur between a male and a female or between two females, although this model may be used to assess interfemale aggression as well. Intermale aggression shares some characteristics with the resident–intruder model, but differs from the resident–intruder model because the interaction does not take place within one subject's territory.
7. predatory—a form of aggression that is directed by a predator toward its natural prey object

Note that both irritable and predatory forms of aggressive behavior can be elicited by electrical stimulation of the medial or lateral regions of the hypothalamus and dorsal and ventral aspects of the midbrain periaqueductal gray matter, respectively, in particular in the cat (Siegel & Pott, 1988; Siegel & Brutus, 1990), but also in the rat (Kruk et al., 1983).

Other models for studying aggression do not easily fit into Moyer's classification scheme, including competition for a limited resource and infanticide. It may be argued that competition for a limited resource may be viewed as a form of territorial aggression. The competition model places subjects in a situation in which they must interact to obtain necessary substances such as food and water, or to obtain access to wanted objects such as a mate or a better territory. However, this model differs from the resident–intruder model in several ways. First, in the resident–intruder model, the interaction takes place within the subject's territory whereas, in the competition model, the interaction takes place within a common area. Second, in the resident–intruder model no necessity to compete exists since the resources available are not limited. In the infanticide model, the subject has access to pups and the variable measured is the death of the pups.

These classifications of aggressive behavior may be reduced to two general categories—defensive aggression and predatory attack. In this sense, forms of aggression such as fear-induced, maternal, irritable, territorial, and intermale may constitute part of the same or related processes. The primary distinctions may comprise the environmental or experimental conditions by which the responses are elicited rather than the nature of the response.

III. Testosterone and Aggressive Behavior

In attempting to develop a better understanding of the role of testosterone in aggressive behavior, a number of the models described have been employed. This discussion summarizes some of the more recent data obtained with the use of these models.

A. Experimental Manipulation of Testosterone Levels

One way to examine the effects of testosterone is to manipulate the subject's testosterone levels experimentally. This goal is accomplished most easily through systemic administration of testosterone or by castration. Intracerebral administration also can be used. In the following discussion, systemic administration was used unless otherwise noted. Data are summarized in Table 1.

1. Resident–Intruder Model

The capacity of hormones to increase or decrease aggression is not simply the result of the level of hormone. The directionality of the effects of testosterone on aggression is the product of the interdependence of the neuroendocrine status of both the intruder and the resident, the environmental conditions present at the time

TABLE I
Effects of Experimental Manipulation of Testosterone on Aggression in Animals

Reference	Species	Results obtained
Resident–intruder model		
Haug & Brain (1983) Whalen & Johnson (1987) Haug et al. (1986)	Male mice	Castration increased female typical aggression
Haug & Brain (1983) Whalen & Johnson (1987)	Male mice	T[a] suppressed castration-induced aggression
Whalen & Johnson (1988)	Female mice	T increased female-typical aggression
Whalen & Johnson (1987)	Male mice	Castration decreased male-typical aggression T increased male-typical aggression
Simon et al. (1984) Simon & Masters (1987) Whalen & Johnson (1988)	Female mice	T increased male-typical aggression
Albert et al. (1987a)	Male rat	Hypothalamic implant increased aggression
Albert et al. (1987b)	Male rat	Hypothalamic lesion decreased aggression
Competition model		
Albert et al. (1989)	Male rat	T increased aggression
Intermale aggression model		
Bermond et al. (1982)	Male rat	Hypothalamic stimulation increased aggression Castration increased and T decreased hypo-thalamic current thresholds for aggression
Infanticide model		
Brown (1986)	Male rat	T increased aggression but castration did not affect aggression

[a]T, Testosterone.

at which the pairing is investigated, as well as the gender of the pair. For example, attack on a lactating female is considered a female-typical form of aggressive behavior, whereas attack on an olfactory bulbectomized male is considered a male-typical form of aggressive behavior. In male mice, castration leads to an increase in attack on lactating females (i.e., female-typical aggression) (Haug & Brain, 1983; Haug, Spets, Ouss-Schlegel, Benton, & Brain, 1986; Whalen & Johnson, 1987) that is suppressed by testosterone treatment (Haug & Brain, 1983; Whalen & Johnson, 1987). In contrast, testosterone treatment increases the in-cidence of attack on lactating female targets in female ovariectomized mice (Whalen & Johnson, 1988). In male mice, castration leads to decreased attack of olfactory bulbectomized targets (i.e., male-typical aggression) and testosterone treatment leads to an increase in this type of attack (Whalen & Johnson, 1987). Several investigators have demonstrated that testosterone given to ovariectomized females leads to increased male-typical aggression toward olfactory bulbectomized males (Simon, Gandelman, & Gray, 1984; Simon & Masters, 1987; Whalen & Johnson, 1988). Thus, when comparing the effects of testosterone on these two models of aggression, it is evident these effects are governed by both the target stimulus under study and the gender of the resident. Specifically, testosterone increases male-typical aggression in both male and female residents, whereas it decreases female-typical aggression in male residents and increases female-typical aggression in female residents.

Alcohol may induce aggressive behavior and therefore can be used as an additional approach to the study of hormones on this process. Aggressive responses to alcohol, such as attack bites, offensive sideways threats, and tail rattles, are thought to be dose dependent. In fact, at higher doses, alcohol may decrease aggression. DeBold and Miczek (1985) demonstrated the facilitatory effects of testosterone by showing that male subjects receiving high doses of testosterone in connection with a low alcohol dosage displayed elevated levels of aggression. This form of aggression could be reduced somewhat by the administration of a higher dose of alcohol. With higher levels of testosterone, more alcohol was needed to reduce aggressive behavior. More recently, Lisciotto, DeBold, and Miczek (1990) demonstrated a dimorphic response to alcohol and testosterone. These researchers showed that sham-castrated males had increased levels of aggression after alcohol administration, whereas neither androgenized females nor neonatally gonadectomized males showed this response. Therefore, not just the level of circulating testosterone but also the gender is important as a determinant of aggressive behavior.

In the studies described here, testosterone was administered systemically. The effects of testosterone on the resident–intruder model of aggression also has been measured after intracerebral administration of testosterone. The rationale for this approach is that peripherally administered testosterone presumably acts through receptors present in the central nervous system (CNS), and that the possible sites of action can be determined by observing the effects of implanting the hormone at such sites. Albert, Dyson, and Walsh (1987a) demonstrated increased aggression toward a male intruder when testosterone was implanted into the medial hypothalamus of castrated male hooded rats. This effect was not obtained when testosterone was implanted either dorsal or anterior to this site, nor when cholesterol was implanted into the medial hypothalamus. This result suggests a site-specific process. Additionally, Albert, Dyson, Walsh, and Gorzalla (1987b) demonstrated that male hooded rats with electrolytic lesions of the medial hypothalamus showed lowered aggressiveness as measured by deficits in attack, biting, and piloerection than sham-operated controls when presented with an unfamiliar male intruder. Collectively, these studies suggest that the medial hypothalamus is a central site critical to the development of aggressive behavior and that testosterone acts as a neuromodulator in this region of the hypothalamus.

2. Competition Model

Evidence suggests that testosterone also can affect aggressive behavior when tested in the competition model. In a study by Albert, Petrovic, and Walsh (1989), rats that were subjected to a competitive experience for palatable food were shown to exhibit aggressive behavior that was characterized by lateral attacks and pioerection. Such attacks were increased after peripheral testosterone implants.

3. Intermale Aggression

Similar to its function in the resident–intruder model, the hypothalamus appears to play an important role in intermale aggression. One of the clearest illustrations of the facilitatory effects of testosterone was demonstrated by Bermond, Mos, Meelis, van der Poel, and Kruk (1982) using intermale fighting induced by electrical stimulation of the hypothalamus. This effect was evidenced by the fact that the threshold current required for elicitation of biting behavior was higher in castrated subjects, which lack androgens in their general circulation, than in normal subjects. In addition, this increase in threshold was lowered after intramuscular testosterone propionate treatment. These findings suggest that circulating testosterone is necessary for the induction of aggressive behavior generated by electrical stimulation of the hypothalamus. The authors suggest that the effects of testosterone on intermale aggression are a direct result of the modulatory action of the hormone on the attack mechanism that acts, perhaps, at the level of the hypothalamus, and are not the result of a secondary response to pain.

4. Infanticide

Perrigo, Bryant, and vom Saal (1989) demonstrated that infanticide in the male could be inhibited by each of the following variables: (1) the birth of its own pups; (2) castration; or (3) hypophysectomy. This result suggests that testosterone is necessary for males to display infanticide and that, in intact males, endogenous levels of testosterone decrease at the time of birth of their pups. Partial support for this view was obtained by Brown (1986), who observed that systemic testosterone implants increased the occurrence of infanticide. However, this investigator also observed that castration had no effect on infanticide. Other data also seem to question the role of testosterone with respect to infanticide. Svare and co-workers (Svare et al., 1983) showed that age-linked decreases in aggression and infanticide were not dependent on testosterone levels and that testosterone supplementation did not lead to recovery of aggression. Nevertheless, the two studies may not be contradictory since down-regulation or loss of testosterone receptors could occur with aging. Clearly, further work needs to be conducted to answer this question.

B. Effects of Natural Variations in Testosterone Levels on Aggression

Data on natural variations in testosterone levels are summarized in Table 2.

1. Seasonal Variations

Seasonal variations are associated with the occurrence of aggressive behavior, suggesting that the annual rhythms may be thought of as inducers of aggressive behavior. Caldwell, Glickman, and Smith (1984) demonstrated that seasonal increases in aggression are independent of testosterone levels. These researchers castrated male wood rats postpubertally and demonstrated that seasonal increases in aggression were maintained in the castrates. This finding replicates an earlier result found in rhesus monkeys, in which annual changes in male aggression occurred in both castrated and intact subjects (Michael & Zumpe, 1981).

Seasonal variations of testosterone and aggression have been studied extensively in lizards. Evidence suggests that changes in testosterone levels in male mountain spiny lizards correlate well with changes in aggressive territorial behavior. Specifically, when testosterone levels are low, territorial aggression levels are low; when testosterone levels are high, territorial aggression levels are high (Moore, 1986). As an additional test of this notion, Moore and Marler (1987) manipulated testosterone levels by administering implants to animals in the non-

TABLE 2
Effects of Natural Variations in Testosterone Levels on Aggression in Animals

Reference	Species	Results obtained
Seasonal variations		
Caldwell et al. (1984)	Male wood rats	T^a levels not correlated with aggression
Moore (1986) Moore & Marler (1987) Moore (1988)	Lizard	T levels positively correlated with aggression
Schlinger (1987) Archawaranon & Wiley (1988) Wingfield (1984)	Sparrow	T levels positively correlated with aggression
Developmental effects		
Gandelman & Graham (1986) Mann & Svare (1983) Rines & Vom Saal (1984)	Female mice	T *in utero* increased adult aggression
Shrenker et al. (1985)	Male mice	Castration at 30 days postnatally decreased aggression whereas castration at 50 days was not effective

[a]T, Testosterone.

breeding season so hormonal levels would be equivalent to those present during the breeding season. This procedure led to an increase in aggression but not to the level observed during the breeding season. When testosterone implants were given to castrated subjects during the breeding season, this procedure resulted in a recovery to normal breeding season levels of territorial aggression (Moore, 1988). The authors concluded that the territorial aggression seen in the breeding season is controlled to a large extent by testosterone, but presumably also includes a number of additional unknown mechanisms. During the breeding season, other neuroendocrine factors are likely to interact with increased levels of testosterone to yield high levels of territorial aggression.

In addition to studying rodents and lizards, several investigators examined the effects of seasonal testosterone levels on aggression in birds. Seasonal variations in aggression in the white-throated sparrow can be correlated with androgen levels (Schlinger, 1987). For instance, in this species, both androgen levels and aggression are increased in November but not in January or March. The authors suggest that the increases in both androgen and aggression help establish flock formation and dominance hierarchies. Further elaboration of this phenomenon in this species was carried out by Archawaranon and Wiley (1988). These investigators sought to determine whether the effects of androgens on aggression and dominance were caused by testosterone or by its metabolites. The subjects were subcutaneously implanted with testosterone or its metabolites—androstenedione, 5α-dihydrotestosterone (DHT), androsterone, or estradiol. The investigators noted that the most potent effect on aggression and dominance resulted from testosterone administration. A second experiment asked whether aggressive behavior was affected more significantly by testosterone treatment alone or by the combined treatment with only its metabolites. The results indicated that similar effects were obtained under both conditions. Moreover, these subjects displayed higher aggression and dominance scores than those administered either estradiol or DHT treatment alone. Additionally, when subjects were administered compounds that blocked the conversion of testosterone to both DHT and estradiol, aggression and dominance scores were lower than in animals subjected to the blockade of conversion of testosterone to either DHT or estradiol. This difference occurred despite the fact that animals subjected to the blockade of a single metabolite had elevated testosterone levels. Thus, these findings suggest that the androgenic effects on aggression are significantly dependent on the conversion of testosterone to both androgenic and estrogenic metabolites.

The effects of variations in testosterone levels during the breeding season on aggression also were tested in a study by Dufty (1989). In this study, testosterone levels in a free-living avian species were adjusted to maintain breeding season levels after implantation with testosterone capsules. Subjects that received these implants showed greater injury and had lower survival rates to the following year. Prolonged elevated testosterone levels were thought to produce a risk factor by maintaining aggressive intermale interactions that are normally evident only during the breeding season. An earlier study by Wingfield (1984) suggests that this conclusion may be true. Wingfield subcutaneously implanted free-living adult male song sparrows with testosterone so plasma testosterone levels were maintained at the springtime peak. Each bird was challenged with tape-recordings of conspecifics intruding into their territory. The sparrows with the testosterone implants showed higher levels of aggression than did controls.

C. Role of Testosterone in the Development of Aggressive Behavior

In mammals, sexual differentiation begins during fetal life and continues through puberty. During development, masculinization and defeminization occur in males. Androgens are thought to play an important role in inducing prenatal changes that occur in the structure and function of various tissues, including the gonads and the brain. The presence or absence of androgens continues to be an important developmental variable through puberty and may effect long-term physiological and behavioral manifestations in the organism. *In utero* position is one aspect of development that appears to be significant in determining the hormonal status of rodents. This result establishes the importance of the microenvironment of the developing fetus as a possible factor that governs a later propensity to display aggressive behavior. Rodent fetuses are positioned randomly *in utero*, so they may develop next to either male or female fetuses. Fetuses that develop between two male fetuses have higher blood testosterone levels during gestation than do fetuses that develop between two females (Perrigo et al., 1989). A fetus adjacent to a male fetus receives stimulation from the androgens produced by that male fetus *in utero*. Therefore, a subject that develops between two females *in utero* would be expected to be somewhat different endocrinologically from one that develops between two males *in utero*. Further, a subject developing alone would be expected to be endocrinologically different from one that develops with the rest of the litter because of the microenvironment present during gestation.

Gandelman and Graham (1986) made use of the idea of differing microenvironments to examine the effect of testosterone treatment in ovariectomized female mice. These researchers compared the response of female subjects with an *in utero* position between two females with the response of female subjects that developed as single fetuses because of the removal of their litter mates during gestation. As adults, the females that developed between two females attacked olfactory bulbectomized males when treated with testosterone. The single fetus females did not show aggressive behavior and were unresponsive to hormonal treatment. The authors concluded that, in intact litters that contain both male and female pups, some masculinization occurs *in utero* to female fetuses because of the presence of male fetuses, even if the fetuses are not contiguous during development. Mann and Svare (1983) also demonstrated the importance of *in utero* testosterone in adult aggression. In this study, female subjects whose mothers received testosterone during the gestational period showed increased aggression toward male intruders.

The question of an *in utero* position effect on adult aggressive behavior was explored further by Rines and vom Saal (1984). To control for possible differential levels of hormones postnatally, the subjects were ovariectomized at birth and given hormone replacement. These investigators found that young females that were in an *in utero* position between two males (2M females) were more aggressive than young females that were in an *in utero* position between zero males (0M females). However, old 0M females were aggressive, a result that suggests an age related difference in which 0M females develop a greater sensitivity to testosterone later in life than do 2M females.

Similarly, Perrigo and co-workers (1989) demonstrated that *in utero* position is also important in the incidence of infanticide in mice. In their experiment, males that developed between two females *in utero*, and thus had lower circulating testosterone levels during gestation, were shown to display increased infanticide, both before and after mating. Thus, these studies indicate the important of developmental factors that contribute to the organization of aggressive behavior during the postnatal period.

With respect to postnatal development, the intermale aggression model has been used to demonstrate the importance of the temporal relationship between castration and the tendency for expression of aggressive behavior. In one study, Shrenker, Maxson, and Ginsburg (1985) demonstrated that male mice castrated at 30 days after birth were less aggressive than those castrated at 50 days after birth. The subjects castrated at 50 days showed levels of aggression similar to those of sham-operated controls. Therefore, the timing of this procedure appears to be of importance for the later development of aggressive behavior. Consistent with these findings, Simon, Gandelman, and Gray (1984) reported that testosterone administered to females at birth produced subjects that behaved more aggressively than did subjects given estradiol neonatally when treated with testosterone as adults.

The studies described suggest that a critical time period exists during development when a given level of circulating testosterone must be present for appropriate interactions of neural structures associated with the expression of aggressive reactions to take place, and thus insure that such responses will remain present in the adult organism. Further, once this critical developmental period has passed, the presence or absence of circulating testosterone becomes less critical. This statement would seem to contradict studies that use adult rodents in which testosterone has been shown to be effective in modulating aggression. However, although critical periods exist during which changes have the most dramatic and long-lasting effects, additional manipulation of the system can occur at a later time as well. Whether such changes can alter brain function in adulthood significantly has yet to be determined.

Early postnatal effects of testosterone also have been shown to be important with respect to the early expression of aggressive behavior in hyenas. The data from this species stem from a study by Frank, Glickman, and Light (1991), who examined the role of androgens in fatal sibling aggression in neonatal spotted hyenas. These researchers demonstrated that androstenedione was elevated in both sexes at birth and remained high in females for the first month but fell in males. However, testosterone remained higher in males than in females in the first month following birth. Of particular significance is the fact that sibling fighting, which frequently results in death for one of the combatants, was most pronounced during this time period. The authors suggested that this unusual aggressive phenomenon was related directly to high androgen levels.

D. Studies of Testosterone in Humans

In the studies just described, we assessed the possible relationship between testosterone and aggressive behavior in a variety of animal species and indicated, whenever appropriate, how factors such as development and seasonal changes may affect both testosterone levels and aggressive behavior. In this section, we assess the evidence in support of a relationship between testosterone and aggressive behavior in humans. Data of testosterone variations in humans are summarized in Table 3.

Over the past two decades, several studies were designed to assess the relationship between testosterone levels and the probability of exhibiting aggressive behavior. Several methodological difficulties are apparent with respect to our capacity to compare the findings of one investigation with those of another, including differences in: (1) the nature of the populations sampled; (2) the ages of the sample; (3) the size of the sample considered; and (4) the kinds of behavioral processes investigated that are interpreted as aggressive in nature. Nevertheless, in spite of these methodological considerations, a pattern appears to emerge from these studies that, although not entirely consistent, is supportive of the findings obtained from the animal literature that demonstrate a positive correlation between testosterone levels and the presence of aggressive behavior.

The most convincing data in support of this view are derived from studies involving subjects who were incarcerated in either prisons or mental institutions. In one of the earliest studies in which clear-cut evidence was obtained, Ehrenkranz, Bliss, and Sheard (1974) categorized prison inmates on the basis of their level of aggressiveness (as determined from a battery of psychological tests) and thus were able to distinguish a chronically aggressive group from socially dominant and nonaggressive individuals. The principal finding noted by these investigators was that testosterone levels were consistently higher in individuals displaying chronic aggressive behavior than in those inmates who were classified as either socially dominant or nonaggressive. In a related study, Dabbs, Frady, Carr, and Besch (1987) also categorized prison inmates in terms of whether or not they had committed violent crimes. These authors obtained results similar to those obtained by Ehrenkranz and colleagues (Ehrenkranz et al., 1974). Using free

TABLE 3
Human Studies Correlating Testosterone Levels and Aggressive Behavior

Reference	Population sample	Results obtained
Positive relationship between T[a] levels and aggression		
Booth et al. (1989)	College men	T levels rose before a tennis match and were highest in the winners
Dabbs et al. (1987)	Prison inmates	High T levels associated with violent criminal acts and low T values associated with nonviolent crimes
Olweus et al. (1980, 1988)	Adolescent males	Subjects with high T levels showed increased readiness to respond to threat and increased propensity to engage in aggressive destructive behavior (Olweus Aggression Inventory)
Christiansen & Knussmann (1987)	20- 30-yr-old males	Subjects with high T showed self-ratings of spontaneous aggression
Susman et al. (1987)	10- 14-yr-old boys	Higher androstenedione related to acting out of behavioral problems
Rada et al. (1983)	Rapists, child molesters	T levels were higher in rapists than in child molesters or nonviolent control subjects
Scaramella & Brown (1978)	Hockey players	T levels were higher in response to threat but other measures of aggression were not positively correlated with T levels
Rada et al. (1976)	Rapists	T values were higher in rapists who were violent than in those who raped but were not otherwise violent
Ehrenkranz et al. (1974)	18- to 45-yr-old prison inmates	T levels were higher in a chronically aggressive group relative to a socially dominant or nonaggressive group
Persky et al. (1971)	College students	In younger men, T levels were higher for hostility inventory and aggression but not in older men (Buss–Durkee Hostility Inventory)
Absence of a positive relationship between T levels and aggression		
Sourial & Fenton (1988)	Case study [31-yr-old 48 XXYY Klinefelter's syndrome]	Aggressive and sexual fantasies and impulses declined with T treatment in individual with low endogenous T levels
Raboch et al. (1987)	Case study [Klinefelter's syndrome]	Low T levels associated with sexually motivated murder
Worthman & Konner (1987)	!Kung San men	Pattern of change in T levels is associated with prolonged exercise rather than the success or failure of the hunt
Bradford & McLean (1984)	Sex offenders	No correlation observed between high levels of violence and high levels of T
Kreuz & Rose (1972)	Young prisoners	Little correlation observed between T levels of verbally aggressive people who frequently fight and nonaggressive individuals

[a]T, Testosterone.

testosterone values obtained from saliva, positive correlations were obtained between individuals who displayed high testosterone levels and those who committed violent crimes. Moreover, those individuals who had the lowest testosterone levels were noted to have committed nonviolent crimes. High testosterone levels also were observed in those individuals who were rated as "tougher" by their peers, as well as in individuals who had committed nonviolent crimes but had been incarcerated for the longest period of time and had received punishment for disciplinary infractions.

Rada (1976) and Rada, Laws, Kellner, Stivastava, and Peake (1983) initially selected patients at the Atascadero State Hospital in California and classified them according to the degree of violence exhibited during the commission of a rape. Subjects also were classified according to their scores on Buss–Durkee and Megargee overcontrolled-hostility inventories; plasma testosterone levels were obtained from each of the subjects. The results were consistent with the findings just described and revealed that rapists who were judged to be most violent (i.e., those who beat their victims during the rape) had significantly higher plasma testosterone levels than child molesters or normal subjects. In addition, hostility rating scores were also higher for rapists than for normals, but individual correlations between testosterone levels and hostility ratings were not demonstrated.

Similar conclusions supporting the notion of a positive correlation between the propensity to express aggressive behavior and high testosterone levels were reached by a number of other investigators whose samples included normal adolescents or young adults. Christiansen and Knussmann (1987) administered a battery of standardized and projective tests to young adult males in their study and observed a positive correlation between serum testosterone levels and self-ratings of spontaneous aggression and dominance. This finding paralleled the earlier findings by Persky, Smith, and Basu (1971), who observed that testosterone levels were higher in college students who also scored high for the factors of hostility and aggression as measured on a hostility inventory scale. Several investigators sampled college athletes in order to define further the relationship between testosterone levels and aggressive behavior. In one study, Scaramella and Brown (1978) demonstrated that testosterone levels were correlated positively with the response to threat in college hockey players. Booth, Shelley, Mazur, Tharp, and Kittok (1989) observed that testosterone levels rose just prior to a tennis match and were highest in the winners of the matches. Findings consistent with those described here were reported in two papers describing studies of adolescent males. Olweus, Mattsson, Schalling, and Low (1980, 1988) compared testosterone levels in individuals who were also administered an Olweus Aggression Inventory scale. These authors observed that individuals who had high testosterone levels responded more vigorously to provocations and threats. In addition, such individuals were also more irritable, which appeared to increase their likelihood of engaging in aggressive behavior.

Fewer studies conducted during the past two decades have failed to provide data in support of a positive relationship between high testosterone levels and the propensity to express aggressive behavior. Bradford and McLean (1984) examined 50 males incarcerated for sex offenses that ranged from exhibitionism, fetishism, and pedophilia to more violent behavior (i.e., rape). These authors did not observe a significant correlation between high levels of sexual violence and elevated testosterone levels. This result, in itself, may not be in conflict with the findings described earlier since Rada and co-workers (Rada et al., 1983) showed that such a correlation depends on whether or not other violent acts are committed on the victim. In the Bradford and McLean study, many of the individuals classified as rapists may not have, in fact, committed other violent acts on their victims. Another negative report was provided by Kreuz and Rose (1972), who examined young prisoners and observed little correlation between testosterone levels of verbally aggressive people who frequently fought and of those who were nonaggressive. More recently, Worthman and Konner (1987) followed the pattern of change in testosterone levels of !Kung San men during the course of their hunt. The authors reported changes in testosterone levels that were correlated more closely with the length of exercise than with the success or failure of the hunt. Finally, several investigators provided case study reports of individuals diagnosed with Klinefelter's syndrome, which is characterized in part by lower levels of testosterone and by the relative absence of secondary sex characteristics. In one individual, testosterone treatment was observed to result in a decline of aggressive impulses and sexual fantasies (Sourial & Fenton, 1988). In this instance, the authors suggest that the reduction in aggressive tendencies probably was associated with the psychological benefit derived from the development of secondary sex characteristics following testosterone treatment, which allowed the patient to feel better about himself. Obviously the psychological impact of testosterone treatment far outweighed any countereffect that increased testosterone levels might have had in facilitating the occurrence of aggressive behavior. In the second study, Raboch, Cerna, and Zemek (1987) reported that an individual who had committed a sexually motivated murder had relatively low testosterone levels.

Thus, despite the presence of negative data, the overall observations obtained from human studies seem to lend some support to the presence of a relationship between endogenous testosterone levels and the propensity to commit aggressive acts. These data, however, must be treated with considerable caution since other variables such as environmental conditions, the psychological state of the individual studied, and the type of aggressive act committed appear to play more critical roles in determining whether or not a positive correlation will be obtained.

The role of testosterone in aggressive behavior constitutes the single largest concentration of research endeavors in this area of investigation. However, the possible role of other hormones in this process has been explored as well. In the following sections, we summarize some of the main findings concerning the effects of estrogen, progesterone, and adrenal steroids on aggressive behavior.

IV. Estrogen

A. Experimental Manipulation of Estrogen Levels

The overwhelming majority of studies described use systemic administration of the hormone in question to assess its possible effects on aggressive behavior. However, we also report studies that have considered the effects of intracerebral administration of this hormone as well as the effects of *in utero* exposure to estrogen on the development of aggression.

1. Resident–Intruder Model

As noted earlier, the neuroendocrine status of both the resident and the intruder are important variables in determining the level of aggression of the encounter. DeBold and Miczek (1984) systematically varied the neuroendocrine status of the intruder to determine how this procedure affected aggression as measured by the frequency of biting attack toward the neck, back, or flank, or a nip to the snout or face. Overall, these researchers found that residents were more likely to attack intruders of the same sex than those of the opposite sex. Although castration of the male intruder led to a decrease in attack from male residents, ovariectomy of female intruders did not change the frequency of attack by female residents. Further, castrated males with testosterone replacement were attacked by male residents in a manner similar to their attacks on intact intruders, but estrogen (or progesterone) replacement in female intruders did not affect the attack response. Thus, manipulation of the neuroendocrine status of the intruders by gonadectomy further demonstrated that this form of aggressive behavior is gonadally dependent in males but not in females.

In an attempt to determine how modification of the neuroendocrine status of the resident would alter the propensity to elicit attack, Meisel, Sterner, and Diekman (1988) examined the effect of chronic estradiol treatment on aggression in female hamsters. The subjects were ovariectomized, treated with varying doses of estradiol, and retested 3, 7, 10, and 14 days postimplantation. Although baseline levels of aggression were relatively high, estradiol treatment failed to alter attack responses. This finding is consistent with DeBold and Miczek's results because aggression was not affected by changes in estrogen levels in female subjects. However, when subsequently injected with progesterone, the estradiol-treated subjects showed a reduction in aggression relative to the cholesterol-treated controls, which remained aggressive after progesterone treatment. The authors concluded that inhibition of aggression in female subjects is likely to be dependent on the combined effects of estradiol and progesterone.

In these studies, estrogen was administered systemically. Since estrogen has been suggested to act via the CNS to affect aggressive reactions, Takahashi, Lisk, and Burnett (1985) implanted estradiol into both the ventromedial hypothalamus and the medial preoptic area or anterior hypothalamus of female golden hamsters

where estrogen receptors are believed to be located. Controls consisted of cholesterol injected into the ventromedial nucleus and estradiol placed into the anterior
hypothalamus or preoptic area, or cholesterol injected into both sites. Decreased
agonistic responses toward a male partner were observed after both single and dual
estradiol implantation. This result suggests that estrogen may act at receptor sites
within the anterior hypothalamus or preoptic region to modulate agonistic behavior.

2. Maternal Aggression

Other laboratories have examined the effects of systematic variations in
estrogen levels on maternal aggression against intruders. Mayer and Rosenblatt
(1987) examined the hypothesis that hormonal factors that mediate maternal care
also mediate maternal aggression. In their study, resident subjects were either
pregnant or virgin female rats. All subjects were hysterectomized, ovariectomized,
injected with estrogen or sham operated, and exposed to pups continuously.
Nonfamiliar male intruders were introduced at several different times during the
process of induction of maternal behavior. Aggressive behavior was evident in
groups with elevated estrogen levels prior to their sensitization with the pups.
Once maternal behavior was initiated, pregnant and pregnancy-terminated females
showed increased aggression but such tendencies were not displayed by females
that had never been impregnated. Females whose pregnancies had been terminated
without estrogen treatment also became aggressive after the initiation of maternal
behavior. This study therefore suggests that, although elevated estrogen levels
may have some modulatory effects on nonmaternal aggression, factors other than
estrogen may be critical for the expression of maternal aggression. In particular,
conception itself or the cascade of physiological processes that follows conception
may cause long-lasting neuroendocrine changes that affect ensuing response patterns, including maternal aggression.

Mayer, Ahdieh, and Rosenblatt (1990) further examined the role of sex
steroids in maternal aggression. These researchers employed nonpregnant ovariectomized rats and treated them with levels of estrogen and progesterone normally
seen during pregnancy, which caused a short-latency maternal behavior in the
nonpregnant subjects. The subjects given hormonal treatment at pregnancy levels
displayed aggressive behavior even when they did not express maternal behavior,
whereas vehicle controls were not aggressive. In a second experiment, some of the
subjects were hypophysectomized in addition to the other treatments. The results
of this experiment were similar to those of the first experiment. Hormonal treatment led to heightened aggression, whereas hypophysectomy had no effect on this
response. Overall, these studies demonstrated that a hormonal regimen that induced short-latency maternal behavior is associated with an increase in aggression
and that this response does not seem to be pituitary dependent.

Lisk and Nachtigall (1988) obtained results that appear contradictory to the findings just discussed. These investigators examined the role of estrogen in agonistic behavior of a female hamster toward a male. The pair, including a sexually active male, was acclimated in separate halves of a two-compartment test chamber for 2 days. The partition separating the pair was removed and the frequency of attack and chase was recorded for 15 min. Lisk and Nachtigall found a higher level of attack–chase when endogenous estrogen levels were low than when estrogen levels were naturally high in the estrous cycle. Ovariectomy produced an increase in attack–chase and the addition of exogenous estrogen decreased attack–chase. Thus, this finding also suggests that estrogen plays an inhibitory role in the regulation of aggressive behavior but further questions its precise role in this process.

B. Role of Estrogen in the Development of Aggressive Behavior

As previously indicated, sexual differentiation begins during fetal life and continues through puberty in mammals. In addition to androgens, estrogens are thought to be important in inducing changes that occur in the structure and function of various tissues, such as brain and gonads. These changes may be important in the manifestation of aggressive forms of behavior.

Several investigators have examined the effects of fetal exposure to estrogen on aggressive behavior. In one study, Gandelman, Peterson, and Hauser (1982) administered estrogen during days 12–18 of gestation. As adults, the subjects whose mothers received this treatment were given chronic testosterone treatments. Prenatal exposure to estrogen did not appear to alter the expression of aggressive behavior as adults.

The effects of endogenous prenatal exposure to estrogen also was examined using an *in utero* fetal position model (Vom Saal, Grant, McMullen, & Laves, 1983). Males that developed between two female fetuses had higher levels of estradiol in their amniotic fluid than males located between two male fetuses. Possible differences in postnatal exposure to gonadal hormones were controlled by castrating the subjects at birth and giving hormone replacement treatment. Male mice that developed between two female fetuses were observed to be less aggressive as adults. The authors suggested that the variations in levels of aggression were the result of developmental differences in brain tissue that occurred prenatally, due in part to the action of estrogen. These findings may not contradict those of Gandelman and colleagues (Gandelman et al., 1982), who showed that prenatal exposure to estrogen had no effect on adult aggression, if we assume that a critical time period exists during gestation for the development of aggression. Gandelman and co-workers perhaps failed to administer estrogen during the critical time window for this hormone to be effective. Note also that these findings are not likely to be attributed solely to differences in estrogen levels, but probably are due to the interaction of the constellation of endocrine microchanges that result from *in utero* position effects. Such effects also would include changes in androgen levels as well as in any neuroendocrine factors that interact with estrogens and androgens. Therefore, although a component of *in utero* position effects may, in fact, be due to estrogen differences present in the microenvironment of the organism, one cannot clearly separate the effects of this variable from those associated with the entire constellation of changes evident under different *in utero* positions.

Simon and Whalen (1987) examined the effect of early postnatal exposure to estrogen (or testosterone) in female mice on their probability of eliciting aggressive behavior as adults. Between 90 and 140 days, the subjects were ovariectomized and administered androgen or estrogen. Aggression was measured 48 hr after surgery, using biting or chasing of an olfactory bulbectomized male intruder as a response measure. In neonatally oil-treated controls, androgens produced increased aggression whereas estrogen was not effective. The neonatal androgen treatment also led to an enhanced effectiveness of subsequent androgen, but not estrogen, treatment in eliciting aggression as adults. Similarly, neonatal estrogen exposure resulted in the expression of aggression when the subjects subsequently were given estrogen stimulation as adults, but this procedure was less effective in enhancing adult responsiveness to androgens. Thus, this study lends further support to the view that responses to these compounds found in adult subjects are dependent on the entire neuroendocrine history of the subject. Moreover, the differential effects of perinatal androgens and estrogens on aggression in adulthood are confounded by the fact that androgens such as testosterone are converted metabolically (aromatized) to estradiol in the developing and adult mammalian brain (Naftolin et al., 1975). Factors that affect this conversion have been shown to modify the organization (McEwen, Lieberburg, Chaptal, & Krey, 1977) as well as the activation (Harding, 1986) of sexual and sexually dimorphic behavior.

V. Progesterone

A. Experimental Manipulation of Progesterone Levels

1. Maternal Aggression

Aggressive behavior such as attack on a male intruder progresses through pregnancy, increasing in magnitude during its course (Mann & Svare, 1982). Similarly, circulating levels of progesterone have been shown to change through the course of pregnancy. Mann, Konen, and Svare (1984) examined whether these changes in progesterone paralleled the increases seen in aggressive behavior. These researchers measured attack on a male intruder and found that, early in pregnancy, progesterone levels were correlated somewhat with aggression, in which both variables were low on days 6 and 10 and high on day 14. By day 18 in pregnancy, a dissociation occurred between progesterone levels and aggressive behavior. At this time, progesterone values were at their nadir while aggressive behavior was still high. A further dissociation was found between progesterone levels and aggressive behavior when supplemental administration of progesterone did not yield an early onset of aggression. However, progesterone treatment in virgin mice with intact ovaries yielded an increase in aggressive behavior, but this response was weaker than the one evident in pregnant females. Further, when progesterone levels were reduced, aggressiveness in virgin mice was attenuated whereas the levels of aggressive behavior remained high under this hormonal condition in pregnant mice.

Svare, Miele, and Kinsley (1986) examined the role of progesterone in late pregnancy on pregnancy-induced aggression against male intruders. These investigators demonstrated that hysterectomy on day 15 decreased aggression toward male intruders and that this decrease could be attenuated by progesterone implanted under the nape of the neck. Aggressive behavior obtained with progesterone-treated hysterectomized females was inhibited by estradiol, whereas estradiol treatment alone did not affect aggression in hysterectomized mice. This study supports the data of Mann and co-workers (Mann et al., 1984) that progesterone is associated with enhancement of aggressive encounters, but the overall aggressive reaction is modified further by the interactions with other hormones such as estrogen, as well as with other aspects of the internal milieu of the pregnant animal.

That this conclusion may be quite limited is suggested by a subsequent study by Svare (1988). This author observed that the effects of progesterone on pregnancy-induced aggression are strain specific. In this study, female DBA/2J mice were more likely to show pregnancy-induced aggressive behaviors than C57BL/6J mice. Further, virgin DBA/2J mice that received subcutaneously implanted progesterone exhibited aggressive behavior whereas such responses were not evident in C57BL/6J mice. These effects could not be attributed to differences in circulating progesterone since similar levels of circulating progesterone during pregnancy and following implantation were observed in both strains. Therefore, genotypic differences may cause differential aggressive responses to progesterone that may result from variations in CNS tissue sensitivity to progesterone.

2. Resident–Intruder Model

Although progesterone has been shown, under some conditions, to increase pregnancy-induced aggression, some support exists for the notion that it may inhibit other forms of aggression. Fraile, McEwen, and Pfaff (1987) used a same-sex resident–intruder model to examine the effects of progesterone on aggressive behavior. Castrated male hamsters showed decreased aggressive behavior after progesterone treatment and ovariectomized female hamsters treated with progesterone showed decreased frequency of attack.

In contrast to the decreases in aggression found following progesterone treatment, Meisel and Sterner (1990) reported an increase in aggression following this treatment. Here, ovariectomized female hamsters were treated with estradiol and given subcutaneous progesterone injections 2 and 3 days later. Subjects that received injections of progesterone showed higher levels of aggression than control subjects.

To address the effects of central administration of progesterone, Takahashi and Lisk (1985) implanted progesterone into the medial preoptic area, anterior hypothalamus, or ventromedial hypothalamus in female hamsters. Aggression was measured by placing a sexually experienced male into the home-cage of the subject. Subjects with implants in the ventromedial nucleus and medial preoptic area but not in the anterior hypothalamus showed decreased aggression toward the male, thus demonstrating a site-specific central effect of progesterone on aggressive behavior.

B. Role of Progesterone in Development of Aggressive Behavior

In one study, prenatal exposure to progesterone was shown to modulate aggressive behavior in adulthood (Wagner, Kinsley, & Svare, 1986). Female subjects whose mothers received progesterone injections during days 12–16 of gestation showed increased postpartum aggression as adults. The author suggested that prenatal exposure may have affected prenatal brain differentiation, which could have led to a masculinization of the brain, as shown by an increase in aggression.

VI. Adrenal Steroids and Aggressive Behavior

Many investigators have attempted to examine the possible role that adrenal steroids may play during the process of aggression. Several authors have used the models of resident–intruder, shock-induced aggression, and maternal aggression, and have focused their attention on altered corticosterone levels in an animal defeated after fighting rather than on the aggressor. More pertinent to this chapter, however, are the data concerning how adrenal steroids may alter the propensity to commit aggressive acts. This subject has been examined but the findings are sparse at best. The following paragraphs briefly summarize the findings.

Studies that have attempted to determine how adrenal steroids affect aggressive reactions have been reported in the lizard, rat, and human. Two studies have been conducted in separate species of lizards. In one study, Tokarz (1987) reported that animals pretreated with corticosterone pellets showed a reduction in aggressive tendencies such as approach and biting behavior as well as in aggressive postural responses. In a related study, Moore (1987) raised the question of whether corticosterone levels are altered after aggressive responses. In this experiment, blood samples that were collected during regular periods during aggressive encounters revealed no changes in circulating levels of corticosterone. Thus, at the level of the lizard along the phylogenetic scale, no clear-cut conclusions can be drawn concerning whether or not aggressive reactions are mediated by changes in circulating levels of corticosterone.

In the rat, the data provide some suggestive evidence that adrenal steroids may play an inhibitory modulating role in aggressive behavior. Meaney, Stewart, and Beatty (1982) observed play fighting in Norway rat pups and noted that neonates treated with corticosterone within the first 4 days of life eventually fought less frequently than did control animals. Moreover, corticosterone treatment at a later period (days 9 or 10) had little effect on fighting, thus suggesting that a critical period exists with respect to the time at which corticosterone treatment can be effective in modulating aggressive behavior.

Consistent with these findings are the observations of Severyanova (1988). In this study, in which a shock-induced model of aggression was employed, administration of low doses of the corticosterone precursor deoxycorticosterone was reported to result in a decrease in fighting as well as in an increase in the current threshold (delivered to an electrified grid) required to initiate fighting behavior.

Another approach was employed by Tazi and colleagues (Tazi et al., 1987). These investigators, who also used the model of shock-induced fighting, observed that microinjections of corticotropin-releasing factor (CRF) administered intracerebroventricularly facilitated the occurrence of this response. Additional support for this finding was obtained by these authors, who showed that delivery of the CRF antagonist alpha-helical CRF-(9–41) could block shock-induced fighting. Relating this finding to the previous results of studies in which corticosterone was employed is difficult for several reasons. One possibility is that CRF, in addition to acting on the anterior pituitary, may act as a transmitter or neuromodulator at sites in the brain that are associated with the expression of aggressive behavior. Second, CRF is likely to have widespread effects on the adrenal system and alter the release of other hormones, which might include aldosterone as well as androgens that could modify aggressive reactions in a manner different from that of corticosterone.

Other investigators have studied the effects of adrenal corticotropic hormone (ACTH) on aggressive behavior. One study has suggested that ACTH administration can lead to an increase in aggressive behavior (Brain & Evans, 1977), possibly by acting directly on brain mechanisms that govern this response or by acting through a corticosterone mechanism. However, these findings seem inconsistent with an earlier report (Leshner, 1975) in which ACTH treatment was proposed to reduce aggressiveness in both intact mice and mice with controlled corticosterone levels. Thus, our current understanding of the role of adrenal cortical steroids in aggression in rodents remains unclear.

Several studies conducted in humans used either male violent offenders or male substance users and attempted to correlate cortisol levels with aggression, impulsivity, or otherwise antisocial behavior. Again, these studies were inconsistent. In one investigation, cortisol levels were reported to be low among habitually violent offenders who maintained antisocial personalities in comparison with a control group of people who had antisocial personalities but did not display violent behavior (Virkkunen, 1985). In contrast, cortisol levels were reported to be higher in individuals characterized with higher levels of aggressiveness and impulsivity following administration of the indirect serotonin agonist fenfluramine (Fishbein, Lozovsky, & Jaffe, 1989).

Thus, at different levels along the phylogenetic scale, studies aimed at identifying a relationship between the pituitary–adrenal axis and aggressive behavior have been inconsistent and do not permit any firm and clear-cut conclusions to be drawn. These data would suggest, instead, that adrenal steroids function primarily in response to acts of violence or other forms of stress. Along these lines, several investigators have sought to examine how the adrenocortical system responds when the organism is subjected to aggression. In studies involving either mice (File, 1984) or monkeys (Scallet, Suomi, & Bowman, 1981; Martensz et al., 1987) cortisol levels were reported to be elevated following an attack (or defeat) by an aggressor of the same species. In a parallel fashion, 3α, 5α-tetrahydroxycorticosterone, a metabolite of the corticosterone precursor deoxycorticosterone, when administered peripherally to intruder mice, has been demonstrated to hasten the time for defeat to occur (Kavalliers, 1988). The data suggest that this naturally occurring steroid can have a potent effect on how the organism reacts to aggressive encounters.

VII. Comments and Conclusions

The studies reviewed in this chapter provide substantive evidence that sex hormones play significant roles in the expression of aggressive behavior. However, this relationship is a highly complex one in which the effects of hormonal treatment on aggressive behavior appear to be dependent on interactions with such factors as gender, species, and model of aggressive behavior under consideration.

From the literature we have reviewed, the following conclusions are drawn. The studies conducted with animal subjects have, for the most part, demonstrated that testosterone is the most important hormonal variable involved in the induction of aggressive behavior. In some instances, this response is gender specific and selective to the particular form of aggression measured, which would suggest that the effects may not be totally generalizable, but such findings do not detract from their significance. Further, specific times, developmentally, have been demonstrated to be important for the manifestation of the effects of testosterone on the induction of aggressive behavior. The human data are less consistent, but still seem to suggest that testosterone affects aggressive behavior. Possibly, at different levels along the phylogenetic scale, other psychogenic variables begin to override the relationship between testosterone and aggression.

With respect to other sex hormones, estrogen appears to oppose the effects of testosterone because it may inhibit aggressive responses, whereas the effects of progesterone are less clear cut and seem to be dependent on interaction with other neuroendocrine factors. The literature on adrenal steroids fails to demonstrate conclusively that these hormones play a role in the regulation of aggressive behavior. Instead, we suggest that adrenal steroid hormones serve a more significant function with respect to processes associated with the defeat of an organism following an aggressive bout.

A question of central concern is the possible mechanism by which sex hormones control of aggressive reactions. Such a mechanism is likely to involve the action of hormones, which pass through the blood–brain barrier, on selective neuronal pools in the brain. In such instances, hormones presumably act to modify the release of specific transmitters and may affect the activity of second messengers as well. On the basis of our knowledge of the neuroanatomy of aggressive behavior, we may suggest on which structures in the brain hormones serve as neuromodulators to alter the expression of aggressive behavior. Since many of the studies considered involve models that may be classified as defensive aggression, two of the most critical structures would seem to be the medial hypothalamus and the midbrain periaqueductal gray matter. These structures are of significance because they constitute regions from which this form of aggression can be elicited readily by electrical or chemical stimulation (Fuchs, Edinger, & Siegel, 1985a,b; Shaikh, Barrett, & Siegel, 1987; Siegel, & Pott, 1988; Siegel, & Brutus, 1990). In particular, the medial hypothalamus plays a unique role in this process. This conclusion is derived from the facts that: (1) implantation of hormones into this structure clearly can modify the propensity for the occurrence of an attack response; (2) lesions of the medial hypothalamus result in an attenuation of aggressive behavior such as maternal aggression (Hansen, 1989), which has been shown to be modified by hormonal treatment; and (3) sex hormone concentrating cells can be found in high densities within the medial hypothalamus (Pfaff, 1968a,b; Pfaff & Keiner, 1973). Although the medial hypothalamus and possibly the periaqueductal gray would appear to be primary candidates for areas where hormonal actions would modify aggressive behavior most effectively, other structures also should be considered. These regions include the limbic system—amygdala, hippocampal formation, septal area, and prefrontal and cingulate cortices—which is characterized by its modulatory actions on aggressive behavior (Siegel & Edinger, 1981, 1983; Siegel & Brutus, 1990). It is of interest to note that several recent studies have shown that substance P neurons situated in the medial amygdala—a structure known to facilitate the occurrence of defensive rage behavior and suppress predatory attack in the cat (Shaikh, Steinberg & Siegel, 1992; Siegel & Brutus, 1990; Steinberg, Shaikh & Siegel, 1992) are increased in numbers following testosterone treatment (Swann & Macchione, 1992; Swann & Newman, 1992). Thus, sex hormone regulation of aggressive behavior may result from its combined effects on both the medial hypothalamus and limbic structures. This notion is depicted in Figure 1. A potential heuristic line of research would be one that is directed toward elucidating the precise mechanism that underlies sex hormone modulation of the hypothalamus and limbic system and how such modulation affects the attack process.

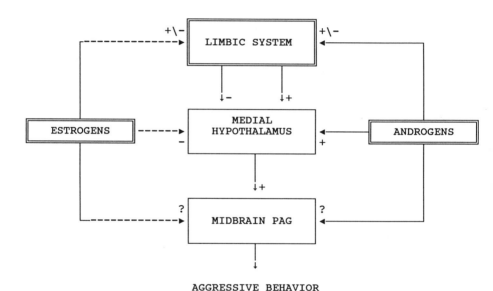

Figure 1. Schematic representation of the possible mechanism by which sex hormones regulate aggressive behavior. That the expression of aggressive responses that are defensive in nature is mediated by a pathway that arises from the medial hypothalamus and descends directly into the midbrain periaqueductal gray (PAG) Neurons from the PAG, in turn, pass caudally to lower regions of the brainstem and ultimately to the spinal cord, where contact is made with somatomotor and autonomic cell groups that collectively constitute the final common paths for the elicitation of aggressive responses. Output neurons in the medial hypothalamus and PAG also are modulated directly by various component nuclei of the limbic system. In this scheme, circulating levels of androgens and estrogens, which easily can pass through the blood–brain barrier, alter the discharge patterns of neurons within the medial hypothalamus, limbic system, and possibly the midbrain PAG. This modification of the firing patterns results in a change in the propensity for aggressive reactions. Note that the evidence suggests that androgens normally serve to facilitate aggressive responses (+), whereas estrogens appear to play an inhibitory role (−). That these hormones may act directly on neurons within the PAG remains a possibility that has yet to be tested experimentally.

Acknowledement

The authors gratefully acknowledge Barbara Fadem for her helpful suggestions and criticisms of the manuscript.

References

Albert, D. J., Dyson, E. M., & Walsh, M. L. (1987a). Intermale social aggression: Reinstatement in castrated rats by implants of testosterone propionate in the medial hypothalamus. *Physiology and Behavior, 39,* 555–560.

Albert, D. J., Dyson, E. M., Walsh, M. L., & Gorzalka, B. B. (1987b). Intermale social aggression in rats: Suppression by medial hypothalamic lesions independently of enhanced defensiveness of decreased testicular testosterone. *Physiology and Behavior, 39,* 693–698.

Albert, D. J., Petrovic, D. M., & Walsh, M. L. (1989). Competitive experience activates testosterone-dependent social aggression toward unfamiliar males. *Physiology and Behavior, 45,* 723–727.

Archawaranon, M., & Wiley, R. H. (1988). Control of aggression and dominance in white-throated sparrows by testosterone and its metabolites. *Hormones and Behavior, 22,* 497–517.

Barfield, R. J. (1984). Reproductive hormones and aggressive behavior. *Progress in Clinical and Biological Research, 169,* 105–134.

Bermond, B., Mos, J., Meelis, W., VanDerPoel, A. M., & Kruk, M. R. (1982). Aggression induced by stimulation of the hypothalamus: Effects of androgens. *Pharmacology, Biochemistry, and Behavior, 16,* 41–45.

Booth, A., Shelley, G., Mazur, A., Tharp, G., & Kittok, R. (1989). Testosterone, and winning and losing in human competition. *Hormones and Behavior, 23,* 556–571.

Bouissou, M.-F. (1983). Androgens, aggressive behavior and social relationships in higher mammals. *Hormone Research, 18,* 43–61.

Bradford, J. M. W., & McLean, D. (1984). Sexual offenders, violence and testosterone: A clinical study. *Canadian Journal of Psychiatry, 29,* 335–343.

Brain, P. F. (1977). *Hormones and aggression.* First ed. Montreal, Canada: Eden Press.

Brain, P. F. (1978). Hormones and aggression. Second ed. Montreal, Canada: Eden Press.

Brain, P. F., & Evans, A. E. (1977). Acute influences of some ACTH-related peptides on fighting and adrenocortical activity in male laboratory mice. *Pharmacology, Biochemistry, and Behavior, 7,* 425–433.

Brown, R. E. (1986). Social and hormonal factors influencing infanticide and its suppression in adult male Long-Evans rats (*Rattus norvegicus*). *Journal of Comparative Psychology, 100,* 155–161.

Caldwell, G. S., Glickman, S. E., & Smith, E. R. (1984). Seasonal aggression independent of seasonal testosterone in wood rats. *Proceedings of the National Academy of Sciences, 81,* 525–527.

Christiansen, K., & Knussmann, R. (1987). Androgen levels and components of aggressive behavior in men. *Hormones and Behavior, 21,* 170–180.

Coe, C. L., & Levine, S. (1983). Biology of aggression. *Bulletin of the American Academy of Psychiatry and the Law, 11(2),* 131–148.

Dabbs, J. M., Frady, R. L., Carr, T. S., & Besch, N. F. (1987). Saliva testosterone and criminal violence in young adult prison inmates. *Psychosomatic Medicine, 49,* 174–182.

DeBold, J. F., & Miczek, K. A. (1984). Aggression persists after ovariectomy in female rats. *Hormones and Behavior, 18,* 177–190.

DeBold, J. F., & Miczek, K. A. (1985). Testosterone modulates the effects of ethanol on male mouse aggression. *Psychopharmacology, 86,* 286–290.

Dufty, A. M. (1989). Testosterone and survival: A cost of aggressiveness? *Hormones and Behavior,* *23,* 185–193.

Ehrenkranz, J., Bliss, E., & Sheard, M. H. (1974). Plasma testosterone: Correlation with aggressive behavior and social dominance in man. *Psychosomatic Medicine, 36,* 469–475.

File, S. E. (1984). The stress of intruding: Reduction by chlordiazepoxide. *Physiology and Behavior, 33,* 345–347.

Fishbein, D. H., Lozovsky, D., & Jaffe, J. H. (1989). Impulsivity, aggression, and neuroendocrine responses to serotonergic stimulation in substance abusers. *Biological Psychiatry, 25,* 1049–1066.

Fraile, I. G., McEwen, B. S., & Pfaff, D. W. (1987). Progesterone inhibition of aggressive behaviors in hamsters. *Physiology and Behavior, 39,* 225–229.

Frank, L. G., Glickman, S. E., & Light, P. (1991). Fatal sibling aggression, precocial development, and androgens in neonatal spotted hyenas. *Science, 252,* 702–704.

Fuchs, S. A. G., Edinger, H. M., & Siegel, A. (1985a). The role of the anterior hypothalamus in affective defense behavior elicited from the ventromedial hypothalamus of the cat. *Brain Research, 330,* 93–107.

Fuchs, S. A. G., Edinger, H. M., & Siegel, A. (1985b). The organization of the hypothalamic pathways mediating affective defense behavior in the cat. *Brain Research, 330,* 77–92.

Gandelman, R., & Graham, S. (1986). Singleton female mouse fetuses are subsequently unresponsive to the aggression-activating property of testosterone. *Physiology and Behavior, 37,* 465–467.

Gandelman, R., Peterson, C., & Hauser, H. (1982). Mice: Fetal estrogen exposure does not facilitate later activation of fighting by testosterone. *Physiology and Behavior, 29,* 397–399.

Hansen, S. (1989). Medial hypothalamic involvement in maternal aggression of rats. *Behavioral Neuroscience, 103,* 1035–1046.

Harding, C. (1986). The role of androgen metabolism in the activation of male behavior. In B. Komisaruk, H. I. Siegel, M. F. Cheng and H. Feder (Eds.), *Reproduction: A behavioral and neuroendocrine perspective,* (Vol. 4, pp. 371–378). New York: New York Academy of Science.

Haug, M., & Brain, P. F. (1983). The effects of differential housing, castration and steroidal hormone replacement on attacks directed by resident mice towards lactating intruders. *Physiology and Behavior, 30,* 557–560.

Haug, M., Spetz, J. F., Ouss-Schlegel, M. L., Benton, D., & Brain, P. F. (1986). Effects of gender, gonadectomy and social status on attack directed towards female intruders by resident mice. *Physiology and Behavior, 37,* 533–537.

Kavaliers, M. (1988). Inhibitory influences of the adrenal steroid, 3-alpha,5-alpha-tetrahydroxycorticosterone on aggression and defeat-induced analgesia in mice. *Psychopharmacology, 95,* 488–492.

Kreuz, L. E., & Rose, R. M. (1972). Assessment of aggressive behavior and plasma testosterone in a young criminal population. *Psychosomatic Medicine, 34,* 321–332.

Kruk, M. R., VanDerPoel, A. M., Meelis, W., Hermans, J., Mostert, P. G., Mos, J., & Lohman, A. H. M. (1983). Discriminant analysis of the localization of aggression-inducing electrode placements in the hypothalamus of male rats. *Brain Research, 260,* 61–79.

Leshner, A. I. (1975). A model of hormones and agonistic behavior. *Physiology and Behavior, 15,* 225–235.

Lisciotto, C. A., DeBold, J. F., & Miczek, K. A. (1990). Sexual differentiation and the effects of alcohol on aggressive behavior in mice. *Pharmacology, Biochemistry, and Behavior, 35,* 357–362.

Lisk, R. D., & Nachtigall, M. J. (1988). Estrogen of agonistic and proceptive responses in the golden hamster. *Hormones and Behavior, 22,* 35–48.

McEwen, B. S., Lieberburg, I., Chaptal, C., & Krey, L. C. (1977). Aromatization: Important for sexual differentiation of the neonatal rat brain. *Hormones and Behavior, 9,* 249–263.

Mann, M. A., Konen, C., & Svare, B. (1984). The role of progesterone in pregnancy-induced aggression in mice. *Hormones and Behavior, 18,* 140–160.

Mann, M. A., & Svare, B. (1982). Factors influencing pregnancy-induced aggression in mice. *Behavioral and Neural Biology, 36,* 242–258.

Mann, M. A., & Svare, B. (1983). Prenatal testosterone exposure elevates maternal aggression in mice. *Physiology and Behavior, 30,* 503–507.

Martensz, N. D., Vellucci, S. V., Fuller, L. M., Everitt, B. J., Keverne, E. B., & Herbert, J. (1987). Relation between aggressive behavior and circadian rhythms in cortisol and testosterone in social groups of talapoin monkeys. *Journal of Endocrinology, 115,* 107–120.

Mayer, A. D., Ahdieh, H. B., & Rosenblatt, J. S. (1990). Effects of prolonged estrogen–progesterone treatment and hypophysectomy on the stimulation of short-latency maternal behavior and aggression in female rats. *Hormones and Behavior, 24,* 152–173.

Mayer, A. D., & Rosenblatt, J. S. (1987). Hormonal factors influence the onset of maternal aggression in laboratory rats. *Hormones and Behavior, 21,* 253–267.

Meaney, M. J., Stewart, J., & Beatty, W. W. (1982). The influence of glucocorticoids during the neonatal period on the development of play-fighting in norway rat pups. *Hormones and Behavior, 16,* 475–491.

Meisel, R. L., & Sterner, M. R. (1990). Progesterone inhibition of sexual behavior is accompanied by an activation of aggression in female Syrian hamsters. *Physiology and Behavior, 47,* 415–417.

Meisel, R. L., Sterner, M. R., & Diekman, M. A. (1988). Differential hormonal control of aggression and sexual behavior in female syrian hamsters. *Hormones and Behavior, 22,* 453–466.

Michael, R. P., & Zumpe, D. (1981). Relation between the seasonal changes in aggression, plasma testosterone and the photoperiod in male rhesus monkeys. *Psychoneuroendocrinology, 6,* 145–158.

Moore, M. C. (1986). Elevated testosterone levels during nonbreeding season territoriality in a fall-breeding lizard, *Sceloporus jarrovi. Journal of Comparative Physiology, 158,* 159–163.

Moore, M. C. (1987). Circulating steroid hormones during rapid aggressive responses of territorial male mountain spiny lizards, *Sceloporus jarrovi. Hormones and Behavior, 21,* 511–521.

Moore, M. C. (1988). Testosterone control of territorial behavior: Tonic-release implants fully restore seasonal and short-term aggressive responses in free-living castrated lizards. *General and Comparative Endocrinology, 70,* 450–459.

Moore, M. C., & Marler, C. A. (1987). Effects of testosterone manipulations on nonbreeding season territorial aggression in free-living male lizards, *Sceloporus jarrovi. General and Comparative Endocrinology, 65,* 225–232.

Moyer, K. E. (1976). *The psychology of aggression.* New York: Harper & Row.

Naftolin, F., Ryan, K. J., Davies, I. J., Reddy, V. V., Flores, F., Petro, Z., & Kuhn, M. (1975). The formation of estrogens by central neuroendocrine tissues. *Recent Progress in Hormone Research, 31,* 295–315.

Olweus, D., Mattsson, A., Schalling, D., & Low, H. (1980). Testosterone, aggression, physical, and personality dimensions in normal adolescent males. *Psychosomatic Medicine, 42,* 253–269.

Olweus, D., Mattsson, A., Schalling, D., & Low, H. (1988). Circulating testosterone levels and aggression in adolescent males: A causal analysis. *Psychosomatic Medicine, 50,* 261–272.

Perrigo, G., Bryant, W. C., & vom Saal, F. S. (1989). Fetal, hormonal and experiential factors influencing the mating-induced regulation of infanticide in male house mice. *Physiology and Behavior, 46,* 121–128.

Persky, H., Smith, K. D., & Basu, G. K. (1971). Relation of psychologic measures of aggression and hostility to testosterone production in man. *Psychosomatic Medicine, 33,* 265–277.

Pfaff, D. W. (1968a). Autoradiographic localization of radioactivity in the rat brain after injection of tritiated sex hormones. *Science, 161,* 1355–1356.

Pfaff, D. W. (1968b). Uptake of estradiol-17-beta-H3 in the female rat brain: An autoradiographic study. *Endocrinology, 82*, 1149–1155.

Pfaff, D. W., & Keiner, M. (1973). Atlas of estradiol-concentrating cells in the central nervous system of the female rat. *Journal of Comparative Neurology, 151*, 121–158.

Raboch, J., Cerna, H., & Zemek, P. (1987). Sexual aggressivity and androgens. *British Journal of Psychiatry, 151*, 398–400.

Rada, R. T. (1976). Alcoholism and the child molester. *Annals of the New York Academy of Sciences, 273*, 492–496.

Rada, R. T., Laws, D. R., Kellner, R., Stivastava, L., & Peake, G. (1983). Plasma androgens in violent and nonviolent sex offenders. *Bulletin of the American Academy of Psychiatry and the Law, 11(2)*, 149–158.

Rines, J. P., & Vom Saal, F. S. (1984). Fetal effects on sexual behavior and aggression in young and old female mice treated with estrogen and testosterone. *Hormones and Behavior, 18*, 117–129.

Scallet, A. C., Suomi, S. J., & Bowman, R. E. (1981). Sex differences in adrenocortical response to controlled agonistic encounters in rhesus monkeys. *Physiology and Behavior, 26*, 385–390.

Scaramella, T. J., & Brown, W. A. (1978). Serum testosterone and aggressiveness in hockey players. *Psychosomatic Medicine, 40*, 262–265.

Schlinger, B. A. (1987). Plasma androgens and aggressiveness in captive winter white-throated sparrows (*Zonotrichia albicollis*). *Hormones and Behavior, 21*, 203–210.

Severyanova, L. A. (1988). Neuromodulator mechanism of the inhibitory influence of deoxycorticosterone on the aggressive–defensive behavior of rats. *Neuroscience and Behavioral Physiology, 18(6)*, 486–492.

Shaikh, M. B., Barrett, J. A., & Siegel, A. (1987). The pathways mediating affective defense and quiet biting attack behavior from the midbrain central gray of the cat: An autoradiographic study. *Brain Research, 437*, 9–25.

Shaikh, M. B., Steinberg, A. & Siegel, A. (1992). Substance P pathway from the medial amygdala to the medial hypothalamic defensive rage sites, *Soc. Neurosci. Abstr., 18*, 356.

Shrenker, P., Maxson, S. C., & Ginsburg, B. E. (1985). The role of postnatal testosterone in the development of sexually dimorphic behaviors in DBA/1Bg mice. *Physiology and Behavior, 35*, 757–762.

Siegel, A., & Brutus, M. (1990). Neurosubstrates of aggression and rage in the cat. In A. N. Epstein & A. R. Morrison (Eds.), *Progress in psychobiology and physiological psychology* (pp. 135–233). San Diego: Academic Press.

Siegel, A., & Edinger, H. (1981). Neural control of aggression and rage. In P. J. Morgane & J. Panksepp (Eds.), *Handbook of the hypothalamus* (pp. 203–240). New York: Marcel Dekker.

Siegel, A., & Edinger, H. M. (1983). Role of the limbic system in hypothalamically elicited attack behavior. *Neuroscience and Biobehavioral Reviews, 7*, 395–407.

Siegel, A., & Pott, C. B. (1988). Neural substrate of aggression and flight in the cat. *Progress in Neurobiology, 31*, 261–283.

Simon, N. G., Gandelman, R., & Gray, J. L. (1984). Endocrine induction of intermale aggression in mice: A comparison of hormonal regimens and their relationship to naturally occurring behavior. *Physiology and Behavior, 33*, 379–383.

Simon, N. G., & Masters, D. B. (1987). Activation of male-typical aggression by testosterone but not its metabolites in C57BL/6J female mice. *Physiology and Behavior, 41*, 405–407.

Simon, N. G., & Whalen, R. E. (1987). Sexual differentiation of androgen-sensitive and estrogen-sensitive regulatory system for aggressive behavior. *Hormones and Behavior, 21*, 493–500.

Sourial, N., & Fenton, F. (1988). Testosterone treatment of an XXYY male presenting with aggression: A case report. *Canadian Journal of Psychiatry, 33*, 846–850.

Steinberg, A., Shaikh, M. B., & Siegel, A. (1992). The role of substance P in medial amygdaloid facilitation of defensive rage behavior in the cat, *Soc. Neurosci. Abstr., 18*, 356.

Susman, E. J., Inoff-Germain, G., Nottelmann, E. D., Loriaux, D. L., Cutler, G. B., & Chrousos, G. P. (1987). Hormones, emotional dispositions, and aggressive attributes in young adolescents, *Child Development, 58* 1114–1134.

Svare, B. B. (Ed.) (1983). *Hormones and aggressive behavior.* New York: Plenum Press.

Svare, B. (1988). Genotype modulates the aggression-promoting quality of progesterone in pregnant mice. *Hormones and Behavior, 22*, 90–99.

Svare, B., Mann, M., Broida, J., Kinsley, C., Ghiraldi, L., Miele, J., & Konen, C. (1983). Intermale aggression and infanticide in aged C57BL/6J male mice: Behavioral deficits are not related to serum testosterone (T) levels and are not recovered by supplemental T. *Neurobiology of Aging, 4*, 305–312.

Svare, B., Miele, J., & Kinsley, C. (1986). Mice: Progesterone stimulates aggression in pregnancy-terminated females. *Hormones and Behavior, 20*, 194–200.

Swann, J. M., & Macchione, N. (1992). Photoperiodic regulation of substance P immunoreactivity in the mating behavior pathway of the male golden hamster, *Brain Research, 590*, 29–38.

Swann, J. M., & Newman, S. W. (1992). Testosterone regulates substance P within neurons of the medial nucleus of the amygdala, the bed nucleus of the stria terminalis and the medial preoptic area of the male golden hamster, *Brain Research, 590*, 18–28.

Takahashi, L. K., & Lisk, R. D. (1985). Diencephalic sites of progesterone action for inhibiting aggression and facilitating sexual receptivity in estrogen-primed golden hamsters. *Endocrinology, 116*, 2393–2399.

Takahashi, L. K., Lisk, R. D., & Burnett, A. L. (1985). Dual estradiol action in diencephalon and the regulation of sociosexual behavior in female golden hamsters. *Brain Research, 359*, 194–207.

Tazi, A., Dantzer, R., Le Moal, M., Rivier, J., Vale, W., & Koob, G. F. (1987). Corticotropin-releasing factor antagonist blocks stress-induced fighting in rats. *Regulatory Peptides, 18*, 37–42.

Tokarz, R. R. (1987). Effects of corticosterone treatment on male aggressive behavior in a lizard (*Anolis sagrei*). *Hormones and Behavior, 21*, 358–370.

Virkkunen, M. (1985). Urinary free cortisol secretion in habitually violent offenders. *Acta Psychiatrica Scandinavica, 72*, 40–44.

Vom Saal, F. S., Grant, W. M., McMullen, C. W., & Laves, K. S. (1983). High fetal estrogen concentrations: Correlation with increased adult sexual activity and decreased aggression in male mice. *Science, 220*, 1306–1308.

Wagner, C. K., Kinsley, C., & Svare, B. (1986). Mice: Postpartum aggression is elevated following prenatal progesterone exposure. *Hormones and Behavior, 20*, 212–221.

Whalen, R. E., & Johnson, F. (1987). Individual differences in the attack behavior of male mice: A function of attack stimulus and hormonal state. *Hormones and Behavior, 21*, 223–233.

Whalen, R. E., & Johnson, F. (1988). Aggression in adult female mice: Chronic testosterone treatment induces attack against olfactory bulbectomized male and lactating female mice. *Physiology and Behavior, 43*, 17–20.

Wingfield, J. C. (1984). Environmental and endocrine control of reproduction in the song sparrow, *Melospiza melodia.* II. Agonistic interactions as environmental information stimulating secretion of testosterone. *General and Comparative Endocrinology, 56*, 417–424.

Worthman, C. M., & Konner, M. J. (1987). Testosterone levels change with subsistence hunting effort in !Kung San men. *Psychoneuroendocrinology, 12*, 449–458.

CHAPTER

5

Anabolic Steroids: Misuse or Abuse?

Lynn H. O'Connor and Theodore J. Cicero

Department of Psychiatry
Washington University School of Medicine
St. Louis, Missouri 63110

I. Introduction

The purpose of this chapter is to discuss the prevalence of the use of anabolic androgenic steroids (AAS), whether these compounds satisfy the criteria for substances of abuse, the physiological actions and side effects of these steroids, and, finally, whether they influence the response to other psychotrophic drugs. This chapter is divided into three sections: first, the prevalence of AAS abuse and the reasons most often given for AAS use; second, a critical discussion of the physiological effects of anabolic steroids and their side effects; and, finally, a commentary on whether anabolic steroid use may influence the response to other psychoactive drugs.

At the outset, several considerations should be borne in mind: First, researchers have little doubt that AAS are misused and that such use has significant adverse effects, particularly on prepubertal and pubertal adolescents who are unusually susceptible to the disruptive effects of AAS. (For reviews, see Haupt and Rovere, 1984; Taylor, 1987; Wilson, 1988; Lombardo, 1990.) However, the issue of whether these compounds have abuse liability in the classical pharmacological sense remains a matter of debate and will be one focus of this chapter. Second,

129

very little data are available that are relevant to the abuse potential of anabolic steroids in humans and even less information is at hand using animal models. Moreover, the absence of well-controlled double-blind studies using placebo controls severely limits the conclusions that can be drawn. Specifically, distinguishing between "expectancy" effects of AAS use and the immediate reinforcing properties of the drugs is virtually impossible from the studies currently available. Finally, AAS abusers rarely if ever abuse only a single AAS. Rather, use of a number of AAS, other illicit and prescription drugs, and hormones (e.g., growth hormone)—a practice often referred to as "stacking"—appears to be the normal pattern (Taylor, 1987; Wilson, 1988). Compounding this problem, the patterns of AAS use by regular users, including doses and frequency of administration, are highly variable, making well-controlled studies in an experimental situation difficult at best and perhaps of somewhat limited value with respect to an examination of "normal" AAS misuse in humans. Moreover, since these compounds generally are obtained illicitly on the black market, their purity is frequently suspect, which makes self-reported estimates of the doses and types of compounds administered of questionable validity. As a result of these considerations, assessing the dependence liability of a given AAS in multi-drug abusing individuals with highly irregular patterns of self-administration has proven to be extremely difficult.

II. Prevalence of and Factors Responsible for Anabolic Androgenic Steroid Misuse

In a recent survey of 853 male adolescents (Johnson, Jay, Shoup, & Rickert, 1989), 11.1% reported past or current use of anabolic steroids. Of these users, over 80% were involved in organized sports activities. These rates are alarming and appear to be growing. Estimating the prevalence of AAS misuse by college or professional athletes or in the general population is impossible, but anecdotal data suggest that such use is quite high (Burkett & Falduto, 1984; Taylor, 1985). Use among weight lifters and body builders is estimated at 80% and among competitors of all sports at 50% (Bell & Doege, 1987). Consequently, the misuse of AAS appears to be reaching epidemic proportions; recent publicity and overall increased awareness of the use and misuse of these compounds apparently has done little to discourage their ingestion.

Anabolic androgenic steroids are reported by regular users to produce euphoria, a sense of well-being, increased libido, aggression, improved self-image, and, most commonly, an enhancement of athletic performance and muscular development. (For reviews, see Wilson and Griffin, 1980; Haupt and Rovere, 1984; Taylor, 1987; Wilson, 1988; Lombardo, 1990). Assessing whether these self-reports represent immediate discernible effects of the AAS or the expectation that the use of AAS over time will produce these effects is difficult because of the inherent limitations in the design of most studies carried out to date (see subsequent discussion).

In addition to the commonly described reasons for AAS misuse in athletes, a number of side effects are undesirable consequences of ingesting suprapharmacological doses of AAS (Haupt and Rovere, 1984; Taylor, 1987; Pope and Katz, 1988; Wilson, 1988), including most prominently endocrine disturbances (particularly in adolescents and women), cardiac and liver problems, stunted growth, testicular atrophy, psychiatric disturbances, and violent expressions of hostility (frequently termed "roid rage" by chronic misusers). Despite the alarming incidence of these side effects, chronic users still apparently are willing to risk them to obtain what they believe will be a long-term desirable effect. Conversely, many users take steps to avoid the side effects of AAS by ingesting other substances to counteract some of these adverse reactions, for example, growth hormone and various prescription drugs obtained legally or illicitly. Regrettably, however, the results of several surveys of large numbers of AAS abusers indicated that although many of them were aware of the side effects of AAS misuse, substantial numbers had little appreciation for the possible adverse consequences of chronic misuse (Fuller & La Fountain, 1987; Johnson et al., 1989; Wilson, 1988; Komoroski & Rickert, 1992).

One final factor that must be considered when evaluating the effects of AAS consists of the reasons most often given by individuals who take AAS on a regular basis. Several studies have examined this issue (Fuller & La Fountain, 1987; Taylor, 1987; Johnson et al., 1989; Wilson, 1988; Komoroski & Rickert, 1992). Not surprisingly, most users reported that, with a regular program of exercise and proper diet, AAS improved their physique and athletic performance and thereby their self-image, despite the fact that the efficacy of the drugs in this regard is not clear. Interestingly, some AAS misusers stated that, although they were not sure that the AAS had any direct athletic-enhancing performance effects, they used the drugs because others against whom they were competing were suspected of using AAS and, hence, these individuals could not risk the possibility of losing the "competitive edge." In addition, a relatively large number of adolescents reported that peer pressure was a significant factor in their decision to use AAS; some reported that other presumed actions of AAS (e.g., enhanced sexual performance) were strong motivating factors. These data clearly point out the need for educational efforts, since regular users apparently are not well versed in the full spectrum of actions of AAS, including their efficacy in producing a desired effect and, most importantly, in the side effects associated with such misuse. Indeed, in all surveys examining the factors responsible for the use of AAS, the individuals uniformly reported that virtually all their information about the "efficacious" effects of AAS and their side effects came from friends, magazines, "experts" at gyms, and, unfortunately, coaches.

III. Physical Effects and Side Effects of Anabolic Androgenic Steroids

Testosterone, the principal steroid secreted by the testes, is of limited medical use because it is metabolized rapidly, making effective physiological levels difficult to sustain. In the 1950s, many synthetic derivatives were developed to make testosterone suitable for oral administration, to prolong its action by decreasing the rate of absorption and catabolism, and to enhance its potency. These efforts were undertaken with the goal of finding a derivative in which the masculinizing (androgenic) and muscle building (anabolic) effects of testosterone could be separated. The anabolic and androgenic effects of testosterone are not known not to be mediated by different mechanisms. Rather, the muscle building or masculinizing actions of androgens depend on the type of target issue in which the receptors that bind these hormones are located. Thus, the term "anabolic steroid" is a misnomer: all AAS are simply androgens. The actions of androgens are influenced by metabolizing enzymes. In some tissues, a high content of the enzyme 5α-reductase reduces certain steroids to compounds that have a higher affinity for the androgen receptor than does the parent compound. Paradoxically, many actions of AAS depend on their aromatization to estrogens (Naftolin et al., 1975). For example, testosterone is aromatized to estradiol or reduced to the more potent derivative 5α-dihydrotestosterone, and exerts direct effects on its own.

The number of legitimate medical uses for androgens is quite small. The extent to which the mechanisms of action of androgens is understood is rather limited in comparison with other classes of steroid hormones such as estrogens and glucocorticoids. Finding toxicological and pharmacological data relevant to AAS abuse by athletes is particularly difficult because athletes typically self-administer these drugs in dosages 10 to 100 times higher than clinical dosages (Wilson, 1988), and they administer these drugs in step-wise combinations (stacking). Classical endocrinological studies of steroid action generally keep drug dosages within the physiological range or as close to physiological range as possible. The doses of AAS self-administered by athletes are far in excess of doses used in replacement therapy for hypogonadism and far in excess of the dose required to saturate androgen receptors completely. For example, 15 mg methandrostenolone a day is sufficient to restore androgen levels in hypogonadal men, but athletes have been reported to take 100 to 300 mg a day (Friedl, 1990). At such ridiculously high doses, the full array of effects of AAS is unlikely to be mediated solely by the androgen receptor. At pharmacological doses, synthetic AAS interact with progestin, estrogen, and glucocorticoid receptors in addition to androgen receptors (see Janne, 1990, for review). For example, certain synthetic androgens have a higher affinity for the progesterone receptor than does progesterone and can induce biological effects that are characteristic of progesterone (Janne, 1990). At pharmacological doses, steroid hormones also bind directly to neurotransmitter receptors. (For review, see La Bella, Kim, & Templeton, 1978; McEwen, Krey, & Luine, 1978; Su, London, & Jaffe, 1988; Baulieu & Robel, 1990.)

Several mechanisms have been proposed to account for potential effects of AAS on athletic performance. These mechanisms include physiological effects at the neuromuscular junction and psychological effects via the central nervous system (CNS). Potential actions of AAS on muscle tissue or on the CNS include (1) counteraction of catabolic effects of glucocorticoids by increased protein synthesis or inhibition of genomic and membrane effects of glucocorticoids; (2) central neural effects; and (3) placebo effects. The process of determining the biological significance and physiological mechanisms of action of AAS promises to be complicated because of the lack of information about the pharmacological effects of AAS at the high doses employed by athletes, and because of the wide spectrum of effects of AAS at suprapharmacological doses. With these caveats in mind, some of the physiological effects of AAS are summarized.

A. Athletic Performance and Muscle Development

Most investigations of AAS action have focused on whether AAS enhance athletic performance and muscle development. The general public, particularly many young athletes, is convinced that AAS enhance physique and athletic performance. However, the results of the relatively few available double-blind studies using placebo controls indicate that this staunchly held belief may have little scientific basis (see Table 1). Many well-controlled studies show no effect on muscular strength or athletic performance. (See Cicero & O'Connor, 1990, and Elashoff, Jacknow, Shain, & Brownstein, 1991, for recent reviews.) Thus, no compelling evidence exists that AAS have dramatic effects on muscular strength. Although several studies have shown, under appropriately designed conditions, that AAS enhance athletic performance and that subjects can discern AAS from a placebo over an extended time period, an equal or perhaps greater number of studies has failed to demonstrate such effects. (For reviews, see Wilson & Griffin, 1980; Haupt & Rovere, 1984; Lamb, 1984; Wilson, 1988.)

Inadequacies of many studies for example, the lack of a placebo group and no blinding, contribute to the confusion in the literature (Cicero & O'Connor, 1990; Elashoff et al., 1991). Blinding is difficult because weight gain, increased acne, and beard growth frequently allow the AAS being tested to be discriminated. Typically, substantial improvements in performance also occur in the placebo groups (Ariel & Saville, 1972). Ethical considerations do not permit the testing of AAS in the megadose range abused by athletes. The dose range tested is probably too narrow to see a dose–response effect, even if one does exist (Elashoff et al., 1991). If effects of high doses of AAS are small, effects beyond those achievable

TABLE I

Summary of Well-Controlled, Double-Blind Studies That Have
Reported That Anabolic Androgenic Steroid Use Improves Athletic
Performance or, Conversely, Has No Effect

Positive studies	Negative studies
O'Shea (1971)	Samuels et al. (1942)
Johnson et al. (1972)	Fowler et al. (1965)
Ariel (1973, 1974)	Ariel and Saville (1927)
Ward (1973)	Fahey and Brown (1973)
Freed et al. (1975)	Golding et al. (1974)
Win-May and Mya-Tu (1975)	Stomme et al. (1974)
Hervey et al. (1981)	Johnson et al. (1975)
	Hervey et al. (1976)
	Loughton and Ruhling (1977)
	Crist et al. (1983)

with an intense training and exercise program would be difficult to demonstrate on a statistical basis. However, a competitive athlete for whom the difference between winning and losing might be measured in fractions of a second, a 4–5% increase in strength or performance capability would be of substantial personal advantage. Without extensive animal experimentation, excluding the possibility that under the appropriate conditions—including specific training regimens in combination with protein supplementation—AAS might have some minimal effects on athletic performance capacity is nearly impossible.

High doses of AAS have a tendency to increase body weight in animals and in humans. However, does increased body weight due to an anabolic effect on protein synthesis lead to increased protein deposition as muscle and increased muscle strength? In this area, animal studies might usefully complement clinical studies since stringent controls, higher doses of AAS, and detailed biochemical analyses of muscle tissue are feasible in animals experiments. Three types of muscle activity have been examined for sensitivity to AAS. The first type of muscle activity is compensatory muscle hypertrophy or functional overload after complete or partial ablation of synergistic muscles or after severing a tendon. The second type of muscle activity is isometric or strength training, in which rats carry weighted belts. The third type of muscle activity, endurance training, is designed to increase muscle capacity for aerobic metabolism rather than to improve muscle strength or muscle mass. In an endurance exercise program, rats are trained to run a treadmill.

Despite the variety of models and exercise programs tested, the assumption that AAS increase muscle strength generally is not supported by animal experiments. Most studies show no effect (Exner, Staudte, & Pette, 1973b; Hickson et al., 1976; Stone & Lipner, 1978; Richardson, 1977; Young, Crookshank, & Ponder, 1977; Morano, 1984; Danhaive & Rousseau, 1988), although a few demonstrate improvement (Exner, Staudter, & Pette, 1973a; Viru & Korge, 1979). For example, Richardson (1977) found no effect of high doses of AAS on muscular strength in primates receiving a high protein diet and undergoing a regular program of weight training.

Some evidence suggests that AAS influence muscle fiber diameter, muscle protein synthesis, RNA polymerase activity, and the activity of enzymes associated with muscle mitochondria and the sarcotubular system in athletes (Alèn, Hakkinen, & Komi, 1984; Griggs et al., 1989) and in animals (Rogozkin, 1976, 1979; Saborido, Vila, Molano, & Megias, 1991). On the other hand, Boissonneault and colleagues (Boissonneault, Gaagnon, Ho-Kin, & Tremblay, 1987) were unable to find an effect of AAS on the hypertrophy-induced changes in specific mRNAs of skeletal muscle. Reported changes in protein synthesis and enzyme activity might be interpreted to imply that AAS increase muscular strength, but results gathered to date do not favor this possibility. For example, 12 months of treatment with AAS increased muscle protein synthesis and muscle mass in patients with muscular dystrophy but did not improve strength (Griggs, Halliday, Kingston, & Moxley, 1986). One explanation for these findings is that AAS might increase protein synthesis in noncontractile tissue contained in skeletal muscle (Griggs et al., 1989). Saborido, Vila, Molano, and Megias (1991) found that AAS increased the content in the sarcotubular fraction from slow-twitch muscle in rats. These results are in agreement with the results of Griggs and co-workers (1989) and Kuhn and Max (1985), and support the idea that the noncontractile elements are affected by AAS. Female rats might be predicted to be more sensitive to the effects of AAS because they have lower endogenous testosterone levels and higher concentrations of androgen receptors, but the few results in female rats have been conflicting. Kuhn and Max (1985) discovered no effect of high doses of AAS in hypertrophic muscles but Exner and colleagues (1973a) determined an effect of moderate doses of AAS using a program of isometric training. In castrated male rats, AAS induced significant increases in body weight and muscle mass (see Kochakian, 1990, for review).

Several investigators favor the possibility that the primary mechanism of action of high doses of AAS in skeletal muscle is by antagonism of the catabolic effects of glucocorticoids. Glucocorticoids cause protein degradation and muscle atrophy (Goldberg & Goodman, 1969). Fast-twitch muscle fibers undergo a greater degree of weakness than do slow-twitch fibers (Robinson & Clamann, 1988). Natural fluctuations in glucocorticoid and androgen levels occur after a period of physical exercise in athletes and in animals. Based on the limited number of studies available, the nature of the fluctuation of hormone levels appears to depend on the duration of strenuous exercise. After a single short bout of exercise, testosterone levels increase whereas, after longer periods of exercise, testosterone levels decrease and glucocorticoid levels increase (Mather, Toriola, & Dada, 1986; Tchaikovsky, Astratenkovan, & Basharina, 1986; Villaneuva et al., 1986; Urhausen & Kindermann, De Souza et al., 1991). Results of the limited number of studies on the ability of AAS to reverse the muscular atrophy effects of glucocorticoids in rats are inconsistent. Danhaive and Rousseau (1988) reported that AAS treatment attenuated the effects of glucocorticoids on muscle atrophy but Capaccio and colleagues (1987) found no effect of AAS on glucocorticoid-induced muscle wasting. Interestingly, Danhaive and Rousseau (1988) reported that adrenalectomy abolished the effects of AAS in females. The balance between the anabolic effects of glucocorticoids also is highlighted by work showing that castration-induced muscular atrophy is prevented partially by administration of an antiglucocorticoid (Konagaya & Max, 1986). Haupt and Rovere (1984) proposed that intense weight lifting might force athletes into a chronic catabolic state and that the anticatabolic effects of AAS might be more evident in these individuals. These researchers suggested that AAS shift the nitrogen balance from negative to positive because of better protein utilization and combat the catabolic effects of increased release of glucocorticoids during stressful training. A consistent finding in the literature is that exercise counteracts the effects of glucocorticoids on muscle atrophy (see Hickson, Czerwinski, Falduto, & Young, 1990, for review). Whether AAS in combination with exercise produces any greater effect than exercise alone on muscle sparing is an important area that must be investigated.

The molecular mechanisms that underlie the anticatabolic effects of AAS are unclear at present. The antagonism could occur at several different levels, including direct antagonism of binding to glucocorticoid receptors or physiological antagonism of glucocorticoid function at the gene level that would involve inhibition of glucocorticoid receptor gene expression, glucocorticoid synthesis, or glucocorticoid release. The relatively poor affinity of AAS for glucocorticoid receptors is well documented (for review, see Hickson et al., 1990; Janne, 1990) but, in spite of their poor affinity, pharmacological doses of AAS have been shown to inhibit the binding of glucocorticoids to glucocorticoid receptors in muscle (Mayer & Rosen, 1975; Barbieri Lee & Rean, 1979). Attempts to explore the relationship between glucocorticoid receptor content, androgen receptor content, and physical exercise have yielded results that are still inconclusive, partly because some investigators measured total androgen receptor content and others measured unoccupied cytosol receptors. Muscle androgen receptor content was unchanged by endurance exercise-induced muscle sparing (Hickson, Kurowski, Capaccio & Chattertos, 1984) but was reported to be increased by a short swim (Tchaikovsky, Astratenkova & Basharida, 1986). A long training program of endurance exercise had no effect on muscle glucocorticoid receptor binding (Hickson, Kurowski, Cappaccio & Chatterton, 1984), but a shorter endurance training program increased glucocorticoid receptor binding (Czerwinski, Kurowski, Falduto, Zak, & Hickson, 1987). The significance of these findings is unclear at present.

B. Virilization and Hypogonadism

AAS administered in doses that produce blood levels of androgens above the normal physiological range (e.g., 5–10 mg/day testosterone; Caminos-Torres, Ma, & Snyder, 1977) produce hypogonadism that is characterized by decreased testosterone production by the testes (see Wilson, 1988, for review). Luteinizing hormone releasing hormone (LHRH), luteinizing hormone (LH), and follicle stimulating hormone (FSH) levels decrease through negative feedback of AAS on the hypothalamic–pituitary–gonadal axis (Mooradian, Morley, & Korenman, 1987). The size of the testes and the production of sperm also are decreased (Johnson, Fisher, Silvester, & Hofheins, 1972; Alèn & Suomenen, 1984). These effects may persist for several months after the drugs are discontinued, presumably because of testicular refractoriness (Martikainen et al., 1986). Increased acne is common and male pattern baldness is accelerated in those individuals who are susceptible to baldness. Some investigators are concerned that AAS might increase the risk of prostrate cancer (Roberts & Essenhigh, 1986; Bardin, Swerdloff, & Santen, 1991).

In females, permanent virilizing effects of AAS include hirsutism, clitoral hypertrophy, and deepening of the voice; these changes are considered permanent. Menstrual irregularities, breast atrophy, increased aggressiveness, and increased libido also have been reported (Salmon & Geist, 1943; Strauss, Ligyetta, & Lanese, 1985; Eliot, Goldberg, Kuehl, & Catlin, 1987).

In children, AAS may cause precocious puberty and may stunt growth permanently through premature skeletal maturation and closure of the epiphyses (Rogol, Martha, & Blizzard, 1990).

C. Feminization

The feminizing effects of AAS observed in some male athletes are related to increased levels of estrogen that approach levels normally seen only in women (Alèn, Reinila, & Vihko, 1985). The ability of testosterone and certain other synthetic AAS to be aromatized to estrogens is presumed to be the mechanism responsible for the relatively high levels of estrogen in these athletes. As a consequence of increased levels of estrogen, some athletes develop gynecomastia (the growth of small lumps of breast tissue under the nipples, occasionally accompanied by secretion of fluid from the nipples) (Strauss, Wright, Finerman, & Catlin, 1983). This problem is particularly prevalent in athletes engaged in sports in which a "stacking" regimen of AAS abuse is common, for example, weight lifting (Wilson, 1988). The incidence of gynecomastia is also high in children because of a higher rate of aromatization (Wilson, 1988).

D. Cardiovascular Disease

The most commonly reported cardiovascular change associated with AAS abuse is a striking suppression of high density lipoproteins (HDL) and an increase in low density lipoproteins (LDL) (Allen & Fraser, 1981; Taggert, Applebaum-Bowden, & Haftner, 1982; Hurley, Seals, & Hagberg, 1984; Webb, Laskarzewski, & Glueck, 1984; Kantor et al., 1985; Lenders et al., 1988; Glazer, 1991). Thus, one of the major worries is that AAS abuse will raise cholesterol levels and predispose athletes to cardiovascular disease. The data suggest that changes in lipoproteins may be related to a particular class of synthetic androgens. The 17α-alkylated steroids such as stanozol cause adverse alterations in serum lipoprotein profiles, whereas testosterone does not (Thompson, Cullinane, Sady, & Chenevert, 1989). The differences among the effects of different androgens might be related to a combination of factors, including the hepatotoxic effects of 17α-alkylated steroids, the aromatization of testosterone to estradiol (which counteracts androgen effects on lipids), and the route of administration of the AAS (Thompson et al., 1989; Bardin et al., 1991).

Several case reports of acute myocardial infarction and stroke in athletes using AAS suggest the possibility that AAS use increases the risk of thrombosis (Frankel, Eichberg, & Zachariah, 1988; McNutt, Ferenchik, Kirlin, Hamlin, 1988; Glazer, 1991). In animal studies (see Ferenchick, 1991, for review), androgens potentiate the effects of thrombotic stimuli (Uzunova, 1978), increase platelet aggregation (a mediator of acute myocardial infarction, unstable angina, and sudden death) (Johnson, 1974), and increase the synthesis of procoagulant factors (see Ferenchick, 1990, for review).

In view of the well-documented effects of AAS on serum lipoproteins, platelet activity, and coagulation factors, it is perplexing that more cases of accelerated arteriosclerosis, acute myocardial infarction, and stroke have not been reported. Although more time may be needed before these changes become manifest, uncontrolled AAS use also might induce changes in the cardiovascular system that counteract each other (Bardin, 1991). For example, 17α-alkylated androgens increase several anticoagulant and fibrinolytic proteins, in addition to their adverse effects on serum lipoproteins (Blamey et al., 1984; Kluft et al., 1984; Small et al., 1984; Ahn et al., 1986). The risk of coronary heart disease has been estimated to be increased 3–6 times by AAS (Glazer, 1991). The effects of AAS on insulin levels may magnify the risk of heart disease several fold (Glazer, 1991). Further studies using a variety of synthetic androgens will be necessary to determine the extent to which uncontrolled AAS use increases the risk of arteriosclerosis, acute myocardial infarction, and stroke.

E. Liver Disease

Liver tumors (Overly, Dankoff, Wang, & Singh, 1984; Goldman, 1985) and kidney tumors (Pratt, Gray, Stolley, & Coleman, 1977) have been reported in athletes who abused AAS, but whether AAS use is linked causally to tumor growth in these particular individuals is not clear. However, benign and malignant liver tumors have been reported in individuals receiving long-term androgen treatment for hypogonadism (Boyd & Mark, 1977) and aplastic anemia (see Wilson, 1988, for review). There have been 23 cases of peliosis hepatitis, a rare form of hepatitis characterized by small blood-filled lesions in the liver, reported in conjunction with AAS use (see Haupt & Rovere, 1984; Wilson, 1988, for review).

F. Psychiatric Disturbances

Despite the fact that effects of AAS on athletic performance have been difficult to document scientifically, many athletes are convinced that these drugs enhance physique and athletic performance. Moreover, even among athletes who are not sure that AAS have any performance enhancing effects, use of the drugs is widespread because of the belief that their competitors use them. Hence, these athletes do not want to risk losing the competitive edge. Thus, the use of AAS by professional athletes is overwhelming. Their use by college athletes and by adolescents who seek to improve their physique is increasing. In view of the difficulty documenting marked effects of AAS on muscular strength, mood altering effects of AAS could be a substantial contributor to the motivational aspects of improved athletic performance.

Considering the widespread use of these substances, the growing number of reports of psychological and emotional effects of uncontrolled AAS use is particularly disturbing and may prove to be the most serious adverse effect of AAS misuse. To date, systematic studies of psychological effects of high doses of AAS have not been performed. The preponderance of evidence for their psychological effects comes from case reports and from incidental mention of side effects of AAS in studies designed to examine the influence of these substances on athletic performance.

In a review of the literature encompassing 155 athletes from 13 studies, Haupt and Rovere (1984) found the most commonly reported side effects to be increased aggressive behavior and increased libido (Johnson et al., 1972; Stamford & Moffatt 1974, Strauss, Wright, & Finerman, 1982). Subsequent case reports confirmed the possibility of increased aggressiveness (Pope & Katz, 1988; Brower, Blow, Beresford, & Fuelling, 1989; Katz & Pope, 1990; Hannan, Friedl, Zold, Kettler & Plymate, 1991) and violent crime (Conacher, Ch, & Workman, 1989) during the period of AAS abuse. Anecdotal information from long-term users of AAS indicates that these individuals believe that violent expressions of hostility (termed "roid rage" by steroid misusers) are caused by AAS (Taylor, 1987).

Many of the known clinical consequences of androgen therapy were discovered in studies designed to evaluate the potential use of very low doses of AAS as male contraceptive agents or in studies designed to restore testosterone levels to normal in hypogonadal men. In a review of this literature, Katz and Pope (1990) reported inconsistent effects of physiological doses of AAS on tension, irritability, and aggression in hypogonadal men. Establishing a dose–response relationship between levels of aggressiveness and AAS in humans with normal gonadal function has been difficult; the evidence is somewhat contradictory (Bancroft, 1978; Dabbs, Frady, Carr, & Besch, 1987; Doering, Brodie, Kramer, Becker, & Hamburg, 1974; Ehrenkranz, Bliss, & Sheard, 1974; Kreuz & Rose, 1972; Mayer-Bahlburg, Nat, Bon, Sharma, & Edwards, 1974; Persky, Smith, & Basu, 1971; Scaramella & Brown, 1978; Sturup, 1968). Moreover, because of ethical considerations involved in administering varying doses of AAS to humans, studies are correlational in nature, making it nearly impossible to determine whether alterations in levels of endogenous AAS are a cause or a consequence of aggressive behavior.

Little evidence suggests that surges in endogenous AAS are associated with athletic success. In one study (Booth, Shelley, Mazur, Tharp, & Kittok, 1989), eventual losers in a competitive athletic event had higher prematch AAS levels than eventual winners. The study also reported that, in all competitors, AAS levels were higher just prior to and just after a match than at other times. Interestingly, the experiment suggested that mood and attitude may be a better predictor of AAS levels than competitive success. Winners who were elated after a match experienced a greater postmatch rise in AAS than winners with a moderate or negative mood (Booth et al., 1989). Worthman and Konner (1987) reported that hunting success was not correlated with AAS levels in !Kung San men on a 6-day hunt. Scaramella and Brown (1978) found a positive correlation between degree of aggressive response to threat and AAS levels in hockey players, but no correlation between AAS levels and any other measures of aggressiveness.

In contrast to the literature on humans, in animals little doubt remains that castration decreases aggression and moderate levels of AAS restore aggressive behavior in castrated male mice (see Leshner, 1978; Dixson, 1980, for review). Unfortunately, the degree to which supplemental doses of AAS potentiate aggression in intact animals has received very little study. A threshold for aggressiveness is known to exist; also, as the dose of testosterone is increased, the percentage of castrated male mice that will fight is increased also (Edwards, 1969). Banarjee (1971) showed that testosterone increased aggressive behavior in relatively non-aggressive intact male mice. However, the more important study, which has not been done, would establish whether a dose–response relationship exists between AAS and intensity of aggressive responding. In this regard, note that, in one study, treatment of castrated male mice with high pharmacological doses of AAS did not restore the aggressive behavior whereas more moderate doses did restore aggressiveness (Bevan, Levy, Whitehouse, & Bevan, 1957; Bevan, Bevan, & Williams, 1958). Thus, currently the potential for increased aggressive behavior following long-term use of high doses of AAS receives little support from the limited animal literature.

In view of the finding that high doses of AAS fail to induce aggressive behavior in castrated male mice (Bevan et al., 1957; Bevan et al., 1958), considering whether the primary mechanism of action of high doses of AAS on aggression is counteracting the effects of glucocorticoids is worthwhile. During an aggressive encounter, animals display two types of behavior (1) fighting or aggressive behavior and (2) nonaggressive or submissive behavior. In mice, glucocorticoids have effects on aggressive behavior but the relationship is complicated because adrenal corticotropic hormone (ACTH) also influences aggressive behavior, in the opposite direction (see Leshner, 1978, for review). Also readiness of animal to avoid aggressive encounters with an opponent that it knows to be a superior fighter can be examined. In this type of situation, the critical hormone controlling the tendency of the animal to behave submissively appears to be corticosterone (Leshner, Moyer, & Walker, 1975; Moyer & Leshner, 1976). In contrast, testosterone does not affect the ability of an animal to learn to avoid aggressive encounters (Leshner & Moyer, 1975). If the normal increase in glucocorticoid levels that occurs during an aggressive encounter with a superior opponent is prevented, a mouse will fight longer and take more time to become submissive (Nock & Leshner, 1976). The normal elevation in ACTH and glucocorticoid levels that occurs during an aggressive encounter appears to potentiate the behaviorally appropriate decrease in aggressive behavior and increase in submissiveness by a mouse that ultimately will be beaten. When the normal increase in glucocorticoids is blocked, the animal behaves inappropriately for the situation and fights longer. If the data on aggressive encounters between mice are extrapolated to humans, certain predictions about the mechanisms underlying the effects of high doses of AAS can be made. The available evidence suggests that high doses of AAS do not facilitate aggressive behavior, raising the possibility that pharmacological doses of AAS act to oppose the actions of the hypothalamo–pituitary–adrenocrotical system on aggression and oppose the actions of glucocorticoids on the avoidance of aggressive encounters. If this is true, high doses of AAS would counteract the effects of glucocorticoids in aggressive situations and cause a behaviorally inappropriate display of aggression. Further studies of the roles of glucocorticoids and pharmacological doses of AAS in aggressive and nonaggressive behaviors are needed to test this possibility.

In men and women, a commonly reported side effect of AAS is increased libido (Johnson et al., 1972; Stamford & Moffatt, 1974; Strauss et al., 1983, 1985). Decreased libido also occasionally has been reported (Johnson et al., 1972; Freed, Banks, Longson, & Burley, 1975; Strauss et al., 1982, 1983). In men, AAS are considered an important contributor to normal libido (O'Carrol & Bancroft, 1984; Mooradian et al., 1987). In women, AAS also are considered to be the primary hormones involved in libido (Salmon & Geist, 1943).

Additional effects mentioned by users of AAS are increased irritability and nervous tension, diminished fatigue, and euphoria (Sammuels, Henschel, & Keys, 1942; Johnson & O'Shea, 1969; Johnson, Fischer, Silvester, & Hofheirs, 1972; Stamford & Moffatt, 1974; Freed et al., 1975; Hervey, 1975; Wilson & Griffin, 1980; Lucking, 1982; Strauss et al., 1983; Lamb, 1984; Taylor, 1984; Tennant, Black, & Voy, 1988). The idea that AAS makes athletes "feel better" receives some support from the clinical effects of testosterone in depressed patients. In the 1940s, large doses of testosterone were used with some success to treat depression (Miller, Hubert, & Hamilton, 1938; Danziger & Blank, 1942; Danziger, Schroeder, & Unger, 1944; Altschule & Tillotson, 1948). Vogel, Klaiber, and Broverman (1978, 1985) reported that AAS, at doses up to 10 times those used to treat hypogonadism (400 vs 40 mg methyltestosterone/day), were an effective antidepressant in chronically depressed men. Yesavage, Davidson, Widrow, and Berger (1985) found a correlation between testosterone levels and severity of depression, but whether the reduced levels of testosterone were a consequence of depression or a causative factor is unclear. Thus, the limited use of AAS to alleviate symptoms of depression and the correlation between fluctuations in levels of AAS and severity of depression raise the possibility that alterations in mood might be a factor that contributes to the misuse of AAS.

More serious psychological side effects of AAS abuse include symptoms of mania and psychosis (Wilson et al., 1974; Annitto & Layman, 1980; Freinhaar & Alvarez, 1985; Pope & Katz, 1987, 1988). Pope and Katz (1988) reported that 41 of 45 AAS users interviewed meet criteria for psychotic symptoms, including paranoid delusions and hallucinations. Of the subjects, 4 met criteria for manic episodes and 5 had major depression.

G. Are Anabolic Androgenic Steroids Abused or Misused?

Several investigators have hypothesized that long-term AAS use may develop into a dependency disorder similar to that accompanying long-term cocaine, alcohol, or opioid abuse (Tennant et al., 1988; Brower et al., 1989; Kashkin & Kleber, 1989; Goldstein, 1990; Yesalis, Vicary, Buckley, Smeit, Katrard, Wright, 1990b). Two cases have been reported of patients who met certain criteria for psychoactive substance dependence. Among the criteria met were (1) continued use despite substantial psychological problems such as paranoia, suicidal tendency, and increased aggression; (2) unsuccessful attempts to stop using the drug; and (3) evidence of withdrawal symptoms on discontinuation of the drug (Tennant et al., 1988; Brower et al., 1989). Pope and Katz (1988) reported that 12% of AAS users developed a depressive episode during a period of AAS withdrawal.

Many examples provide strong support for the idea that a drop in endocrine hormones can precipitate physical and psychological disturbances that are strongly suggestive of a withdrawal syndrome. Patients undergoing corticosterone withdrawal show a withdrawal syndrome characterized by anorexia, nausea, lethargy, arthralgia, weakness, desquamation, and weight loss. This syndrome occurs in patients who are adrenalectomized during treatment for Cushings disease, as well as in those who experience abrupt cessation of prolonged corticosterone therapy for a variety of other disorders (Woodbury, 1958; Amatrudo, Hurst, & D'Esopo, 1965; Dixon & Christy, 1980; Sherwin & Gelfand, 1985). Another example of a potentially harmful drop in steroid levels is the precipitous drop in progesterone that occurs postpartum (Hamburg, 1966). The postpartum period is associated with an increased incidence of depression (Hamburg, 1966; Hamilton, Parry, & Blumenthal, 1988). After bilateral oophorectomy, an abrupt drop in estrogen and testosterone levels occurs. In a well-controlled study of effects of estrogen and testosterone administration on the physical and psychological symptoms following surgical menopause, energy level, well-being, and appetite were found to be enhanced by AAS administration (Sherwin & Gelfand, 1985). Anabolic androgenic steroids had a more profound effect on these symptoms than did combined therapy with AAS and estrogen or estrogen alone. These results suggest that abrupt withdrawal of AAS secreted by the ovaries plays a critical role in the development of psychological disturbances after surgical menopause. In a clinical report, a withdrawal syndrome was produced in an AAS abuser by injection of the opiate antagonist naloxone (Tennant et al., 1988). Endogenous opioid peptides play a key role in the regulation of the hypothalamo–pituitary–gonadal axis. Moreover, considerable evidence suggests that AAS modulate the endocrine responses to opiates and alcohol (see Section IV).

From the brief review provided here, clearly the use of AAS is a goal-oriented behavior, which some researchers claim provides evidence that AAS are drugs of abuse (Taylor, 1987; Kashkin & Kleber, 1989). However, whether self-administration of these substances is based on the intrinsic pharmacological or reinforcing properties of the AAS or on some anticipated effect that is either real or perceived remains to be determined. As a result, whether chronic users simply misuse these compounds or whether AAS have significant abuse liability as appropriately defined in a strict pharmacological sense of (1) self-administration for the immediate rewarding physiological or pharmacological effects; (2) discrimination of AAS from other steroids or placebo; (3) tolerance so more drug is required to produce the desired effect; and (4) physical dependence, in which a withdrawal syndrome ensues on abrupt withdrawal of the drug, is unclear. At this time, insufficient information is available to determine whether AAS possess intrinsic abuse liability as defined by these criteria (see Cicero & O'Connor, 1990, for a review). Additional systematic studies would be required to determine the abuse potential of AAS. Moreover, animal models of AAS use and abuse must be developed since they ultimately will provide valuable insights into the mechanisms involved in the abuse liability of AAS. Animal models have proven extremely useful in studies of the abuse potential of other drugs. Their development in this area would be extremely valuable.

IV. Role of Anabolic Androgenic Steroids in the Abuse of Other Substances

Increasing evidence suggests that gonadal steroids may participate in the acute and chronic effects of many abused substances. Drug-induced alterations in the levels of testosterone have been postulated to contribute to the mood changes associated with the acute effects of drugs, particularly alcohol (e.g., Mendelson & Mello, 1974, 1976, 1979). However, these conclusions have not gained widespread acceptance for several reasons. For example, alcohol intoxication in some individuals produces aggression or hostility, whereas in others it generates passivity. In both groups of individuals, alcohol reduces serum testosterone levels; hence, essentially no correlation appears to exist between the mood altering effects of alcohol and serum testosterone levels. In addition, establishing correlations between testosterone levels and mood shifts and/or passive–aggressive behavior has been difficult under normal drug-free conditions in humans or animals (Sturup, 1968; Persky et al., 1971; Doering et al., 1974; Mayer-Bahlburg et al., 1974; Bancroft, 1978). Note that other studies (Winslow, Ellingbor, & Miczek, 1988; Winslow & Niczek, 1988) have shown a strong positive correlation between acute changes in serum testosterone levels and mood shifts (passive–aggressive behavior) in nonhuman primates. However, these effects were dependent on several factors such as social context and endocrine status at the time of alcohol administration. Despite these reservations and somewhat contradictory evidence regarding a relationship between alcohol-induced changes in serum testosterone levels and the behavioral effects of the drug, dismissing the possibility of a link between AAS and the mood altering properties of alcohol would be unwise. Specifically, because of the slow-onset genomic effects of AAS, poor correlations between the acute behavioral effects of alcohol and drug-induced modifications in AAS levels might be expected. Rather, focusing on long-term correlations between serum testosterone levels and behavioral disturbances induced by chronic alcohol abuse might be more valid.

In the female, attempts have been made to ask a somewhat different question. Do the marked alterations in sex steroids throughout the menstrual (human) or estrous (animal) cycle modify the acute response to abused drugs? The literature is now replete with evidence that the effects of opiate agonists and antagonists on reproductive hormones are influenced markedly by the stage of the estrous cycle during which they are administered (for reviews see Cicero, 1980a,b, 1984, 1987). Similarly, the effects of alcohol on reproductive hormones also are affected by the stage of the estrous or menstrual cycle (Dees & Kozlowski, 1984; Mello &

Mendelson, 1985; Mello, Mendelson, Bree, and Skupny, 1986; Gibeau, Hosobu-chi, & Lee, 1986; Mello, Mendelson, Bree, & Skupny, 1986; Mello, 1988; Men-delson, Mello, Tech, & Ellingboe, 1989). However, virtually nothing is known about whether other pharmacological or physiological actions of alcohol or opiates are affected or whether the effects of other commonly abused drugs (e.g., mari-juana or cocaine) are modulated by intrinsic fluctuations in female sex hormones.

In addition to the foregoing studies in males, a limited number of studies has been done in females. With respect to the issue of whether alterations in drug seeking behavior occur as a function of the menstrual or estrous cycle, little information is at hand. Although not conclusive, a limited number of studies suggest that the use of marijuana (Mello & Mendelson, 1985) and alcohol (Podol-sky, 1963; Belfer, Shader, Carroll, & Hermatz, 1971; Belfer & Shader, 1976) may increase during various phases of the menstrual cycle that appear to correlate with changes in moods. Clearly, these findings should be extended to determine wheth-er a causal relationship exists between the self-administration of marijuana or alcohol and normal fluctuations in ovarian and pituitary hormones during the menstrual cycle in humans. At present, the data suggest a correlation, but cause–effect relationships are unclear since an enhanced self-administration of a psycho-active compound at various points in the menstrual cycle simply could be related to alterations in mood, independent of steroid fluctuations themselves. Never-theless, these findings are exciting and determining whether the self-administra-tion of other drugs is affected similarly during the menstrual cycle would be of substantial interest. We are aware of no systematic reports in the literature that would suggest that the self-administration of drugs of abuse is influenced by the menstrual cycle in nonhuman primates or the estrous cycle in animals. This absence of data is difficult to understand since such studies are clearly feasible and would provide valuable insights into the role of AAS in the abuse liability of commonly used drugs. Animal models would permit a degree of control of experimental variables that is clearly not feasible in humans and, hence, in the long-run will provide much more definitive data.

In the male, the role of AAS in mediating the effects of substances of abuse also has been examined. Considerable evidence suggests that AAS modulate the endocrine responses to opiates and to alcohol (for reviews, see Cicero, 1980, 1982a,b, 1984, 1987). For example, studies in this laboratory have shown that, in the testosterone-depleted castrated male rat, the effects of morphine and alcohol on serum LH levels are diminished so the normally suppressive effect of the drugs on serum LH levels virtually was abolished in long-term castrated male rats (Cicero, 1982b; Cicero, Greenwald, Nock, & O'Connor, 1990). Interestingly, in these studies we found that the castration-induced insensitivity to morphine and alcohol appeared to be selective to LH secretion because long- and short-term castrated and sham-operated rats were equally or more sensitive to other CNS effects of the two drugs. However, note that we have not assessed the full pharmacological profiles of morphine or alcohol in the castrated animal. Determining whether other acute effects of these two drugs are affected in animals in which steroid concentrations are altered systematically would be of substantial interest. Given the discrete localization of steroids in the CNS, designing studies to examine the effects of alcohol and morphine in castrated animals in which specified AAS levels are maintained and to examine how specific behavioral or physiological measures thought to be mediated by brain areas richly innervated by steroid-concentrating neuronal systems are affected should be possible. The results discussed here clearly indicate that, in male animals, AAS may mediate at least some of the acute effects of alcohol and opiates. This area of research is in its very early stages of development, but promises to be rewarding in terms of defining the possible role of AAS in mediating the actions and, consequently, the use of psychoactive compounds. In particular, the foregoing studies raise the possibility that the pharmacological response to other commonly abused substances, as well as to legitimately prescribed medications, could be altered by chronic misuse of suprapharmacological doses of AAS. The misuse of AAS also could lead to polydrug abuse and unpredictable responses to drug combinations.

We are unaware of any studies in human males in which the possible role of AAS in mediating the effects of psychoactive compounds has been investigated, undoubtedly because of the absence in males of the very large and prominent rhythms in serum steroid levels that are observed in females. In addition, relatively few physiological conditions display persistent increases or decreases in serum testosterone levels, which imposes restrictions on the types of experiments that can be carried out. Nevertheless, circadian rhythms of testosterone release exist in males (Keating & Tcholokian, 1979). Consequently, determining whether these fluctuations in any way modify the acute effects of a variety of commonly abused substances should be possible. Currently, however, the castrated male rodent seems to provide an extremely valuable model to examine the role of testosterone in mediating the acute and chronic effects of abused substances, since the investigator has complete control of circulating serum testosterone levels throughout the course of the experiments.

Although the foregoing results suggest that AAS may play an important role in the acute effects of alcohol and opiates on reproductive function in the male rat, several extremely interesting questions remain to be examined. Do AAS participate in the nonendocrine effects of alcohol and morphine? Are the effects of other abused substances similarly affected by manipulations in the steroid milieu? Is the development of tolerance to and physical dependence on morphine and alcohol altered by AAS? Are the reinforcing properties of these drugs affected by AAS? We are unaware of any systematic studies addressing these important issues. Clearly, this important area of research deserves much greater attention.

One extremely interesting question, not discussed here, should be singled out for immediate intensive study in view of the currently available evidence regarding an interaction between AAS and the effects of abused substances. Specifically, does the misuse of AAS by young athletes for prolonged periods of time result in any modifications in the effects of alcohol or other commonly abused drugs? Of perhaps greater interest is whether the use of AAS leads to an enhanced self-administration of other abused substances. The literature reviewed in this section suggests that this possibility should be explored, but currently we are unaware of any literature relevant to this extremely important question. Clearly, this idea should be investigated. Such studies can be carried out easily, initially by appropriately designed epidemiological studies.

V. Concluding Remarks

Little doubt remains that anabolic androgenic steroids are misused by an increasing number of individuals for a variety of anticipated effects, including most prominently an enhancement of physique and athletic performance. Despite the fact that the evidence concerning the efficacy of AAS in this regard is not compelling, users believe that these compounds produce such effects. The misuse of AAS is, consequently, becoming an important problem in society that cannot be ignored. Hopefully, this chapter will serve as a stimulus to investigators to carry out more extensive studies to examine more fully the incidence and motivating factors responsible for the use of AAS, their efficacy in producing the anticipated effects, the distinction between expectancy and the immediate reinforcing properties of the AAS, and, finally, whether these compounds have abuse liability in the true sense of this term.

In addition to studies addressing the important problems related to the misuse or abuse liability of AAS, a growing body of data suggests that AAS may be involved in mediating the acute effects of substances of abuse on at least certain parameters, and could be involved in those processes associated with tolerance and physical dependence. Although this area of research is very new it is very promising and is worthy of much further investigation. Three particularly fruitful areas of research are (1) whether intrinsic changes in sex steroids modulate the acute or chronic effects of abused substances; (2) whether drug-induced alterations in the steroid milieu in humans and animals are involved in mediating the acute and, particularly, chronic effects of substances of abuse; and (3) whether the acute effects of commonly abused drugs and their self-administration are modified by long-term AAS misuse in chronic users.

Acknowledgment

This chapter was adapted in part from the authors' chapter "Abuse Liability of Anabolic Steroids and Their Possible Role in the Abuse of Alcohol, Morphine, and Other Substances" in *Anabolic Steroid Abuse*, edited by G. C. Lin and L. Erinoff.

References

Ahn, Y. S., Harrington, W. J., Simon, S. R., Mylvaganam, R., Pall, L. M., & So, A. G. (1986). Danazol for the treatment of idiopathic thrombocytopenic purpura. *New England Journal of Medicine, 314*, 645–650.

Alèn, M., Hakkinen, K., & Komi, P. V. (1984). Changes in neuromuscular performance and muscle fiber characteristics of elite power athletes self-administering androgenic and anabolic steroids. *Acta Physiologica Scandinavica, 122*, 535–544.

Alèn, M., Reinila, M., & Vihko, R. (1985). Response of serum hormones to androgen administration in power athletes. *Medical Science, Sports, and Exercise, 17*, 354.

Alèn, M., & Suominen, J. (1984). Effect of androgenic and anabolic steroids in spermatogenesis in power athletes. *International Journal of Sports Medicine, 5*, 189–192.

Allen, J. K., & Fraser, I. S. (1981). Cholesterol, high density lipoprotein and danazol. *Journal of Endocrinology and Metabolism, 53*, 149–152.

Altschule, M. D., & Tillotson, K. J. (1948). The use of testosterone in the treatment of depressions. *New England Journal of Medicine, 239*, 1036–1038.

Amatrudo, T. T., Jr., Hurst, M. M., & D'Esopo, N. D. (1965). Certain endocrine and metabolic facets of the steroid withdrawal syndrome. *Journal of Clinical Endocrinology, 25* 1207–1217.

Annitto, W., & Layman, W. A. (1980). Anabolic steroids and acute schizophrenic episode. *Journal of Clinical Psychiatry, 41(4)*, 154–144.

Ariel, G. (1972). The effect of anabolic steroid (methandrostenolone) upon selected physiological parameters. *Athletic Training, 7*, 190–200.

Ariel, G. (1973). The effect of anabolic steroid upon skeletal muscle contractile force. *Journal of Sports Medicine, 13*, 187.

Ariel, G. (1974). Residual effect of an anabolic steroid upon isotonic muscular force. *Journal of Sports Medicine, 14*, 103.

Ariel, G., & Saville, W. (1972). Anabolic steroids: The physiological effects of placebos. *Medical Science, Sports, and Exercise, 4*, 124–126.

Arnold, A. P., & Breedlove, S. M. (1985). Organizational and activational effects of sex steroid on brain and behavior: A reanalysis. *Hormones and Behavior, 19*, 469–498.

Banarjee, U. (1971). The influence of some hormones and drugs on isolation-induced aggression in male mice. *Communications in Behavioral Biology, 6*, 163–170.

Bancroft, J. (1978). The relationship between hormones and sexual behavior in humans. In J. B. Hutchinson (Ed.), *Biological determinants of sexual behavior* (pp. 493–520). Wiley, New York.

Barbieri, R. L., Lee, H., & Ryan, K. J. (1979). Danazol binding to rat androgen, glucocorticoid, progesterone, and estrogen receptors: Correlation with biologic activity. *Fertility and Sterility, 31*, 182.

Bardin, C. W., Swerdloff, R. S., & Santen, R. J. (1991). Androgens: Risks and benefits. *Journal of Clinical Endocrinology and Metabolism, 73(1)*, 4.

Baulieu, E.-E., & Robel, P. (1990). Neurosteroids: A new brain function. *Journal of Steroid Biochemistry and Molecular Biology, 37(3)*, 395–403.

Belfer, M. L., & Shader, R. I. (1976). Premenstrual factors as determinants of alcoholism in women. In M. Greenblatt, & M. A. Schuckit (Eds.), *Alcohol problems in women and children* (pp. 97–102). New York.

Belfer, M., Shader, R. I., Carroll, M., and Hermatz, J. S. (1971). Alcoholism in women. *Archives of General Psychiatry, 25*, 540–544.

Bell, J. A., & Doege, T. C. (1987). Athletes' use and abuse of drugs. *Physiology and Sports Medicine, 15*, 99–108.

Bevan, J. M., Bevan, W., & Williams, B. F. (1958). Spontaneous aggressiveness in young castrate C3H mice treated with three dose levels of testosterone. *Physiological Zoology, 31,* 284–288.

Bevan, W., Levy, G. W., Whitehouse, J. M., & Bevan, J. M. (1957). Spontaneous aggressiveness in two strains of mice castrated and treated with one of three androgens. *Physiological Zoology, 30,* 341–349.

Blamey, S. L., Lowe, G. D. O., Bertina, R. M., Kluft, C., Sue-Ling, H. M., Davies, J. A., & Forbes, C. D. (1984). Protein C antigen levels in major abdominal surgery: Relationships to deep vein thrombosis, malignancy and treatment with stanozolol. *Thrombosis and Haematology, 54,* 622–625.

Boissonneault, G., Gagnon, J., Ho-Kin, M. A., & Tremblay, R. R. (1987). Lack of effect of anabolic steroids on specific mRNAs of skeletal muscle undergoing compensatory hypertrophy. *Molecular and Cellular Endocrinology, 51,* 19–24.

Booth, A., Shelley, G., Mazur, A., Tharp, G., & Kittok, R. (1989). Testosterone, and winning and losing in human competition. *Hormones and Behavior, 23,* 556–57.

Boyd, P. R., & Mark, G. J. (1977). Multiple hepatic adenoma and hepatocellular carcinoma in a man on oralmethyl testosterone for eleven years. *Cancer, 40,* 1765–1770.

Brower, K. J., Blow, F. C., Beresford, T. P., & Fuelling, C. (1989). Anabolic–androgenic steroid dependence. *Journal of Clinical Psychiatry, 50,* 31–32.

Burkett, L. N., & Falduto, M. T. (1984). Steroid abuse by athletes in a metropolitan area. *Physician and Sports Medicine, 12,* 69–74.

Caminos-Torres, R., Ma, L., & Snyder, P. J. (1977). Testosterone-induced inhibition of the LH and FSH responses to gonadotropin-releasing hormone occurs slowly. *Journal of Clinical Endocrinology and Metabolism, 44,* 1142.

Capaccio, J. A., Kurowski, T. T., Czerwinski, S. M., Chatterton, R. T., Jr., & Hickson, R. C. (1987).

Cicero, T. J. (1980a). Common mechanisms underlying the effects of ethanol and narcotics on neuroendocrine function. In N. K. Mello (Ed.), *Advances in substance abuse, behavioral and biological research* (Vol. I, pp. 201–254). Greenwich, Connecticut: JAI Press.

Cicero, T. J. (1980b). Sex differences in the effects of alcohol and other psychoactive drugs on endocrine function: Clinical and experimental evidence. In O. J. Kalant (Ed.), *Alcohol and drug problems in women* (pp. 545–593). New York: Plenum Press.

Cicero, T. J. (1982a). Neuroendocrinological effects of alcohol. In W. P. Creger (Ed.), *Annual review of medicine* (pp. 123–142). Palo Alto, California: Annual Reviews.

Cicero, T. J. (1982b). Involvement of hormones in the development of tolerance to and physical dependence on ethanol. In R. Collu (Ed.), *Brain peptides and hormones* (pp. 379–390). New York: Raven Press.

Cicero, T. J. (1984). Opiate-mediated control of luteinizing hormone in the male: physiological implications. In G. Delitala (Ed.), *Opioid modulation of endocrine function* (pp. 211–222). New York.

Cicero, T. J. (1987). Basic endocrine pharmacology of opioid agonists and antagonists. In Wakeling and Furr (Eds.), *Pharmacology and clinical uses of inhibitors of hormone secretion and action* (pp. 518–537). England: Bailliere Tindall.

Cicero, T. J., Greenwald, J., Nock, B., & O'Connor, L. S. (1990). Castration-induced changes in the response of the hypothalamic-pituitary axis to alcohol in the male rat. *Journal of Pharmacology and Experimental Therapeutics, 252(2),* 456–461.

Cicero, T. J., & O'Connor, L. H. (1990). Abuse liability of anabolic steroids and their possible role in the abuse of alcohol, morphine and other substances. In G. C. Lin and L. Erinoff (Eds.), *Anabolic steroid abuse. pp. 1–28.* Washington, D.C.: National Institutes of Drug Abuse.

Conacher, G. N., & Workman, D. G. (1989). Violent crime possibly associated with anabolic steroid use. *American Journal of Psychiatry, 146,* 5.

Testosterone fails to prevent skeletal muscle atrophy from glucocorticoids. *Journal of Applied Physiology, 63,* 328–334.

Crist, D. M., Stackpole, P. J., & Peake, G. T. (1983). Effects of androgenic-anabolic steroids on neuromuscular power and body composition. *Journal of Applied Physiology, 54,* 366–370.

Czerwinski, S. M., Kurowski, T. T., Falduto, M. T., Zak, R., & Hickson, R. C. (1987). Myosin heavy chain turnover in the prevention of glucocorticoid-induced muscle atrophy by exercise. *Journal of Applied Physiology, 63,* 1504–1510.

Dabbs, J., Frady, R., Carr, T., & Besch, N. (1987). Saliva testosterone and criminal violence in young adult prison inmates. *Psychosomatic Medicine, 49,* 174–182.

Danhaive, P. A., & Rousseau, G. G. (1988). Evidence for sex-dependent anabolic response to androgenic steroids mediated by muscle glucocorticoid receptors in the rat. *Journal of Steroid Biochemistry, 29(6),* 575–581.

Danziger, L., & Blank, H. R. (1942). Androgen therapy of agitated depressions in the male. *Medical Annals of the District of Columbia, 11,* 181–183.

Danzinger, L., Schroeder, H. T., & Unger, A. A. (1944). Androgen therapy for involutional melancholia. *Archives of Neurology and Psychiatry, 51,* 457–461.

Dees, W. L., & Kozlowski, G. P. (1984). Differential effects of ethanol on luteinizing hormone, follicle stimulating hormone and prolactin secretion in the female rat. *Alcohol, 1,* 429–433.

Dees, W. L., Rettori, V., Kozlowski, G. P., & McCann, S. M. (1985). Ethanol and the pulsatile release of luteinizing hormone, follicle stimulating hormone and prolactin in ovariectomized rats. *Alcohol, 2,* 641–646.

De Souza, M. J., Maguire, S., Maresh, C. M., C. M., Kraemer, W. J., Rubin, K. R., & Loucks, A. B. (1991). Adrenal activation and the prolactin response to exercise in eumenorrheic and amenorrheic runners. *American Physiological Society, 70(6),* 2378–2387.

Dixon, R. B., & Christy, N. P. (1980). On the various forms of corticosteroid withdrawal syndrome. *American Journal of Medicine, 68,* 224.

Dixson, A. (1980). Androgens and aggressive behavior in primates: A review. *Aggressive Behavior, 6,* 37–67.

Doering, C. H., Brodie, H. K., Kramer, H., Becker, H., & Hamburg, D. A. (1974). Plasma testosterone levels in psychologic measures in men over a two month period. In R. C. Frideman, R. M. Richard, and R. O. L. VandeWiele (Eds.), *Sex differences in behavior* (pp. 413–431). New York: Wiley.

Edwards, D. A. (1969). Early androgen stimulation and aggressive behavior in male and female mice. *Physiology and Behavior, 4,* 333–338.

Ehrenkranz, J., Bliss, E., & Sheard, M. (1974). Plasma testosterone: Correlation with aggressive behavior and social dominance in man. *Psychosomatic Medicine, 36,* 469–475.

Elashoff, J. D., Jacknow, A. D., Shain, S. G., & Braunstein, G. D. (1991). Effects of anabolic-androgenic steroids on muscular strength. *Annals of Internal Medicine, 115,* 387–393.

Eliot, D. L., Goldberg, L., Kuehl, K. S., & Catlin, D. H. (1987). Characteristics of anabolic-androgenic steroid-free competitive male and female bodybuilders. *Physician and Sports Medicine, 15(6),* 169.

Exner, G. U., Staudte, H. W., & Pette, D. (1973a). Isometric training of rats—Effects upon fast and slow muscle and modification by an anabolic hormone (nandrolone decanoate). *Pflugers Archives, 345,* 1–14.

Exner, G. U., Staudte, H. W., & Pette, D. (1973b). Isometric training of rats—Effects upon fast and slow muscle and modification by an anabolic hormone (nandrolone decanoate). II. Male rats. *Pflugers Archives, 345,* 15–22.

Fahey, T. D., & Brown, C. H. (1973). The effects of an anabolic steroid on the strength, body composition, and endurance of college males when accompanied by a weight training program. *Medicine and Science in Sports, 5,* 272–276.

Ferenchick, G. S. (1991). Anabolic/androgenic steroid abuse and thrombosis: Is there a connection? *Medical Hypotheses, 35,* 27–31.

Fowler, W. M., Gardner, G. W., & Egstrom, G. H. (1965). Effect of an anabolic steroid on physical performance of young men. *Journal of Applied Physiology, 20,* 1038–1040.

Frankel, M. A., Eichberg, R., & Zachariah, S. B. (1988). Anabolic androgenic steroid and a stroke in an athlete: Case report. *Archives of Physiology and Medical Rehabilitation, 69,* 632.

Freed, D. L. J., Banks, A. J., Longson, D., & Burley, D. M. (1975). Anabolic steroids in athletes: Crossover double-blind trial on weight lifters. *British Medical Journal, 2,* 471–473.

Freinhar, J., & Alvarez, W. (1985). Androgen-induced hypomania. *Journal of Clinical Psychiatry, 46(8),* 354–355.

Friedl, K. E. (1990). Reappraisal of the health risks associated with the use of high doses of oral and injectable androgenic steroids. In G. C. Lin and L. Erinoff (Eds.), *Anabolic steroid abuse* (pp. 142–177). Washington, D.C.: National Institutes of Drug Abuse.

Fuller, J. R., & LaFountain, M. J. (1987). Performance-enhancing drugs in sport: A different form of drug abuse. *Adolescence, 22,* 969–976.

Gibeau, P. M., Hosobuchi, Y., & Lee, N. M. (1986). Dynorphin effects on plasma concentrations of anterior pituitary hormones in the nonhuman primates. *Journal of Pharmacology and Experimental Therapeutics, 238,* 974–97.

Glazer, G. (1991). Atherogenic effects of anabolic steroids on serum lipid levels. *Archives of Internal Medicine, 151,* 1925–1933.

Goldberg, A., & Goodman, H. (1969). Relationship between cortisone and muscle work in determining muscle size. *Journal of Physiology (London), 200,* 667–675.

Golding, L. A., Freydinger, J. E., & Fishel, S. S. (1974). Weight, size and strength: Unchanged with steroids. *Physician and Sports Medicine, 2,* 39–43.

Goldman, B. (1985). Liver carcinoma in an athlete taking anabolic steroids. *Journal of the American Osteopathic Association, 85,* 25.

Goldstein, P. J. (1990). Anabolic steroids: An ethnographic approach. In G. C. Lin and L. Erinoff (Eds.), *Anabolic steroid abuse* pp. 74–96. Washington, D.C.: National Institutes of Drug Abuse.

Griggs, R. C., Halliday, D., Kingston, W., & Moxley, R. T., III (1986). Effect of testosterone on muscle protein synthesis in myotonic dystrophy. *Annals of Neurology, 20,* 590–596.

Griggs, R. C., Kingston, W., Jozefowicz, R. F., Herr, B. E., Forbes, G., & Halliday, D. (1989). Effect of testosterone on muscle mass and muscle protein synthesis. *Journal of Applied Physiology, 66,* 498–503.

Hamburg, D. A. (1966). Effects of progesterone on behavior. *Research Publications: Association for Research in Nervous and Mental Disease, 43,* 251.

Hamilton, J. A., Parry, B. L., & Blumenthal, S. J. (1988). The menstrual cycle in context, I: Affective syndromes associated with reproductive hormonal changes. *Journal of Clinical Psychiatry, 49(12),* 474–480.

Hannan, C. J., Jr., Friedl, K. E., Zold, A., Kettler, T. M., & Plymate, S. R. (1991). Psychological and serum homovanillic acid changes in men administered androgenic steroids. *Psychoneuroendocrinology, 16,* 335–343.

Haupt, H. A., & Rovere, G. D. (1984). Anabolic steroids: A review of the literature. *American Journal of Sports Medicine, 12(6),* 469–484.

Hervey, G. R. (1975). Are athletes wrong about anabolic steroids? *British Journal of Sports Medicine, 9,* 74–77.

Hervey, G. G., Hutchinson, I., Knibbs, A. V., Burkinshaw, L., Jones, P. R. M., Norgan, N. G., & Lewell, M. J. (1976). Anabolic effects of methandienone in men undergoing athletic training. *Lancet, 2,* 699.

Hervey, G. R., Knibbs, A. V., Burkinshaw, L., Morgan, D. B., Jones, P. R. M., Chettle, D. R., &

Vartsky, D. (1981). Effects of methandienone on the performance and body composition of men undergoing athletic training. *Clinical Science, 60*, 457–461.

Hickson, R. C., Kurowski, T. T., Capaccio, J. A., & Chatterton, R. T., Jr. (1984). Androgen cytosol binding in exercise-induced sparing of muscle atrophy. *American Journal of Physiology, 247*, E597–E603.

Hickson, R. C., Czerwinski, S. M., Falduto, T., & Young, A. P. (1990). Glucocorticoid antagonism by exercise and androgenic-anabolic steroids. *Medical Science of Sports and Exercise, 22(3)*, 331–340.

Hickson, R. C., Heusner, W. W., Van Huss, W. D., Jackson, D. E., Anderson, D. A., Jones, D. A., & Psaledas, A. T. (1976). Effects of dianabol and high-intensity sprint training on body composition of rats. *Medical Science of Sports, 8(3)*, 191–195.

Hurley, B. F., Seals, D. R., & Hagberg, J. M. (1984). High-density-lipoprotein cholesterol in body-builders vs powerlifters. *Journal of the American Medical Association, 252*, 507–513.

Janne, O. A. (1990). Androgen interaction through multiple steroid receptors. In G. C. Lin and L. Erinoff (Eds.), *Anabolic steroid abuse* (pp. 178–186). Washington, D.C.: National Institutes of Drug Abuse.

Johnson, L. C., & O'Shea, J. P. (1969). Anabolic steroids: Effects on strength development. *Science, 165*, 957–959.

Johnson, L. C., Fisher, G., Silvester, L. J., & Hofheins, C. C. (1972). Anabolic steroid: effects on strength, body weight, oxygen uptake and spermatogenesis upon mature males. *Medicine and Science in Sports, 4*, 43–45.

Johnson, L. C., Roundy, E. D., Allsen, P. E., Fisher, A. G., & Silvester, L. J. (1975). Effect of anabolic steroid treatment on endurance. *Medicine and Science in Sports, 7*, 287–289.

Johnson, M. D., Jay, M. S., Shoup, B., & Rickert, V. I. (1989). Anabolic steroid use by male adolescents. *Pediatrics 83(6)*, 921–9249.

Kantor, M. A., Bianchini, A., Bernier, D., Sady, S. P., & Thompson, P. D. (1985). Androgens reduce HDL2-cholesterol and increase hepatic triglyceride lipase activity. *Medical Science of Sports and Exercise, 17*, 462–465.

Kashkin, K. B., & Kleber, H. D. (1989). Hooked on hormones? An anabolic steroid addiction hypothesis. *Journal of the American Medical Association, 262(22)*, 31669.

Katz, D. L., & Pope, H. G., Jr. (1990). Anabolic-androgenic steroid-induced mental status changes. In G. C. Lin and L. Erinoff (Eds.), *Anabolic steroid abuse* (pp. 215–223). Washington, D.C.: National Institutes of Drug Abuse.

Keating, R. J., & Tcholokian, R. K. (1979). In vivo patterns of circulating steroids in adult male rats. I. Variations in testosterone during 24- and 48-hour standard and reverse light/dark cycles. *Endocrinology, 104*, 184–188.

Kluft, C., Bertina, R. M., Preston, F. E., Malia, R. G., Blamey, S. L., Lowe, G. D. O., & Forbes, C. D. (1984). Protein C, an anticoagulant protein, is increased in healthy volunteers and surgical patients after treatment with stanozolol. *Thrombosis Research, 33*, 297–304.

Kochakian, C. D. (1990). History of anabolic-androgenic steroids. In G. C. Lin and L. Erinoff (Eds.), *Anabolic steroid abuse* (pp. 29–00). Washington, D.C.: National Institutes of Drug Abuse.

Komoroski, E. M., & Rickert, V. I. (1992). Adolescent body image and attitudes to anabolic steroid use. *American Journal of Diseases of Children, 146*, 823–828.

Konagaya, M., & Max, S. R. (1986). A possible role for endogenous glucocorticoids in orchiectomy-induced atrophy of the rat levator ani muscle: Studies with RU 38486, a potent and selective antiglucocorticoid. *Journal of Steroid Biochemistry, 25(3)*, 305–308.

Kreuz, L., & Rose, R. (1972). Assessment of aggressive behavior and plasma testosterone in a young criminal population. *Psychosomatic Medicine, 34*, 321–332.

Kuhn, F. E., & Max, S. R. (1985). Testosterone and muscle hypertrophy in female rats. *Journal of Applied Physiology, 59*, 24–27.

LaBella, F. S., Kim, R. S. S., & Tempelton, J. (1978). Opiate receptor binding activity of 17-alpha estrogenic steroids. *Life Sciences, 23*, 1797–1804.

Lamb, D. R. (1984). Anabolic steroids in athletics: How well do they work and how dangerous are they? *American Journal of Sports Medicine, 12*, 31–38.

Lenders, J. W. M., Demacker, P. N. M., Vos, J. A., Jansen, P. L. M., Hoitsma, A. J., van't Laar, A., & Thien, T. (1988). Deleterious effects of anabolic steroids on serum lipoproteins, blood pressure, and liver function in amateur body builders. *International Journal of Sports Medicine, 9*, 19–23.

Leshner, A. I. (1978). *Agonistic behavior: An introduction to behavioral endocrinology.* New York: Oxford Press.

Leshner, A. I., & Moyer, J. A. (1975). Androgens and agonistic behavior in mice: Relevance to aggression and irrelevance to avoidance-of-attack. *Physiology and Behavior, 15*, 695–699.

Leshner, A. I., Moyer, J. A., & Walker, W. A. (1975). Pituitary-adrenocortical activity and avoidance-of-attack in mice. *Physiology and Behavior, 15*, 689–693.

Lombardo, J. A. (1990). Anabolic/androgenic steroids. In *Anabolic Steroid Abuse* G. L. Lin and L. Erinoffleds) Washington, D.C.: National Institutes of Drug Abuse, pp. 60–73.

Loughton, S. J., & Ruhling, R. O. (1977). Human strength and endurance response to anabolic steroid and training. *Journal of Sports Medicine, 17*, 285–296.

Lucking, M. T. (1982). Steroid hormones in sports: Sex hormones and their derivatives. *International Journal of Sports Medicine, 3*, 65–67.

McEwen, B. S., Krey, L. C., & Luine V. N. (1978). Steroid hormone action in the neuroendocrine system: When is the genome involved? In S. Reichlin, R. J. Baldessarini, and J. B. Martin (Eds.), *The hypothalamus* (p. 255–268). New York: Raven Press.

McNutt, R. A., Ferenchick, G. S., Kirlin, P. C., & Hamlin, N. J. (1988). Acute myocardial infarction in a 22-year-old world class weight lifter after using anabolic steroids. *American Journal of Cardiology, 62*, 1648.

Martikainen, H., Alén, M., Rahkila, P., & Vihko, R. (1986). Testicular responsiveness to human chorionic gonadotropin during transient hypogonadotrophic hypogonadism induced by androgenic/anabolic steroids in power athletes. *Journal of Steroid Biochemistry, 25*, 109–112.

Mather, D. N., Toriola, A. L., & Dada, O. A. (1986). Serum cortisol and testosterone levels in conditioned male distance runners and nonathletes after maximal exercise. *Journal of Sports Medicine, 26*, 245–250.

Mayer, M., & Rosen, F. (1975). Interaction of anabolic steroids with glucocorticoid receptor sites in rat muscle cytosol. *American Journal of Physiology, 229(5)*, 1381–1386.

Mayer-Bahlberg, H. F., Nat, R., Boon, D. A., Sharma, M., & Edwards, J. A. (1974). Aggressiveness and testosterone measures in man. *Psychosomatic Medicine, 36*, 269–274.

Mello, N. K. (1988). Effects of alcohol abuse on reproductive function in women. In M. Galanter (Eds.), *Recent developments in alcoholism* (Vol. 6, pp. 253–276). New York: Plenum Press.

Mello, N. K., & Mendelson, J. H. (1985). Operate acquisition of marijuana by women. *Journal of Pharmacology and Experimental Therapeutics, 235*, 162–1715.

Mello, N., Mendelson, J. H., Bree, M. P., & Skupny, A. S. T. (1986). Alcohol effects on luteinizing hormone-releasing hormone stimulated luteinizing hormone and follicle-stimulating hormone in ovariectomized female rhesus monkeys. *Journal of Pharmacology and Experimental Therapeutics, 239*, 693–700.

Mendelson, J. H., & Mello, N. K. (1974). Alcohol, aggression and androgens. *Research Publications: Association for Research in Nervous and Mental Disease, 52*, 225–247.

Mendelson, J. H., & Mello, N. K. (1976). Behavioral and biochemical interrelations in alcoholism. *Annual Reviews of Medicine, 27*, 321–333.

Mendelson, J. H., & Mello, N. K. (1979). Biologic concomitants of alcoholism. *New England Journal of Medicine, 301*, 912–929.

Mendelson, J. H., Mello, N. K., Teoh, S. K., & Ellingboe, J. (1989). Alcohol effects on luteinizing hormone releasing hormone stimulated anterior pituitary and gonadal hormones in women. *Journal of Pharmacology and Experimental Therapeutics, 250,* 902–909.

Miller, N. E., Hubert, G., & Hamilton, J. B. (1938). Mental and behavioral changes following male hormone treatment of adult castration, hypogonadism and psychic impotence. *Proceedings of the Society of Experimental Biology and Medicine, 38,* 538–540.

Mooradian, A. D., Morley, J. E., & Korenman, S. G. 91987). Biological actions of androgens. *Endocrinology Review, 8(1),* 1.

Morano, I. (1984). Influence of exercise and dianabol on the degradation rate of myofibrillar proteins of the heart and three fiber types of skeletal muscle of female guinea pigs. *International Journal of Sports Medicine, 5,* 317–319.

Moyer, J. A., & Leshner, A. I. (1976). Pituitary-adrenal effects on avoidance-of-attack in mice: Separation of the effects of ACTH and corticosterone. *Physiology and Behavior, 17,* 297–301.

Naftolin, F., Ryan, K., Davies, I., Reddy, V. V., Flores, F., Petro, Z., Kuhn, W., White, R. J., Takoaka, Y., & Wolin, L. (1975). The formation of estrogens by central neuroendocrine tissues. *Recent Progress in Hormone Research, 31,* 295.

Nock, B. L., & Leshner, A. I. (1976). Hormonal mediation of the effects of defeat on agonistic responding in mice. *Physiology and Behavior, 17,* 111–119.

O'Carrol, R., & Bancroft, J. (1984). Testosterone therapy for low sexual interest and erectile dysfunction in men: A controlled study. *British Journal of Psychiatry, 145,* 146–151.

O'Shea, J. P. (1971). The effects of an anabolic steroid on dynamic strength levels of weightlifters. *Nutrition Reports International, 4,* 363–370.

Overly, W., Dankoff, J. A., Wang, B. K., & Singh, U. D. (1984). Androgens and hepatocellular carcinoma in an athlete. *Annals of Internal Medicine, 100(1),* 158.

Persky, H., Smith, K. D., & Basu, G. K. (1971). Relation of psychologic measures of aggression and hostility to testosterone production in man. *Psychosomatic Medicine, 33,* 265–277.

Podolsky, E. (1963). Women alcoholics and premenstrual tension. *Journal of the American Medical Womens Association, 18,* 816–818.

Pope, H. B., & Katz, D. L. (1987). Bodybuilders psychosis. *Lancet,* April 11, 1(8573) 863.

Pope, H. B., & Katz, D. L. (1988). Affective and psychotic symptoms associated with anabolic steroid use. *American Journal of Psychiatry, 145,* 487–490.

Pratt, J., Gray, G. F., Stolley, P. D., & Coleman, J. W. (1977). Wilms tumor in an adult associated with androgen abuse. *Journal of the American Medical Association, 237(21),* 2322.

Richardson, J. H. (1977). A comparison of two drugs on strength increase in monkeys. *Journal of Sports Medicine, 17,* 251.

Roberts, J. T., & Essenhigh, D. M. (1986). Adenocarcinoma of prostate in 40-year-old body-builder. *Lancet, 2,* 742.

Robinson, A. J., & Clamann, H. P. (1988). Effects of glucocorticoids on motor units in cat hindlimb muscles. *Muscle and Nerve, 11,* 703–713.

Rogol, A. D., Martha, P. M., Jr., & Blizzard, R. M. (1990). Anabolic-androgenic steroids profoundly affect growth at puberty in boys. In G. L. Lin and L. Erinoff (Eds.), *Anabolic steroid abuse* (pp. 187–195). Washington, D.C.: National Institutes of Drug Abuse.

Rogozkin, V. A. (1976). The role of low molecular weight compounds in the regulation of skeletal muscle genome activity during exercise. *Medical Science and Sports, 8,* 104.

Rogozkin, V. A. (1979). Anabolic steroid metabolism in skeletal muscle. *Journal of Steroid Biochemistry, 11,* 923–926.

Saborido, A., Vila, J., Molano, F., & Megías, A. (1991). Effect of anabolic steroids on mitochondria and sarcotubular system of skeletal muscle. *American Physiology Society, 70(3),* 1038–1043.

Salmon, U. J., & Geist, S. H. (1943). Effect of androgens upon libido in women. *Journal of Clinical Endocrinology, 3,* 235–238.

Samuels, L. T., Henschel, A. F., & Keys, A. (1942). Influence of methyl testosterone on muscular work and creatine metabolism in normal young men. *Journal of Clinical Endocrinology and Metabolism, 2*, 649–654.

Scaramella, T., & Brown, W. (1978). Serum testosterone and aggressiveness in hockey players. *Psychosomatic Medicine, 40*, 262–265.

Sherwin, B. B., & Gelfand, M. M. (1985). Transactions of the fortieth annual meeting of the society of obstetrics and gynaecologists of Canada. *American Journal of Obstetrics and Gynecology, 151(2)*, 153–605.

Small, M., MacLean, J. A., McArdle, B. M., Bertina, R. M., Lowe, G. D. O., & Forbes, C. D. (1984). Haemostatic effects of stanozolol in elderly medical patients. *Thrombosis Research, 35*, 353–358.

Stamford, B. A., & Moffatt, R. (1974). Anabolic steroid: Effectiveness as an ergogenic aid to experienced weight trainers. *Journal of Sports Medicine, 14*, 191–197.

Stone, M. H., & Lipner, H. (1978). Responses to intensive training and methandrostenelone administration. I. Contractile and performance variables. *Pflugers Archives, 375*, 141–146.

Strauss, R. H., Liggett, M. T., & Lanese, R. R. (1985). Anabolic steroid use and perceived effects in ten weight-trained women athletes. *Journal of the American Medical Association, 253(19)*, 2871.

Strauss, R. H., Wright, J. E., & Finerman, G. A. M. (1982). Anabolic steroid use and health status among forth-two egith-trained male athletes. *Medical Science of Sports, 14*, 119.

Strauss, R. H., Wright, J. E., Finerman, G. A. M., & Catlin, D. H. (1983). Side effects of anabolic steroids in weight-trained men. *The Physician and Sports Medicine, 11(12)*, 87.

Stromme, S. B., Meen, H. D., & Aakvaag, A. (1974). Effects of androgen-anabolic steroid on strength development and plasma testosterone levels in normal males. *Medicine and Science in Sports, 6*, 203–208.

Sturup, G. K. (1968). Treatment of sexual offenders in Herstedvester, Denmark. *Acta Psychiatrica Scandinavica (Supplementum), 204*, 5–62.

Su, T. P., London, E. D., & Jaffe, J. H. (1988). Steroid binding at s receptors suggests a link between endocrine, nervous, and immune systems. *Science, 240*, 219–221.

Taggart, H., M. A., Applebaum-Bowden, D., Haftner, S. (1982). Reduction in high density lipoproteins by anabolic steroid (stanozolol) therapy for postmenopausal osteoporosis. *Metabolism, 311*, 1147–1152.

Taylor, W. N. (1984). Are anabolic steroids for the long distance runner? *Annals of Sports Medicine, 2*, 51–52.

Taylor, W. N. (1987). Synthetic anabolic-androgenic steroids: A plea for controlled substance status. *The Physician and Sports Medicine, 15(5)*, 140–50.

Taylor, W. N., & Black, A. B. (1987). Pervasive anabolic steroid use among health club athletes. *Annals of Sports Medicine, 3*, 155–159.

Tchaikovsky, V. S., Astratenkova, B., & Basharina, O. B. (1986). The reflection of exercises on the content and reception of the steroid hormones in rat skeletal muscles. *Journal of Steroid Biochemistry, 24*, 251.

Tennant, F., Black, D. L., & Voy, R. O. (1988). Anabolic steroid dependence with opiod-type features. *New England Journal of Medicine, 319*, 578.

Thompson, P. D., Cullinane, E. M., Sady, S. P., & Chenevert, C. (1989). Contrasting effects of testosterone and stanozolol on serum lipoprotein levels. *Journal of the American Medical Association, 261(8)*, 1165.

Urhausen, A., & Kindermann, W. (1987). Behavior of testosterone, sex hormone binding globulin (SHBG), and cortisol before and after a triathlon competition. *International Journal of Sports Medicine, 8*, 305–308.

Uzunova, A. D., Ramey, E. R., & Ramwell, P. W. (1978). Gonadal hormones and pathogenesis of occlusive arterial thrombosis. *American Journal of Physiology, 234(4)*, 4454–4459.

Villaneuva, A. L., Schlosser, S., Hopper, B., Liu, H. J., Hoffman, D. I., & Rebar, R. W. (1986). Increased corticsol production in women runners. *Journal of Clinical Endocrinology and Metabolism, 63(1)*, 113–136.

Viru, A., & Korge, P. (1979). Role of anabolic steroids in the hormonal regulation of skeletal muscle adaptation. *Journal of Steroid Biochemistry, 11*, 931–932.

Vogel, Klaiber, E. L., & Broverman, D. M. (1978). Roles of the gonadal steroid hormones in psychiatric depression in men and women. *Progress in Neuropsychopharmacology, 2*, 487–503.

Vogel, W., Klaiber, E. L., & Broverman, D. M. (1985). A comparison of the antidepressant effects of a synthetic androgen (mesterolone) and amitriptyline in depressed men. *Journal of Clinical Psychology, 46*, 6–8.

Ward, P. (1973). The effects of anabolic steroid on strength and lean body mass. *Medicine and Science in Sports, 5*, 277–282.

Webb, O. L., Laskarzewski, P. M., & Glueck, C. J. (1984). Severe depression of high-density lipoprotein cholesterol levels in weight lifters and body builders by self-administered exogenous testosterone and anabolic-androgenic steroids. *Metabolism, 33*, 971–975.

Wilson, J. D. (1988). Androgen abuse by athletes. *Endocrine Review, 9*, 181–199.

Wilson, J. D., & Griffin, J. E. (1980). The use and misuse of androgens. *Metabolism, 29*, 1278–1295.

Wilson, I. C., Prange, A. J., Jr., & Lara, P. P. (1974). Methyltestosterone with imipramine in men: Conversion of depression to paranoid reaction. *American Journal of Psychiatry, 131*, 21–24.

Win-May, M., & Mya-Tu, M. (1975). The effect of anabolic steroids on physical fitness. *Journal of Sports Medicine and Physical Fitness, 15*, 266–271.

Winslow, J. T., Ellingboe, J., & Miczek, K. A. (1988). Effects of alcohol on aggressive behavior in squirrel monkeys: influence of testosterone and social context. *Psychopharmacology, 95*, 356–363.

Winslow, J. T., & Miczek, K. A. (1988). Androgen dependency of alcohol effects on aggressive behavior: A seasonal rhythm in high-ranking squirrel monkeys. *Psychopharmacology, 95*, 92–98.

Woodbury, D. M. (1958). Relation between the adrenal cortex and the central nervous system. *Pharmacology Review, 10*, 275–357.

Worthman, C. M., & Konner, M. J. (1987). Testosterone levels change with subsistence hunting effort in !Kung San men. *Psychoneuroendocrinology, 12*, 449–458.

Yesalis, C. E., Anderson, W. A., Buckley, W. E., & Wright, J. E. (1990a). Incidence of the nonmedical use of anabolic-androgenic steroids. In G. C. Lin and L. Erinoff (Eds.), *Anabolic steroid abuse* (pp. 97–112). Washington, D.C.: National Institutes of Drug Abuse.

Yesalis, C. E., Vicary, J. R., Buckley, W. E., Streit, A. L., Katz, D. L., & Wright, J. E. (1990b). Indications of psychological dependence among anabolic-androgenic steroid abusers. In G. C. Lin and L. Erinoff (Eds.), *Anabolic steroid abuse* (pp. 196–214). Washington, D.C.: National Institutes of Drug Abuse.

Yesavage, J. A., Davidson, J., Widrow, L., & Berger, P. A. (1985). Plasma testosterone levels, depression, sexuality, and age. *Biological Psychiatry, 20*, 199–228.

Young, M., Crookshank, H. R., & Ponder, L. (1977). Effect of anabolic steroid on selected parameters in male albine rats. *Research Quarterly of the American Association of Health, Physical Education, and Recreation, 48*, 653–656.

6

Adrenal Steroid Effects on the Brain: Versatile Hormones with Good and Bad Effects

Bruce S. McEwen, Randall R. Sakai, and Robert L. Spencer

Laboratory of Neuroendocrinology
The Rockefeller University
New York, New York 10021

I. Introduction

A. Environment and Genome

The genome provides the limits and possibilities for each person to express the characteristics of an individual human being; the environment is responsible for calling forth many of those individual characteristics. This idea is illustrated clearly by identical twins, who have the same genome and yet, despite their many similarities, do not show complete concordance for many inherited characteristics, including behavior (Plomin, 1990). This difference is caused by the effects of differing environmental factors starting with intrauterine position, to which each twin is exposed. This phenomenon is also evident for disorders such as Alzheimer's disease, Type I diabetes, schizophrenia, and depressive illness, for which the concordance in twins is below 50% (Davies, 1986; Plomin, 1990). Therefore, the environment (referring to all influences, both internal and external, from conception onwards) plays an enormous role in shaping individual characteristics during normal development as well as in the expression of abnormal states.

What is the environment? Among the most pervasive and important of all environmental influences are the hormones of the pituitary, adrenal, and thyroid glands and the gonads. Hormone secretion is directed by the brain via the hypothalamus; hormone secretions act throughout the body to coordinate and signal changing environmental conditions as well as a changing mental state. These hormones, in turn, feed back on the brain and pituitary to regulate hormone secretion and to affect brain function and behavior. Many actions of hormones involve modulation of gene expression via receptors that bind to DNA and alter transcription of DNA information into messenger RNA. The hormones then direct the formation of proteins in the cytoplasm of the cells. Cells in the brain, as well as cells and tissues in many organs of the body, contain the types of receptors for hormones that mediate these changes in gene expression.

B. Stress

One of the most inevitable experiences in modern life is "stress." "Stress" represents an interaction between various elements of the environment—including the "external environment" of the organism, other living organisms, and the results of a range of experiences that challenge homeostasis and activate hormone secretion. Restoration of physiological balance is the final goal of the body's response to stressors. Heat, cold, trauma, physical exertion, and hunger are examples of "external" stressors; internal experiences or stressors include fear, joy, anger, and frustration. Social interactions are among the most powerful of all "stressors" (Sapolsky, 1990b). However, the most important aspect of stress is the "stress response." The stress response involves a high degree of individual variability in terms of whether an event is interpreted by the individual as threatening to physiological or psychological balance and thus will lead to a measurable physiological response.

C. The Stress Response

The secretion of adrenal corticotropic hormone (ACTH) by the pituitary gland, under control of the hypothalamus, and the stimulation of adrenal glucocorticoid secretion by ACTH is among the most ubiquitous of the physiological stress responses. However, adrenal steroids also are involved in many basic metabolic and homeostatic functions and are under the control of the internal clock in the suprachiasmatic nucleus (SCN) (Richter, 1965). The adrenal steroids act on the brain and on many tissues of the body by altering gene expression. These steroids and their diverse actions will be the main focus of this chapter.

In 1968, we discovered receptors for adrenal steroids in the hippocampal formation of the brain (McEwen, Weiss, & Schwartz, 1968). Subsequent studies showed that two types of receptors for adrenal steroids exist throughout the body: Type I (Mineralocorticoid receptors) and Type II (glucocorticoid receptors) (reviewed by McEwen, DeKloet, & Rostene, 1986b). These two receptor types are products of different genes (Evans, 1988) and can coexist within the same cells. Both types of receptors occur in the hippocampus in high concentrations but are present also in other brain areas and in other tissues of the body. Many other brain regions contain Type II receptors; scattered groups of neurons throughout the brain contain Type I receptors as well. Much effort is being devoted to relating these two receptor types to the actions of adrenal steroids on the brain.

Because adrenal steroid secretion is driven both by external events and experiences and by internal events, as well as by the biological clock, the fact that adrenal steroids influence brain function in many ways is not surprising. These compounds alter food intake, modulate the consumption of salt, and influence locomotor activity and blood pressure. Adrenal steroids alter neurotransmitter systems of the brain, often in a biphasic and dose-dependent manner, which involves modulation of biosynthesis of transmitter or modulation of the transmitter receptors. This effect occurs for the serotonin and benzodiazepine systems, as described in a later section. A variety of neuropeptide systems also are regulated by adrenal steroids, including corticotropin releasing hormone (CRH), vasopressin, angiotensinogen, enkephalin, and preprotachykinin (Sawchenko, Swanson, & Vale, 1984; Chao & McEwen, 1990, 1991; Angulo, Riftina, Ledoux, & McEwen, 1991; Imaki, Nahan, Rivier, Sawchenko, & Vale, 1991). Finally, adrenal steroids exacerbate destructive effects of excitatory amino acids on pyramidal neurons in the hippocampus, while protecting neurons of the dentate gyrus from destruction. Many of these effects appear to be mediated by Type I and Type II receptors, but some actions also may involve membrane receptors that respond to metabolites of these steroids that are generated systemically or locally in the brain (see subsequent text).

This chapter reviews the beneficial and destructive actions of adrenal steroids and attempts to relate them to normal and abnormal brain function and behavior and to the broader issue of how we develop and maintain our individuality.

II. Neural and Neuroendocrine Control of Adrenocortical Function

The two major classes of adrenocortical hormones are the glucocorticoids and the mineralocorticoids. Although structurally similar and derived from similar sterol precursors, the regulation of their secretion and their physiological actions differ substantially. The hypothalamus regulates glucocorticoid secretion via the release of CRH as well as vasopressin and oxytocin, which act synergistically on the pituitary to cause the release of ACTH; the secretion of vasopressin and oxytocin may be governed by different stressors and possibly by different neural pathways (Gibbs, 1986). Together, this responding system is referred to as the hypothalamo–

pituitary–adrenal (HPA) axis. Glucocorticoids feed back at the level of the pituitary, as well as at the paraventricular nuclei where CRH, vasopressin, and oxytocin are produced (Sawchenko et al., 1984; Imaki et al., 1991). Type II receptors appear to be the primary mediators of the feedback actions of glucocorticoids. Type I receptors are found in both the pituitary and the paraventricular nuclei, as well as in the hippocampus and other limbic brain structures (Arriza, Simerly, Swanson, & Evans, 1988; Sakai & Epstein, 1990); yet their role is not clear.

The limbic system of the brain plays an important role in modulating the HPA axis, resulting in both increases and decreases in glucocorticoid secretion (for review, see McEwen, 1977), whereas the SCN is the site of an endogenous clock that maintains the diurnal rhythm of waking and pituitary–adrenal activity (Richter, 1965). Lesion studies of the hippocampus revealed that it plays an important role as an inhibitor of the secretion of glucocorticoids and in shutting off their secretion after stress. (For reviews, see Sapolsky, Krey, & McEwen, 1986; McEwen et al., 1990; Jacobson & Sapolsky, 1991.)

The secretion of aldosterone, the primary mineralocorticoid, is largely independent of the HPA axis and is under the control of more than a dozen factors, both positive and negative, that modulate its synthesis and release (Quinn & Williams, 1988). ACTH influences aldosterone secretion to a limited extent, but its stimulatory effects are complemented and largely overwhelmed by those of angiotensin, potassium ions, vasopressin, catecholamines, and serotonin operating through most of the known second-messenger transduction systems; on the negative side, atrial natriuretic peptides and dopamine inhibit aldosterone secretion (Quinn & Williams, 1988).

III. Adrenal Steroid Effects on Target Tissues

A. Genomic versus Nongenomic Mechanisms

Intracellular receptors mediate many of the effects of adrenal steroids on many target tissues throughout the body. These receptors are DNA binding proteins that enhance, or in some cases inhibit, the transcription of genes that are adjacent to the DNA sites at which they bind. Genomic effects are delayed in onset and sustained in duration, although the onset latency may be on the order of minutes and the duration also may be only a matter of minutes (McEwen, 1991).

Also, for many years steroids have been recognized to act at the cell membrane level. Until recently, these effects were relegated to pharmacology because they appeared to be elicited only by high steroid concentrations in a relatively specific fashion (McEwen, 1991). This perception has changed with the discovery of highly specific sites on the chloride channel of the γ-aminobutyric acid (GABA)a–benzodiazepine receptor complex, which interacts with physiological concentrations of natural steroids that are generated *in vivo* from naturally occurring steroid hormones. Now these membrane-active steroids are referred to as "neuroactive steroids." (See McEwen, 1991; McEwen et al., 1992, for summaries.) The relevance of this mechanism to the actions of adrenal steroids was shown by the demonstration that stress-induced secretion of adrenal steroids leads to the accumulation in blood and brain of THDOC, the 3α,5α-pregnane metabolite of desoxycorticosterone, as well as 3α,5α-pregnenolone (Purdy, Morrow, Moore, & Paul, 1991). THDOC is a potent agonist of the steroid site on the GABAa–benzodiazepine receptor and potentiates the influx of chloride ions (for review, see Simmonds, 1990). As might be expected from this action, THDOC is an anxiolytic agent with a short onset of action when administered to rats (Crawley, Majewska, Glowa, & Paul, 1986). Other membrane-mediated effects of adrenal steroids on neuronal electrical activity and behavior have been less well characterized with respect to mechanism of action (see Simmonds, 1990, for review).

B. Type I and Type II Receptor Properties and Occupancy *in Vivo*

Except for the actions of THDOC on GABAa–benzodiazepine receptors, the membrane actions of adrenal steroids and other steroids, such as the progestins, are not well characterized in terms of receptor pharmacology (McEwen, 1991). On the other hand, as noted earlier, the genomic effects of adrenal steroids involve two well-characterized types of receptor that are found in the brain (Reul & deKloet, 1985), as well as in other tissues and organs (Funder, Feldman, & Edelman, 1973). Studies *in vitro* have shown that Type I receptors have a high affinity for corticosterone and aldosterone, whereas Type II receptors have a lower affinity for corticosterone and a much lower affinity for aldosterone (see McEwen et al., 1986a, for review). Synthetic steroids such as Ru28362 (agonist) and Ru38486 (antagonist) are very useful tools for manipulating Type II receptors, whereas aldosterone (agonist) and Ru28318 (antagonist) are the most selective tools for manipulating Type I receptors. These two receptor types account for the two-level recognition of corticosterone by the brain; low levels of hormone concentrate primarily in the hippocampus and higher levels of hormone bind to sites throughout the brain. This difference was shown initially by studies involving radioimmunoassay of corticosterone in brain cell nuclei isolated from rats with different circulating levels of corticosterone (McEwen, Stephenson, & Krey, 1980). Subsequently, Type I receptors were identified; these have the higher affinity for corticosterone and are concentrated in the hippocampus (Reul & DeKloet, 1985). Type II receptors were identified also. These have a lower affinity than Type I receptors for corticosterone and tend to recognize stress levels of hormone; they are found in hippocampus as well as in many other brain regions (for review, see McEwen et al., 1986a).

The affinity and specificity of Type I and Type II adrenal steroid receptors based on *in vitro* studies does not predict their behavior completely in various brain regions and tissues of the body in relation to levels of steroids circulating *in vivo*. This story has two parts, one dealing with the mineralocorticoid aldosterone and the other dealing with resting and stress levels of corticosterone. With respect to aldosterone, since Type I receptors have a high affinity for aldosterone, these receptors would seem to mediate the sodium-conserving effects of this important steroid. Indeed, in aldosterone-sensitive tissues such as the kidney, this appears to be the case, but the question is how these receptors are made available to bind aldosterone in the face of 1000-fold higher circulating levels of glucocorticoids, which should occupy them completely. The most likely explanation is that an enzyme, 11β-hydroxysteroid dehydrogenase (11-DH), converts corticosterone (in the rat) and cortisol (in the human) to 11-dehydro metabolites that do not bind avidly to Type I or Type II receptors. Since aldosterone is not metabolized by this enzyme, it is able to gain access to the Type I receptors. The kidney appears to use this mechanism to allow preferential access of aldosterone to the Type I receptor (Edwards et al., 1988; Funder, Pearce, Smith, & Smith, 1988). Although the brain contains 11-DH, 11-DH has not yet been demonstrated to work in the brain in the same way as in the kidney. However, specific effects of aldosterone on salt appetite in the brain are blocked by the specific antagonist Ru28318 (Sakai, Nicolaidis, & Epstein, 1986), which must be explained by some mechanism that allows this hormone to gain access to some of the Type I receptors that regulate salt appetite. Studies of this phenomenon in the brain are currently underway.

Substantial differences exist between the brain and peripheral tissues, as do regional differences within the brain, in the estimated proportion of Type I and Type II receptors that are occupied and activated under basal (low corticosterone) and acute stress (high corticosterone) conditions (Miller, Spencer, Stein, & McEwen, 1990; Spencer, Young, Choo, & McEwen, 1990). Within a particular tissue or brain region, the population of higher affinity Type I receptors becomes nearly saturated before significant occupation of the lower affinity Type II receptors occurs. However, across tissues, considerable heterogeneity in the relative occupancy and activation of Type I and Type II receptors by diurnal and stress levels of corticosterone may occur.

In general, more adrenal steroid receptor activation by endogenous corticosterone appears to occur in brain than in peripheral tissues; within the brain, Type I and Type II receptors of the hippocampus are activated to a greater degree by basal and acute stress levels of glucocorticoids than are other brain regions. An illustration of this concept is that, under conditions of low basal (A.M.) secretion of glucocorticoids in the rat, only Type I receptors in brain areas such as the hippocampus and the cortex exhibit significant receptor occupation and activation whereas Type II receptors in the brain and Type I and Type II receptors in the pituitary and peripheral tissues are predominantly unoccupied. During acute stress,

Type I receptors in the pituitary also become activated. Type II receptors in the brain also become activated significantly during acute stress, but the proportion of occupied and activated Type II receptors only approaches 50%. Interestingly, very little activation of Type II receptors occurs in the pituitary or peripheral immune tissues during acute stress.

In the evening, when basal plasma corticosterone levels of the rat reach their diurnal peak, the pattern of Type I and Type II receptor activation in the brain resembles that present during acute stress: saturation of Type I receptors occurs and activation of Type II receptors is only slightly less than that seen during acute stress (Spencer, Miller, Kang, Stein, & McEwen, 1991). Thus, in the rat brain, the difference between basal and stress levels of Type I and Type II receptor activation is much greater in the morning than in the evening.

These results suggest that unknown factors may regulate the concentration of free glucocorticoids available to bind to receptors in different cells and tissues of the body. One possible factor modulating free corticosterone levels is corticosteroid binding globulin, which is present in much higher concentrations in the periphery than in the brain. Another factor regulating the local availability of corticosterone may be tissue-associated metabolic enzymes, such as 11-DH described earlier. Heterogeneity among tissues in the proportion of Type I and Type II receptors that are activated may be a physiological means of fine tuning the impact that glucocorticoids have on various tissues during basal and stress conditions. For example, during times of acute stress, adrenal steroid receptors in peripheral tissues such as the immune tissues may be largely buffered from an acute rise in the potent glucocorticoids, whereas adrenal steroid receptors in regions in the brain that mediate negative feedback effects of glucocorticoids appear to be very sensitive to elevations of glucocorticoids.

C. Type I and Type II Receptor Function

The initial descriptions of the cloning of the Type I (mineralocorticoid) receptor drew attention to the fact that the DNA binding domain of this receptor shares more than 90% homology with the DNA binding domain of the Type II glucocorticoid receptor (Arriza et al., 1987). Indeed, these two receptors also were shown to regulate the same genes when present in the same cell type (Arriza et al., 1988). Curiously, demonstrating such overlapping or synergistic interactions between Type I and Type II receptors *in vivo* has not been so easy. One example of a possible synergy concerns the induction of beta amylase in the pancreas; neither corticosterone nor aldosterone alone prevented the decline in this enzyme activity after adrenalectomy, but the combination of the two steroids prevented the decline. This effect was blocked partially by a Type I receptor antagonist spironolactone, but unfortunately no attempt was made in this study to block Type II receptors with Ru38486; the involvement of Type II receptors remains unresolved (Alliet et al., 1989). Another example in which some synergy may occur is the regulation of beta adrenergic receptors in arterial smooth muscle, which is stimulated by aldosterone yet can be blocked by a combination of Ru38486 and spironolactone but not by Ru38486 alone (Jazayeri & Meyer, 1989). Since spironolactone was not used by itself, the exclusive involvement of Type I receptors remains unproven. A third example demonstrates an interaction of Type I and Type II receptor-mediated effects. In this study on the rat proximal colon, both aldosterone and the Type II agonist Ru28362 increased net sodium ion and chloride ion absorption. Whereas the aldosterone effect was blocked by spironolactone, the effect of Ru28362 was not. On the other hand, aldosterone stimulated active potassium secretion by the colon but Ru28362 did not (Turnamian & Binder, 1990).

Attempts to determine the cellular effects that are mediated by the Type I and Type II receptors in brain have produced a similar array of complex interactions that are not explained by the overlap of the DNA binding domains of the two receptor types. In several cases, specific effects of mineralocorticoids appear to be mediated by Type I receptors. For example, for induction of the glucose transporter (Glut-1) in neurons and astrocytic glial cells in culture, aldosterone was an effective inducer and dexamethasone, a Type II receptor agonist, was not (Rydzewski, Sumner, Shen-orr, & Raizada, 1991). Aldosterone and desoxycorticosterone acetate also induce angiotensin II receptors in rat brain and in cultured neurons; at least the effects in culture are blocked by Type I receptor antagonists (Wilson, Sumners, Hathaway, & Fregly, 1986; Sumners & Fregly, 1989). However, not all effects on brain cells are mediated by Type I receptors. For example, potassium and sodium ion flux in rat C6 glioma cells is unaffected by aldosterone treatment, whereas potassium uptake is inhibited by treatment with a synthetic Type II receptor agonist, triamcinolane, as well as by dexamethasone and corticosterone (Beaumont, Vaughn, & Fanestil, 1987).

For other effects of adrenal steroids in the brain, Type I and Type II receptors appear to mediate opposing effects, even if by different mechanisms or on different groups of cells. One example deals with the death and survival of neurons in the hippocampal formation of the rat. The survival of dentate gyrus neurons is maintained via low levels of adrenal steroids acting on Type I receptors (Woolley, Gould, Sakai, Spencer, & McEwen, 1991). On the other hand, the destruction of pyramidal neurons of CA1 and CA3 by high levels of circulating corticosterone in conjunction with the actions of excitatory amino acids and calcium ions appears to involve mediation by Type II receptors (Sapolsky, 1990a). This effect is not a question of receptor distribution because dentate gyrus and pyramidal neurons contain both Type I and Type II receptors. Type I and Type II receptors also mediate opposing effects of corticosterone on hippocampal neuronal excitability; Type I receptors mediate an excitatory effect involving suppression of neuronal inhibition by serotonin and Type II receptors mediate an inhibitory action of the signal-enhancing effects of norepinephrine (Joels & deKloet, 1989, 1991). These effects appear to be biphasic with respect to both dose and time. Adrenal steroids secreted under stress also permissively regulate serotonin formation in the brain, Type II receptors appear to mediate these effects in many forebrain regions, whereas Type I receptors may be involved in the hippocampus.

How can Type I and Type II receptors mediate such different effects when they share over 90% homology in their DNA binding regions? We do not know, but studies indicate that the Type II receptors, at least, possess an activity other than DNA binding, namely, the ability to bind to the protooncogene c-*fos* and render it unable to bind to its AP-1 transcription regulatory site (Jonat et al., 1990; Schulle et al., 1990; Yang-Yen et al., 1990). Through this mechanism, Type II receptors can interfere with transcriptional enhancement mediated by *fos–jun* actions on the AP-1 transcription site. Insofar as Type I receptors differ in this capability or possess other, similar, activities, we may know the answer to the puzzle of the specificity of Type I and Type II actions.

IV. Type I and Type II Adrenal Steroid Receptor Involvement in Physiological Processes

As noted earlier, the adrenal steroids have been shown to be important in modulating a diverse array of behavioral and physiological processes, including sodium homeostasis, blood pressure regulation, circadian rhythmicity, energy metabolism, and locomotor activity (see Figure 1). In this section, we discuss the role of adrenal steroids in these processes and consider the relative contributions of Type I and Type II receptors. Because of the large collection of literature on adrenal steroid effects on behavior, we focus on particular studies that use specific receptor agonist or antagonist ligands and thereby specify Type I or Type II receptor effects.

A. Sodium Appetite

Mineralocorticoids and glucocorticoids influence the intake of sodium-rich solutions (see also Schulkin and Fluharty, Introduction). The biphasic effects of mineralocorticoids on sodium intake are illustrated best by the sodium appetite aroused in the adrenalectomized rat. The adrenalectomized rat displays an avid appetite for sodium that is caused in part by increased angiotensin II activity acting on the brain, as a consequence of renal sodium loss because of the removal of the adrenal steroids (Sakai & Epstein, 1990). Mineralocorticoid administration that returns steroid levels to the normal physiological range decreases the appetite for sodium by restoring sodium balance and by normalizing angiotensin activity.

Administration of the glucocorticoid corticosterone by itself neither decreased nor increased the sodium intake of the adrenalectomized rat (McEwen, DeKloet,

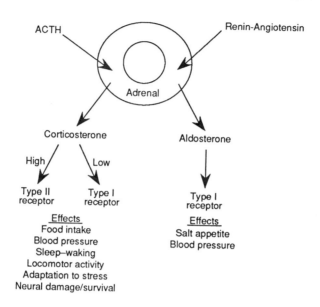

Figure 1. The separate control of corticosterone and aldosterone secretion by different endocrine pathways is shown, with the existence of separate receptor-mediated pathways for the effects of adrenal steroids on neural processes and behavior, as described in the text. High levels of corticosterone occupy Type II receptors found throughout the brain, whereas low levels of corticosterone can occupy Type I receptors found in hypothalamic and limbic system neurons, and especially in the hippocampus. We have postulated, on the basis of behavioral evidence from studies of salt intake, that Type I receptors also exist in some brain regions that recognize and bind circulating aldosterone in spite of 100- to 1000-fold higher levels of corticosterone that also can bind to Type I receptors. As noted in the text, a key enzyme, 11-β hydroxysteroid dehydrogenase may prevent corticosterone from reaching Type I receptors in certain cells of the brain, as is already known to be the case in kidney collecting tubules. Reprinted by permission of Elsevier Science Publishing Company, Inc. from "Paradoxical effects of adrenal steroids on the brain: Protection versus degeneration," by McEwen et al., *Biological Psychiatry, 31*, 177–199. Copyright 1992 by the Society of Biological Psychiatry.

& Rostene, 1986b). Going beyond the physiological range, administration of either aldosterone or deoxycorticosterone to adrenalectomized rats at pharmacological doses causes a mineralocorticoid-induced sodium intake that is induced independently of brain angiotensin (Sakai & Epstein, 1990) and is similar to that seen in adrenal-intact rats given pharmacological doses of mineralocorticoids (Wolf, 1965; Fregly & Waters, 1966; Wolf & Handel, 1966; Sakai et al., 1986; Schulkin, Marini, & Epstein, 1989).

Similar to rats under adrenal steroid manipulations, adrenal-intact rats that are depleted of sodium by various natural or pharmacological treatments also express a sodium appetite. This sodium appetite occurs through the synergistic interaction of aldosterone and the central renin–angiotensin system (Epstein, 1982; Fluharty

& Epstein, 1983; Epstein, 1991). When the Type I receptor antagonist Ru28318 is infused into the cerebral ventricles, the synergistic interaction with brain angiotensin is blocked and sodium appetite is suppressed (Sakai et al., 1986).

Collectively, these studies of adrenalectomized, aldosterone treated, and sodium-depleted animals suggest that the steroid regulation of sodium appetite involves mineralocorticoid actions. Moreover, these findings suggest that a population of aldosterone-preferring Type I receptor sites exists somewhere in the brain, through which the mineralocorticoids produce at least some of their behavioral and physiological effects. However, growing evidence suggests that synergistic interactions between glucocorticoids and mineralocorticoids also affect the expression of sodium appetite as well as thirst. Low doses of aldosterone that are normally ineffective in arousing sodium intake become effective when combined with glucocorticoids such as dexamethasone or corticosterone (Coirini, Schulkin, & McEwen, 1988). Further, Type I and Type II adrenal steroid receptor activation has been shown to promote induction of angiotensin II receptors in neurons (Wilson et al., 1986; Castren & Saavedra, 1989; Sumners & Fregly, 1989; Chow, Sakai, Reagan, McEwen, & Fluharty, 1991) or synergistically increase angiotensin's ability to elicit drinking (Sumners, Gault, & Fregly, 1991). In these studies, differences exist in the degree to which glucocorticoids and mineralocorticoids produce increases in receptor density or sensitivity.

Another complication of the brain angiotensin system is that the messenger RNA for angiotensinogen is found in glial cells (Lynch, Simnad, Ben-Ari, & Garrison, 1986; Stornetta, Hawelu-Johnson, Guyenet, & Lynch, 1988), as is angiotensinogen immunoreactivity (Campbell, Sernia, Thomas, & Oldfield, 1991), whereas angiotensin II immunoreactivity is detected in neuronal processes (Lind, Swanson, & Ganten, 1984, 1985). How, or at what step in processing of the angiotensinogen precursor protein, transfer to neurons occurs is not clear. Angiotensinogen mRNA is under positive control by adrenal steroids in liver (Sernia, Clemens, & Funder, 1989) and in preoptic area (Angulo et al., 1991) via a Type II receptor mechanism. Whether aldosterone has any effect on angiotensinogen mRNA levels in brain is not clear.

B. Blood Pressure

Adrenal steroids have a modulatory effect on blood pressure regulation by interacting with central vasoactive peptidergic systems. Mineralocorticoid infusion into the cerebral ventricles or deoxycorticosterone given systemically increases blood pressure, whereas glucocorticoid infusion has the opposite effect (Gomez-Sanchez, Fort, & Gomez-Sanchez, 1990; Kalimi, Opoku, Agarwal, & Corley, 1990; van den Berg, DeKloet, van Dijken, & DeJong, 1990). These opposing effects on blood pressure are thought to be mediated by Type I receptors to promote increased blood pressure, whereas Type II receptors mediate the opposite effect. The central infusion of Type I receptor antagonist Ru28318 can reduce basal blood pressure and reverse the hypertension induced by systemic deoxy-corticosterone or centrally administered aldosterone (Janiak & Brody, 1988; Go-mez-Sanchez et al., 1990; van den Berg et al., 1990). In contrast, Type II receptors promote decreased blood pressure, that is, central infusions of the Type II receptor agonist Ru28362 decrease blood pressure and of the Type II receptor antagonist Ru38486 increase blood pressure (van den Berg et al., 1990).

C. Food Intake, Obesity, and Circadian Behaviors

The adrenal steroids are secreted in a circadian pattern. This secretion helps coordinate daily activities such as sleep–wake cycles and ingestive and locomotor behavior. Prior to waking and the beginning of the daily cycle of eating and drinking, plasma ACTH and corticosterone levels reach their diurnal peak. This daily rise in corticosterone occurs at a time when energy stores are low, and helps promote both gluconeogenesis and appetite for food. Therefore, the rise in glu-cocorticoids precedes, and may act as a signal for feeding. This relationship is demonstrated best in food-restricted animals, in which the peak of circadian corticosterone precedes the time of restricted food availability and schedule-induced behaviors such as wheel-running and polydipsia (see subsequent text). These schedule-induced behaviors are attenuated by adrenalectomy and return with adrenal steroid replacement. The best studied of these behaviors is food intake.

An important aspect of adrenal steroid action in normal physiology is its ability to activate food intake and specific hunger for carbohydrates at the time of waking during the diurnal rest–activity cycle in animals with normal body weight (Tempel & Leibowitz, 1989). Mineralocorticoids and glucocorticoids modulate food intake by integrating multiple neurotransmitter systems that include the alpha adrenergic system, neuropeptide Y, and galanin (Leibowitz, 1986, 1989). For example, the alpha-2 adrenergic receptors in the paraventricular nucleus (PVN) of the hypothalamus show a diurnal rhythm, in which the peak of receptor binding coincides with the plasma corticosterone peak; noradrenaline seems to control the rat's first meal of carbohydrate in the early dark, the time of the rat's normal arousal during the 24-hr clock period (Jhanwar-Uniyal, Roland, & Leibowitz, 1985). As the levels of corticosterone begin falling after the first meal, the rats begin to shift their macronutrient selection to a mixture of fats and carbohydrates and fats and protein toward the end of the dark period. As mentioned earlier, these effects on food intake and body weight gain can be abolished by adrenalectomy and can be reinstated by selective replacement with corticosterone, aldosterone, or the Type II agonist Ru28362 (King, 1988; Devenport, Knehans, Sundstrom, & Thomas, 1989a; Devenport, Knehans, Thomas, & Sundstrom, 1989b; Tempel &

Leibowitz, 1989; Devenport & Thomas, 1990; Green, Woods, & Wilkinson, 1991).

As noted, diurnal patterns of food intake are modulated by glucocorticoids and mineralocorticoids. High levels of glucocorticoids suppress food intake and reduce body weight (Devenport et al., 1989, 1990), except when endogenous energy reserves have already been depleted by prior food deprivation and body weight loss (Dulloo, Seydoux, & Gerardier, 1990; Green et al., 1991). Interestingly, in animal models of inherited obesity, treatment with the Type II receptor antagonist Ru38486 is effective (as is adrenalectomy) in preventing the onset of obesity; however, the genetically related lean animals are insensitive to the receptor antagonist (Langley & York, 1990a).

Type I adrenal steroid receptors are implicated in the stimulatory effects of aldosterone and low doses of corticosterone on food intake (Tempel & Leibowitz, 1989; Devenport & Thomas, 1990), whereas Type II receptors are implicated in the inhibitory effects of high levels of corticosterone as well as of synthetic steroids such as dexamethasone and Ru28362 on food intake (Devenport & Thomas, 1990). Type I receptors have been found in low levels in PVN (Sakai et al., 1990), as have Type II receptors (Fuxe et al., 1985). Although the PVN plays a prominent role in these actions, other brain regions including the hippocampus also may be involved in mediating steroid sensitivity (King, 1988). This inference is based in part on the distribution of Type I and Type II receptors, in which the hippocampus figures prominently as the major repository of Type I receptors (for review, see McEwen et al., 1986a). Whether the Type I receptors found in the parvocellular region of the PVN (Sakai et al., 1990) play a role in food intake or the diurnal release in ACTH secretion remains to be seen.

Increased food intake and body weight, leading to obesity, occur naturally in several strains of rats and mice; obesity also can be produced experimentally by lesions of the paraventricular and ventromedial nuclei of the hypothalamus. Both natural and lesion-induced obesity are dependent, to a very large degree, on the presence of circulating adrenal steroids, since maintenance of obesity requires adrenal steroids and their removal by adrenalectomy attenuates obesity (Bray, 1985; King, 1988; Tokuyama & Himms-Hagen, 1989b). Genetically obese mice and rats show enhanced basal secretion of ACTH and corticosterone as well as enhanced sensitivity to glucocorticoid effects on target tissues (Tokuyama & Himms-Hagen, 1989a; Guillaume-Gentil et al., 1990; Langley & York, 1990a). Type II receptors are prominent in producing these effects, since the Type II antagonist Ru38486 blocks the development of obesity in the obese fa/fa Zucker rat (Langley & York, 1990b) and the Type II receptor levels and Type II receptor-mediated effects are larger in obese fa/fa Zucker rats than in lean rats of the same strain (Langley & York, 1990a). This elevation of Type II receptor levels was found in hippocampus as well as in hypothalamus, and also manifested an altered K_d, suggesting a lower binding affinity for corticosterone (Langley & York,

1990a). At the same time, although brain and serum levels of corticosterone were not significantly higher in these particular obese rats compared with lean rats, adrenalectomy revealed larger effects of endogenous corticosteroids on glyceraldehyde phosphate dehydrogenase activity (Langley & York, 1990a). Collectively, these data show that the effects of glucocorticoids on maintenance of obesity occur through both changes in glucocorticoid receptor properties (i.e., increased receptor number and reduced K_d) and glucocorticoid modulation of glucose metabolism via its regulation of glyceraldehyde phosphate dehydrogenase.

D. "Food-Shift" Effect

In addition to obesity, the other striking connection between glucocorticoids and food intake is the "food-shift" effect, in which the adrenal steroids play an important role in synchronizing food intake with food availability (Krieger, 1974). When food is available for restricted periods, an anticipatory elevation of glucocorticoids occurs in conjunction with a shift in body temperature; this shift is independent of the SCN (Krieger & Hauser, 1977) or ventromedial nucleus (Honma, Honma, Nagasaka, & Hiroshige, 1987); blinding or constant light do not prevent it (Takahashi et al., 1977a; Takahashi, Inoue, & Takahashi, 1977b). Thus, the cycle of fasting and eating appears to be a more potent synchronizer of the HPA axis than the light–dark cycle (Morimoto, Arisue, & Yamamura, 1977) and may involve a neural timing system that lies outside the SCN clock (Mistlberger, Houpt, & Moore-Ede, 1990).

Food and water deprivation also lead to increased corticosterone secretion, particularly in anticipation of access to food and water or when a high level of uncertainty is associated with its availability (Levine, Weinberg, & Brett, 1979; Coover, 1984).The consummatory act leads to a reduction in glucocorticoid levels (Levine et al., 1979), sometimes in conjunction with a biphasic elevation of corticosterone blood levels (Honma, Honma, & Hiroshige, 1984). However, neither salt hunger (Rosenwasser, Schulkin, & Adler, 1988) nor water deprivation is as potent a synchronizer of HPA activity as food deprivation, as described earlier (Armario & Jolin, 1986; Honma, Honma, Hirai, Katsuro, & Hiroshige, 1986).

Although the hypothalamus is implicated most directly in glucocorticoid actions on food intake, the anticipation of food availability and whether or not food actually becomes available appears to involve other brain structures, notably the hippocampal formation. Rats with damage to the fornix or ablation of the hippocampus fail to show elevations of plasma corticosterone when eating is prohibited (Osborne, Sirakumaran, & Black, 1979; Osborne, 1986) or during the initial phase of extinction of lever response for food reward (Coover, Goldman, & Levine, 1971). Consummatory behavior of rats also is disrupted by fornix lesions, with shorter meals and increased trips away from the food cup (Osborne, 1986; Osborne & Dodek, 1986). These behavioral changes and the lack of a corticosterone response to an expected reward are consistent with hippocampal involvement in matching actual events to expected behavior (Levine, Goldman, & Coover, 1972; Osborne, 1986). How hippocampal sensitivity to circulating adrenal steroids plays a role in this process of meal anticipation remains to be seen.

E. Locomotor Activity

Rats have a diurnal rhythm not only of food intake but also of locomotor activity (i.e., increased wheel-running). Adrenocortical secretions appear to have a synchronizing and facilitative role in activity as well as in food intake. Adrenalectomy decreases spontaneous exploratory activity in rats; corticosterone restores this activity (Veldhuis, DeKloet, Van Zorst, & Bohus, 1982). Evidence also exists for involvement of the hippocampal formation in this type of locomotor activity. Hippocampal lesions increase locomotor activity; these effects are reduced by chemical adrenalectomy with metyrapone (Ryan, Springer, Hannigan, & Isaacson, 1985). Clearly, from these results the hippocampus is, if anything, an inhibitor of locomotor activity and not the primary site of action of glucocorticoids to facilitate locomotion. However, the hippocampus may be involved more importantly in timing of motor events and matching them to the appropriate cues.

Specific neurochemical systems under glucocorticoid control may play a role in the synchronization of locomotor activity. A role for enkephalin in locomotor activity and its diurnal variation is made attractive by the finding that enkephalin administration into both ventral tegmentum and nucleus accumbens stimulates locomotor activity (Kalivas, 1985). Glucocorticoids elevate preproenkephalin mRNA in caudate–putamen and nucleus accumbens (Chao & McEwen, 1990). The involvement of the glucocorticoid rhythm in the facilitation of locomotor activity is suggested by the fact that a rhythm of enkephalin release occurs, with an increase as darkness approaches (Bayon & Anton, 1986); a diurnal rhythm of preproenkephalin mRNA has been discovered, with an increase during the dark period (Chao & McEwen, 1990, 1991), suggesting that genomic activation follows the increased activation of enkephalin release in the caudate–putamen.

Tachykinins also are implicated in the regulation of locomotor activity. For example, substance P exerts stimulatory effects on locomotor activity when administered into the nigrostriatal system (Stinus, Kelly, & Iversen, 1978; Kelly, Cador, & Stinus, 1985). Adrenal steroids increase tachykinin mRNA levels in the striatum (Chao & McEwen, 1991); therefore the link between adrenal steroids and locomotor activity may involve more than one regulatory neuropeptide system.

F. Amphetamine-Induced Stereotypy

Amphetamine-induced stereotypy is an abnormal drug-induced behavior, so named because it is repetitive and serves no apparent useful purpose to the animal. Stereotypy frequently is stimulated by drugs of abuse, or is the delayed consequence of drug administration for therapeutic purposes. Adrenalectomy increases stereotypy involving activation of the dopamine system of the nigrostriatal and mesolimbic neural systems; corticosterone decreases it, even in the presence of the adrenals (Faunt & Crocker, 1988). Enkephalin also inhibits this stereotypy (Broderick, Gardner, & van Progg, 1984), as does neurotensin (Stoessel & Szcuzutkowski, 1991). A role of enkephalin in mediating corticosterone suppression of stereotypy is consistent with the positive effects of corticosterone on preproenkephalin mRNA (Chao & McEwen, 1990).

Once again, as for locomotor activity, the hippocampus also may be involved as may the nigrostriatal and mesolimbic systems. Note also that behavioral sensitization of locomotor activity to amphetamine administration is blocked by fimbria–fornix lesions (Yoshikawa, Shibuya, Kaneno, & Toru, 1991), indicating that the hippocampus does play a role in some of these effects.

G. Learned Immobility during Forced Swimming

A step beyond locomotor activity is the tailoring of such activity to environmental demands. Rats placed into a tank of water swim rapidly at the beginning, but with time they adapt as they discover they cannot escape. Thus, they slow down their activity and begin to float. Such immobility appears to be an adaptive response to conserve energy (Hawkins, Hicks, Phillips, & Moore, 1978). This adaptive process is dependent on the presence of glucocorticoids during training (Jefferys, Copolov, Irby, & Funder, 1983; Veldhuis, DeKorte, & DeKloet, 1985; Jefferys, Boublik, & Funder, 1985; Jefferys & Funder, 1987; Mitchell & Meaney, 1991). To what extent this response depends on the mesolimbic dopaminergic system or the hippocampus is unknown but, on the basis of what is described here, the participation of both brain areas is likely. However, the relationships between hormones, metabolism, and the adaptive process is undoubtedly very complicated. Also glucocorticoids are involved in forced swimming in mice. One study has shown that involvement of glucocorticoids is dependent on the water temperature in which the mice swim; adrenal involvement is demonstrated at 25°C, but not at 20°, 30°, or 35°C (Peeters, Smets, & Broekkamp, 1991).

H. Schedule-Induced Behaviors

Schedule-induced polydipsia (SIP) is repetitive drinking behavior induced by schedules of reinforcement in operant learning paradigms. Adrenal steroids appear to control this behavior negatively, because SIP is elevated by adrenalectomy (Devenport, 1978). Moreover, in the presence of the adrenals, adrenal secretions are part of a negative feedback loop, that is, adrenal steroid levels are elevated in rats that have acquired SIP but glucocorticoid secretion is reduced during the actual behavior (Brett & Levine, 1979, 1981). Whether glucocorticoid effects on

SIP involve regulatory neuropeptides such as enkephalins and substance P that are implicated in locomotor activity and that synergize with the mesolimbic and nigrostriatal dopamine systems (see previous section) remains to be established.

The negative feedback loop between adrenal steroids and SIP is consistent with decreased glucocorticoid secretion as a result of consummatory behaviors in rats that are food deprived and show elevated anticipatory glucocorticoid levels. Such a link to expectation returns our attention to another brain structure, the hippocampus. As noted earlier in the discussion on the "food-shift" effect, evidence from these latter situations suggests that the hippocampus plays a role in the recognition that expectations are either violated or established (Osborne, 1986).

I. Sleep

Adrenal steroids have been shown to influence sleep in humans and in experimental animals. The diurnal rhythm of glucocorticoid secretion involves a peak at the end of the sleeping period. Do adrenal steroids play a regulatory role in the timing or subtypes of sleep? The most consistent observation in humans is that glucocorticoids, both synthetic and natural, as well as ACTH infusions produce a reduction in total time in rapid eye movement (REM) sleep (Gillin, Jacobs, Fram, & Snyder, 1972; Gillin, Jacobs, Snyder, & Henkin, 1974; Feinberg, Carroll, King, & Greden, 1984; Fehm et al., 1986; Born, Zwick, Roth, Fehm-Wolfsdorf, & Fehm, 1987; Born, Spath-Schualbe, Schwakenhofer, Kern, & Fehm, 1989; Born, DeKloet, Wenz, Kern, & Fehm, 1991). Consistent with their daily secretion at the end of the sleeping period, synthetic and natural glucocorticoids tend to increase intermittent wakefulness (Fehm et al., 1986). However, differences exist among these treatments in relation to the other forms of sleep. Note also that cortisol infusions increase slow-wave sleep time, whereas synthetic glucocorticoids and ACTH infusions are noneffective on slow-wave sleep (Gillin et al., 1974; Feinberg et al., 1984; Fehm et al., 1986; Born et al., 1987, 1989).

Clinical conditions involving adrenocortical hypo- and hyperfunction also are associated with sleep abnormalities. Addison's disease patients (patients with low or absent adrenal steroids) respond poorly to ACTH infusions and show less reduction in REM sleep time than normal subjects receiving ACTH; however, their actual sleep is poorer and discomfort is greater than that of normal subjects because they are withdrawn temporarily from replacement therapy to receive ACTH infusions (Gillin et al., 1974). Cushing's disease patients, who have excess adrenal steroid secretion, have much reduced levels of total sleep and REM sleep (Krieger & Glick, 1972), which is consistent, at least in part, with the effects of exogenous glucocorticoids just described.

Studies on rats, cats, and rabbits tend to support these findings in humans. ACTH infusions suppressed REM sleep in intact cats (Koranyi, Beyer, & Guzman-Flores, 1971) and rabbits (Kawakami, Negoro, & Terasawa, 1965). Moreover, these studies indicate that timing is a very important aspect of adrenal influence on sleep. Adrenalectomy abolished the circadian distribution of REM sleep in rats subjected to an artificial 1-hr lighting schedule involving 30-min lights on and 30-min lights off periods. Moreover, cortisol infusion into adrenalectomized rats at the time of normal glucocorticoid elevation (i.e., late afternoon), reestablished a circadian rhythm; however, this rhythm was out of phase with the original rhythm (Johnson & Sawyer, 1971). Nevertheless, in this study, total REM sleep time was not altered by adrenalectomy or by cortisol, which is a result somewhat different from the findings in humans. However, the paradigm used to elicit sleep was different from measuring the distribution of sleep in the natural 24-hr day. Indeed, in experiments on rats involving the normal 24-hr day, hypophysectomy increased slow-wave sleep time and decreased REM sleep duration by approximately 30% (Zhang, Valatx, & Jouuet, 1988). Unfortunately, no attempt was made in this study to replace hormones in hypophysectomized rats to rule out the role of ACTH, so completely assessing the contribution of adrenal steroids is impossible.

The possibility of distinguishing Type I and Type II receptor involvement in sleep was investigated in a study on human subjects (Born et al., 1991). Induction of increased slow-wave sleep by cortisol was opposed by a Type I receptor blocker, potassium canrenoate, whereas the cortisol effect to reduce REM sleep duration was not counteracted by canrenoate, suggesting that the effect was not Type I receptor mediated but might be mediated by Type II receptors (Born et al., 1991). These conclusions fit with the known specificity of Type II receptors for synthetic steroids such as dexamethasone and another synthetic glucocorticoid, prednisone, both of which have been used to cause reduced REM sleep time (see previous section). Collectively, these examples illustrate how sleep frequency and sleep type can be influenced markedly by normal and abnormal secretion of adrenal steroids.

V. Adrenal Steroid Receptor Involvement in Pathophysiological Processes

A. Destructive Effects of Glucocorticoids on Neural Tissue Related to Aging, Hypoxia, and Severe Stress

Adrenal steroids also mediate pathological effects on the brain as associated with transient ischemia, aging, and severe social stress (Sapolsky et al., 1986; Sapolsky, 1990a). All these effects appear to involve a synergism between excitatory amino acids produced in the brain, the calcium ions that they mobilize to exert their transmitter effects, and circulating glucocorticoids present at the time of the ischemic event (McEwen & Gould, 1990; Sapolsky, 1990a). Glucocorticoids produce their destruction-potentiating effects via Type II receptors (Sapolsky, 1990a). These effects apparently stem from the action of glucocorticoids to decrease glucose uptake and reduce available energy stores at a time when neurons are excited maximally by released endogenous excitatory amino acids (Sapolsky, 1990a). The hippocampus is one of the most vulnerable brain structures, so inhibitory glucocorticoid effects on glucose uptake are detected most readily in hippocampal neurons and glial cells (Horner, Packan, & Sapolsky, 1990).

The hippocampus is not uniformly sensitive to the destructive effects of transient ischemia, aging, and severe social stress. Ischemic effects are most pronounced in CA1 and subiculum, whereas aging and social stress effects are larger in the CA3 region of the hippocampus. Chronic glucocorticoid treatment produces larger destructive effects in the CA3 region. We have found that, after treatment of young adult rats with corticosterone for 3 weeks, dendrites in CA3 but not in CA1 or in dentate gyrus were atrophied and branched less extensively (Woolley, Gould, & McEwen, 1990) (see Figure 2). These effects were particularly evident on the apical dendrites of the CA3 pyramidal neurons, which strongly suggests that the heavy mossy fiber input to these dendrites somehow is involved (Woolley et al., 1990). Indeed, stimulation of the perforant pathway, which activates the mossy fiber pathway, causes seizures that lead to damage of the CA3 pyramidal neurons (Sloviter, 1983). In keeping with these results, kainate damage of CA3 is prevented by prior destruction of the mossy fiber input (Nadler & Cuthbertson, 1980).

B. Protective Effects of Adrenal Steroids on the Dentate Gyrus

Surprisingly, the dentate gyrus is not affected adversely by many of the treatments, such as ischemia, high glucocorticoids, and seizures, which destroy pyramidal neurons. Instead, adrenalectomy leads to cell death of granule neurons within the dentate gyrus (see Figure 2). This effect can be detected within 3–7 days (Gould, Woolley, & McEwen, 1990) and can result, after several months, in destruction of the dentate gyrus (Sloviter et al., 1989; Roy, Lynn, & Benn, 1990). The rats that show large-scale destruction of the dentate gyrus are animals that lack the accessory adrenal tissue that can produce some adrenal steroids and therefore protect against further cell loss (Sloviter et al., 1989; Roy et al., 1990). The importance of Type I receptor activation to protect against the loss of dentate gyrus neurons after adrenalectomy has been demonstrated. Woolley et al. (1991) have shown that a low dose of aldosterone, which only occupies Type I receptors in the hippocampus of adrenalectomized rats, prevented the loss of granular cells whereas the Type II receptor agonist Ru28362 was ineffective in preventing cell loss in the same brain area. The vulnerability of the dentate gyrus to cell death is not seen in cerebellar granule neurons, nor in pyramidal cells of the hippocampus.

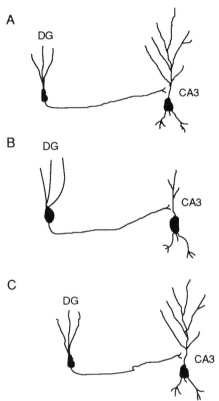

Figure 2. Mossy fiber projections from dentate gyrus (DG) granule neurons to CA3 pyramidal neurons connect two regions where opposing effects of adrenal steroids are manifested. (A) CA3 pyramidal neurons receive the mossy fiber input from the granule neurons of the DG. (B) Chronic corticosterone treatment or repeated restraint stress promotes atrophy of the apical dendrites of CA3 pyramidal neurons. (C) Adrenalectomy causes neurons of the DG to die and be replaced by new neurons, whereas pyramidal neurons are unaffected. Low levels of adrenal steroids prevent this and stabilize the DG granule neuron population via Type I adrenal steroid receptors. Reprinted with permission from McEwen (1991).

C. Role of Adrenal Steroids in Affective Illness

The association between disregulation and hypersecretion of cortisol and depressive illness (Carroll, Martin, & Davies, 1968; Sachar, Hellmann, Fukushima, & Gallagher, 1970) raises the question of whether adrenal steroids play a role in the etiology or symptomatology of the illness (Quarton, Clark, Cobb, & Bauer, 1955; Von Zerssen, 1976). Although moderate to high levels of glucocorticoids can affect mood adversely (Persky, Smith, & Basu, 1971; Von Zerssen, 1976), the absence of adrenal steroids produced by adrenalectomy reveals increased anxiety that is reversed by corticosterone replacement (Weiss, McEwen, Silva, & Kalkut, 1970; File, Vellucci, & Wendlandt, 1979). These anxiolytic effects of corticosterone argue for a protective adaptive influence of adrenocortical secretions under normal circumstances, during which elevations in endogenous corticosterone as well as THDOC can work together as anxiolytic agents (Crawley et al., 1986; Purdy et al., 1991). This inference is supported by studies showing that learned helplessness behavior elicited by inescapable electric footshock is increased in rats that are adrenalectomized and can be reduced by corticosterone replacement (Edwards, Harkins, Wright, & Henn, 1990).

A neurochemical explanation for the protective actions of glucocorticoids on affective state has been proposed, based on an analogy with the anti-inflammatory actions of adrenal steroids in the periphery, where glucocorticoids appear to exert counterregulatory effects on a number of neurochemical systems that are activated by stress. This scheme is summarized in Figure 3. One example of this phenomenon involves the noradrenergic innervation of the cerebral cortex. The effect of stress is to increase noradrenergic activity; repeated stress leads to an induction of the enzyme tyrosine hydroxylase (TOH), which can be detected by increases in the mRNA in the cell bodies of the locus coeruleus that project directly to the cerebral cortex. Adrenal steroids appear to have very little effect on TOH induction. To counteract increased noradrenergic tone during chronic stress, glucocorticoids mediate a suppression of the postsynaptic second messenger response (i.e., the activation of cAMP formation), which is mediated by beta adrenergic receptors. However, the effect of glucocorticoids is an indirect one and is mediated by an alpha-1 adrenergic receptor mechanism that normally potentiates the beta receptor response via calcium ions. Calcium ions in nervous tissue are involved centrally in cAMP generation because, with calmodulin, they activate a form of adenylate cyclase that in turn affects cAMP formation. We have found that calcium–calmodulin-stimulated adenylate cyclase activity is suppressed by chronic stress via a glucocorticoid-dependent mechanism in the cerebral cortex (Gannon & McEwen, 1990). Collectively, these actions of glucocorticoids secreted during stress are consistent with an action of stress hormones to reduce arousal mediated through the noradrenergic systems, and thereby prevent overactivation of this system. These glucocorticoid actions are similar to the effects of tricyclic antidepressant drugs and suggest that glucocorticoids may be, in this sense, one of the body's own antidepressant substances.

Glucocorticoids have parallel effects on other neurotransmitter actions. Serotonin actions via 5HT1A receptors, which can be anxiogenic, are suppressed by glucocorticoids in the hippocampal formation but not in the cerebral cortex (Mendelson & McEwen, 1992). Moreover, indications are that central actions of ser-

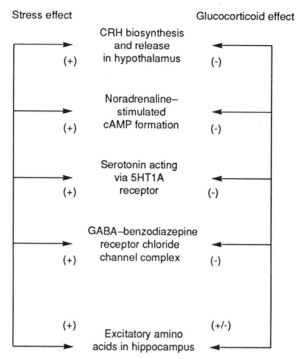

Stress effect   Glucocorticoid effect

CRH biosynthesis
and release
in hypothalamus

Noradrenaline–
stimulated
cAMP formation

Serotonin acting
via 5HT1A
receptor

GABA–benzodiazepine
receptor chloride
channel complex

Excitatory amino
acids in hippocampus

Figure 3.   Counterregulation by glucocorticoids. Stress activation of four neurotransmitter systems in the brain is opposed by the actions of glucocorticoids, at least in some brain areas, as described in the text. For excitatory amino acids (EAA) in hippocampus, glucocorticoids potentiate damaging effects arising from EAA release in pyramidal neurons while stabilizing the granule cell population and exerting other counterregulatory effects as well. Reprinted with permission of Elsevier Science Publishing Company Copyright 1992 by the Society of Biological Psychiatry, from McEwen et al. (1992).

otonin are attenuated by glucocorticoid treatment (Dickinson, Kennett, & Kurzon, 1985). The serotonin system's interactions with glucocorticoids are complicated by the fact that circulating glucocorticoids are permissive agents that acutely activate serotonin formation during stress (Azmitia & McEwen, 1969, 1974, 1976; Neckers & Sze, 1976; Singh, Corley, Phan, & Boodle-Biber, 1990). Thus, the acute effects of circulating steroid are counterbalanced, at least in part, by their ability to regulate negatively the actions of serotonin via 5HT1A receptors in some brain regions.

Adrenal secretions also have been shown to reduce the number of benzodiazepine receptors in a number of brain areas (DeSouza, Goeders, & Kuhar, 1986; Acuna, Fernandez, Gomar, Aquila, & Castillo, 1990). These effects, which require behavioral studies to determine their functional relevance, imply that glucocorticoids can dampen even the anxiolytic actions of endogenous substances such as steroid metabolites. As noted earlier, metabolites of progesterone and desoxycorticosterone, such as the pregnanediols and THDOC, that are generated rapidly *in vivo* during stress (Purdy et al., 1991) activate the chloride channel of the GABAa–benzodiazepine receptor and lead to anxiolytic actions (Crawley et al., 1986). Thus, like the serotonin system, adrenal steroids are involved in both the acute activation of a receptor system as well as, putatively, in the regulation of the same receptor type in the opposite direction chronically.

VI. Conclusions

We have seen that adrenocortical secretions have multiple effects on the brain, many of which are related to the regulation of normal physiological processes—including food and sodium intake, blood pressure, locomotion, and sleep—that are manifested during the diurnal cycle of waking and sleeping. Other actions of adrenal steroids are involved in the response of the brain to stress. Some of these actions are adaptive and protective, whereas other effects are associated with damage and disregulation.

This diversity of effects is a reflection of the multiple functions of adrenal secretions and the different modes of control of adrenocortical activity, involving separate regulation of aldosterone and glucocorticoid secretion as well as two distinct modes of neurally driven ACTH and glucocorticoid secretion, namely, a diurnal rhythm and response to stressors. The diversity of adrenal steroid actions is also a reflection of the separate, and sometimes opposing, roles that the two types of intracellular adrenal steroid receptors play. Despite their similarity in DNA binding domains, these receptors perform rather different functions in the living cell. In addition to genomic effects, nongenomic actions of adrenal steroids at the membrane level have been recognized; these actions exert other types of regulatory influence over brain function, but on a much shorter time scale than the genomic effects.

One of the principal actions of hormones on the brain is regulation and coordination of the response of the brain to changes in the environment. Cyclical processes such as seasonal and diurnal variations in light intensity result in cyclical patterns of hormone secretion that, in turn, alter brain structure and neurochemistry in a cyclical fashion, enabling the brain to respond optimally and appropriately at each stage of the cycle. In contrast, the experiences of individuals with other creatures and with events in an often unpredictable world also evoke changes in endocrine function that feed back on the brain to enable it to adjust and to function as normally as possible.

The secretion of glucocorticoids in response to stressors represents a means of promoting adaptation and restoring balance in physiological systems in the aftermath of challenges to homeostasis; the role of adrenal steroids in this regard has been likened to a second line of defense that buffers the organism from responding too strongly to its primary line of defense (Munck, Guyre, & Holbrook, 1984). Prime examples in the body are the anti-inflammatory effects of adrenal steroids and the inhibitory effects of adrenal steroids over autoimmune or allergic processes that represent overreactions of the cellular immune system. Similar counterregulatory processes are now recognized in the brain. Thus, we have seen that, for each primary response of a neurotransmitter system to stress, a delayed counterregulatory action of glucocorticoids exists at the level of a neurotransmitter receptor or second-messenger system.

Whereas the counterregulatory systems in the brain act to prevent over-reactions of neurochemical systems, under some circumstances the protective mechanisms fail and damage ensues. The actions of stressors on the hippocampus illustrate the negative side of dysregulation. Exactly how much and what types of stress can cause the beginnings of damage to the hippocampal formation most readily remains to be established. Initial information suggests that severe social stress is capable of causing hippocampal neuronal loss (Uno, Ross, Else, Suleman, & Sapolsky, 1989; Uno, Glugge, Thieme, Johren, & Fuchs, 1991), and that ordinary laboratory stressors such as restraint cause atrophy of hippocampal neurons within only a few weeks; such damage may be reversible if stress is terminated at that time (Watanabe, Gould, & McEwen, 1992). Thus, considerable protection against this type of brain damage probably exists, except under extreme circumstances of certain types of potent stressors. However, the very existence of such damage reveals that the environment can produce wear and tear on the brain. Conceivably, such wear and tear may be an inevitable and unavoidable part of living and responding to a changing environment (McEwen, 1991b).

Beginning shortly after conception, each individual experiences a unique set of events that begins to have a cumulative effect on brain and body development and the operation of adaptive mechanisms. As the organism matures and develops a functioning hypothalamo–pituitary–adrenal axis, such experiences act in part through the reactions of the individual of secreting adrenal steroids and other associated hormones that we have not discussed in this chapter. Through the interplay of experience and endocrine response we can see, in principle, one of the ways in which the environment produces a unique individual.

Acknowledgment

Research in the authors' laboratory that is described in this chapter was supported by NIMH Grants MH 41256 and MH 43787.

References

Acuna, D., Fernandez, B., Gomar, M., Aguila, C., & Castillo, J. (1990). Influence of the pituitary-adrenal axis on benzodiazepine receptor binding to rat cerebral cortex. *Neuroendodrinology, 51,* 97–103.

Alliet, P., Lu, R., de la Garza, M., Santer, R., Lebenthal, E., & Lee, P. (1989). Response of exocrine pancrease to corticosterone and aldosterone after adrenalectomy. *Journal of Steroid Biochemistry, 33,* 1097–1102.

Angulo, J., Riftina, F., Ledoux, M., & McEwen, B. S. (1991). Adrenal glucocorticoids regulate the angiotensinogen gene expression in the rat brain: In vivo and in situ hybridization studies. *Abstracts of the Society of Neuroscience, 17,* 165.15.

Armario, A., & Jolin, T. (1986). Effects of water restriction on circadian rhythms of corticosterone, growth hormone and thyroid stimulating hormone in adult male rats. *Physiology and Behavior, 38,* 327–330.

Arriza, J., Simerly, R., Swanson, L., & Evans, R. (1988). The neuronal mineralocorticoid receptor as a mediator of glucocorticoid response. *Neuron, 1,* 887–900.

Arriza, J., Weinberger, C., Cerelli, G., Glaser, T., Handelin, B., Housman, D., & Evans, R. (1987). Cloning of human mineralocorticoid receptor complementary DNA: Structural and functional kinship with the glucocorticoid receptor. *Science, 237,* 268–275.

Azmitia, E., & McEwen, B. S. (1969). Corticosterone regulation of tryptophan hydroxylase in rat midbrain. *Science, 166,* 1274–1276.

Azmitia, E., & McEwen, B. S. (1974). Adrenocortical influence on rat brain tryptophan hydroxylase activity. *Brain Research, 78,* 291–302.

Azmitia, E., & McEwen, B. S. (1976). Early response of rat brain tryptophan hydroxylase activity to cycloheximide, puromycin and corticosterone. *Journal of Neurochemistry, 27,* 773–778.

Bayon, A., & Anton, B. (1986). Diurnal rhythm of the in vivo release of enkephalin from the globus pallidus of the rat. *Regulatory Peptides, 15,* 63–70.

Beaumont, K., Vaughn, D., & Fanestil, D. (1987). Effect of adrenocorticoid receptors on potassium and sodium flux in rat C6 glioma cells. *Journal of Steroid Biochemistry, 28,* 593–598.

Born, J., DeKloet, R., Wenz, H., Kern, W., & Fehm, H. (1991). Gluco- and antimineralocorticoid effects on human sleep: A role of central corticosteroid receptors. *American Journal of Physiology, 260,* E183–E188.

Born, J., Spath-Schwalbe, E., Schwakenhofer, H., Kern, W., & Fehm, H. (1989). Influences of corticotropin-releasing hormone (CRH) adrenocorticotropin (ACTH), and cortisol on sleep in normal man. *Journal of Clinical Endocrinology and Metabolism, 68,* 904–911.

Born, J., Zwick, A., Roth, G., Fehm-Wolfsdorf, G., & Fehm, H. (1987). Differential effects of hydrocortisone, fluocortolone and aldosterone on nocturnal sleep in humans. *Acta Endocrinologica, 116,* 129–137.

Bray, G. (1985). Autonomic and endocrine factors in the regulation of food intake. *Brain Research Bulletin, 14,* 505–510.

Brett, L. P., & Levine, S. (1979). Schedule-induced polydipsia suppressed pituitary-adrenal activity in rats. *Journal of Comparative Physiology and Psychology, 93,* 946–956.

Brett, L. P., & Levine, S. (1981). The pituitary–adrenal response to "minimized" schedule-induced drinking. *Physiology and Behavior, 26,* 153–158.

Broderick, P., Gardner, E., & van Pragg, H. (1984). In vivo electrochemical and behavioral evidence for specific neural substrates modulated differently by enkephalin in rat stimulant stereotypy and locomotion. *Biological Psychiatry, 19,* 45–54.

Campbell, D., Sernia, C., Thomas, W., & Oldfield, B. (1991). Immunocytochemical localization of angiotensinogen in rat brain: Dependence of neuronal immunoreactivity on method of tissue processing. *Journal of Neuroendocrinology 3,* 653–660.

Carroll, B., Martin, F., & Davies, B. (1968). Resistance to suppression by dexamethasone of plasma 11-O.H.C.S. levels in severe depressive illness. *British Medical Journal, 3,* 285–287.

Castren, E., & Saavedra, J. (1989). Angiotensin II receptors in paraventricular nucleus, subfornical organ, and pituitary gland of hypophysectomized, adrenalectomized, and vasopressin-deficient rats. *Proceedings of the National Academy of Science, U.S.A., 86,* 725–729.

Chao, H., & McEwen, B. S. (1990). Glucocorticoid regulation of preproenkephalin messenger ribonucleic acid in the rat striatum. *Endocrinology, 126,* 3124–3130.

Chao, H., & McEwen, B. S. (1991). Glucocorticoid regulation of neuropeptide mRNAs in the rat striatum. *Molecular Brain Research, 9,* 307–311.

Chow, S., Sakai, R., Reagan, L., McEwen, B. S., & Fluharty, S. (1991). Interactions between angiotensin II and corticosteroid receptors in neuroblastoma cells. *Abstracts of the Society for Neuroscience, 17,* 321.11.

Coirini, H., Schulkin, J., & McEwen, B. S. (1988). Behavioral and neuroendocrine regulation of mineralocorticoid and glucocorticoid action. *Abstracts of the Society for Neuroscience, 14,* 528.15.

Coover, G. (1984). Plasma corticosterone and meal expectancy in rats: Effects of low probability cues. *Physiology and Behavior, 33,* 179–184.

Coover, R. G., Goldman, L., & Levine, S. (1971). Plasma corticosterone levels during extinction of a lever-press response in hippocampectomized rats. *Physiology and Behavior, 7,* 727–732.

Crawley, J., Majewska, J., Glowa, M., & Paul, S. (1986). Anxiolytic activity of endogenous adrenal steroid. *Brain Research, 339,* 382–386.

Davies, P. (1986). The genetics of Alzheimer's disease: A review and discussion of the implications. *Neurobiology and Aging, 7,* 459–466.

De Kock, S., & DeKloet, R. (1987). Neurotrophic peptide ACTH-(4–10) permits glucocorticoid-facilitated retention of acquired immobility response of hypophysectomized rats. *European Journal of Pharmacology, 141,* 461–466.

De Souza, E., Goeders, N., & Kuhar, M. (1986). Benzodiazepine receptors in rat brain are altered by adrenalectomy. *Brain Research, 381,* 176–181.

Devenport, L. (1978). Schedule-induced polydipsia in rats: Adrenocortical and hippocampal modulation. *Journal of Comparative Physiology and Psychology, 92,* 651–660.

Devenport, L., Knehans, A., Sundstrom, A., & Thomas, T. (1989a). Corticosterone's dual metabolic actions. *Life Sciences, 45,* 1389–1396.

Devenport, L., Knehans, A., Thomas, T., & Sundstrom, A. (1989b). Macronutrient intake and utilization by rats: Interactions with type I adrenocorticoid receptor stimulation. *Life Sciences, 45,* 1389–1396.

Devenport, L., & Thomas, T. (1990). Acute, chronic and interactive effects of type I and II corticosteroid receptor stimulation on feeding and weight gain. *Physiology and Behavior, 47,* 1221–1228.

Dickinson, S. L., Kennett, G. A., & Curzon, G. (1985). Reduced 5-hydroxytryptamine-dependent behavior in rats following chronic corticosterone treatment. *Brain Research, 345,* 10–18.

Dulloo, A., Seydoux, J., & Girardier, L. (1990). Role of corticosterone in adaptive changes in energy expenditure during refeeding after low calorie intake. *American Journal of Physiology, 259,* E658–E664.

Edwards, C., Burt, D., McIntyre, M., DeKloet, R., Stewart, P., Brett, L., Sutanto, W., & Monder, C. (1988). Localization of 11-beta-hydroxysteroid dehydrogenase-tissue specific protector of the mineralocorticoid receptor. *Lancet, 2,* 986–989.

Edwards, E., Harkins, K., Wright, G., & Henn, F. (1990). Effects of bilateral adrenalectomy on the induction of learned helplessness behavior. *Neuropsychopharmacology, 3,* 109–114.

Epstein, A. N. (1982). Mineralocorticoids and cerebral angiotensin may act together to produce sodium appetite. *Peptides, 3,* 493–494.

Epstein, A. N. (1991). Neurohormonal control of salt intake in the rat. *Brain Research Bulletin, 27,* 315–320.

Evans, R. (1988). The steroid and thyroid hormone receptor superfamily. *Science, 240,* 889–895.

Faunt, J., & Crocker, A. (1988). Adrenocortical hormone status affects responses to dopamine receptor agonists. *European Journal of Pharmacology, 152,* 255–261.

Fehm, H., Benkowtisch, R., Kern, W., Fehm-Wolfsdorf, G., Pauschinger, P., & Born, J. (1986). Influences of corticosteroids, dexamethasone and hydrocortisone on sleep in humans. *Neuropsychobiology, 16,* 198–204.

Feinberg, M., Carroll, B., King, D., & Greden, J. (1984). The effect of dexamethasone on sleep: Preliminary results in eleven patients. *Biological Psychiatry, 19,* 771–775.

File, S., Vellucci, A., & Wendlandt, S. (1979). Corticosterone—Anxiogenic or anxiolytic agent? *Journal of Pharmaceutics and Pharmacology, 31,* 300–305.

Fluharty, S., & Epstein, A. (1983). Sodium appetite elicited by intracerebroventricular infusion of angiotensin II in the rat: Synergistic interaction with systemic mineralocorticoids. *Behavioral Neuroscience, 97,* 746–758.

Fregly, M., & Waters, W. (1966). Effect of mineralocorticoids on spontaneous sodium chloride appetite of adrenalectomized rats. *Physiology and Behavior, 1,* 65–74.

Funder, J., Feldman, D., & Edelman, I. (1973). The roles of plasma binding and receptor specificity in the mineralocorticoid action of aldosterone. *Endocrinology, 92,* 994–1004.

Funder, J., Pearce, P., Smith, R., & Smith, A. (1988). Mineralocorticoid action: Target tissue specificity is enzyme, not receptor, mediated. *Science, 242,* 583–585.

Fuxe, K., Wikstrom, A., Okret, S., Agnati, L., Harfstrand, A., Yu, Z.-Y., Granholm, L., Zoli, M., Vale, W., & Gustafsson, J.-A. (1985). Mapping of glucocorticoid receptor immunoreactive neurons in the rat tel- and diencephalon using a monoclonal antibody against rat liver glucocorticoid receptor. *Endocrinology, 117,* 1803–1812.

Ganesan, R., & Sumners, C. (1989). Glucocorticoids potentiate the dipsogenic action of angiotensin II. *Brain Research, 499,* 121–130.

Gannon, M., & McEwen, B. S. (1990). Calmodulin involvement in stress- and corticosterone-induced down-regulation of cyclic AMP-generating systems in brain. *Journal of Neurochemistry, 55,* 276–284.

Gibbs, D. (1986). Vasopressin and oxytocin: hypothalamic modulators of the stress response: A review. *Psychoneuroendocrinology, 11,* 131–140.

Gillin, J., Jacobs, L., Fram, D., & Snyder, F. (1972). Acute effect of a glucocorticoid on normal human sleep. *Nature (London), 237,* 398–399.

Gillin, J., Jacobs, L., Snyder, F., & Henkin, R. (1974). Effects of ACTH on the sleep of normal subjects and patients with Addison's disease. *Neuroendocrinology, 15,* 21–31.

Gomez-Sanchez, E., Fort, C., & Gomez-Sanchez, C. (1990). Intracerebroventricular infusion of Ru28318 blocks aldosterone-salt hypertension. *American Journal of Physiology, 258,* E482–E484.

Gould, E., Woolley, C., & McEwen, B. S. (1990). Short-term glucocorticoid manipulations affect neuronal morphology and survival in the adult dentate gyrus. *Neuroscience, 37,* 367–375.

Green, P., Woods, C., & Wilkinson, S. (1991). Activation of central type II adrenal steroid receptors enhances rate of weight gain in underweight adrenalectomized rats. *Abstracts of the Society for Neuroscience, 17,* 196.13.

Guillaume-Gentil, C., Rohner-Jeanrenaud, F., Abramo, F., Besteti, G., Rossi, G., & Jeanrenaud, B. (1990). Abnormal regulation of the hypothalamo-pituitary-adrenal axis in the genetically obese fa/fa rat. *Endocrinology* **126,** 1873–1879.

Hawkins, J., Hicks, R., Phillips, N., & Moore, J. (1978). Swimming rats and human depression. *Nature (London), 274,* 512–513.

Honma, K.-I., Honma, S., Hirai, T., Katsuno, Y., & Hiroshige, T. (1986). Food ingestion is more important to plasma corticosterone dynamics than water intake in rats under restricted daily feeding. *Physiology and Behavior, 37,* 791–795.

Honma, K.-I., Honma, S., & Hiroshige, T. (1984). Feeding-associated corticosterone peak in rats under various feeding cycles. *American Journal of Physiology, 246,* R721–R726.

Honma, S., Honma, K.-I., Nagasaka, T., & Hiroshige, T. (1987). The ventromedial hypothalamic nucleus is not essential for the prefeeding corticosterone peak in rats under restricted daily feeding. *Physiology and Behavior, 39,* 211–215.

Horner, H., Packan, D., & Sapolsky, R. (1990). Glucocorticoids inhibit glucose transport in hippocampal neurons and glia. *Neuroendocrinology, 52,* 57–64.

Imaki, T., Nahan, J.-L., Rivier, C., Sawchenko, P., & Vale, W. (1991). Differential regulation of corticotropin-releasing factor mRNA in rat brain regions by glucocorticoids and stress. *Journal of Neuroscience, 11,* 585–599.

Jacobson, L., & Sapolsky, R. (1991). The role of the hippocampus in feedback regulation of the hypothalamic-pituitary-adrenocortical axis. *Endocrine Review, 12,* 118–134.

Janiak, P., & Brody, M. (1988). Central interactions between aldosterone and vasopressin on cardiovascular system. *American Journal of Physiology, 24,* R166–R173.

Jazayeri, A., & Meyer, W. (1989). Mineralocorticoid-induced increase in beta-adrenergic receptors of cultured rat arterial smooth muscle cells. *Journal of Steroid Biochemistry, 33,* 987–991.

Jefferys, D., Boublik, J., & Funder, J. (1985). A K-selective opioidergic pathway is involved in the reversal of a behavioral effect of adrenalectomy. *European Journal of Pharmacology, 107,* 331–335.

Jefferys, D., Copolov, D., Irby, D., & Funder, J. (1983). Behavioral effect of adrenalectomy: Reversal by glucocorticoids or [D-Ala2-Met5]enkaphlinamide. *European Journal of Pharmacology, 92,* 99–103.

Jefferys, D., & Funder, J. (1987). Glucocorticoids, adrenal medullary opioids, and the retention of a behavioral response after stress. *Endocrinology, 121,* 1006–1009.

Jhanwar-Uniyal, M., Roland, C., & Leibowitz, S. (1985). Diurnal rhythm of α2-noradrenergic receptors in the paraventricular nucleus and other brain areas: Relation to circulating corticosterone and feeding behavior. *Life Sciences, 38,* 473–482.

Joels, M., & DeKloet, E. R. (1989). Effects of glucocorticoids and norepinephrine on the excitability in the hippocampus. *Science, 245,* 1502–1505.

Joels, M., & DeKloet, E. R. (1991). Control of neuronal excitability by corticosteroid hormones. *Trends in Neurosciences, 15,* 25–30.

Johnson, J., & Sawyer, C. (1971). Adrenal steroids and the maintenance of a circadian distribution of paradoxical sleep in rats. *Endocrinology, 89,* 507–512.

Jonat, C., Rahmsdorf, H., Park, K.-K., Cato, A., Gebel, S., Ponta, H., & Herrlich, P. (1990). Antitumor promotion and anti-inflammation: Down-modulation of AP-1 (Fos/Jun) activity by glucocorticoid hormone. *Cell, 62,* 1189–1204.

Kalimi, M., Opoku, J., Agarwal, M., & Corley, K. (1990). Effects of antimineralocorticoid Ru26752 on steroid-induced hypertension in rats. *American Journal of Physiology, 258,* E737–E739.

Kalivas, P. (1985). Interactions between neuropeptides and dopamine neurons in the ventromedial mesencephalon. *Neuroscience and Behavioral Reviews, 9,* 573–587.

Kawakami, M., Negoro, H., & Terasawa, E. (1965). Influence of immobilization stress upon the paradoxical sleep (EEG after-reaction) in the rabbit. *Journal of Physiology, 15,* 1–16.

Kelly, A., Cador, M., and Stinus, L. (1985). Behavioral analysis of the effect of substance P injected into the ventral mesencephalon on investigatory and spontaneous motor behavior in the rat. *Psychopharmacology (Berlin), 85,* 37–46.

King, B. (1988). Glucocorticoids and hypothalamic obesity. *Neuroscience and Biobehavioral Reviews, 12,* 29–37.

Koranyi, L., Beyer, C., & Guzman-Flores, C. (1971). Multiple unit activity during habituation, sleep wakefulness cycle, and the effects of ACTH and corticosteroid treatment. *Physiology and Behavior, 7,* 321–329.

Krieger, D. (1974). Food and water restriction shifts corticosterone, temperature, activity and brain amine periodicity. *Endocrinology, 95,* 1195–1201.

Krieger, D., & Glick, S. (1972). Growth hormone and cortisol responsiveness in Cushing's syndrome. Relation to a possible central nervous system etiology. *American Journal of Medicine, 52,* 25–40.

Krieger, D., & Hauser, H. (1977). Suprachiasmatic nuclear lesions do not abolish food-shifted circadian adrenal and temperature rhythmicity. *Science, 197,* 398–399.

Langley, S., & York, D. (1990a). Effects of antiglucocorticoid RU486 on development of obesity in obese fa/fa/ Zucker rats. *American Journal of Physiology, 259,* R539–R544.

Langley, S., and York, D. (1990b). Increased type II glucocorticoid-receptor numbers and glucocorticoid-sensitive enzyme activities in the brain of the obese Zucker rat. *Brain Research, 533,* 268–274.

Leibowitz, S. (1986). Brain monoamines and peptides: Role in the control of eating behavior. *Federation Proceedings, 45,* 1396–1403.

Leibowitz, S. (1989). Hypothalamic neuropeptide Y and galanin: Functional studies of coexistence with monoamines. In V. Mutt (Ed.), *Neuropeptide* (pp. 267–281). New York: Raven Press.

Levine, S., Goldman, L., & Coover, G. (1972). Expectancy and the pituitary-adrenal system. In R. Porter and J. Knight (Eds.), *Physiology, emotion and psychosomatic illness* (pp. 281–296). Ciba Foundation Symposium. Amsterdam: Elsevier.

Levine, S., Weinberg, J., & Brett, L. (1979). Inhibition of pituitary–adrenal activity as a consequence of consummatory behavior. *Psychoneuroendocrinology, 4,* 275–286.

Lind, R. W., Swanson, L., & Ganten, D. (1984). Angiotensin II immunoreactivity in the neural afferents and efferents of the subfornical organ of the rat. *Brain Research, 321,* 209–215.

Lind, R. W., Swanson, L., & Ganten, D. (1985). Organization of angiotensin II immunoreactive cells and fibers in the rat central nervous system. *Neuroendocrinology, 40,* 2–24.

Lynch, K., Simnad, V., Ben-Ari, E., & Garrison, J. (1986). Localization of preangiotensinogen messenger RNA sequences in the rat brain. *Hypertension, 8,* 540–543.

McEwen, B. S. (1977). Adrenal steroid feedback on neuroendocrine tissues. *Annals of the New York Academy of Sciences, 297,* 568–579.

McEwen, B. S. (1991a). Steroids affect neural activity by acting on the membrane and the genome. *Trends in Pharmacological Science, 12,* 141–147.

McEwen, B. S. (1991b). Our changing ideas about steroid effects on an ever-changing brain. *Seminars in the Neurosciences, 3,* 497–507.

McEwen, B. S., Angulo, J., Cameron, H., Chao, H., Daniels, D., Gannon, M., Gould, E., Mendelson, S., Sakai, R., Spencer, R., & Woolley, C. (1992). Paradoxical effects of adrenal steroids on the brain: Protection versus degeneration. *Biological Psychiatry, 31,* 177–199.

McEwen, B. S., Brinton, R., Chao, H., Coirini, H., Gannon, M., Gould, E., O'Callaghan, J., Spencer, R., Randall, S., & Woolley, C. (1990). The hippocampus: A site for modulatory interactions between steroid hormones, neurotransmitters and neuropeptides. In E. Muller and R. MacLeod (Eds.), *Neuroendocrine perspectives* (pp. 93–131). New York: Springer Verlag.

McEwen, B. S., DeKloet, E. R., & Rostene, W. (1986b). Adrenal steroid receptors and actions in the nervous system. *Physiology Reviews, 66,* 1121–1188.

McEwen, B. S., & Gould, E. (1990). Adrenal steroid influences on the survival of hippocampal neurons. *Biochemistry and Pharmacology, 40,* 2393–2402.

McEwen, B. S., Lambdin, L., Rainbow, T., & DeNicola, A. (1986a). Aldosterone effects on salt appetite in adrenalectomized rats. *Neuroendocrinology, 43,* 38–43.

McEwen, B. S., Stephenson, B. S., & Krey, L. (1980). Radioimmunoassay of brain tissue and cell nuclear corticosterone. *Journal of Neuroscience Methods, 3,* 57–65.

McEwen, B. S., Weiss, J., & Schwartz, L. (1968). Selective retention of corticosterone by limbic structures in rat brain. *Nature (London), 220,* 911–912.

Mendelson, S., & McEwen, B. S. (1991). Autoradiographic analyses of the effects of restraint-induced stress on 5-HT_{1A}, 5-HT_{1C} and 5-HT_2 receptors in the dorsal hippocampus of male and female rats. *Neuroendocrinology, 54,* 454–461.

Mendelson, S., & McEwen, B. S. (1992). Autoradiographic analyses of the effects of adrenalectomy and corticosterone on 5-HT_{1A} and 5-HT_{1B} receptors in the dorsal hippocampus and cortex of the rat. *Neuroendocrinology, 55,* 444–450.

Miller, A., Spencer, R., Stein, M., & McEwen, B. S. (1990). Adrenal steroid receptor binding in spleen and thymus after stress or dexamethasone. *American Journal of Physiology, 259,* E405–E412.

Miller, A., Spencer, R., Trestman, R., Kim, B., McEwen, B. S., & Stein, M. (1991). Adrenal steroid receptor activation in vivo and immune function. *American Journal of Physiology, 261,* E126–E131.

Mistlberger, R., Houpt, T., & Moore-Ede, M. (1990). Food-anticipatory rhythms under 24 hour schedules of limited access to single macronutrients. *Journal of Biological Rhythms, 5,* 35–46.

Mitchell, J., & Meaney, M. (1991). Effects of corticosterone on response consolidation and retrieval in the forced swim test. *Behavioral Neuroscience, 105,* 798–803.

Morimoto, Y., Arisue, K., & Yamamura, Y. (1977). Relationship between circadian rhythm of food intake and that of plasma corticosterone and effect of food restriction on circadian adrenocortical rhythm in the rat. *Neuroendocrinology, 23,* 212–222.

Munck, A., Guyre, P., & Holbrook, N. (1984). Physiological functions of glucocorticoids in stress and their relation to pharmacological actions. *Endocrine Reviews, 5,* 25–44.

Nadler, J., & Cuthbertson, G. (1980). Kainic acid neurotoxicity toward hippocampal formation: Dependence on specific excitatory pathways. *Brain Research, 195,* 47–56.

Neckers, L., & Sze, P. (1975). Regulation of 5-hydroxytryptamine metabolism in mouse brain by adrenal glucocorticoids. *Brain Research, 93,* 123–132.

Nitabach, M., Schulkin, J., & Epstein, A. (1989). The medial amygdala is part of a mineralocorticoid-sensitive circuit controlling NaCl intake in the rat. *Behavioral Brain Research, 35,* 127–134.

Osborne, B. (1986). Behavioral and adrenal responses and meal expectancy in rats with fornix transection. *Physiology and Behavior, 37,* 499–502.

Osborne, B., & Dodek, A. (1986). Disrupted patterns of consummatory behavior in rats with fornix transections. *Behavioral and Neural Biology, 45,* 212–222.

Osborne, B., Sivakumaran, T., & Black, A. (1979). Effect of fornix lesions on adrenocortical responses to changes in environmental stimulation. *Behavioral Neuroscience and Biology, 25,* 227–241.

Peeters, B., Smets, R., & Broekkamp, C. (1991). The involvement of glucocorticoids in the acquired immobility response is dependent on the water temperature. *Physiology and Behavior, 51,* 127–129.

Persky, H., Smith, K., & Basu, G. (1971). Effect of corticosterone and hydrocortisone on some indicators of anxiety. *Journal of Clinical Endocrinology, 33,* 467–473.

Plomin, R. (1990). The role of inheritance in behavior. *Science, 48,* 183–188.

Purdy, R., Morrow, L., Moore, P., & Paul, S. (1991). Stress-induced elevations of gamma-aminobutyric acid type A receptor-active steroids in the rat brain. *Proceedings of the National Academy of Sciences, U.S.A., 88,* 4553–4557.

Quarton, G., Clark, L., Cobb, S., & Bauer, W. (1955). Mental disturbances associated with ACTH and cortisone: A review of explanatory hypotheses. *Medicine, 34,* 13–50.

Quinn, S., & Williams, G. (1988). Regulation of aldosterone secretion. *Annual Review of Physiology, 50,* 409–426.

Reul, J. M., & DeKloet, E. R. (1985). Two receptor systems for corticosterone in rat brain: Micro-distribution and differential occupation. *Endocrinology, 117,* 2505–2511.

Richter, C. (1965). *Biological clocks in medicine and psychiatry.* Springfield, Illinois: Charles C. Thomas.

Rosenwasser, A., Schulkin, J., & Adler, N. (1988). Anticipatory appetitive behavior of adrenalectom-ized rats under circadian salt-access schedules. *Animal Learning and Behavior, 16,* 324–329.

Roy, E., Lynn, D., & Bemm, C. (1990). Individual variations in hippocampal dentate degeneration following adrenalectomy. *Behavioral Neuroscience and Biology, 54,* 330–336.

Ryan, J., Springer, J., Hannigan, J., & Isaacson, R. (1985). Suppression of corticosterone synthesis alters the behavior of hippocampally lesioned rats. *Behavioral Neuroscience and Biology, 44,* 47–59.

Rydzewski, B., Sumner, C., Shen-orr, Z., & Raizada, M. (1991). Mineralocorticoids influence steady-state levels of glucose transporter-1 mRNA in neuronal and astrocytic glial cells in primary culture. *Abstracts of the Endocrine Society,* 1311.

Sachar, R., Hellman, J., Fukushima, D., & Gallagher, T. (1970). Cortisol production in depressive illness. *Archives of General Psychiatry, 23,* 289–298.

Sakai, R., & Epstein, A. (1990). Dependence of adrenalectomy-induced sodium appetite on the action of angiotensin II in the brain of the rat. *Behavioral Neuroscience, 104,* 167–176.

Sakai, R., Lakshmi, V., Monder, C., Funder, J., Kvozowski, Z., & McEwen, B. (1990). Co-localization of 11β-hydroxysteroid dehydrogenase and mineralocorticoid receptor in rat brain. *Abstracts Soc. Neurosci., 16,* #541.1 1309.

Sakai, R., Nicolaidis, S., & Epstein, A. (1986). Salt appetite is suppressed by interference with angiotensin II and aldosterone. *American Journal of Physiology, 251,* R762–R768.

Sapolsky, R. (1990a). Glucocorticoids, hippocampal damage and the glutamatergic synapse. *Progress in Brain Research, 86,* 13–23.

Sapolsky, R. (1990b). Stress in the wild. *Scientific American, 262,* 116–123.

Sapolsky, R., Krey, L., & McEwen, B. S. (1986). The neuroendocrinology of stress and aging: The glucocorticoid cascade hypothesis. *Endocrine Reviews, 7,* 284–301.

Sawchenko, P., Swanson, L., & Vale, W. (1984). Co-expression of CRF- and vasopressin-immun-oreactivity in parvocellular neurosecretory neurons of adrenalectomized rats. *Proceedings of the National Academy of Sciences, U.S.A., 81,* 1883–1887.

Schulkin, J., Marini, J., & Epstein, A. (1989). A role for the medial region of the amygdala in mineralocorticoid-induced salt hunger. *Behavioral Neuroscience, 103,* 178–185.

Schulle, R., Rangarajan, P., Kliewer, S., Ransone, L., Bolado, J., Yang, N., Verma, I., & Evans, R. (1990). Functional antagonism between oncoprotein c-*Jun* and the glucocorticoid receptor. *Cell, 62,* 1217–1226.

Sernia, C., Clements, J., and Funder, J. (1989). Regulation of liver angiotensinogen mRNA by glucocorticoids and thyroxine. *Molecular and Cellular Endocrinology, 61,* 147–156.

Simmonds, M. (1990). *Steroids and neuronal activity.* Chichester, England: John Wiley and Sons.

Singh, V., Corley, K., Phan, T.-H., & Boadle-Biber, M. (1990). Increases in the activity of tryptophan hydroxylase from rat cortex and midbrain in response to acute or repeated sound stress are blocked by adrenalectomy and restored by dexamethasone treatment. *Brain Research, 516,* 66–76.

Sloviter, R. (1983). "Epileptic" brain damage in rats induced by sustained electrical stimulation of the perforant path. I. Acute electrophysiological and light microscopic studies. *Brain Research Bulletin, 10,* 675–697.

Sloviter, R., Valiquette, G., Abrams, G., Ronk, E., Sollas, A., Paul, L., & Neubort, S. (1989). Selective loss of hippocampal granule cells in the mature rat brain after adrenalectomy. *Science, 243,* 535–538.

Spencer, R., Miller, A., Kang, S., Stein, M., & McEwen, B. S. (1991). Diurnal comparison of adrenal steroid receptor activation in brain, pituitary and immune tissue. *Abstracts Soc. Neurosci., 17*, 829.

Spencer, R., Young, E., Choo, P., & McEwen, B. S. (1990). Adrenal steroid type I and type II receptor binding: Estimates of in vivo receptor number, occupancy and activation with varying levels of steroid. *Brain Research, 514*, 37–48.

Stinus, L., Kelly, A., & Iversen, S. (1978). Increased spontaneous activity following substance P infusion into A10 dopaminergic area. *Nature (London), 276*, 616–618.

Stoessl, A., and Szczutkowski, E. (1991). Neurotensin and neurotensin analogs modify the effects of chronic neuroleptic administration in the rat. *Brain Research, 558*, 289–295.

Stornetta, R., Hawelu-Johnson, C., Guyenet, P., & Lynch, K. (1988). Astrocytes synthesize angiotensinogen in brain. *Science, 242*, 1444–1446.

Sumners, C., & Fregly, M. (1989). Modulation of angiotensin II binding sites in neuronal cultures by mineralocorticoids. *American Journal of Physiology, 256*, C121–C129.

Sumners, C., Gault, T., & Fregly, M. (1991). Potentiation of angiotensin II-induced drinking by glucocorticoids is a specific glucocorticoid Type II receptor (GR)-mediated event. *Brain Research, 552*, 283–290.

Sze, P., Neckers, L., & Towle, A. (1976). Glucocorticoids as a regulatory factor for brain tryptophan hydroxylase. *Journal of Neurochemistry, 26*, 169–173.

Takahashi, K., Inoue, K., Kobayashi, K., Hayafuji, C., Nakamura, Y., & Takahashi, Y. (1977a). Effects of food restriction on circadian adrenocortical rhythm in rats under constant lighting conditions. *Neuroendocrinology, 23*, 193–199.

Takahashi, K., Inoue, K., & Takahashi, Y. (1977b). Parallel shift in circadian rhythms of adrenocortical activity and food intake in blinded and intact rats exposed to continuous illumination. *Endocrinology, 100*, 1097–1107.

Tempel, D., & Leibowitz, S. (1989). PVN steroid implants: Effect on feeding patterns and macronutrient selection. *Brain Research Bulletin, 23*, 553–560.

Tokuyama, K., & Himjs-Hagen, J. (1989a). Enhanced acute response to corticosterone in genetically obese (ob/ob) mice. *American Journal of Physiology, 257*, E133–E138.

Tokuyama, K., & Himms-Hagen, J. (1989b). Adrenalectomy prevents obesity in glutamate-treated mice. *American Journal of Physiology, 257*, E129–E144.

Turnamian, S., & Binder, H. (1990). Aldosterone and glucocorticoid receptor-specific agonists regulate ion transport in rat proximal colon. *American Journal of Physiology, 258*, G492–G498.

Uno, H., Glugge, G., Thieme, C., Johren, O., & Fuchs, E. (1991). Degeneration of the hippocampal pyramidal neurons in the socially stressed tree shrew. *Abstracts of the Society for Neuroscience, 17*, 52.20.

Uno, H., Ross, T., Else, J., Suleman, M., & Sapolsky, R. (1989). Hippocampal damage associated with prolonged and fatal stress in primates. *Journal of Neuroscience, 9*, 1705–1711.

Van den Berg, D., DeKloet, R., van Dijken, H., & De Jong, W. (1990). Differential central effects of mineralocorticoid and glucocorticoid agonists and antagonists on blood pressure. *Endocrinology, 126*, 118–124.

Veldhuis, H., DeKloet, E. R., Van Zoest, I., & Bohus, B. (1982). Adrenalectomy reduces exploratory behavior activity in rat: A specific role of corticosterone. *Hormones and Behavior, 16*, 191–198.

Veldhuis, H., DeKorte, C., & DeKloet, E. R. (1985). Glucocorticoids facilitate the retention of acquired immobility during forced swimming. *European Journal of Pharmacology, 115*, 211–217.

Von Zerssen, D. (1976). Mood and behavioral changes under corticosteroid therapy. In T. Util, G. Landahn, and Herrman (Eds.), *Psychotropic action of hormones* (pp. 195–222). New York: Spectrum.

Watanabe, Y., Gould, E., & McEwen, B. S. (1992). Stress induces growth of apical dendrites of hippocampal CA3 pyramidal neurons. *Brain Research, 588*, 341–345.

Weiss, J., McEwen, B. S., Silva, M., & Kalkut, M. (1970). Pituitary-adrenal alterations and fear responding. *American Journal of Physiology, 218,* 864–868.

Wilson, K., Sumners, C., Hathaway, S., & Fregly, M. (1986). Mineralocorticoids modulate central angiotensin II receptors in rats. *Brain Research, 382,* 87–96.

Wolf, G. (1965). Effect of deoxycorticosterone on sodium appetite of intact and adrenalectomized rats. *American Journal of Physiology, 208,* 1281–1285.

Wolf, G., & Handel, P. (1966). Aldosterone induced sodium appetite: Dose–response and specificity. *Endocrinology, 6,* 1120–1124.

Woolley, C., Gould, E., & McEwen, B. S. (1990). Exposure to excess glucocorticoids alters dendritic morphology of adult hippocampal pyramidal neurons. *Brain Research, 531,* 225–231.

Woolley, C., Gould, E., Sakai, R., Spencer, R., & McEwen, B. S. (1991). Effects of aldosterone or RU28362 treatment on adrenalectomy-induced cell death in the dentate gyrus of the adult rat. *Brain Research, 554,* 312–315.

Yang-Yen, H., Chambard, J.-C., Sun, Y.-L., Smeal, T., Schmidt, T., Drouin, J., & Karin, M. (1990). Transcriptional interference between c-*Jun* and the glucocorticoid receptor: Mutual inhibition of DNA binding due to direct protein-protein interaction. *Cell, 62,* 1205–1215.

Yoshikawa, T., Shibuya, H., Kaneno, S., & Toru, M. (1991). Blockade of behavioral sensitization to methamphetamine by lesion of hippocampo-accumbal pathway. *Life Sciences, 48,* 1325–1332.

Zhang, J., Valatx, J.-L., & Jouvet, M. (1988). Effects of hypophysectomy on the sleep of neonatally monosodium glutamate-treated rats. *Brain Research Bulletin, 21,* 897–903.

CHAPTER

7

Psychoimmunology: The Missing Links

David Saphier

Departments of Pharmacology and Therapeutics, and Psychiatry
Louisiana State University Medical Center
Shreveport, Louisiana 71130

I. Introduction

Biomedical science has used an increasingly reductionist approach over the centuries to understand the intricacies of life. This approach has been successful in many ways but we are all familiar with the scenario inside the office of the physician. In addition to the physical examination and the taking of specimens for analysis, the medical practitioner almost inevitably will ask questions of a personal nature to find out whether any circumstances exist that may contribute to the state of health of the patient. This approach has not been learned as a part of the reductionist science taught in the first preclinical years of medical school; rather, it is a result of millennia of human evolution. Indeed the Greek philosopher–physicians were well aware, for example, that "melancholic women were more prone to unhealthy swellings in the breast," as observed by Galen more than 2000 years ago. This type of observation has been combined with the fields of psychiatry, psychology, physiology, endocrinology, neurosciences, and immunology to create a new division of scientific endeavor that has been termed variously as psychoimmunology, neuroimmunology, immunoendocrinology, psychoneuroendocrine-immunology, and so on (Solomon & Moos, 1964; Besedovsky & Sorkin, 1977; Solomon, 1987; Pierpaoli & Maestroni, 1988; Ader, Felten, & Cohen, 1991). All these neologisms have their uses and all have been used to describe the study of the interactions now known to exist between the immune and central nervous systems, which had, until now, been considered separate entities in terms of maintenance of the state of well-being of the organism. The lengthy names employed clearly indicate that the "reduced" sciences are being reintegrated in a more holistic fashion and that the "missing links" are being defined to enable a better understanding of the bidirectional connections between the psyche and the immune system. Some of the basic principles, combined with specific examples, of the multiplicity of these interactions form the substance of this chapter.

II. Evidence for Psychosomatic Interactions: Specific Hunger

As the reader progresses through the various sections of this chapter, an impression of a somewhat diverse bank of information will emerge, but the connections that exist between the various different constituents that fall under the general heading of "psychoimmunology" also should be apparent. The fundamental principles and mechanisms to be discussed are not intended to be assimilated into a series of therapeutic tenets. Instead, these studies should serve as portals to the future. In this light, perhaps showing how a number of studies performed over the last few decades can be integrated to provide a consolidated view of another somatic process intimately related to the central nervous system and behavior will be useful.

In 1940, Curt Richter made an important discovery in the field of taste perception, which led to the coining of the term "specific hunger." This discovery followed studies in which he observed that a child whose adrenal cortex had been removed because of a tumor developed a specific hunger for salt. After adrenalectomy, the child was no longer able to secrete adrenocortical hormones (including aldosterone). As a result, he lost much of his salt through urinary excretion and consequently had low plasma concentrations of sodium. Given free access to food, the child compensated for the deficiency by deliberately selecting salt or salty foods. The studies were taken to the laboratory, where rats also were found to select salty foods after surgical adrenalectomy (Richter, 1942). These observations of a clear neural influence on food selection led to further investigations attempting to unravel this apparently simple process of which we are hardly aware and take for granted. Thus, removal of the adrenal gland is now known to result in a desensitization of the salt receptors in the tongue to salt (Contreras, 1977). Obviously, this story is not complete, since clearly the tongue is not the final organ responsible for the modification of behavior (craving for salty food). The messages for such selection must arise at the level of the brain (Chapters 1 and 6). How this perception and selection is integrated is not clear, but a number of other experiments provided clues. In one study, the increasing recruitment of hypothalamic vasopressin-secreting neurons into phasic electrical activity patterns of firing was observed after increasing periods of water deprivation in the rat (Wakerley, Poulain, & Brown, 1978). Vasopressin, or antidiuretic hormone, is responsible for water resorption at the level of the kidney. In the conscious monkey, the stimulus of drinking was found to inhibit the activity of such neurons and to decrease vasopressin secretion (Vincent, Arnauld, & Bioulac, 1972). We also have shown that the activity of such neurons is increased by application of hyperosmotic salt solutions and decreased by application of distilled water to the tongue and oropharynx of the rat, indicating the presence of an immediate and specific response (Saphier & Feldman, 1990).

Vasopressin is known to serve as a neurotransmitter than modulates aspects of behavior that include memory acquisition and retrieval (de Wied, 1984). Perhaps a component of the salt craving discussed here is in some way modulated by processes of memory that are strengthened by vasopressin. Further, the vasopressin neurons are sensitive to adrenocortical hormones (Saphier & Feldman, 1988), and are located in an area of the hypothalamus into which microinjections of adrenergic agents have been shown to cause a preferential increase in carbohydrate intake (Leibowitz, Weiss, Yee, & Tretter, 1985). These observations do not serve as a complete and direct explanation of the originally observed phenomenon, but several other observations will doubtless one day be integrated into a more complete picture. The results obtained simply represent a part of the complex of events designed to maintain homeostasis, but much effort has been required to identify the mechanisms involved in such regulation. Similar principles almost certainly will apply in psychoimmunology, although the completed picture will be very complex indeed.

III. Effects of the Psyche on Immune Function

Clearly psychological factors, such as the stress associated with bereavement, may lead to a decreased immunocompetence and increased incidence of disease. In light of such observations, the brain seems able to influence activity of the immune system. Apparently, any such effects of the central nervous system (CNS) on immunological parameters must have physical bases; these must be identified if we are to manage such effects effectively. However, first we must identify the wide variety of neural effects on the immune system and characterize those effects in terms of immune function. This section describes some of the more important observations that have been made.

Some of the earliest reports were those cited by Smith and Salinger (1933), who noted that Osler had described an allergic patient who experienced an asthma attack when presented with an artificial rose. Another similar example was that of hay fever attacks that could be induced in very sensitive patients when shown a picture of a hay field (Hill, 1930). Of course, these examples show conditioned responses that were not entirely new observations, since earlier animal studies by Russian investigators had demonstrated that immune responses could be conditioned classically according to Pavlovian paradigms (Metal'nikov & Chorine, 1926). After a number of studies were done that primarily supported the observations of these investigators, a long period of scientific unresponsiveness ensued during which only sporadic reports appeared in the scientific press. However, a resurgence of interest has been evident in the last 30 years or so; now such responses have been shown to be immunopharmacologically manipulable and behaviorally conditioned suppression of immune responses has been demonstrated clearly (Ader & Cohen, 1975).

Asthma is a good example of a potentially fatal clinical disorder that appears to have an important psychological component. Allergists have known for many years that asthmatic children in particular frequently have underlying emotional problems. Now we know that the bronchoconstriction typical of asthma also may be caused by activation of autonomic pathways. Indeed, some asthmatic patients may have lower levels of plasma IgE (the class of immunoglobulin antibodies associated with allergic responses, mast cell activation, and histamine release) than normal subjects. Thus, in addition to the assessment of allergic status, examination of the emotional and psychological profile of asthmatic patients is important (Mrazek & Klinnert, 1988).

Psychological factors also appear to be of importance in the etiology of a number of other diseases, particularly those of an autoimmune nature such as rheumatoid arthritis (RA), systemic lupus erythematosus (SLE), myasthenia gravis (MG), multiple sclerosis (MS), ulcerative colitis (Engel's colitis), and even Graves' disease. In diseases such as these, so-called "forbidden" T lymphocyte clones seem to be permitted to develop and work in cooperation with B lymphocytes to produce the autoimmune antibodies that attack the host and cause the disease. The deselection of forbidden clonal T-cell lines is a complex phenomenon that is understood incompletely, but nevertheless recognized as being of crucial importance to the ability of the immune system to discriminate between self and nonself (Tomer & Shoenfeld, 1988). Several factors are clearly of importance in the deselection process; genetic predisposition is perhaps of prime importance, but overlaid on the DNA substrate are a number of nebulous psychosocial elements. These elements have proven difficult to identify, but the classical studies of Solomon and co-workers have provided some standards that can be applied to analyses of the interactions between psychological factors and somatic disease processes. Among the identified psychological components are those that appear, to the lay person, to be somewhat abnormal. These elements include depression, apathy, anxiety, alienation, general "psychotic" symptoms, and particular personality traits such as prejudice and dependency (Solomon, 1981). Many of these factors represent long-term stressful situations; acutely stressful events are of much less relevance, although these may exacerbate the disease process.

Genetic factors also may be superimposed on the scenario described here. As an indication of the importance of genetic components in diseases such as RA, Solomon extended his studies to include the siblings of a patient cohort with RA. Differences in psychological and personality indices were found not only between the diseased subjects and their healthy siblings, but also between healthy individuals that were positive on immunoglobulin tests for the FII rheumatoid factor and those that were FII seronegative. The conclusions drawn from the study were that the physically healthy individuals in the FII-positive group were in better psychological equilibrium than individuals in the other two groups, the investigators suggested that, had they not been so, they may have developed the disease. The classical psychological factors associated with the physically ill patients also were represented more in the FII-seronegative group, suggesting that only with a genetic predisposition to RA would the negative psychological constituents result in expression of the disease state. Note that RA is more common in women and that the foregoing studies were done with female patients. The sex and reproductive status of the individual are known to play a role in development and regulation of disease processes, which will be discussed in greater detail in a subsequent section.

Additional clinical evidence for the existence of psychoneuroimmunomodulatory mechanisms arises from studies in which abnormal immunoglobulin levels were detected in schizophrenic patients (Solomon, Moos, Fessel, & Morgan, 1966). Similar observations were made in association with other forms of mental illness and a variety of personality factors (Solomon, 1981). These observations now have been shown to be of functional significance in a number of controlled experimental situations. In an attempt to relate the clinical observations to laboratory models, Solomon and co-workers demonstrated that opposing effects of opposing stimuli, which may be termed stressing and gentling, were effective in decreasing and increasing immunocompetence, respectively (Solomon, Levine, & Kraft, 1968; Solomon, 1969). Repeated exposure to a stressor may result in adaptation and sometimes an enhanced immune response (Gisler, 1974; Monjan & Collector, 1977).

IV. Cerebral Lateralization

Cerebral lateralization is an additional factor, not essentially related to the psyche, that has been suggested to be involved in the modulation of immune processes. Lateralization is known to be of importance in the regulation of a number of biological functions. Anatomical, biochemical, and toxicopharmacologic bases appear to exist for the described functional asymmetries (Glick, 1985). One such difference is an increased binding of the lipophilic immunostimulatory agent diethyldithiocarbamate in the right neocortex (Guillaumin, Lepape, & Renoux, 1986), correlated with a greater fatty acid content on the right side of the brain (Pediconi & Rodriguez de Turco, 1984). Renoux and Biziere (1991) demonstrated the importance of cerebral lateralization on immune responses, that is, lesions of the left cerebral neocortex in mice depressed T-cell responses without affecting B-cell responses. These workers also showed that the effects of the T-cell-specific immunopotentiator diethyldithiocarbamate do not require an intact neocortex, but may be subject to a negative influence by the left hemisphere, which is in turn inhibited by the right hemisphere (Renoux & Biziere, 1991). The clinical findings of Geschwind and Behan (1982) also support the view that cerebral lateralization may be correlated with altered immunocompetence. These researchers found that immune disorders such as atopic diseases, autoimmune thyroiditis, and migraine were more frequent in left-handers and their relatives than in right-handers. Collectively, these clinical and animal studies indicate that a neuroanatomical basis is likely to exist for the neural and psychological modulation of immune responses under consideration.

V. Neural Substrates for Immunomodulation

To determine the CNS sites that may be of importance in the modulation of immune responses, various brain manipulations such as electrolytic lesions and electrical stimulation have been employed. Such approaches have provided some details of the central neural circuitry involved in the regulation of immune responses. Most studies have focused on the hypothalamus and limbic structures, although some have examined the role of neocortical structures, as discussed in the previous section. Anterior hypothalamic lesions decrease the number of nucleated spleen cells and thymocytes; decrease natural killer cell activity, lymphoproliferative blastogenic responses to mitogens, and antibody production; and also inhibit the development of anaphylactic responses, autoimmune disease, and experimental allergic encephalomyelitis (Korneva & Khay, 1963; Tyrey & Nalbandov, 1972; Jankovic & Isakovic, 1973; Stein, Schleyer, & Keller, 1981; Cross, Markesbery, Brooks, & Roszman, 1984; Roszman, Cross, Brooks, & Marksbery, 1985). Blockade of electrical activity in the anterior hypothalamus with tetrodotoxin has the same effect as lesions, whereas electrical stimulation of the same region results in opposing modulation of antibody responses (D. Saphier, H. Ovadia, A. Maimon, G. Mor, and O. Abramsky, unpublished observations). Electrical stimulation of various brain regions, including the hypothalamus, has been shown by a number of other authors to alter immune responses; the changes generally oppose those induced by lesions of the same brain structures (Kanda,

1959; Korneva & Khay, 1963). Such lesioning and stimulation studies have provided some insight into the neural substrates for immunomodulation but do not provide any mechanistic solutions.

The fact that some effects of hypothalamic lesions can be reversed by hypophysectomy (Cross et al., 1984; Tyrey & Nalbandov, 1972) indicates that the observed immunomodulation may have been a result of altered neuroendocrine activity. The hypothalamus is the primary regulatory structure controlling pituitary function; the anterior hypothalamus is of importance as an integrative "driving" center regulating neural activity in the endocrine hypothalamus and, consequently, adenohypophyseal hormonal secretory activity (Dyer & Saphier, 1981; Saphier & Feldman, 1986). Therefore, some of the influences of this structure (and those of the limbic system) on immune responses are likely to be mediated by hormones of the pituitary gland. The principal role of the hypothalamus is in maintaining homeostasis, which frequently involves altering activity in response to changes in the internal or external environment. This function is, in some ways, similar to that of the immune system because the function of the immune system is responding to infection to return the organism to health. In addition, the hypothalamus, by way of its effects on neuroendocrine activity, is responsible for the regulation of temperature, circadian rhythms, sexual and reproductive function, and some aspects of ontogeny and senescence. The effects of some of these parameters on immune function, and evidence for specific modulatory effects of pituitary hormones, will be detailed elsewhere in this chapter.

The hypothalamus has been described as the head ganglion of the autonomic nervous system. Stimulation of the anterior hypothalamus results in autonomic activation that is associated with increased circulating concentrations of catecholamines, which have been suggested to alter immune function. Dense catecholaminergic (and other) innervation of various components of the immune system is known, thus providing an additional neuroanatomical basis of potential importance in the modulation of immune activity. The bone marrow is innervated by sympathetic fibers from the femoral artery and vein, in addition to sensory innervation from the sciatic nerve. The function of the autonomic catecholaminergic innervation is unclear, but is believed to be associated with the regulation of blood flow and fat mobilization. The release of reticulocytes following stimulation of the posterior hypothalamus is likely to be mediated by these sympathetic nerves (Feldman, Rachmilewitz, & Izak, 1966).

Innervation of the spleen, thymus gland, and other lymphoid tissue has been the subject of much interest in the past decade. Many studies have arisen from the laboratories of Felten and co-workers and Bulloch and co-workers. These studies have been reviewed by Felten and Felten (1991) and by Bulloch (1985), so the interested reader should refer to these texts for more detailed information. Briefly, sympathetic innervation of the spleen, thymus, lymph nodes, and gut-associated lymphoid tissues has been demonstrated. Evidence also exists for sensory innervation of these tissues by myelinated fibers. Noradrenergic and cholinergic innervation has been demonstrated and a number of peptide-containing nerve fibers have been shown using immunocytochemical techniques (Felten, Felten, Carlson, Olschowka, & Livnat, 1985). Peptides demonstrated include neuropeptide Y, vasoactive intestinal polypeptide, substance P, calcitonin gene related peptide, somatostatin, and even the immune system cytokine interleukin 1 (IL-1). The physiological significance of such innervation remains largely speculative. However, the close synapse-like apposition of such nerve terminals with reticular fibers and cells, macrophages, and IgG-positive B lymphocytes, in conjunction with the observations of nerve fiber retraction and recrudescence under different conditions of immune activity, strongly suggests direct neural influences on the function of cellular components of the immune system. Further, the presence of specific membrane transmitter receptors on the surface of immunocytes supports this concept, as detailed subsequently. In addition to the innervation of immune tissues, local innervation of other tissues such as skin may be able to modify immunological activity, as demonstrated by the slower growth of tumors in mouse skin deprived of sympathetic innervation by unilateral superior third cervical ganglionectomy (Romeo et al., 1991). Clinical studies have not provided data to demonstrate direct neural–immune interactions, but innervation of the bone marrow, thymus, spleen, and other lymphoid tissues of humans (Kudoh, Hoski, & Murakami, 1979; Ghali, Abdel-Rahman, & Nagib, 1980; Yamashita et al., 1984; Fink & Weihe, 1988) suggests that the situation in humans will turn out to be similar to that in animals. What is clear, however, is that, in humans as in other species, the nervous system is able to influence activity of the immune system by way of neuroendocrine and endocrine transduction.

VI. Neuroendocrine Effects on the Immune System

In addition to the effects on immune function of the gonadal steroid hormones described in the next two sections, many studies have demonstrated the potential role played by hypothalamic and pituitary peptide hormones in the modulation of immune responses. Hypophysectomy and peptide hormone injections have been used to evaluate the effects of pituitary hormones on the immune system, and considerable influences have been demonstrated (Berczi, 1986). The effects of pituitary grafts, as a model of hyperprolactinemia, also have been studied extensively. The primary effect of hypophysectomy is a suppression of antibody production and lymphocyte proliferation and differentiation in response to immunogenic challenges. Animals so treated have small thymus glands and spleens. These observations indicate that, under normal conditions, the pituitary gland apparently exerts trophic stimulatory effects on the immune system. Since the pituitary is usually responsible for the maintenance of homeostasis and the well-being of the organism, this result is not surprising. However, identifying the endocrine factors secreted by the gland that mediate positive effects on the immune system is important.

Growth hormone and prolactin are pituitary hormones that have arisen by gene duplication from a common ancestral hormone but have distinct effects, mediated via different receptors, in mammalian species. The principal effect of a lack of growth hormone in early life is dwarfism, whereas hypersecretion results in gigantism (or acromegaly, if it occurs in later life). In hypophysectomized rats, growth hormone administration augments antibody and DNA synthesis; this effect also is observed in normal animals, albeit to a lesser extent. Growth hormone exerts mitogenic effects on the thymus gland and is believed to stimulate thymic epithelial cells to produce greater quantities of thymic hormones. Growth hormone deficiency in children does not seem to lead to an increased incidence of infection; these children usually have normal T- and B-cell distributions, antibody levels, and proliferative responses to mitogenic or antigenic stimulation. However, some studies in humans have reported a decreased immunocompetence as reflected by thymic hypoplasia, reduced antibody and cell-mediated immune responses, and decreased natural killer cell activity. Growth hormone secretion decreases with age, which is associated with an increased thymic involution. Such involution may be decreased by growth hormone in experimental animals, and the reduced immunocompetence associated with aging also may be reduced during growth hormone therapy. Human mononuclear lymphocytes each have several thousand high affinity receptors for growth hormone. *In vitro* studies generally have demonstrated positive effects of this polypeptide on lymphoproliferative responses. Growth hormone also appears to stimulate macrophage phagocytosis and production of cytokines such as tumor necrosis factor α, IL-1, and the growth factor somatomedin C. The overall effects of growth hormone on activity of various components of the immune system thus appear to be stimulatory. The results of some studies do not entirely support this view. Clearly further work is required to determine the precise interrelationships between growth hormone and the immune system (Kelley, 1989).

Prolactin usually is associated with lactation and reproductive cyclicity, although numerous other effects have been ascribed to this hormone, not the least of which are effects on the immune system. Prolactin is also a stress-responsive hormone like adrenal corticotropic hormone (ACTH), which will be considered separately. The regulation of prolactin secretion from the anterior pituitary gland is well characterized; factors derived from the immune system, for example, IL-1 and thymic factors, may modulate prolactin secretion (Bernton, Bryant, & Holaday, 1991). Prolactin probably interacts with glucocorticoid hormones *in vivo* in an antagonistic manner to stimulate immune activity, which normally is suppressed by increased secretion of such steroids. This effect has not been demonstrated *in vitro* and requires further substantiation. The specific effects of prolactin on immune system function have been reviewed extensively by Berczi and Nagy (1988) and by Bernton, Bryant, and Holaday (1991). Prolactin exerts effects similar to those of growth hormone in terms of increasing proliferative activity in the liver and thymus gland. Only about one-tenth as many receptors for prolactin exist on lymphocytes, as for growth hormone, and actions on immune cells have been difficult to characterize. Although prolactin appears to be stimulatory *in vivo,* it has not been possible to demonstrate such effects *in vitro.* The studies performed have been complicated by the fact that lymphocytes themselves are able to synthesize prolactin or a prolactin-like protein (Montgomery et al., 1987). Clearly, further work is required to characterize any effects of prolactin on immune system activity, although an immunopotentiating role seems probable.

Corticotropin releasing factor (CRF) is another peptide that has received considerable attention as a modulator of immunological activity. CRF is a hypothalamic peptide responsible for activation of pituitary–adrenocortical secretion in response to stress, and also is considered to act as a neurotransmitter in the CNS. Receptor sites for CRF have been found on human blood lymphocytes and monocytes, and on mouse splenic macrophages. Exposure of human peripheral blood mononuclear cells to this peptide induces the release of β-endorphin, and has been shown to increase proliferation of human lymphocytes *in vitro.* In addition, CRF increases both basal and induced synthesis of the cytokines IL-1 and IL-2 by such cells, and stimulates expression of IL-2 receptors on human T cells (Singh, Warren, White, & Leu, 1990).

VII. Sex, Reproduction, and Immune Function

Much of the data available concerning the effects of sex and reproductive status on immune function is somewhat anecdotal. Despite the plethora of reports available in the scientific literature, the results of many of the systematic studies have been contradictory and confusing. However, certain details are apparent, and are summarized here. The reader is referred to the excellent reviews by Grossman (1984), Schuurs and Verheul (1990), and McCruden and Stimson (1991) for further information and particulars.

Generally speaking, the findings in humans and in animals are comparable. This correlation is likely to continue in future studies. Female rodents and humans have higher IgM, IgG, and IgA plasma concentrations and are able to mount larger antibody responses of various Ig classes to both T-cell dependent and T-cell independent antigens. Before puberty, boys tend to be more prone to allergy than girls, but this situation appears to reverse at puberty when testosterone secretion in males and estrogen secretion in females increases, suggesting that the allergy phenomenon is sex-hormone dependent. Female rats and mice have been reported to exhibit either lower or higher cell-mediated immune responses, and show a higher resistance to tumors and parasites. Human females appear to have decreased cell-mediated immune responses compared with males, although this result does not correlate with a possible reduced incidence of tumors and greater resistance to viral and parasitic infections. Macrophages from mature female rats produce more IL-1 than those from mature males or immature females; ovariectomy causes a reduction that can be reversed by estradiol. IL-1 production by human monocytes *in vitro* is stimulated by low concentrations and inhibited by high concentrations of estradiol (or progesterone).

Castration of male rodents increases immunoglobulin levels and humoral and cell-mediated immune responses, and also increases the weight of the thymus, spleen, and lymph nodes, with a concomitant increase in T cells in the last two tissues. Gonadectomy of mature females does not alter humoral immune responses, but does increase cell-mediated immune responses and thymus weight. The increase in thymic weight following castration is believed to result from a delay in the involution of the thymus and in the hypertrophy of cortical and medullary thymocytes and reticuloendothelial cells. Data obtained from studies of genetically hypogonadal mice reveal apparent differences between the sexes. The splenocyte count in normal males was lower than that in normal females and hypogonadal males and females. Thymus weights and thymocyte counts in males were higher in hypogonadal animals, whereas in females they were lower. Treatment with high doses of androgens decreases humoral and cell-mediated immune responses and decreases thymus weight. In addition, mouse strains with high androgen responsiveness show low immune responses. In a clinical study, administration of human chorionic gonadotropin to prepubertal boys with incomplete testicular descent concomitantly increased plasma testosterone concentrations, while significantly decreasing the CD4/CD8 cell ratio. On the other hand, treatment of intact and castrated male or female rats with estradiol increased antibody responses to various T-dependent and T-independent antigens. High doses of estradiol decrease cell-mediated responses; cyclic exposure (mimicking the estrous cycle) has a greater effect on antibody formation than chronic exposure. The antiestrogen, tamoxifen has opposing effects, inhibiting antibody responses and cell-mediated immunity. *In vitro,* physiological doses of estradiol stimulate pokeweed mitogen-induced Ig synthesis by B lymphocytes, whereas testosterone inhibits such synthesis or has no effect. Overall, testosterone appears to be important as a physiological (down-)regulator of immune system function, whereas estrogens appear to have a positive, immunopotentiating effect.

Receptors for sex hormones have been identified in thymic tissue and apparently are localized in the epithelial cells, suggesting that the effects of estradiol and testosterone on immune responses are, at least partly, indirect. Estradiol receptors also have been identified in lymphoid cells, CD8-positive T cells, peripheral blood mononuclear cells, and thymic cells. Although androgen receptors have been demonstrated in human thymus, they are reported to be absent in peripheral blood cells.

The progestogens constitute an additional class of sex steroids that may exert important influences on immune activity in the female. Menopause is accompanied by a significant increase in IL-1 release from monocytes; this rise may be reversed by estrogen or medroxyprogesterone hormone replacement therapy. Progesterone increases production of immunosuppressive factors by endometrial tissue during proliferation; interestingly, premenstrual exacerbations of asthma may be ameliorated by administration of progesterone. *In vitro*, the progestogen lynestrenol stimulates active T-cell rosetting, leukocyte adherence inhibition, and monocyte phagocytosis. Intraarticular injection of progesterone reduces local inflammation, probably by inhibiting monocyte IL-1 production. Women taking oral contraceptives containing progestogen and estrogen have been reported to show altered responses to some immunogenic stimuli, the recorded effects apparently are due to the progestogen, although these effects may require the presence of the estrogen. Natural killer cell activity fluctuates during the estrous cycle and during pregnancy in the mouse, and is decreased during the periovulatory period in women, again indicating influences of sex steroids on immune activity.

The hormonal changes during pregnancy are profound and complex. Clearly this state is able to exert important effects on immune function. Mitogenic and cell-mediated immune responses are decreased during pregnancy, whereas skin homografts survive longer than in nonpregnant women. Natural killer cell activity is depressed during pregnancy, in negative correlation with estrogen levels, and is increased significantly 1 month postpartum. Although the concentrations and proportions of sex hormones are altered substantially during pregnancy, this is clearly not the only reason for the observed changes in immune reactivity.

The influences of gender and sex hormones on cell-mediated and humoral immune responses is incomplete, but the available evidence suggests that physiological concentrations of estrogens stimulate such responses and that male hormones do the opposite. The presence of receptors for both estrogens and androgens in lymphoid cells or in the thymus gland may explain how these steroid hormones are able to modify structure and function of these organs and interact with and modulate activity of the immune system. In summary, estrogens may have direct effects on B cells, inhibit B suppressor function, facilitate T helper maturation, and influence macrophage maturation and function. On the other hand, androgens may interfere with maturation processes of B cells and T suppressor cells of the bone marrow, resulting in increased T suppressor function. In view of the popularity of anabolic steroids in athletes to increase muscle mass and strength, it is relevant to note that such steroids may result in altered immunocompetence. One of the few reports available has indicated that the use of anabolic steroids has potent immunomodulatory effects, and that autoantibodies are more prevalent in strength-trained men, even in the absence of anabolic steroid use (Calabrese et al., 1989).

Gender and sex hormones have been suggested to influence the course of autoimmune diseases in humans. Women show a distinct preponderance for several autoimmune diseases, such as SLE, Sjögren's syndrome, RA, and MS. The greater incidence of SLE and RA in females is even higher during the reproductive years than before or after; premenstrual flare-ups are frequent, suggesting that these diseases are at least partly dependent on the sex hormone milieu. Patients with Klinefelter's syndrome (XXY genotype) have increased estrogen and decreased testosterone levels, and also suffer from an increased incidence of these diseases. Another indication that sex hormones can influence autoimmune disease is that pregnancy can alter the symptomatology of certain diseases, perhaps because of the steadily increasing circulating estrogen and progesterone concentrations. Clinical symptoms in RA, MS, and MG are reduced during pregnancy; however, exacerbations of SLE commonly are observed during the first trimester. In addition, RA, MS, and SLE all have been observed to deteriorate postpartum. Clinical activity of Graves' and Hashimoto's disease and the levels of autoantibodies found in these conditions are reduced during pregnancy. Transient postpartum thyroiditis has been reported and may recur after subsequent pregnancies, sometimes progressing to a permanent hypothyroidism. In MS, exacerbations are less common during pregnancy but occur more often 3 months postpartum.

A reduced incidence of RA has been reported in women taking oral contraceptives that contain an estrogen and a progestin; a significantly reduced incidence of thyroid disease also has been reported. The severity of RA also may be reduced by oral contraceptives and, in some patients, the dose of glucocorticoids required to control RA can be reduced when they receive concomitant oral contraceptives. In SLE patients, the findings are different: oral contraceptives containing estrogen and progestogen induce exacerbations that are suppressed when oral contraceptives are discontinued or when progestogen-only oral contraceptives are used. A reduction of RA incidence has been found among users of noncontraceptive hormones, in particular estrogens, and among previous oral contraceptive users; the protective effects are particularly apparent in seropositive RA. All these observations suggest that estrogens and progesterone have a positive effect on a number of autoimmune diseases, although not in SLE. Interestingly, some male patients with autoimmune disease have an increased incidence of autoantibodies against the estrogen receptor.

Androgens also may regulate the expression of autoimmune diseases. Some studies have suggested beneficial effects of androgens in RA, SLE, and Sjögren's syndrome. The weak androgen danazol has some favorable effects in SLE, discoid lupus, and idiopathic thrombocytopenia. On the other hand, the antiandrogen cyproterone acetate may suppress exacerbations of SLE, perhaps by reducing estradiol levels. However, antiestrogens such as tamoxifen have not been found to be useful in the management of SLE. Thus, the data regarding the effects of androgens (as well as estrogens and progestogens) in autoimmune diseases are confusing and, at times, contradictory. Some of the confusion may arise from the multiplicity of metabolic effects exerted by the drugs examined, but overall androgenic hormones appear to restrain certain autoimmune diseases.

Not only do the gonadal steroid hormones exert influences on (autoimmune) diseases, but abnormalities in the production and metabolism of such hormones have been reported in SLE and RA patients. These effects range widely and may be linked to some genetic factors controlling such diseases. For example, increased 16α-hydroxylation of estrogens, leading to increased levels of 16α-hydroxy-estrone and estriol, occurs in SLE. Also, increased oxidation of testosterone is seen during disease exacerbation; low plasma concentrations of testosterone, dihydro-testosterone, dihydroepiandrosterone, and dihydroepiandrosterone sulfate, with increased androstenedione, have been reported in SLE patients. In female patients, levels of all these androgens were lower than normal. Similar reductions in androgens and androgen metabolites are found in male patients with RA. These results suggest abnormalities of adrenal androgen synthesis or metabolism in these diseases, but whether these abnormalities represent primary predisposing or secondary factors in the pathology of these disorders is not known.

The data available to date provide evidence that sex hormones are involved in, and also altered by, the expression of autoimmune diseases in humans, although their effects depend on the disease under consideration. Androgens appear to be correlated negatively with autoimmunity; estrogens seem to have a favorable effect in RA and thyroiditis but may be deleterious in SLE. In animals, as in humans, some autoimmune diseases show an influence of gender and sex hormones, but in some diseases such effects are not apparent.

VIII. Stress, Adrenocortical Secretion, and Disease

Stress is recognized as perhaps the single most important factor in the modulation of immune responses, both in the clinical setting and in the laboratory situation (Solomon, 1987). One of the primary effects of stress is activation of the hypotha-lamo–pituitary–adrenocortical axis, frequently accompanied by release of pituitary β-endorphin. This response involves neural processing of stressful stimuli to cause an increase in the activity of hypothalamic CRF- and vasopressin-secreting neur-ons. Effects of CRF on immune activity have already been addressed (Singh et al., 1990). The release of these, and other factors, into the hypophyseal portal blood vessels then stimulates pituitary ACTH secretion, resulting in elevated plasma con-centrations of adrenal glucocorticoid hormones (Makara, 1985; Antoni, 1986). The principal hormones are cortisol in humans and corticosterone in rats and mice.

Much is known about the central neural pathways and neurotransmitters regulating adrenocortical secretion. Apparently, the limbic system is essential for the integrity of this response (Plotsky, 1985; Antoni, 1986; Feldman & Saphier, 1989). In turn, limbic structures have been shown at the behavioral, anatomical, physiological, and biochemical levels to be responsive to increased plasma levels of glucocorticoid hormones. This sensitivity allows for precise feedback regulation of adrenocortical secretory activity, thus maintaining appropriate concentrations of such hormones for homeostasis. (Chapter 6 provides a detailed examination of the regulation and actions of adrenal glucocorticoid hormone secretion.)

Inappropriate hyperactivity of the adrenocortical axis is known to exist in diseases such as depression or Cushing's disease, which are known to be associated with decreased immunocompetence and increased incidence of disease. On the other hand, hyposecretion of glucocorticoids, as in Addison's disease, has long been associated with an increased production of lymphocytes. The antiinflammatory and inhibitory effects of glucocorticoid hormones on immune responses have been known and utilized extensively in the clinical setting for many years (MacLean & Reichlin, 1981). Increased plasma glucocorticoids may be detected during the course of immune responses, leading to the hypothesis that infection or other immunoactivating stimuli may be considered stressors (Besedovsky, Sorkin, Keller, & Muller, 1975). Such responses prevent immunological defense mechanisms from running out of control and damaging the organism (Munck & Guyre, 1986). Such damage could arise from, for example, excessive hyperthermia.

Glucocorticoid hormones exert a wide variety of effects on immune system activity (Clamen, 1975; Cupps & Fanci, 1982). Most of the identified effects inhibit immune processes, although some notable exceptions exist. Although examining all the known glucocorticoid actions on immune function is beyond the scope of this chapter, the interested reader should examine the reviews by Besedovsky and del Rey (1987) and Munck and Guyre (1986). Glucocorticoids are known to modulate the activity of many of the mediators of immune responses and inflammatory actions, either by direct inhibition of their secretion or by preventing the expression of their effects by interfering with their efficacy. Such effects include a suppression of the synthesis of the cytokines IL-1, IL-2, IL-3, γ-interferon (γ-IFN), and tumor necrosis factor. In addition, the glucocorticoids inhibit many of the inflammatory agents such as the eicosanoids, bradykinins, serotonin, and histamine. The effects of the glucocorticoids are mediated by binding to specific intracellular cytosolic receptors, followed by direct effects on gene expression and protein synthesis. The overall consequences of such effects include suppression of antibody formation, macrophage phagocytosis, antibody-dependent cell-mediated cytotoxicity, anaphylactic responses, and inflammatory processes. As discussed, most studies indicate an inhibitory action of adrenal corticosteroids on immune activity in humans and animals. Even the normal circadian variation in circulating glucocorticoid levels has been correlated with alterations in immune cell traffic in humans and animals (Kawate, Abo, Hinuma, & Kumagai, 1981). The fever associated with infection first becomes apparent, and is greatest, during the evening hours, probably because of the emancipation of immune system activity during the nadir of cortisol secretion. This example demonstrates the potent effects that these hormones may have on immune activity. The administration of many immunosuppressive drugs also can increase plasma glucocorticoid levels, although this is not the primary mechanism by which such drugs exert their immunosuppressive effects. In addition, increased circulating glucocorticoid levels have been observed in rats and mice during primary immune responses, the changes correlating temporally with the elaboration of antibody-forming cells (Besedovsky et al., 1975; Dunn, Powell, & Gaskin, 1987; Saphier, 1989).

The fact that glucocorticoid, and other hormones, are able to exert potent modulating influences on the immune system, coupled with the observed alterations in adrenocortical secretory activity during immune responses, suggests that an interactive immune–neural–endocrine network exists. This concept was proposed originally by Besedovsky, and considers the influences of hormones on the immune system, the "sensory" nature of the immune system, and the effects of immune system-derived factors on neural and neuroendocrine secretory processes (Besedovsky et al., 1975; Besedovsky & del Rey, 1987). These effects form the focus of the remainder of this chapter, but first considering the chemical similarities between components of the nervous, endocrine, and immune systems is relevant.

IX. Chemical Similarities between the Nervous, Endocrine, and Immune Systems

Comprehending the effects of psychological factors and stress on the immune system is perhaps easiest when one considers that such effects may be mediated by hypothalamo–pituitary axes, as discussed earlier. Cells of the immune system contain receptors for certain peptide and steroid hormones, and these substances are able to act on such cells in a classical endocrine fashion. In addition, lymphocytes are capable of synthesizing a prolactin-like hormone (Montgomery et al., 1987). Receptors for a wide variety of neuropeptides have been demonstrated, as previously discussed, and biogenic amine receptors also have been identified on lymphocytes. Catecholaminergic inhibition of lymphocyte proliferation and effector function has been demonstrated (Livnat, Felten, Carlson, Bellinger, & Felten, 1985). Several other chemical similarities exist between the systems that are believed to be of physiological significance, the best studied of which is the proopiomelanocortin (POMC) series of peptides.

POMC is a peptide hormone precursor from which several biologically active peptides are cleaved by posttranslational processing, including ACTH, α-melano-tropic hormone, and β-endorphin (Beaumont & Hughes, 1979). The major site of synthesis of these hormones is the pituitary gland, from which they are released in response to stressful stimuli, as discussed earlier. In addition to synthesis in the pituitary gland, POMC expression has been suggested in a number of other tissues, including the hypothalamus, gonads, placenta, adrenal gland, and neuroendocrine tumors. Cells of the immune system also have been reported to express POMC-derived peptides, as well as other peptides such as proenkephalin-A (Rosen, Behar, Abramsky, & Ovadia, 1984; Blalock, Harbour-McMenamin, & Smith, 1985; Zurawski et al., 1986). Activation of mouse T helper cells has been shown to induce preproenkephalin messenger RNA synthesis, indicating that the gene is responsive to immunological stimuli (Zurawski et al., 1986). Blalock and Smith (1980, 1981) reported the presence of ACTH-like immunoreactivity, with biolog-ical activity, in preparations of the cytokine α-interferon (α-IFN) of human origin. These researchers reported that plasma corticosterone concentrations were in-creased following the infection of hypophysectomized mice with Newcastle Dis-ease Virus. The adrenocortical response was suggested to be due to ACTH produced by lymphocytes in association with α-IFN (Smith, Meyer, & Blalock, 1982). The evidence that cells of the immune system are able to synthesize sufficient amounts of such peptides for this response has not been supported by all studies (Dunn et al., 1987). Evidence suggests that α-IFN, but not β- or γ-IFN, exerts effects in the CNS similar to those of the endogenous opioid peptides; similarities between α-IFN and the structures of ACTH and β-endorphin have been demonstrated (Blalock & Smith, 1981), although not all authorities support this view. Pharmacological studies have shown that α-IFN and these peptides act via common mechanisms of signal transduction after binding to specific receptors, that apparently include opiate binding sites in the mouse brain (Blalock & Smith, 1981). The opioid peptides derived from the POMC precursor molecule also have been shown to exert effects on immune function and, thus, may act as paracrine autoregulators of immune activity, as well as endocrine modulators (Plotnikoff, Miller, & Murgo, 1982; Plotnikoff, Murgo, Miller, Corder, & Faith, 1985). Several studies have now demonstrated physiological similarities between the CNS effects of α-IFN and the endogenous opioid peptides (Dafny, Prieto-Gomez, & Reyes-Vazquez, 1985; Saphier, 1989; Birmanns, Saphier, & Abramsky, 1990).

Other hypothalamo–pituitary hormones also are synthesized by cells of the thymus gland, and perhaps by other cells of the immune system. In particular, the two related posterior pituitary hormones, oxytocin and vasopressin, are secreted by thymic epithelial cells (Geenen et al., 1987). These peptides are known to modulate functional activity of the immune system (Johnson, Farrar, & Torres, 1982); their release seems likely to be regulated by catecholaminergic fibers innervating the thymus. Note that catecholamines are also of prime importance in the central regulation of vasopressin and oxytocin.

Thus, the immune, nervous, and endocrine systems share common receptor sites for a variety of neurotransmitters, neuropeptides, modulators, and (neuro)hormones, as well as the ability to synthesize and secrete a number of these substances. In view of these similarities, secretions from various components of the immune system may be expected to exert influences on neural activity or on endocrine glands as target tissues, as in the case for histamine, ACTH, prolactin, or other neuropeptides synthesized by lymphocytes and other cells of the immune system (Smith & Blalock, 1981). The effects on neural and endocrine activity of some of these soluble factors elaborated by the immune system will be discussed in the next section.

X. Effects of Immune System Activity on the Brain

The previous sections in this chapter illustrated some of the effects that the brain may exert, via neurohumoral transduction mechanisms, on the immune system. The immune system is likely to be able to exert effects on the brain. Such effects may be less readily identifiable but are, nevertheless, of importance in the regulation of host defense responses and modulation of behavior during disease. The expression "I don't feel well" best exemplifies the ability of humans to identify altered immune system function. Such feelings obviously are generated within the brain, and probably within those regions concerned with homeostasis. Thus, targets for investigations of such effects probably lie within the limbic system and hypothalamus, which maintains intimate connections with other brain regions (Isaacson, 1982). Further, the immunoresponsive areas are likely to be those known to modulate immune responses, as discussed earlier (Korneva & Khay, 1963; Stein et al., 1981).

Several studies have been performed in which site-specific temporal changes in brain activity patterns have been analyzed in the hours, days, and weeks following antigenic challenges (Besedovsky, Sorkin, Felix, & Haas, 1977; Grigoriev, 1981; Korneva, 1987; Saphier, Abramsky, Mor, & Ovadia, 1987a,b; Saphier, 1989). The studies have shown that complex changes in the electrical activity of limbic and hypothalamic structures may take place very rapidly following immunization and that these changes alter in a site-specific manner over subsequent days. Besedovsky et al. (1977) demonstrated increases in hypothalamic ventromedial nucleus unit activity that peaked at the time of maximum immune response, as determined by the number of spleen plaque-forming cells. These changes were correlated with increased plasma corticosterone levels and with altered hypothalamic norepinephrine activity (Besedovsky et al., 1975; Besedovsky et al., 1983; Saphier, 1989; Dunn, 1990). Our own studies were designed to monitor sequentially the changes in electrical activity within the preoptic area and anterior hypothalamus and the hypothalamic paraventricular

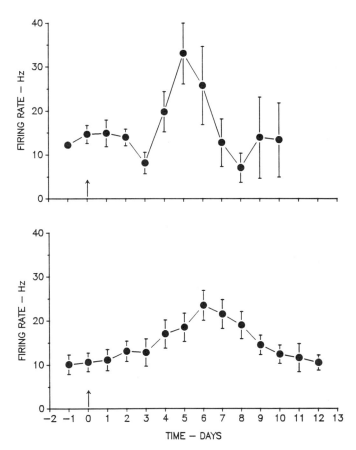

Figure 1. (Top) Preoptic area/anterior hypothalamic multiple unit neuronal electrical activity recorded from adult male rats before and after induction of a peripheral immune response by intraperitoneal injection of a suspension of sheep red blood cells (arrow). Significant changes in neural activity were recorded on days 3–8 following the sensitization, with decreases in activity found on days 3 and 8, whereas substantial increases were found on days 4–6 at the time of maximum antibody production. The recorded changes in electrical activity in this region of the endocrine hypothalamus also have been associated with neurochemical changes and increased adrenocortical secretory activity, as discussed in the text. Control vehicle-injected animals showed no significant alterations in activity. (Bottom) The effect of a second challenge with sheep red blood cells injected 2 wk after the first antigenic challenge evoked similar increases in preoptic area/anterior hypothalamic activity but these were smaller and lasted longer than those following the first injection. In addition, no decreases in activity were recorded. These data demonstrate that the neuronal activity changes during the secondary immune response appear to be related to the different nature of the class of immune activity induced by secondary challenge. Original data from Saphier et al. (1987b), where further details may be obtained.

nucleus during the initial days of antibody generation. Changes in preoptic area and anterior hypothalamic electrical activity were recorded for the 10 days following antigenic stimulation, with threefold increases recorded at the time of initial antibody production. Within the paraventricular nucleus, the site of CRF- and vasopressin-secreting neurons responsible for activation of hypophyseal–adrenocortical secretion, increases in activity were recorded at the time of increased plasma corticosterone secretion. The changes in electrical activity were found to be different in animals that had received a previous antigenic challenge and were undergoing a different (second-set) immune response (Figure 1). Thus, changes in neural activity were recorded in two hypothalamic areas known to be of importance in the regulation of neuroendocrine secretory activity (Dyer & Saphier, 1981; Swanson, Sawchenko, Rivier, & Vale, 1983; Saphier, 1985; Saphier & Feldman, 1985, 1986).

Since antibodies are limited in their ability to cross the blood–brain barrier, the changes in neural activity probably were not related directly to the production of antibodies. The reasons for the increased activity remain unclear, but probably are related to the production, by the immune system, of humoral factors such as lymphokines or, perhaps, POMC-derived peptides. Direct innervation of lymphoid tissue also may play a role in the generation of the recorded central changes (Bulloch, 1985). During the course of immune activation, lymphocytes migrate to the lymph nodes, where mechanical or paracrine effects might cause a flow of sensory information to the brain. To substantiate further the premise that activation of the immune system is able to cause these changes in neural activity by means of chemical messengers, we used cyclophosphamide immunosuppressive drug treatment days prior to, and during, the course of recording. This treatment prevents the generation of antibodies but does not prevent the migration of lymphocytes following antigenic challenge. The treatment was found to be effective in preventing the formation of antibodies and also blocked the changes in neural activity associated with the response. Our results, as well as those discussed previously (Besedovsky et al., 1977; Korneva, 1987), suggest that activation of the immune system indeed is able to modify the activity of the CNS. The changes recorded probably were caused by secretory products released from the activated immune system, although the chemical nature of the proposed humoral factor(s) remain(s) unidentified.

XI. Neurophysiological and Endocrine Effects of Immune System-Derived Factors

In view of the indications that specific soluble chemical factors derived from the activated immune system may be able to alter neural and neuroendocrine activity, reviewing some of the actions of immunomodulatory substances on neurophysiological and endocrine parameters is appropriate. The effects of antibodies on neural activity also has been the focus of some attention, particularly in relation to autoimmune diseases of the CNS (Williams & Schupf, 1987), but this interesting topic will not be covered in this chapter.

Histamine long has been known to exist within components of the immune system (Kazimierczak & Diamant, 1978) and the CNS (Pollard & Schwartz, 1987). This chemical is secreted by mast cells during allergic responses and is located in central neuronal systems in the posterior hypothalamus. The amine exerts central stimulatory effects on arousal and also stimulates adrenocortical secretion (Monnier, Sauer, & Hatt, 1970; Roberts & Calcutt, 1983; Pollard & Schwartz, 1987; Kidron, Saphier, Ovadia, Weidenfeld, & Abramsky, 1989; Saphier, 1989). In both systems, roles have been demonstrated for this amine, but relatively little is known about the mechanism by which it is able to promote arousal and adrenocortical activation. Our data show that histamine is able to cause arousal, as reflected by increases in desynchronization of the cortical electroencephalogram (EEG) while preoptic area and anterior hypothalamic electrical activity is not altered (Kidron et al., 1989). The associated adrenocortical activation, therefore, seems to be mediated at some other site. The increased glucocorticoid levels may, in view of their antiinflammatory actions, serve to limit peripheral allergic responses to the actual site of increased mast cell activity (Saphier, 1989).

The thymus is a gland in which cells produced in the bone marrow undergo the differentiation into T lymphocytes. The thymus is also a source of many peptide factors that are secreted into the peripheral circulation to act as hormones capable of altering the activity of specific target tissues. Thymosin β_4, a component of thymosin fraction 5 (Hooper et al., 1975), has been shown to stimulate reproductive function by stimulating the hypothalamo–pituitary–gonadal axis (Rebar, Miyake, Low, & Goldstein, 1981; Hall et al., 1985), thus substantiating the long-held belief that thymus gland extracts may exert positive (aphrodisiac?) effects on reproductive function. Another factor isolated from thymosin fraction 5, thymosin α_1 also has been shown to stimulate pituitary–adrenocortical secretion without altering CNS biogenic amine turnover (Dunn & Hall, 1987). Peptides of thymosin fraction 5 also are known to be synthesized by other tissues, such as macrophages, and have been found in a wide variety of species. Thymic humoral factor (THF) is another polypeptide synthesized by the thymus gland that exhibits a range of immunostimulatory properties experimentally employed in the treatment of a variety of clinical disorders (Trainin, Handzel, & Pecht, 1985). THF exerts trophic actions on the thymus gland, where glucocorticoid hormones suppress functional activity (Hall & Goldstein, 1981). THF may be able to exert positive feedback effects on its own secretion by suppressing adrenocortical secretory activity, thereby increasing its efficacy as an immunostimulator (Saphier et al., 1988; Kidron et al., 1989; Saphier, Ovadia, & Abramsky, 1990). This effect may be mediated by the CNS, since central administration of THF is able to synchronize the EEG and depress hypothalamic electrical activity (Kidron et al., 1989; Saphier, 1989).

α-IFN, produced by leukocytes after viral stimulation, is known to exert a variety of biological effects apart from its potent antiviral activity, including inhibition of DNA synthesis in the lymphoproliferative response, suppression of antibody synthesis, and enhancement of natural killer cell activity (Friedman & Vogel, 1983). α-IFN has been found to modify behavior by causing central analgesia, catalepsy, and decreased locomotor activity (Blalock & Smith, 1981), effects similar to those of central opiate actions. Central injections of α-IFN have been shown to reduce the abstinence syndrome in morphine tolerant rats, further suggesting an opiate-like mechanism of action (Dafny, 1983; Dafny et al., 1985). In clinical usage, α-IFN exerts diverse neurological effects (Adams, Quesada, & Guttermane, 1984; Kirkwood & Ernstoff, 1984; Rohatiner, Prior, Burton, Balkwill, & Lister, 1985), as well as exerting both acute and long-term positive effects on neurological indices in the disease subacute sclerosing panencephalitis (Steiner, Wirguin, Morag, & Abramsky, 1989), further indicating direct CNS actions of α-IFN. α-IFN also has been found to alter EEG and single unit activity in the rat, probably via endogenous opioid receptor sites (Dafny et al., 1985; Nakashima, Maori, Kuriyama, & Kiyohara, 1987; Kidron et al., 1989; Saphier, 1989; Birmanns et al., 1990). Plasma cortisol concentrations have been found to increase in humans during therapeutic use of α-IFN (Hall et al., 1985; Roosth, Pollard, Brown, & Meyer, 1986), although central administration has been found to decrease plasma corticosterone concentrations in the rat, particularly after chronic administration. As shown in Figure 2, this effect, like that on the EEG, was found to be naloxone sensitive, suggesting a mediation by central μ-opioid receptors (Kidron et al., 1989; Saphier, 1989, 1991; Birmanns et al., 1990). Naloxone-sensitive inhibitory effects of α-IFN on electrical activity of neurons in the preoptic area and anterior hypothalamus also have been shown (Nakashima et al., 1987; Kidron et al., 1989). This effect may be related to decreased adrenocortical secretion, since stimulation of the preoptic area and anterior hypothalamus increases adrenocortical secretion in the rat (Saphier & Feldman, 1986). In addition, our studies have shown that the inhibitory effects of α-IFN on adrenocortical secretion are similarly dependent on specific opioid receptor mechanisms (Figure 2; Saphier, Welch, & Chuluyan, 1993). Thus α-IFN, like THF, decreases adrenocortical secretory activity, an increase in which usually is immunosuppressive, suggesting that α-IFN may exert positive feedback effects on its own efficacy.

IL-1 is a cytokine that is produced mainly by activated macrophages and monocytes, but also is known to exist in other tissues, including glial cells and neurons of the CNS (Dinarello, Renfer, & Wolff, 1977; Fontana, Kristensen, Duba, Gemsa, & Weber, 1982; Bodmer, Tobler, Borbely, & Fontana, 1985). Receptors for IL-1 have been demonstrated in the brain as well as on cells of the immune system (Farrar, Kilian, Ruff, Hill, & Pert, 1987). IL-1 is an endogenous

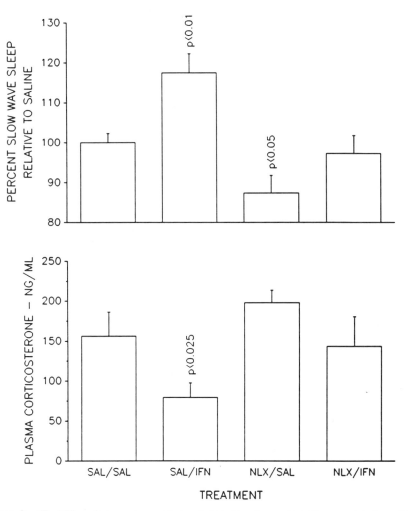

Figure 2. (Top) Effect of intracerebroventricular administration of recombinant human α-interferon (IFN; Inter-Yeda, Israel) on slow wave sleep in conscious freely-moving rats. Sleep/wake activity was assessed from continuous records of the cortical EEG. A significant increase in slow wave sleep was caused by 250 units of α-IFN during the 40-min period immediately following injection, compared with rats given injections of saline (SAL). Intraperitoneal injection of the opiate μ-receptor antagonist naloxone (NLX, 1 mg/kg) caused a decrease in slow wave sleep and was found to prevent the somnogenic effects of α-IFN, suggesting that the immune system-derived cytokine interacts with opiate μ receptors in the central nervous system, providing a communicating link between the two systems. Further details of this study are described in Birmanns et al. (1990). (Bottom) Effects of intracerebroventricular administration of recombinant human α-IFN (Hoffman-La Roche; 100 units) on adrenocortical secretory activity as assessed by measurement of plasma corticosterone concentrations, α-IFN significantly decreased plasma corticosterone concentrations. This effect was antagonized by intraperitoneal pretreatment with naloxone (1 mg/kg), which alone caused a nonsignificant increase in plasma corticosterone. Intraperitoneal injection of α-IFN (1000 units) caused a similar effect, which could be prevented by intracerebroventricular administration of 1 μg naloxone (not shown). These data again suggest an effect of α-IFN on central nervous system opiate receptors, and demonstrate an inverse relationship between arousal (as reflected by the changes in slow wave sleep shown above) and neural regulation of the pituitary–adrenocortical neuroendocrine axis.

pyrogen (Dinarello, 1984; Blatteis, 1985; Besedovsky & del Rey, 1987; Rothwell, 1989), and exerts modulatory influences on the cortical EEG (Krueger, Walter, Dinarello, Wolff, & Chedid, 1984; Blatteis, 1985; Kidron et al., 1989; Saphier, 1989) and on glucose metabolism (Besedovsky & del Rey, 1987). IL-1 also exerts a number of neurochemical (Dunn, 1988) and behavioral (Bluthé, Dantzer, & Kelly, 1989; Uehara, Sekiya, Takasugi, Namiki, & Arimura, 1989) effects, and is a potent activator of hypothalamo–pituitary–adrenocortical secretion (Berkenbosch, van Oers, del Rey, Tilders, & Besedovsky, 1987; Besedovsky & del Rey, 1987; Dunn, 1990; Katsuura, Gottschall, Dahl, & Arimura, 1988; Kidron et al., 1989; Ovadia et al., 1989; Saphier, 1989; Weidenfeld, Abramsky, & Ovadia, 1989). The significant bidirectional interactions between IL-1 and the secretion of glucocorticoid hormones led Besedovsky to develop the theory of an immune-neuroendocrine network (Besedovsky & Sorkin, 1977; Besedovsky, del Rey, Sorkin, & Dinarello, 1986).

In rats and in rabbits, IL-1 usually increases EEG synchronization; such effects last for several hours following a single dose (Krueger et al., 1984; Blatteis, 1985; Bodmer et al., 1985). In our studies, however, a decrease in EEG synchronization was observed in the period immediately following central administration; this increased arousal was associated with increases in plasma corticosterone concentrations (Kidron et al., 1989; Saphier, 1989). We also have shown that IL-1 injected intravenously causes a rapid increase in the electrical activity of putative hypothalamic CRF-secreting neurons (Saphier & Ovadia, 1990; Figure 3). The activity of neighboring electrophysiologically identified vasopressin-secreting neurons was not affected, indicating a lack of involvement of this secretagogue in the adrenocortical response to IL-1. These results support those of other laboratories using different approaches (Berkenbosch et al., 1987) and indicate that IL-1 may exert rapid and specific effects on CNS activity.

XII. Behavioral Consequences of Infection and Immune Activity

Some viral diseases that involve infection of the CNS are associated with frank behavioral symptoms (Steiner et al., 1989). Clearly, other diseases that do not act specifically on the CNS as a target also are able to cause behavioral changes. Indeed, most diseases cause some behavioral changes even if these are minimal, for example, hypersomnia or changes in appetite. We are all familiar with such states; apparently, cytokines such as IL-1 and α-IFN are responsible for these effects. However, other effects of disease states and immune activity on behavior may be identified. In fact, the possibility that mental illness may be due to chronic infection or may represent an autoimmune disease has been raised on many occasions. A classic example of a disease that causes such effects was the epi-

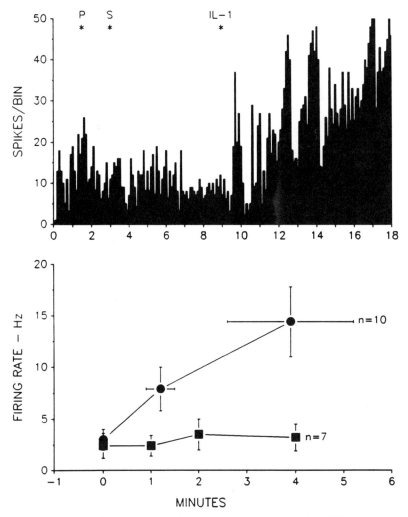

Figure 3. (Top) Peristimulus time histogram showing the activity of a single putative CRF-secreting neuron in the paraventricular nucleus of an adult male rat anesthetized with urethane (1.2 g/kg). The neuron was identified as a tuberoinfundibular neuron by virtue of its antidromic invasion following electrical stimulation of the external zone of the median eminence (Saphier & Ovadia, 1990). Effects of painful somatosensory stimulation (P) consisting of a 1 sec pinch to the contralateral hind limb, intravenous injection of 0.1 ml physiological saline (S), or injection of 50 U recombinant human IL-1β (IL-1) on the activity of the neuron are shown. The mean spontaneous firing rate was 0.9 Hz; somatosensory stimulation evoked a transient increase in firing of the cell. Injection of isotonic saline had no effect on the activity of the cell, but IL-1 caused an increase in firing rate that was sustained for the entire duration of the recording period. (Bottom) Integrated changes for all cells tested as described above. Responses after intravenous administration of saline (■) and of IL-1 (●) are shown. Data points shown represent the means (± standard errors) for the spontaneous activity recorded in the 1 min prior to IL-1 injection, at the time of the response onset, and for the maximum firing rate attained. For the response onset and maximum activity, the times are indicated on the x axis. After saline injections, mean firing rates were determined at 1, 2, and 4 min after injection.

demic of encephalitis lethargica in 1916–1928, which produced psychotic symptoms in adults and intractable hyperactivity in children (Hohman, 1922). In animal studies, neonatal infection with herpes simplex virus type 1 has been shown to cause behavioral alterations that may persist throughout life (Crnic & Pizer, 1988).

Rage responses to trivial stimuli sometimes are observed in brain-damaged patients. These behaviors are a complication of pituitary surgery when inadvertent damage to the hypothalamus occurs. Such responses also are known to follow a number of CNS diseases, including epidemic influenza (and encephalitis), which may destroy neurons in the limbic system and hypothalamus. Other behavioral changes following some viral infections have been determined, and are recognized as discrete clinical entities. Examples of these are the postviral and chronic fatigue syndromes that may have a viral etiology (Goldstein, 1993). Perhaps the best studied virus known to affect higher CNS functions is the human immunodeficiency virus (HIV) associated with acquired immune deficiency syndrome (AIDS). One of the first signs of infection with HIV may be cognitive deficits (Poutiainen, Livanainen, Elovaara, Valle, & Lahdevirta, 1988), that are a prelude to the debilitating dementia often associated with AIDS (Price et al., 1988). Thus, psychosocial factors may influence the development of HIV infection to AIDS, with the direction of effects being either positive or negative (Solomon, Kemeny, & Temoshok, 1991).

XIII. Concluding Comment

A large volume of material from diverse standpoints has been presented in this chapter. However, only relatively few of the salient points have been addressed. What should be clear from the observations presented is that the complex of interactions between sociological, psychological, neurological, endocrine, and immunological factors requires substantial further study to manage the diseases that affect us all in our society today, and in that of the future.

Dedication

This review is written in memory of my father, Ernest Saphier, a physician who inherently recognized the importance of principles of psychoneuroimmunology in clinical practice.

Acknowledgments

The author is grateful to the following people, without whose advice, assistance, and collaboration these studies would not have been possible: Oded Abramsky, Bettina Birmanns, Nissim Conforti,

Shaul Feldman, Ana Itzik, Daphne Kidron, Ada Maimon, Gil Mor, Haim Ovadia, Etty Reinhartz, Jutta Rosenthal, and Joseph Weidenfeld in the Department of Neurology, Hadassah University Hospital and Hadassah-Hebrew University, Jerusalem, Israel. Special thanks also go to Yigal Burstein, Marit Pecht, and Nathan Trainin of the Weizmann Institute for Science, Rehovot, Israel, for the supply of thymic humoral factor (THF). The work was supported by the Lena P. Harvey Endowment Fund for Neurological Research, the Etty and Miguel Meilichson Neuroscience Research Fund, the Reichman Foundation, and by a grant from the Joint Research Fund of the Hebrew University and Hadassah, the National Israeli Council for Research and Development, and the European Economic Community. The cooperation and advice of Adrian J. Dunn in the Department of Pharmacology and Therapeutics, Louisiana State University Medical Center, Shreveport, is also much appreciated. I particularly thank H. Eduardo Chuluyan, Glenn E. Farrar, and Jon E. Welch for their collaboration in the α-IFN studies, which were partly funded by an Institutional Research Grant from the American Cancer Society (# 00171).

References

Adams, F., Quesada, J. R., & Gutterman, J. V. (1984). Neuropsychiatric manifestations of human leukocyte interferon therapy in patients with cancer. *Journal of the American Medical Association, 252*, 938–941.

Ader, R., & Cohen, N. (1975). Behaviorally conditioned immunosuppression. *Psychosomatic Medicine, 37*, 333–340.

Ader, R., Felten, D. L., & Cohen, N. (Eds.) (1991). *Psychoneuroimmunology, 2d ed.* San Diego: Academic Press.

Antoni, F. A. (1986). Hypothalamic control of adrenocorticotropin secretion: Advances since the discovery of 41-residue corticotropin-releasing factor. *Endocrine Reviews, 71*, 351–378.

Beaumont, A., & Hughes, J. (1979). Biology of opioid peptides. *Annual Reviews of Pharmacology and Toxicology, 9*, 245–267.

Berczi, I. (Ed.) (1986). *Pituitary function and immunity.* Boca Raton, Florida: CRC Press.

Berczi, I., & Nagy, E. (1988). The effect of prolactin and growth hormone on hemolymphopoietic tissue and immune function. In I. Berczi & K. Kovacs, (Eds.), *Hormones and immunity* (pp. 145–168). Boston: MTP Press.

Bernton, E. W., Bryant, H. U., & Holaday, J. W. (1991). Prolactin and immune function. In R. Ader, D. L. Felten, & N. Cohen, (Eds.), *Psychoneuroimmunology, 2d ed.* (pp. 403–428). San Diego: Academic Press.

Berkenbosch, F., van Oers, J., del Rey, A., Tilders, F., & Besedovsky, H. (1987). Corticotropin releasing factor producing neurons in the rat activated by interleukin-1. *Science, 238*, 524–526.

Besedovsky, H., & del Rey, A. (1987). Neuroendocrine and metabolic responses induced by interleukin-1. *Journal of Neuroscience Research, 18*, 172–178.

Besedovsky, H., del Rey, A., Sorkin. E., da Prada, M., Burri, R., & Honegger, C. (1983). The immune response evokes changes in brain noradrenergic neurons. *Science, 221*, 564–566.

Besedovsky, H., del Rey, A., Sorkin, E., & Dinarello, C. A. (1986). Immunoregulatory feedback between interleukin-1 and glucocorticoid hormones. *Science, 233*, 652–654.

Besedovsky, H., & Sorkin, E. (1977). Hormonal control of immune processes. *Endocrinology, 21*, 504–513.

Besedovsky, H., & Sorkin, E. (1979). Network of immune-neuroendocrine interactions. *Clinical and Experimental Immunology, 27*, 1–12.

Besedovsky, H., Sorkin, E., Felix, D., & Haas, H. (1977). Hypothalamic changes during the immune response. *European Journal of Immunology, 7*, 325–328.

Besedovsky, H., Sorkin, E., Keller, M., & Muller, J. (1975). Changes in blood hormone levels during immune-response. *Proceedings of the Society for Experimental Biology and Medicine, 150,* 466–470.

Birmanns, B., Saphier, D., & Abramsky, O. (1990). α-Interferon modifies cortical EEG activity: Dose-dependency and antagonism by naloxone. *Journal of the Neurological Sciences, 100,* 22–26.

Blalock, J. E., Harbour-McMenamin, D., & Smith, E. M. (1985). Peptide hormones shared by the neuroendocrine and immunologic systems. *Journal of Immunology, 135,* 858–861.

Blalock, J. E., & Smith, E. M. (1980). Human leukocyte interferon—Structural and biological relatedness to adrenocorticotropic hormone and endorphins. *Proceedings of the National Academy of Sciences, U.S.A., 77,* 5972–5974.

Blalock, J. E., & Smith, E. M. (1981). Human leukocyte interferon (HuIFN-alpha): Potent endorphinlike opioid activity. *Biochemical and Biophysical Research Communications, 101,* 472–478.

Blatteis, C. M. (1985). Central nervous system effects of interleukin-1. In J. Kluger, J. J. Oppenheim, & M. C. Powanda (Eds.), *The physiologic, metabolic, and immunologic actions of interleukin-1* (pp. 107–120). New York: Alan R. Liss.

Bluthé, R. M., Dantzer, R., & Kelly, K. W. (1989). CRF is not involved in the behavioral effects of peripherally injected interleukin-1 in the rat. *Neuroscience Research Communications, 5,* 149–154.

Bodmer, S., Tobler, S., Borbely, A., & Fontana, A. (1985). Interleukin-1 and the central nervous system. In J. Kluger, J. J. Oppenheimer, & M. C. Powanda (Eds.), *The physiologic, metabolic, and immunologic actions of interleukin-1* (pp. 143–149). New York. Alan R. Liss.

Bulloch, K. (1985). Neuroanatomy of lymphoid tissue: A review. In R. Guillemin, M. Cohn, & T. Melnechuk (Eds.), *Neural modulation of immunity* (pp. 111–141). New York: Raven Press.

Calabrese, L. H., Kleiner, S. M., Barna, B. P., Skibinski, C. I., Kirkendale, D. T., Lahita, R. G., & Lombardo, J. A. (1989). The effects of anabolic steroids and weight training on the human immune response. *Medical Science of Sports Exercise, 21,* 386–392.

Clamen, H. N. (1975). How corticosteroids work. *Journal of Allergy and Clinical Immunology, 55,* 145–151.

Contreras, R. J. (1977). Changes in gustatory nerve discharges with sodium deficiency: A single unit analysis. *Brain Research, 121,* 373–378.

Crnic, L. S., & Pizer, L. I. (1988). Behavioral effects of neonatal herpes simplex type 1 infection of mice. *Neurotoxicology and Teratology, 10,* 381–386.

Cross, R. J., Markesbery, W. R., Brooks, W. H., & Roszman, T. L. (1984). Hypothalamic immune interactions: Neuromodulation of natural killer cell activity by lesioning of the anterior hypothalamus. *Immunology, 51,* 399–405.

Cupps, T. R., & Fanci, A. S. (1982). Corticosteroid mediated immunoregulation in man. *Immunology Reviews, 68,* 133–155.

Dafny, N. (1983). Interferon modifies EEG and EEG-like activity recorded from sensory, motor, and limbic system structures in freely behaving rats. *Neurotoxicology, 4,* 235–240.

Dafny, N., Prieto-Gomez, R., & Reyes-Vazquez, C. (1985). Does the immune system communicate with the central nervous system? Interferon modifies central nervous activity. *Neuroimmunology, 9,* 1–12.

de Wied, D. (1984). Neurohypophyseal hormone influences on learning and memory processes. In G. Lynch, J. L. McGaugh, & N. M. Weinberger (Eds.), *Neurobiology of learning and memory* (pp. 289–312). New York: Guilford Press.

Dinarello, C. A. (1984). Interleukin-1 and the pathogenesis of the acute-phase response. *New England Journal of Medicine, 311,* 1413–1418.

Dinarello, C. A., Renfer, L., & Wolff, S. M. (1977). Human leukocytic pyrogen: Purification and development of a radioimmunoassay. *Proceedings of the National Academy of Sciences, U.S.A.* *74*, 4624–4627.

Dunn, A. J. (1988). Systemic interleukin-1 administration stimulates hypothalamic norepinephrine metabolism paralleling the increased plasma corticosterone. *Life Sciences, 43,* 429–435.

Dunn, A. J. (1990). Recent advances in psychoneuroimmunology. *Neuroendocrinimmunology, 16,* 1–5.

Dunn, A. J., & Hall, N. R. (1987). Thymosin extracts and lymphokine-containing supernatant fluids stimulate the pituitary-adrenal axis, but not cerebral catecholamine or indoleamine metabolism. *Brain, Behavior, and Immunity, 1,* 113–122.

Dunn, A. J., Powell, M. L., & Gaskin, J. M. (1987). Virus-induced increases in plasma corticosterone. *Science, 238,* 1423–1424.

Dyer, R. G., & Saphier, D. (1981). Electrical activity of antidromically identified tuberoinfundibular neurones during stimulated release of luteinizing hormone and prolactin in proestrous rats. *Journal of Endocrinology, 89,* 35–44.

Farrar, W. L., Kilian, P. L., Ruff, M. R., Hill, J. M., & Pert, C. B. (1987). Visualization and characterization of interleukin-1 receptors in brain. *Journal of Immunology, 139,* 459–463.

Feldman, S., Rachmilewitz, E. A., & Izak, G. (1966). The effect of central nervous system stimulation on erythropoiesis in rats with chronically implanted electrodes. *Journal of Laboratory and Clinical Medicine, 67,* 713–725.

Feldman, S., & Saphier, D. (1989). Extrahypothalamic neural afferents and the role of neurotransmitters in the regulation of adrenocortical secretion. In F. C. Rose (Ed.), *The control of the hypothalamo-pituitary-adrenocortical axis* (pp. 297–316). Madison, Connecticut: International Universities Press.

Felten, S. Y., & Felten, D. L. (1991). Innervation of lymphoid tissue. In R. Ader, D. L. Felten, & N. Cohen (Eds.), *Psychoneuroimmunology,* 2d ed. (pp. 27–69). San Diego: Academic Press.

Felten, D. L., Felten, S. Y., Carlson, S. L., Olschowka, J. A., & Livnat, S. (1985). Noradrenergic and peptidergic innervation and lymphoid tissue. *Journal of Immunology, 135,* 755–765.

Fink, T., & Weihe, E. (1988). Multiple neuropeptides in nerves supplying mammalian lymph nodes: Messenger candidates for sensory and autonomic neuroimmunomodulation? *Neuroscience Letters, 90,* 39–44.

Fontana, A., Kristensen, F., Duba, R., Gemsa, D., & Weber, E. (1982). Production of prostaglandin E and an interleukin-1 like factor by cultured astrocytes and C-6 glioma cells. *Journal of Immunology, 13,* 685–689.

Friedman, R. M., & Vogel, S. N. (1983). Interferon with special emphasis on the immune system. *Advances in Immunology, 341,* 97–138.

Garcia, J., Hankins, W. G., & Rusiniak, K. W. (1974). Behavioral regulation of the milieu interne in man and rat. *Science, 185,* 824–831.

Geenen, V., Legros, J.-J., Franchimont, P., defresne, M.-P., Boniver, J., Ivell, R., & Richter, D. (1987). The thymus as a neuroendocrine organ: Synthesis of vasopressin and oxytocin in human thymic epithelium. *Annals of the New York Academy of Sciences, 496,* 56–66.

Geschwind, N., & Behan, P. (1982). Left-handedness: Association with immune disease, migraine and developmental learning disorder. *Proceedings of the National Academy of Sciences, U.S.A., 79,* 5097–5100.

Ghali, W. M., Abdel-Rahman, S., & Nagib, M. (1980). Intrinsic innervation and vasculature of pre- and post-natal human thymus. *Acta Anatomica, 108,* 115–123.

Gisler, R. H. (1974). Stress and the hormonal regulation of the immune response in mice. *Psychotherapeutic and Psychosomatic Medicine, 23,* 192–208.

Glick, S. D. (Ed.) (1985). *Cerebral lateralization in nonhuman species.* Orlando, Florida: Academic Press.

Goldstein, J. A. (1993). Chronic fatigue syndromes: The limbic hypothesis. *The Hayworth Library of the Medical Neurobiology of Somatic Disorders, Volume 1*, The Haworth Medical Press, New York.

Grigoriev, V. A. (1981). Dynamics of DC potential of hypothalamic structures in early terms of immune reaction development. *Physiological Journal of the U.S.S.R., 67*, 463–467.

Grossman, C. J. (1984). Regulation of the immune system by sex steroids. *Endocrine Reviews, 5*, 435–455.

Guillaumin, J. M., Lepape, A., & Renoux, G. (1986). Fate and distribution of radioactive sodium diethyldithiocarbamate (Imuthiol®) in the mouse. *International Journal of Immunopharmacology, 8*, 859–865.

Hall, N. R., & Goldstein, A. L. (1981). Neurotransmitters and the immune system. In R. Ader (Ed.), *Psychoneuroimmunology* (pp. 521–544). New York: Academic Press.

Hall, N. R., McGillis, J. P., Spangelo, D. L., Healy, D. L., Chrousos, G. P., Schalte, R. M., & Goldstein, A. L. (1985). Thymic hormone effects on the brain and neuroendocrine circuits. In A. Guillemin, M. Cohn, & T. Melnechuk (Eds.), *Neural modulation of immunity* (pp. 179–193). New York: Raven Press.

Hill, L. E. (1930). *Philosophy of a biologist*. London: Arnold.

Hohman, L. B. (1922). Post-encephalitic behavior disorders in children. *Johns Hopkins Hospital Bulletin, 380*, 372–375.

Hooper, J. A., McDaniel, M. C., Thurman, G. B., Cohen, G. H., Schulof, R. S., & Goldstein, A. L. (1975). Purification and properties of bovine thymosin. *Annals of the New York Academy of Sciences, 249*, 125–144.

Isaacson, R. L. (1982). *The limbic system*. New York: Plenum Press.

Jankovic, B. D., & Isakovic, K. (1973). Neuroendocrine correlates of immune response. I. Effects of brain lesions on antibody production. Arthus reactivity and delayed hypersensitivity in the rat. *International Archives of Allergy and Applied Immunology, 45*, 360–372.

Johnson, H. M., Farrar, W. L., & Torres, B. A. (1982). Vasopressin replacement of interleukin 2 requirement in γ-interferon production: Lymphokine activity of a neuroendocrine hormone. *Journal of Immunology, 129*, 983–986.

Kanda, R. (1959). Studies of the regulation centre on promotion of antibody. II. On the migration and relation of normal precipitin antibody and leucocyte in the peripheral blood by electrical stimuli in the hypothalamus of the rabbit. *Japan Journal of Bacteriology, 14*, 542–545.

Katsuura, G., Gottschall, P. E., Dahl, R. R., & Arimura, A. (1988). Adrenocorticotropin release induced by intracerebroventricular injection of recombinant human interleukin-1 in rats: Possible involvement of prostaglandin. *Endocrinology, 122*, 1773–1779.

Kawate, T., Abo, T., Hinuma, S., & Kumagai, K. (1981). Studies on the bioperiodicity of the immune response, covariations of murine T and B cells and a role of corticosteroid. *Journal of Immunology, 126*, 1364–1367.

Kazimierczak, W., & Diamant, B. (1978). Mechanisms of histamine release in anaphylactic and anaphylactoid reactions. *Progress in Allergy, 24*, 295–365.

Kelley, K. W. (1989). Growth hormone, lymphocytes and macrophages. *Biochemical Pharmacology, 38*, 705–713.

Kidron, D., Saphier, D., Ovadia, H., Weidenfeld, J., & Abramsky, O. (1989). Central administration of immunomodulatory factors alters neural activity and adrenocortical secretion. *Brain, Behavior, and Immunity, 3*, 15–27.

Kirkwood, J. M., & Ernstoff, M. S. (1984). Interferons in the treatment of human cancer. *Journal of Clinical Oncology, 2*, 336–352.

Korneva, E. R. (1987). Electrophysiological analysis of brain reactions to antigen. *Annals of the New York Academy of Sciences, 496*, 318–337.

Korneva, E. R., & Khay, L. M. (1963). Influence of hypothalamic part destruction on the immuno-genesis process (in Russian). *Physiological Journal of the U.S.S.R., 49,* 42–48.

Krueger, J. M., Walter, J., Dinarello, C. A., Wolff, S. M., & Chedid, L. (1984). Sleep-promoting effects of endogenous pyrogen (interleukin-1). *American Journal of Physiology, 246,* 994–999.

Kudoh, G., Hoski, K., & Murakami, T. (1979). Fluorescence microscopic and enzyme histochemical studies of the innervation of the human spleen. *Archivum Histologicum Japonicum, 42,* 169–180.

Leibowitz, S. F., Weiss, G. F., Yee, F., & Tretter, J. B. (1985). Noradrenergic innervation of the paraventricular nucleus: Specific role in control of carbohydrate ingestion. *Brain Research Bulletin, 14,* 561–567.

Livnat, S., Felten, S. Y., Carlson, S. L., Bellinger, D. L., & Felten, D. L. (1985). Involvement of peripheral and central catecholamine systems in neural-immune interactions. *Journal of Neuroimmunology, 10,* 5–30.

McCruden, A. B., & Stimson, W. H. (1991). Sex hormones and immune function. In R. Ader, D. L. Felten, & N. Cohen (Eds.), *Psychoneuroimmunology,* 2d ed. (pp. 475–493). San Diego: Academic Press.

MacLean, D., & Reichlin, S. (1981). Neuroendocrinology and the immune process. In R. Ader (Ed.), *Psychoneuroimmunology* (pp. 475–520). New York: Academic Press.

Makara, G. B. (1985). Mechanisms by which stressful stimuli activate the pituitary-adrenal system. *Federation Proceedings, 44,* 149–154.

Metal'nikov, S., & Chorine, V. (1926). The role of conditioned reflexes in immunity. *Annals of the Pasteur Institute, 40,* 893–900.

Monjan, A. A., & Collector, M. I. (1977). Stress-induced modulation of the immune response. *Science, 196,* 307–308.

Monnier, M., Sauer, R., & Hatt, A. M. (1970). The activating effect of histamine on the central nervous system. *International Review of Neurobiology, 12,* 265–305.

Montgomery, D., Zukoski, C., Shah, G., Buckley, A., Pacholczyk, T., & Russell, D. (1987). Con-A stimulated murine splenocytes produce a factor with prolactin-like bioactivity and immuno-reactivity. *Biochemical and Biophysical Research Communications, 145,* 692–698.

Mrazek, D. A., & Klinnert, M. (1988). The psychological and psychiatric evaluation of asthmatic children. In P. Karoly (Ed.), *Handbook of child health assessment* (pp. 410–433). New York: Wiley.

Munck, A., & Guyre, P. M. (1986). Glucocorticoid physiology, pharmacology, and stress. In G. P. Chrousos, D. L. Loriaux, & M. B. Lipsett (Eds.), *Steroid hormone resistance* (pp. 81–96). New York: Plenum.

Nakashima, T., Maori, T., Kuriyama, K., & Kiyohara, T. (1987). Naloxone blocks the interferon-α induced changes in hypothalamic neuronal activity. *Neuroscience Letters, 82,* 332–336.

Ovadia, H., Abramsky, O., Barak, V., Conforti, N., Saphier, D., & Weidenfeld, J. (1989). Effect of interleukin-1 on adrenocortical activity in intact and hypothalamic deafferentated male rats. *Experimental Brain Research, 76,* 246–249.

Pediconi, M. F., & Rodriguez de Turco, E. B. (1984). Free fatty acid content and release kinetics as manifestation of cerebral lateralization in mouse brain. *Journal of Neurochemistry, 43,* 1–7.

Pierpaoli, W., & Maestroni, G. J. M. (1988). Neuroimmunomodulation: Some recent views and findings. *International Journal of Neuroscience, 39,* 165–175.

Plotnikoff, N. P., Miller, G. C., & Murgo, A. J. (1982). Enkephalins-endorphins: Immunomodulators in mice. *International Journal of Immunopharmacology, 4,* 366–367.

Plotnikoff, N. P., Murgo, A. J., Miller, G. C., Corder, C. N., & Faith, R. E. (1985). Enkephalins: Immunomodulators. *Federation Proceedings, 44,* 118–122.

Plotsky, P. M. (1985). Hypophyseotropic regulation of adenohypophyseal adrenocorticotropin secretion. *Federation Proceedings, 44,* 207–214.

Pollard, H., & Schwartz, J.-C. (1987). Histamine neuronal pathways and their functions. *Trends in Neurosciences, 10,* 86–69.

Poutiainen, E., Livanainen, M., Elovaara, I., Valle, S.-L., & Lahdevirta, J. (1988). Cognitive changes as early signs of HIV infection. *Acta Neurologica Scandinavica, 78,* 49–52.

Price, R. W., Brew, B., Sidtis, J., Rosenblum, M., Schek, A. C., & Cleary, P. (1988). The brain in AIDS: Central nervous system HIV-1 infection and AIDS dementia complex. *Science, 239,* 586–592.

Rebar, R. W., Miyake, A., Low, T. L. K., & Goldstein, A. L. (1981). Thymosin stimulates secretion of luteinizing hormone-releasing factor. *Science, 214,* 669–671.

Renoux, G., & Biziere, K. (1991). Neocortex lateralization of immune function and of the activities of Imuthiol, a T-cell-specific immunopotentiator. In R. Ader, D. L. Felten, & N. Cohen (Eds.), *Psychoneuroimmunology,* 2d ed. (pp. 127–147). San Diego: Academic Press.

Richter, C. P. (1942). Total self regulatory functions in animals and human beings. *Harvey Lectures, 38,* 63–103.

Roberts, F., & Calcutt, C. R. (1983). Commentary: Histamine and the hypothalamus. *Neuroscience, 9,* 721–739.

Rohatiner, A. Z. S., Prior, P., Burton, A., Balkwill, F., & Lister, T. A. (1985). Central nervous system toxicity of interferon. *Progress in Experimental Tumor Research, 29,* 197–202.

Romeo, H. E., Colombo, L. L., Esquifino, A. I., Rosenstein, R. E., Chuluyan, H. E., & Cardinali, D. P. (1991). Slower growth of tumours in sympathetically denervated murine skin. *Journal of the Autonomic Nervous System, 32,* 159–164.

Roosth, J., Pollard, R. D., Brown, S. L., & Meyer, W. J., III. (1986). Cortisol stimulation by re-combinant interferon-α_2. *Journal of Neuroimmunology, 12,* 311–316.

Rosen, H., Behar, O., Abramsky, O., & Ovadia, H. (1989). Regulated expression of proenkephalin A in normal lymphocytes. *Journal of Immunology, 14,* 1–5.

Roszman, T. L., Cross, R. J., Brooks, W. H., & Marksbery, W. R. (1985). Neuroimmunomodulation. Effects of neural lesions on cellular immunity. In R. Guillemin, M. Cohn, & T. Melnechuk (Eds.), *Neural modulation of immunity* (pp. 95–109). New York: Raven Press.

Rothwell, N. J. (1989). Involvement of CRF in the pyrogenic and thermogenic actions of interleukin-1β. *American Journal of Physiology, 256,* E111–E115.

Saphier, D. (1985). Nucleus accumbens and preoptic area stimulation: Tuberoinfundibular single unit responses: Modulation of electrical activity and gonadotrophin secretion. *Experimental Brain Research, 57,* 400–403.

Saphier, D. (1989). Neurophysiological and endocrine consequences of immune activity. *Psychoneuroendocrinology, 14,* 63–87.

Saphier, D. (1990). Vasopressin-secreting neurones of the paraventricular nucleus respond to oro-pharyngeal application of hypertonic saline. *Neuroscience Letters, 109,* 97–101.

Saphier, D. (1992). Electrophysiological effects of cytokines in relation to arousal and adrenocortical secretion. In N. J. Rothwell & R. Dantzer (Eds.), *Interleukin 1 in the brain* (pp. 51–73). Oxford, UK: Pergamon Press.

Saphier, D., Welch, J. E., & Chuluyan, H. E. (1993). α-Interferon inhibits adrenocortical secretion via μ₁-opioid receptors in the rat. *European Journal of Pharmacology, 236,* 186–194.

Saphier, D., Abramsky, O., Mor, G., & Ovadia, H. (1987a). A neurophysiological correlate of an immune response. *Annals of the New York Academy of Sciences, 496,* 354–359.

Saphier, D., Abramsky, O., Mor, G., & Ovadia, H. (1987b). Multiunit electrical in conscious rats during an immune response. *Brain, Behavior, and Immunity, 1,* 40–51.

Saphier, D., & Feldman, S. (1985). Effects of neural stimuli on paraventricular nucleus neurones. *Brain Research Bulletin, 14,* 401–408.

Saphier, D., & Feldman, S. (1986). Effects of stimulation of the preoptic area on hypothalamic paraventricular nucleus unit activity and corticosterone secretion in freely moving rats. *Neuroendocrinology, 42,* 167–173.

Saphier, D., & Feldman, S. (1988). Iontophoretic application of glucocorticoids inhibits identified neurones in the rat paraventricular nucleus. *Brain Research, 453,* 183–190.

Saphier, D., Kidron, D., Ovadia, H., Trainin, N., Pecht, M., Burstein, Y., & Abramsky, O. (1988). Neurophysiological changes in brain following central administration of immunomodulatory factors. *Israel Journal of Medical Sciences, 24,* 261–263.

Saphier, D., & Ovadia, H. (1990). Selective facilitation of putative corticotropin-releasing factor-secreting neurones by interleukin-1. *Neuroscience Letters, 114,* 283–288.

Saphier, D., Ovadia, H., & Abramsky, O. (1990). Neural responses to antigenic challenges and immunomodulatory factors. *Yale Journal of Biology and Medicine, 63,* 109–119.

Schuurs, A. H. W. M., & Verheul, H. A. M. (1990). Effects of gender and sex steroids on the immune response. *Journal of Steroid Biochemistry, 35,* 157–172.

Seligman, M. E. P., & Hager, J. L. (1972). *Biological boundaries of learning.* Englewood Cliffs, New Jersey: Prentice-Hall.

Singh, V. K., Warren, R. P., White, E. D., & Leu, S.-J. C. (1990). Corticotropin-releasing factor-induced stimulation of immune functions. *Annals of the New York Academy of Sciences, 594,* 416–419.

Smith, E. M., & Blalock, J. E. (1981). Human lymphocyte production of corticotropin and endorphin-like substances: Association with leukocyte interferon. *Proceedings of the National Academy of Sciences, U.S.A., 78,* 7530–7534.

Smith, E. M., Meyer, W. J., & Blalock, J. E. (1982). Virus-induced corticosterone in hypophysectomized mice: A possible lymphoid adrenal axis. *Science, 218,* 1311–1312.

Smith, G. H., & Salinger, R. (1933). Hypersensitiveness and the conditioned reflex. *Yale Journal of Biology and Medicine, 5,* 387–402.

Solomon, G. F. (1969). Stress and antibody response in rats. *International Archives of Allergy, 35,* 97–104.

Solomon, G. F. (1981). Immunologic abnormalities in mental illness. In R. Ader (Ed.), *Psychoneuroimmunology* (pp. 259–278). New York: Academic Press.

Solomon, G. F. (1987). Psychoneuroimmunology: Interactions between central nervous system and immune system. *Journal of Neuroscience Research, 18,* 1–9.

Solomon, G. F., Kemeny, M. E., & Temoshok, L. (1991). Psychoneuroimmunologic aspects of human immunodeficiency virus infection. In R. Ader, D. L. Felten, & N. Cohen (Eds.), *Psychoneuroimmunology,* 2d ed. (pp. 1081–1113). San Diego: Academic Press.

Solomon, G. F., Levine, S., & Kraft, K. J. (1968). Early experience and immunity. *Nature (London), 220,* 821–822.

Solomon, G. F., & Moos, R. H. (1964). Emotions, immunity, and disease. A speculative theoretical integration. *Archives of General Psychiatry, 11,* 657–674.

Solomon, G. F., Moos, R. H., Fessel, W. J., & Morgan, E. E. (1966). Globulins and behavior in schizophrenia. *International Journal of Neuropsychiatry, 2,* 20–26.

Stein, M., Schleyer, S. J., & Keller, S. E. (1981). Hypothalamic influences on immune responses. In R. Ader (Ed.), *Psychoneuroimmunology* (pp. 429–448). New York: Academic Press.

Steiner, I., Wirguin, I., Morag, A., & Abramsky, O. (1989). Intraventricular interferon treatment for subacute sclerosing panencephalitis. *Journal of Child Neurology, 4,* 20–24.

Swanson, L. W., Sawchenko, P. E., Rivier, J., & Vale, W. (1983). Organization of ovine corticotropin-releasing factor immunoreactive cells and fibers in the rat brain: An immunohistochemical study. *Neuroendocrinology, 36,* 165–186.

Tomer, Y., & Shoenfeld, Y. (1988). The significance of natural autoantibodies. *Immunological Investigation, 17,* 389–424.

Trainin, N., Handzel, Z. T., & Pecht, M. (1985). Biological and clinical properties of THF. *Thymus, 7*, 137–150.

Tyrey, L., & Nalbandov, A. V. (1972). Influence of anterior hypothalamic lesions on circulating antibody titers in the rat. *American Journal of Physiology, 222*, 179–185.

Uehara, A., Sekiya, C., Takasugi, Y., Namiki, M., & Arimura, A. (1989). Anorexia induced by interleukin-1: Involvement of corticotropin-releasing factor. *American Journal of Physiology, 257*, R613–R617.

Vincent, J. D., Arnauld, E., & Bioulac, B. (1972). Activity of osmosensitive single cells in the hypothalamus of the behaving monkey during drinking. *Brain Research, 44*, 371–384.

Wakerley, J. B., Poulain, D. A., & Brown, D. (1978). Comparison of firing patterns in oxytocin- and vasopressin-releasing neurones during progressive dehydration. *Brain Research, 148*, 425–440.

Weidenfeld, J., Abramsky, O., & Ovadia, H. (1989). Evidence for the involvement of the central adrenergic system in interleukin 1-induced adrenocortical response. *Neuropharmacology, 28*, 1411–1414.

Williams, C. N., & Schupf, N. (1987). The neuropharmacology of immune complex activity in the rat hypothalamus. *Annals of the New York Academy of Sciences, 496*, 250–263.

Yamashita, T., Kumazawa, H., Kozuki, K., Amano, H., Tomoda, K., & Kumazawa, T. (1984). Autonomic nervous system in human palatine tonsil. *Acta Oto-Laryngologica, 416*, 63–71.

Zurawski, G., Benedik, M., Kamb, B. J., Abrams, J. S., Zurawski, S. M., & Lee, F. D. (1986). Activation of mouse T-helper cells induces abundant preproenkephalin MRNA synthesis. *Science, 232*, 772–775.

Oxytocin and the Neuroendocrine Basis of Affiliation

Thomas R. Insel

Laboratory of Neurophysiology
National Institute of Mental Health
Poolesville, Maryland 20837

I. Introduction

Considerable literature exists on the neuroendocrine aspects of social separation and loss. Studies of the hypothalamic–pituitary–adrenal axis (Mendoza, Smotherman, Miner, Kaplan, & Levine, 1978; Hennessy & Ritchey, 1987; Levine & Wiener, 1988), central opioid pathways (Panksepp, Herman, Conner, Bishop, & Scott, 1978; Kalin, Shelton, & Barksdale, 1988), and brain catecholamine concentrations (Kraemer, Ebert, Lake, & McKinney, 1984; Tamborski, Lucott, & Hennessy, 1990; Wiener, Bayart, Faull, & Levine, 1990) all have contributed to developing a picture of the neural responses associated with social separation. In contrast, we know very little about the neural substrates of social attachment. From a behavioral perspective, one might presume that social attachment is simply the inverse of social separation. Indeed, many of the neuroendocrine systems activated by separation are inhibited by social contact (Stanton, Wallstrom, & Levine, 1987). However, if one considers the onset or initiation of social attachment rather than simply the end point of social contact, then the process has an entire set of behavioral and neural factors that the literature on social separation only begins to address.

This chapter focuses on a specific neuroendocrine mechanism involved in the initiation of social bonds. The term "affiliation" is used to describe this process. Although affiliation in an ethologic context usually refers to huddling and grooming behavior (Hinde, 1983), for the purpose of this chapter the term is stretched to include several other behaviors involving nurturing (nonaggressive) social interactions. At different stages of development, affiliation applies to quite different forms of social interaction. For adults, affiliative behaviors include parental care, sexual interaction, pair bounding, allogrooming, and—from a social organization perspective—monogamy and mutual defense. For immature individuals, affiliative behavior might be defined operationally as play, huddling, or reciprocal grooming. Infants of virtually all mammalian species express a drive for social contact by emitting high pitched cries that elicit parental retrieval (Newman, 1988).

Clearly, this list of behaviors is too diverse to be reduced to any single neuroendocrine factor or any specific neural pathway. Indeed, for each of these behaviors, the array of sensory inputs and motor outputs is so complex that any serious neuroscientist interested in describing the circuitry that underlies an affiliative behavior soon would be emitting high pitched cries and huddling by himself in a laboratory corner. However, a moment's reflection about the range of behaviors described as "affiliative" may provide some hope that factors influencing that "great intermediate net" between sensory inputs and motor outputs can be found. Clues to their investigation may be found even in the way these behaviors are described. For instance, the developmental categorization suggests that the emergence of gonadal steroids with sexual maturity may be an important factor in the expression of affiliation. Understanding the neural correlates of puberty—in other words, investigating the effects of gonadal steroids on various neural pathways—may elucidate systems involved in processes such as pair bonding. The ways in which species differ with respect to so many of these affiliative behaviors also may provide important clues. Monogamous species may display high levels of separation distress as infants (Shapiro & Insel, 1990a), marked huddling or side-by-side behavior as juveniles (Dewsbury, 1988), and pair bonding with intense paternal (as well as to maternal) care as adults (Kleinman, 1977). The comparative study of monogamous and polygamous (or highly social versus asocial) species thus may provide another mechanism for investigating the neural substrates of apparently hopelessly complicated behaviors.

This chapter reviews studies of oxytocin as a neuropeptide involved in affiliation. In addition to studies of the central effects of oxytocin, the role of gonadal steroids in the regulation of these actions and comparative studies of oxytocin in monogamous and polygamous rodents support the hypothesis that this neuropeptide may provide a model for understanding a neuroendocrine basis for social attachment.

II. Oxytocin and the Initiation of Affiliation

Oxytocin (OT) is a 9-amino-acid peptide that evolved with the emergence of mammals (Archer, 1974). Perhaps the fact that the best known functions of OT, milk ejection and uterine contraction for viviparity, are two prototypic mammalian traits is no coincidence. OT is synthesized primarily in two hypothalamic nuclei, the paraventricular nucleus (PVN) and the supraoptic nucleus (SON). OT is cleaved from a precursor molecule, neurophysin, that is present both in oxytocin cell bodies and in processes (Brownstein, Russell, & Gainer, 1980). The traditional description of oxytocin focuses on its synthesis in magnocellular secretory neurons of the PVN and SON that project to the posterior pituitary. This view has been extended to include a dense network of nonpituitary projections (Buijs, 1978; Swanson & Kuypers, 1980; Sofroniew & Weindl, 1981). OT fibers, arising principally from small cells (parvocellular neurons) in the PVN, have been found in many areas of the limbic system as well as in several autonomic centers in the brainstem. OT may be, in fact, the predominant PVN peptide with autonomic projections; terminals are evident in both sympathetic and parasympathetic centers (Sawchencko & Swanson, 1982). Ultrastructural studies have demonstrated that these "extrahypothalamic" OT projections make classical synaptic contacts (Voorn & Buijs, 1983) from which OT is released after potassium- or veratridine-induced depolarization (Buijs & van Heerikhuize, 1982). This central OT pool may be considered independent of neurohypophyseal OT release, at least in the sense that cerebrospinal fluid (CSF) and plasma OT responses to various stimuli are not correlated (Perlow et al., 1982; Jones, Robinson, & Harris, 1983; Kendrick, Keverne, Baldwin, & Sharman, 1986).

What is the role of OT in the central nervous system (CNS)? Table 1 summarizes results from diverse studies on the effects of central administration of OT, OT antagonists or antisera, and lesions of the PVN. As can be seen from this profile of behaviors, a pattern exists that matches reasonably well with what has just been described as the realm of affiliative behaviors. Note that other reports of central OT effects, particularly effects on cognitive behavior, are not shown. The point is not that OT affects only affiliative behaviors, but that a pattern of behavioral effects does appear to be consistent. To give the reader a better sense of the nature of these effects, the literature supporting Table 1 is summarized briefly.

A. Parental Behavior

One of the first attempts to investigate OT effects in the CNS tested the hypothesis that OT in the brain would induce maternal behavior correlated with OT effects on labor and lactation. Indeed, during lactation in the rat, OT mRNA increases as much as 3-fold in the SON (van Tol, Bolwerk, Liu, & Burbach, 1988) and 10-fold in the PVN (Lightman & Young, 1987). Plasma concentrations of OT increase during nursing, in association with synchronous firing of PVN and SON cells and the subsequent pulsatile release of OT from the neurohypophysis (Lincoln & Russell, 1985). An ultrastructural reorganization of OT cells in the SON has been reported during lactation, providing an anatomical basis for the enhanced pulsatile release of OT with suckling (Theodosis, Montaguese, Rodriguez, Vincent, &

TABLE I
Central Oxytocin and Social Behaviors[a,b]

Behavior	OT administration	OT antagonist	PVN lx	Species
Maternal behavior	+++	---	---	Rat, sheep, mouse
Sex behavior				
Female	+++	---	?	Rat
Male	++++	---	---	Rat
Infant ultrasounds	---	0	?	Rat
Grooming	++++	?	?	Rat, mouse
Pair-bond formation	+++	0	?	Prairie vole

[a]References are given in the text.
[b]OT, Oxytocin; PVN, Paraventricular nucleus of the hypothalamus; +, increase; −, decrease; 0, no effect; ?, no data available.

Poulain, 1986; Theodosis & Poulain, 1987; Yang, & Hatton, 1988). All these changes may reflect alterations to subserve the peripheral release of the peptide for milk ejection without providing any evidence that OT in the CNS is important for maternal behavior.

Several Pedersen & Prange, 1979; Pedersen, Ascher, Monroe, & Prange, 1982; Fahrbach, Morell, & Pfaff, 1984; Wamboldt & Insel, 1987) but not all (Rubin, Menniti, & Bridges, 1983; Bolwerk & Swanson, 1984) studies have reported that OT given centrally (but not peripherally) to virgin female rats induces full maternal behavior within minutes. Virgin female rats display little interest in infants and, when presented with foster young, will either avoid or cannibalize them (Rosenblatt & Siegel, 1981). At parturition (or following specific steroid regimens to mimic the physiological changes of parturition), a rapid dramatic shift in motivation is seen from a lack of interest to a driven relentless pursuit of nest-building, retrieval, licking, grouping, and protection of pups (Numan, 1988). No other peptide or drug has been shown to induce maternal behavior so quickly in virgin females. However, OT does not work alone. In all studies demonstrating an induction of maternal behavior following central OT administration, the response was dependent on priming with gonadal steroids; no effects of OT were observed in ovariectomized females without estradiol. The sites at which OT might function to induce maternal behavior remain unclear, although the ventral midbrain has been suggested in one preliminary report (Fahrbach, Morrell, & Pfaff, 1985a).

Does OT have a physiological role in the induction of maternal behavior? This question can be answered by blocking central OT pathways using centrally administered antagonists, antisera, or lesions. Studies with all these methods demonstrate that, following either experimentally simulated (Fahrbach, Morrell, & Pfaff, 1985b; Pedersen, Caldwell, & Fort, & Prange, 1985) or natural (van Leengoed, Kerker, & Swanson, 1987; Insel & Harbaugh, 1989) parturition, the onset of maternal behavior can be blocked by OT antagonism. One key feature of these studies is that OT antagonists do not appear to disrupt maternal behavior per se, but block its initiation. The same intervention after parturition, when maternal behavior is established, is without effects.

From these studies in rats, the effects of OT in the CNS appear to influence the initiation of maternal behavior consistent with the role of this peptide in peripheral tissues for the induction of labor and milk ejection. Further evidence for this apparent effect of OT on maternal "motivation" comes from studies in virgin sheep, which resemble nulliparous rats in their normal absence of maternal interest. Vaginal–cervical stimulation, a potent stimulus for both central and peripheral OT release (Kendrick, Keverne, Chapman, & Baldwin, 1988b), induces the rapid onset of maternal behavior in the steroid-primed ewe (Keverne, Levy, Poindron, & Lindsay, 1983). More important, central but not peripheral OT administration increases maternal interest in nulliparous ewes, shifting their behavior toward newborns from avoidance to exploration and caretaking (Kendrick, Keverne, & Baldwin, 1987). At parturition, the concentration of OT in CSF increases to concentrations approximating those found in plasma (Kendrick et al., 1986). Concurrent increases in substantia nigra and olfactory bulb approximate 60 and 30%, respectively (Kendrick, Keverne, Chapman, & Baldwin, 1988a).

B. Reproductive Behavior

Although OT long has been associated with smooth muscle contraction during labor and lactation, the peptide also has a less recognized role in sexual behavior. In human males for instance, plasma OT increases as much as 5-fold with ejaculation (Carmichael et al., 1987; Murphy, Seckl, Burton, Checkley, & Lightman, 1987). Intravenous infusions of OT to male rabbits or rats increase sexual behavior, as measured by decreased time to ejaculation or increased number of ejaculations (Melin & Kihlstrom, 1963; Arletti, Bazzani, Castelli, & Bertolini, 1985; Stoneham, Everitt, Hansen, Lightman, & Todd, 1985; Argiolas, Melis, Stancampiaro, & Gessa, 1989). In female rodents, although peripheral administration of OT does not induce sexual behavior (Rodriguez-Sierra, Crowley, & Komisaruk, 1977), OT mRNA levels increase 1.5- to 2-fold at estrus relative to other stages in the estrous cycle (van Tol et al., 1988). This increase in OT gene expression appears to reflect alterations in OT content and release. Immunoreactive content of OT in the PVN peaks at diestrus (Greer, Caldwell, Johnson, Prange, & Pedersen, 1986), OT pituitary content peaks on the morning of proestrus (Crowley, O'Donahue, George, & Jacobowitz, 1978), and release of peptide into the portal system peaks during the afternoon of proestrus just prior to the onset of sexual receptivity (Sarkar & Gibbs, 1984).

Not only does an OT cycle appear to be associated with female sexual receptivity, but plasma OT concentrations increase with vaginal distension (Roberts & Share, 1968) or mating (in rabbits) (Fuchs, Cubile, & Dawood, 1981). The frequent reports from lactating women that milk ejection accompanies coitus (Fox & Knaggs, 1969; Newton, 1973) suggests that OT is released during sexual activity in human females as well as in males. Although the function of the pituitary release of OT during sexual activity remains unclear, the presence of OT receptors in the contractile tissues of both the male and female genital tract may indicate a role in seminal fluid transport in both sexes (Maggi, Malozowski, Kassis, Guardabasso, & Rodbard, 1987).

Several studies have indicated that OT also may have an important function in the CNS to regulate either the motivation for or the performance of sexual behavior. In the female rat, sexual behavior generally is divided into proceptive (soliciting) and receptive (lordosis) responses, both of which are under the control of gonadal steroids. After low levels of gonadal steroid priming [0.1 µg estradiol benzoate (EB) and 100 µg progesterone], ovariectomized females usually show little proceptive or receptive behavior; central administration of OT (500 ng) induces a 3-fold increase in the time spent in physical contact with the male (Witt & Insel, unpublished observation). At moderate doses of EB (10 µg), OT clearly increases receptive behavior (measured as the amount of lordosis), but only if progesterone is available (Arletti & Bertolini, 1985; Gorzalka & Lester, 1987; Schumacher, Coirini, Frankfurt, & McEwen, 1989). After prolonged priming with EB (3 days), lordosis has been reported to increase following oxytocin (800 ng) even in the absence of progesterone (Caldwell, Prange, & Pedersen, 1986). The neural sites required for OT effects on reproductive behavior appear to include the posterior part of the ventromedial nucleus (VMN) (an area previously implicated in the integration of the lordosis reflex) (Schumacher et al., 1989) and the medial preoptic area (MPOA) (an area showing increased immunoreactive OT following sexual behavior) (Caldwell, Jirikowski, Greer, & Pedersen, 1989).

Although these various experiments indicate that exogenous OT can increase female sexual behavior, none of these results demonstrates that OT is involved physiologically in either proceptive or receptive processes. In preliminary studies, an OT antagonist was ineffective in decreasing lordosis when administered following the onset of sexual receptivity (4 hr postprogesterone). However, as with the antagonist effects on the initiation but not maintenance of maternal behavior, when an OT antagonist was given at the same time as progesterone, prior to the onset of receptivity, lordosis was decreased in a dose-dependent fashion (Witt & Insel, 1991). These results suggest that endogenous OT is important for both proceptive and receptive aspects of sexual behavior in the female rat, consistent with the observations of increased plasma OT just prior to the onset of receptivity (Sarkar & Gibbs, 1984) and peripheral release of OT with sexual behavior (Fuchs,

Cubile, & Dawood, 1981). Further evidence suggesting that OT mediates some aspects of sexual behavior comes from an immunocytochemical study reporting increased OT immunoreactivity in the anterior hypothalamus after sexual activity in ovariectomized females primed with estradiol and progesterone (Caldwell et al., 1989).

Central OT pathways also have been implicated in the mediation of male sex behavior. Penile erection in the rat can be induced with intracerebroventricular OT doses as low as 5 ng (Argiolas, Melis, & Gessa, 1985); even lower doses (3 ng) can elicit this effect with site-specific injections into the PVN (Melis, Argiolas, & Gessa, 1986). As with female sex behavior, central administration of an OT antagonist, $d(CH_2)_5Tyr(Me)$-[Orn[8]]vasotocin, greatly reduces male sexual interest and/or performance as measured by declines in frequency of mounts, intromissions, and ejaculations, although (as with the females) no effects on nonsexual behaviors such as locomotor activity were noted (Argiolas, Bollu, Gessa, Melis, & Serra, 1988). In one study, OT injected centrally prolonged the postejaculatory refractory period, suggesting a possible role in male sexual satiety (Stoneham et al., 1985).

C. Infant Attachment

If OT were important for affiliation, one might expect the peptide to mediate some aspects of infant attachment or separation response. In a very literal sense, OT has been reported to promote pup attachment to the nipple. Washing the ventrum of a rat mother removes an important olfactory cue for pup attachment. Peripheral OT administration to the mother has been reported to reinstate nipple attachment within minutes, even in the absence of milk ejection (Singh & Hofer, 1978). However, whether infant behavior is affected by endogenous or exogenous OT is not clear. OT is synthesized in the fetal rat hypothalamus and, as will be discussed subsequently, brain receptors for OT are more abundant during development than during adulthood (Shapiro & Insel, 1989). Further, OT is concentrated in breast milk relative to plasma (Leake, Weitzman, & Fisher, 1981) and the relative absence of proteolytic activity in the upper gastrointestinal tract of the infant suggests that dietary OT as well as several other large molecules may be absorbed by the infant (Morris & Morris, 1974). The extent to which exogenous OT crosses into the brain has not been investigated in the neonate, but active transport mechanisms have been identified for other larger peptides such as CRF (Insel, 1990) and insulin (Duffy & Pardridge, 1987).

In the 6-day-old pup, social isolation is accompanied by ultrasonic vocalizations (USV) that usually evoke maternal retrieval. These calls or laryngeal cries are in the 35–45 kHz frequency range, well above the range for detection by humans and many predators, but well within a region of high auditory sensitivity for the adult rat. (For a review, see Insel & Winslow, 1991a.) The young of a highly affiliative species produce more calls and exhibit greater glucocorticoid response to isolation than the offspring of nonaffiliative species, suggesting that ultrasonic calls may be a reflection of the distress associated with social isolation (Shapiro & Insel, 1990a). Anxiolytic drugs such as the benzodiazepines and the 5HT1a agonists decrease the number of these calls, whereas anxiogenic compounds such as pentylenetetrazole and 5HT1b agonists increase calling (Insel & Winslow, 1991a). If OT were important for the neural response of social attachment in the pup, one might predict that OT, like the anxiolytics, would decrease the number of ultrasonic calls, essentially conveying the message of attachment in the presence of social isolation.

We have injected OT both centrally and peripherally in 8-day-old rat pups to investigate changes in ultrasonic calls. Following central OT administration (10–1000 ng), the number of ultrasonic calls decreased relative to CSF administration; the greatest effect was observed after 100 ng OT, when USV were decreased approximately 71% from baseline (Insel & Winslow, 1991b). The effects of OT were not due to sedation since changes in locomotor behavior were not significant, nor could they be ascribed to changes in temperature or coordination (i.e., geotaxis). Although an OT antagonist alone did not affect the number of calls, it significantly blocked the effects of 500 ng OT, supporting the notion that agonist effects were mediated by an OT receptor. Moreover, this receptor appears to be in the CNS since peripheral administration of OT did not decrease calling, except at doses 20-fold higher, and peripheral administration of an antagonist failed to block the quieting effects of intraventricular OT injection.

D. Grooming

Grooming increases following OT administration, but the form of grooming that has been observed is primarily autogrooming, not allogrooming as one might expect for an affiliative hormone (Drago, Pedersen, Caldwell, & Prange, 1986). A study of intracerebroventricular OT administration in primates did not report increased allogrooming, but the species used, the squirrel monkey, shows little if any allogrooming under normal conditions (Winslow & Insel, 1991b). In the rat, autogrooming is increased by a number of peptides including (AVP), but central OT administration is characterized by a greater frequency of anogenital grooming (van Wimersma Greidanus, Kroodsma, Pot, Stevens, & Maigret, 1990). Whether OT antagonism would block the initiation of grooming is not yet clear nor is any evidence available for the location of OT effects on the induction of grooming behavior.

E. Summary

In summary, as shown in Table 1, the current literature on the central effects of OT suggests a pattern of what might be called "prosocial" effects. Increases in brain OT are associated with the induction of maternal and reproductive behavior and the reduction of pup isolation calls. Decreases in central OT, from either lesions or antagonist administration, block the initiation of maternal and reproductive behaviors, but have no effect on these behaviors once they are established. This pattern of effects may indicate that endogenous OT promotes the early phases of social interaction, possibly influencing the affective processing of social stimuli. Does a single mechanism exist by which OT alters these various forms of affiliation or does this peptide influence a range of neural circuits, some of which subserve maternal behavior whereas others mediate sex behavior or pup attachment? To answer this question, the cellular mechanisms for OT actions in the brain have been investigated using *in vitro* studies.

III. Cellular Mechanisms of Oxytocin Effects

The investigation of how OT influences social behaviors can begin with an observation from studies of peripheral tissue responses to OT. In both uterine myometrium and mammary myoepithelium, OT effects are mediated by a selective membrane-bound receptor that appears remarkably plastic. In the uterus, for instance, OT receptors increase as much as 10-fold at parturition (Soloff, Alexandrova, & Fernstrom, 1979). OT plasma levels increase only slightly and inconsistently at the onset of parturition (Fuchs, 1985). The increase in uterine contraction during labor may be accounted for partly by this increase in the release of OT, but undoubtedly is due mostly to the amplification of the response at the target organ. What accounts for this dramatic induction of uterine OT receptors? The increase in estrogen and decrease in progesterone that occur physiologically just prior to parturition are the necessary and sufficient stimuli for OT receptor induction in the uterus (Fuchs, Periyasamy, Alexandrova, & Soloff, 1983). This observation from peripheral target tissues may be relevant to central OT function because, as noted earlier in studies of maternal and reproductive behavior, OT behavioral effects are gonadal steroid dependent. A key question for the study of how OT might mediate affiliation, then, is, "How are OT receptors in the CNS regulated?"

A. Brain Oxytocin Receptors

The discovery of OT receptors in the CNS was itself an extremely important finding. Although several immunocytochemical studies already demonstrated extrahypothalamic projections of OT-containing neurons, the autoradiographic demonstration of OT receptors provided compelling evidence that the brain, like the uterus and breast, could be a target organ for OT (Brinton, Wamsley, Gee, Wan, & Yamamura, 1984; De Kloet, Rotteveel, Voorhuis, & Terlou, 1985; Van Leeuwen, Heerikhuize, Van der Meulen, & Wolters, 1985; Insel, 1986; Freund-Mercier et al., 1987). With the development of a highly selective, high affinity, iodinated ligand for OT receptors—[^{125}Id(CH$_2$)$_5$[Tyr(Me)$_2$,Tyr-NH$_2$^{9}]OVT (or ^{125}I-OTA)— rapid study of OT receptor distribution and regulation has become possible (Elands et al., 1987; Elands, Beetsma, Barbaris, & de Kloet, 1988).

OT receptors, identified with ^{125}I-OTA, are present in only a few discrete forebrain regions in the rat (Figure 1). The major loci are the posterior border of the anterior olfactory nucleus (AOP), the taenia tecta (rostral root of the hippocampus), the lateral segment of the bed nucleus of the stria terminalis (BNST), the dorsal–medial aspects of the caudate, the central nucleus of the amygdala (AmC), the VMN, and the ventral subiculum (VS). OT receptors in the forebrain are predominantly in integrative centers rather than in primary sensory or motor areas,

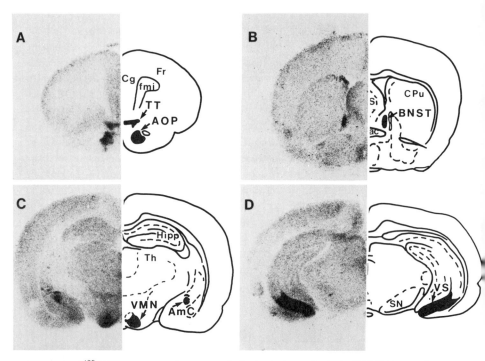

Figure 1. [^{125}I]OTA maps oxytocin receptor distribution in the rat forebrain. Brightfield autoradiograms show distribution of oxytocin receptors in coronal sections from (A) rostral to (D) caudal. Reconstructions provide anatomical landmarks. Abbreviations: AOP, posterior border of anterior olfactory nucleus; Cg, anterior cingulate cortex; fmi, forceps minor of corpus callosum; Fr, frontal cortex; TT, tenia tecta; ac, anterior commissure; BNST, bed nucleus of the stria terminalis; CPu, caudate putamen; S, septum; AmC, amygdala, central nucleus; Hipp, hippocampus; Th, thalamus; VMN, ventromedial nucleus; SN, substantia nigra; VS, ventral subiculum.

as seen for several other neuropeptide receptors such as CRF (De Souza et al., 1984), somatostatin (Reubi, Cortes, Maurer, Probst, & Palacios, 1986), and opioids (Mansour et al., 1987). In addition, OT receptors are found in several brainstem autonomic centers including the dorsal motor nucleus of the vagus and the nucleus of the solitary tract (Tribollet, Audigier, Dubois-Dauphin, & Dreifuss, 1990).

Although the distribution of receptors is identical in male and female brains (Tribollet et al., 1990), OT receptor distribution in the infant is markedly different from the pattern seen in adults. Binding to cingulate cortex, globus pallidus, and midline nuclei of the thalamus is intense only in the infant, whereas certain areas with high levels of binding in the adult brain—BNST and VMN—show virtually no receptors prior to sexual maturity (Shapiro & Insel, 1989; Snijdewint, van Leeuwen, & Boer, 1989; Tribollet, Charpak, Schmidt, Duboid-Dauphin, & Dreifuss, 1989). The significance of this transient expression of OT receptors in development is not understood, but similar patterns of evanescent receptors during ontogeny have been reported previously for several other neuropeptides (Insel, Battaglia, Fairbanks, & DeSouza, 1988; Palacios, Pazos, Dietl, Schlumpf, & Lichtensteiger, 1988; Quirion & Dam, 1988).

B. Brain Oxytocin Receptor Regulation

Now the intriguing question is whether the same mechanism—receptor induction—that serves the onset of labor also serves the initiation of maternal behavior and sexual receptivity. The answer appears to be that the mechanism is the same but the location is very discrete (Figure 2). In most brain regions in which OT receptors are found, no increase in binding is evident at parturition or estrus. However, at parturition, a striking increase in OT receptors occurs in the BNST that persists through at least the first 6 days of lactation but disappears following lactation. Evidence from neurophysiological recording and preliminary lesion studies suggests that this region is important for maternal behavior and milk ejection. During estrus, OT receptors increase in the VMN. This region has been shown previously to subserve the lordosis reflex, local injections of OT increase lordosis in ovariectomized rats primed with gonadal steroids.

What is the mechanism for these regional increases in OT receptors? As in the uterus, these effects in brain appear to be dependent on physiological changes in gonadal steroids (Tribollet et al., 1990). The OT receptor in the VMN and BNST is induced by estrogen (Figure 3). In the VMN, this induction has been reported to be as great as 5-fold with a half-life of approximately 20 hr following gonadectomy (Johnson et al., 1989). This "steroid dependence" may explain the mechanism by which parturition and estrus are associated with OT receptor induction. How can the induction of brain OT receptors be so localized? After all, the AOP and the VS are receiving the same concentrations of circulating estrogen as the

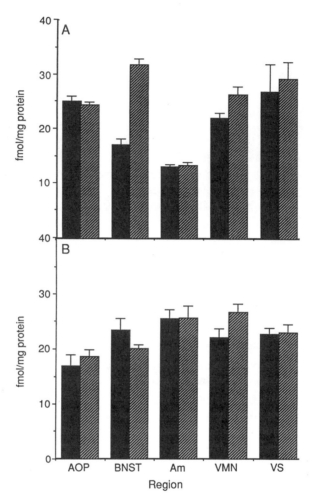

Figure 2. [^{125}I]OTA-specific binding increases in a regionally selective fashion at parturition and estrus. (A) Brains of cycling (n=6, solid bars) and day 1 postpartum (n=6, hatched bars) females were compared for [^{125}I]OTA binding. Quantified optical density measurements demonstrated significant increases in binding in both the BNST (F = 104.3, p < .001) and the VMN (F = 11.46, p = .007). Saturation studies showed that this increased binding was caused by an increase in receptors, not a change in affinity. (B) In a separate study, brains of estrous (n=6, hatched bars) and nonestrous (n=6, solid bars) females were compared for [^{125}I]OTA binding. Estrus was defined behaviorally. In this case, quantified optical density measurements demonstrated significant increases in binding in the VMN (F = 12.3, p = .008) but not in four other regions analyzed. Modified from Insel (1992).

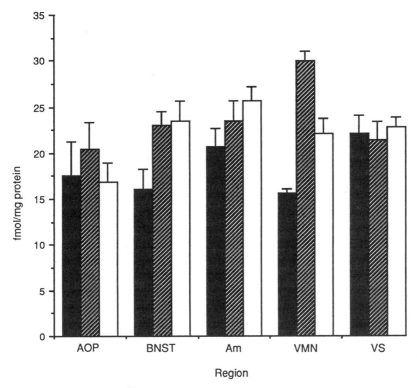

Figure 3. Estradiol induces [^{125}I]OTA-specific binding in BNST and VMN. Ovariectomized female rats received estradiol benzoate (10 μg, sc, $n=4$, hatched bars) or oil (0.1 μl, sc, $n=5$, solid bars) for 2 days. Significant group differences appeared in [^{125}I]OTA binding in BNST ($F = 4.17$, $p = .04$) and VMN ($F = 28.1$, $p < .0001$), but not in other brain regions. Since similar differences from oil treatment group were evident in brains from 6 intact, cycling females (open bars), oxytocin receptors in the BNST and VMN appear to be dependent on physiologic concentrations of gonadal steroids. Modified from Insel (1992).

BNST and the VMN, but receptors in these regions are unaffected by either endogenous or exogenous increases in estrogen. The explanation for this apparent inconsistency is that estrogen receptors are relatively sparse in the AOP and the VS, whereas the BNST and the VMN are among the brain regions with the most intense expression of estrogen receptors.

C. Model of Oxytocin Function in Brain

The mechanism by which OT neurotransmission in the brain may increase at parturition or estrus is remarkably similar to the mechanism believed to operate in peripheral tissues to increase OT effects on smooth muscle contraction. The proposed model is shown in Figure 4. This model rests on three major conceptual points. First, it suggests that the regulation of OT neurotransmission occurs largely at the target organ or receptor in addition to sites of synthesis. Therefore, measuring plasma or CSF OT may be uninformative unless one knows whether OT receptors are increased or decreased. Second, this model posits a key role for gonadal steroids in the regulation of OT pathways. Indeed, describing the OT receptor as a transducer of physiological changes of gonadal steroids would not be an overstatement. Recall that virtually all the behavioral effects of central OT administration were dependent on priming with gonadal steroids. Note that developmental changes in OT receptors in the BNST and the VMN coincide with puberty. Another example of this mechanism was suggested in our studies of OT effects in socially housed primates (Winslow & Insel, 1991b). Dominant and subordinate males not only manifest markedly different circulating concentrations of testosterone, but respond to central administration of OT with a completely different profile of behaviors. Finally, the model shown in Figure 3 suggests parallel processing in which OT release, stimulated by several different environmental factors, can activate brain and peripheral targets simultaneously. In the

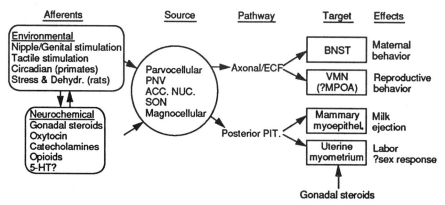

Figure 4. A model of oxytocin physiology posits parallel processing in central and peripheral pathways. Several environmental inputs, acting via a range of neurochemical pathways, are believed to induce oxytocin synthesis in the hypothalamus. Parvocellular neurons, either by direct axonal projections or through exocrine release into extracellular fluid (ECF), affect cells in the BNST and VMN to influence the initiation of maternal (BNST) and reproductive (VMN) behaviors. Concurrent neurohypophyseal release induces milk ejection and uterine contraction. In this model, the key determinant of the functional effects of oxytocin is the induction of receptors at the target organ by circulating gonadal steroids. The regional specificity of gonadal steroid receptor distribution (high in the BNST and VMN) permits the selective induction of receptors. The physiologic variations in estradiol and progesterone may determine whether maternal or reproductive behaviors are expressed. Note that these parallel central and peripheral pathways can be dissociated, resulting in adequate labor and nursing in the absence of normal maternal care. SON, supraoptic nucleus of the hypothalamus; AccNuc, accessory nuclei of the hypothalamus.

physiological condition of normal parturition, one would expect both peripheral and central sites to be stimulated, with labor, nursing, and maternal behavior expressed synchronously. However, these parallel limbs can be dissociated, leading to labor and nursing in a female who in all other ways neglects her young (i.e., maternal neglect) or, conversely, to full maternal behavior in an adoptive mother who has experienced neither labor nor nursing.

In summary, this model suggests an essential mechanism for how OT may alter affiliative behaviors. In the presence of steroid-induced OT receptors, the concurrent CNS release of OT during nursing and coitus may promote social bonding, analogous to the increase in maternal approach following vaginal–cervical stimulation in the steroid-primed ewe (Keverne et al., 1983). Indeed, data support the increased release of OT in the CNS with social experience. At parturition, as OT stores in the PVN decrease, immunoreactive OT in the septum and BNST increase (Caldwell, Greer, Johnson, Prange, & Pedersen, 1987; Jirikowski, Caldwell, Pilgrim, Stumpf, & Pedersen, 1989). Following sexual experience, OT increases in the preoptic area (Caldwell et al., 1989). Priming females with gonadal steroids increases OT immunocytochemically in the ventral hypothalamus, adjacent to the VMN (Schumacher et al., 1989). Even the tactile stimulation of grooming appears sufficient to increase OT release (Stock & Uvnas-Moberg, 1988). In summary, the model originally proposed for peripheral target organs may prove workable in the brain as well. Under physiological conditions, even a modest increase in OT release in the presence of regionally specific enhanced receptor responsivity may be sufficient to generate OT responses that subserve social behaviors.

IV. Comparative Study

At this point, the behavioral data presented implicate OT in the mediation of grooming, maternal and reproductive behaviors, and infant responses to social isolation. The studies of OT regulation demonstrate a remarkable plasticity in OT receptors, suggesting that an important part of the regulation of OT function occurs at the target sites of OT action. Gonadal steroids clearly play an important role in OT function, affecting maternal and reproductive behaviors and OT receptor expression. The remaining question is the extent to which OT mediates affiliation. Although the behavioral studies are suggestive and the receptor studies provide a mechanism, a comparative study provides a test of this hypothesis.

The genus *Microtus* includes several closely related species with an extraordinary range of affiliative patterns. Two of the most dichotomous are *M. ochrogaster,* the prairie vole, and *M. montanus,* the montane vole. Both field and laboratory studies (summarized in Table 2) have demonstrated that these two closely related voles with nearly identical morphology and from relatively similar habitats have opposite patterns of social organization (Dewsbury, 1981; Carter, Williams, & Witt, 1990). The prairie vole is monogamous and highly parental (Getz & Hofman, 1986). Even in the laboratory, adults generally sit side-by-side and young show high levels of ultrasonic calls and glucocorticoid responses to social isolation (Getz, Carter, & Gavish, 1981; Shapiro & Insel, 1990a; Shapiro & Dewsbury, 1993). The montane vole is polygamous, minimally parental, spends little time in contact with conspecifics, and has offspring that show little if any behavioral or physiological response to social isolation (Jannett, 1980; Shapiro & Insel, 1990a; Shapiro & Dewsbury, 1993). Clearly, if OT is important for affiliation, one might expect differences in the OT systems in these two species.

As can be seen in Figure 5, OT receptor maps are virtually complementary between the species. High levels of ^{125}I-OTA binding appeared in the BNST, lateral amygdala, cingulate cortex, and midline thalamus in the prairie vole. In the montane vole, little binding is seen in any of these areas, but high levels of binding are apparent in the lateral septum. When these differences are quantified, the amount of ^{125}I-OTA binding differs between species by at least 300%. Saturation studies, using 500 pM ^{125}I-OTA, essentially replicate these results, confirming a difference in receptor number and not affinity. No sex differences are apparent in either species.

The high density of OT receptors in the BNST and its primary afferent, the lateral amygdala, in the highly parental prairie vole is consistent with a role for this circuit in mediating parental behavior. The significance of the lateral septum OT receptors in the montane vole is not clear, but may relate to the high levels of conspecific aggression observed in this species, since this region has been associated with territorial behaviors such as scent marking and agonistic behavior

TABLE 2
Profiles of Affiliation in Prairie and Montane Voles

Behavior	Prairie	Montane	Reference
Mating system	Monogamous	Polygamous	Dewsbury (1981)
Nest	Shared	Unshared	Getz et al. (1981)
			Jannett (1980)
Maternal care			
Naive	High	Low	Shapiro & Insel (1990a)
Postpartum	High	Mod–High	
Paternal care	High	Low	Shapiro & Insel (1990a)
Side-by-side	High	Low	Shapiro & Dewsbury (1990)
Pup separation			
USV	High	Low	Shapiro & Insel (1990a)
Corticosterone	High	Low	Shapiro & Insel (1990a)

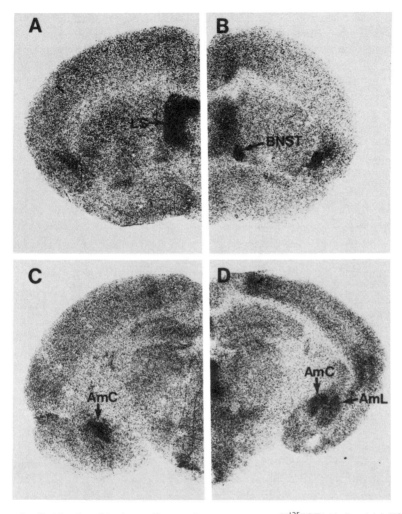

Figure 5.   Social and asocial voles manifest complementary patterns of [^{125}I]OTA binding. Adult (60–90 days old) male prairie voles and montane voles were compared for [^{125}I]OTA binding. A and B show rostral sections at the level of the BNST (level B in Figure 1); C and D show caudal sections at the level of the VMN (level C in Figure 1). (A) Montane vole shows intense binding in the lateral septum (LS), with virtual absence of binding in the BNST. (B) Prairie vole shows relatively little binding in the lateral septum, but intense binding in the BNST. In both species, specific binding occurs in the agranular insular cortex, possibly involved in the processing of gustatory stimuli. (C) Binding in the montane vole amygdala is restricted to the central nucleus. (D) In the prairie vole, binding is high in both the central (AmC) and basolateral (AmL) nuclei of the amygdala, as well as in the midline nuclei of the thalamus. Saturation studies reveal that these contrasting patterns are not caused by species differences in receptor affinity. In each of these same regions, adjacent sections show equivalent numbers and distributions of benzodiazepine and opiate μ receptors in both species. Modified from Insel (1992).

(Gray & McNaughton, 1983). These differences in receptor distribution do appear to be associated with species-typical patterns of social organization, since most of the prairie–montane differences can be replicated with two other microtine rodents, the meadow vole and the pine vole, which are similarly dichotomous for measures of affiliation (Insel & Shapiro, 1992). Moreover, these OT receptor differences do not reflect widespread differences in receptor distribution among microtines. Prairie and montane voles show few if any species differences in the distribution of benzodiazepine and μ opioid receptors, although each of these other systems has been implicated in social behavior (Insel & Shapiro, 1992). In an analogous study of monogamous and polygamous mice (*Peromyscus californicus* and *P. maniculatus*), we previously reported similar species differences, selectively in OT receptor distribution (Insel et al., 1991).

How does the difference in OT receptor distribution affect the response to exogenous OT? The data to answer this question remain incomplete, but one set of studies in prairie voles deserves note. This monogamous vole forms a lasting social bond following extensive mating bouts. In other mammals, OT is released during mating. To determine whether centrally injected OT would confer a mate preference even in the absence of copulatory behavior, male and female voles were housed together without mating for 6 hr while the female received intraventricular OT or CSF by continuous perfusion via osmotic minipump. The infusion was then discontinued and the female was given a choice of a novel male or her "partner" (both of whom were tethered) in a 3-hr test. As shown in Table 2, prairie voles will spend considerable periods sitting side-by-side with their preferred mate. In this experiment, a partner preference was found after OT but not after CSF administration. This effect was found to be dose dependent and blocked by coadministration of an OT antagonist. In the monogamous species, OT appears to be sufficient to induce a partner preference. These results are consistent with the hypothesis that OT released during the long mating bouts of this species provides the neuroendocrine signal (perhaps coordinated with increased estradiol, which is released at the onset of estrus in this species) for pair bond formation. The effects of OT in the montane vole have yet to be elucidated.

The most rigorous test of whether the OT receptor differences are related to the species differences in affiliation would require a change in affiliative behavior to be associated with changes in OT receptors. The behavioral part of this experiment is fraught with problems—lesions that alter the behavior may have nonspecific receptor effects and pharmacological manipulations that alter affiliation are difficult to interpret because of potential differences in pharmacokinetics between the two species. However, under one natural circumstance, the montane vole becomes parental: after parturition, the time spent with pups increases to almost the level observed in the prairie vole. Do OT receptors increase concurrently? ^{125}I-OTA binding appears to increase in one area at parturition in the montane vole—the lateral amygdala. Binding in the prairie vole, which is high in the lateral amygdala in the naive female, does not change at parturition in this or any other region. The lateral amygdala previously has been implicated in the formation of affective associations and the mediation of social affects (Kling, 1972). Since this nucleus is the major afferent for the BNST, these receptor changes resemble the data previously described in the rat which, like the montane vole, shows little interest in pups except during the postpartum period. Of particular note, EB does not alter ^{125}I-OTA binding in the BNST of the prairie vole (Witt, Carter, & Insel, 1991). This brain area in a highly parental species may no longer be steroid dependent for OT receptor expression, just as the behavior is no longer dependent on steroid status.

V. Conclusion

Collectively, these various results of behavioral, cellular, and comparative studies are generally consistent with the hypothesis that central OT pathways are involved in social affiliation. However, these experiments do not address several important issues regarding the mechanism of OT action. The suggestion that OT influences the motivational or affective properties of social interaction has intuitive appeal. More specifically, one might hypothesize that this peptide makes social contact rewarding, permitting an asocial species to become social for the elaboration of parental and reproductive behaviors. A similar (although not identical) proposal has been made previously for opioid peptides, which are the candidate neurotransmitters most frequently associated with affiliation (Panksepp et al., 1978; Fabre-Nys, Meller, & Keverne, 1982). Keverne and colleagues have demonstrated in talapoin monkeys that social contact, such as allogrooming, is associated with a profound increase in CSF β-endorphin (Keverne, Martensz, & Tuite, 1989). The relationship of opioids to telencephalic reward centers has strengthened the very compelling suggestion that these peptides are important for the formation of social bonds (Herman & Panksepp, 1980; Grimm & Bridges, 1983; but see also Winslow & Insel, 1991a). Although opiates alter OT release (Bicknell & Leng, 1982), the possibility that OT affects opioid function has not yet been addressed sufficiently.

Another perspective on the behavioral effects of OT does not rely on an affective or motivational explanation. We suggested previously that OT effects on maternal behavior could be explained by decreased olfactory processing, permitting the neophobic rat to approach its young (Wamboldt & Insel, 1987). Following approach and exploration, other systems might emerge to initiate caretaking behaviors. Does the peptide simply decrease neophobia and increase exploration, similar to a benzodiazepine? Indeed, are OT behavioral effects secondary to stimulation of GABAergic neurons? Currently, viewing OT effects as "disinhibiting" social behavior appears as valid as suggesting explicit "prosocial" actions.

Whatever the ultimate mechanism of OT effects, these experiments support the validity of studying the neuroendocrine basis of social bonds. These behavioral processes frequently have been considered too complex to study experimentally. However, these behaviors—parental care, reproduction, and infant attachment— are of critical survival value, so their neural basis might be expected to be developed exquisitely by natural selection. Perhaps the fact that, at a behavioral level, affiliation displays remarkable diversity even across closely related mammals is no accident. Although this diversity may discourage the rapid generalization from rat to humans, it also provides the opportunity for useful comparative studies in which closely related species that differ markedly on a specific behavioral trait can generate a "natural experiment," yielding insights into both physiology and evolution. From the studies described here, one might suggest that, despite behavioral diversity, specific neuroendocrine mechanisms have been conserved and adapted to facilitate social bond formation across mammalian taxa. Certainly within a single species, OT appears to be involved in the initiation of several different forms of social bonds, including infant–parent and adult–adult interactions. Physiological variation in gonadal steroids appears to direct the regional responsivity of this system, allowing the same receptor to subserve many different behaviors.

In primate species, including our own, parental and sexual behaviors are linked only loosely to gonadal steroid status. The development of an extensive neocortex and the associated emergence of complex social structures in which various individuals contribute to parental care provide a different context in which to assess neuroendocrine regulation of social behavior. Nevertheless, the limbic structures that have OT receptors remain largely unchanged across mammalian phylogeny. Studies of rat maternal behavior demonstrating the dissociation of labor and nursing from maternal attachment may prove useful in understanding the human mother who fails to bond with her infant. The avoidance of sociosexual contact after central OT antagonist administration may provide a model for understanding forms of human psychopathology, for example, the schizoid personality or autism, for which social experience is not rewarding. The recognition that OT affects the distress of social isolation in the rat pup suggests that some of the same neural processes that promote maternal care may signal social contact in the infant.

The brain OT system, whether regulated by gonadal steroids or sculpted by evolution, exhibits remarkable plasticity. The mechanisms described for the heterologous regulation of OT receptors provide a model for neuroendocrine regulation of behavior that may be important in primates as well as in rodents. Although gonadal steroids may not regulate OT receptors in the human brain, understanding what regulates this system in our own species ultimately may shed light on many aspects of human social behavior. In the meantime, exploring the ways in which this system links the brain to behavior in a diverse array of mammals should provide insights into fundamental principles of neuroendocrine function and evolution.

References

Archer, R. (1974). Chemistry of the neurohypophyseal hormones: An example of molecular evolution. In E. Knobil & W. Sawyer (Ed.), *Handbook of physiology* (pp. 119–130). Washington, D.C.: American Physiological Society.

Argiolas, A., Collu, M., Gessa, G. L., Melis, M. R., & Serra, G. (1988). The oxytocin antagonist d(CH_2)$_5$Tyr(Me)-Orn^8-vasotocin inhibits male copulatory behavior in rats. *European Journal of Pharmacology, 149,* 389–392.

Argiolas, A., Melis, M. R., & Gessa, G. L. (1985). Intraventricular oxytocin induces yawning and penile erection in rats. *European Journal of Pharmacology, 117,* 395–396.

Argiolas, A., Melis, M. R., Stancampiano, R., & Gessa, G. L. (1989). Penile erection and yawning induced by oxytocin and related peptides: Structure–activity relationship. *Peptides, 10,* 559–563.

Arletti, R., Bazzani, C., Castelli, M., & Bertolini, A. (1985). Oxytocin improves male copulatory performance in rats. *Hormones and Behavior, 19,* 14–20.

Arletti, R., & Bertolini, A. (1985). Oxytocin stimulates lordosis behavior in female rats. *Neuropeptides, 6,* 247–253.

Bicknell, R. J., & Leng, G. (1982). Endogenous opiates regulate oxytocin but not vasopressin secretion from the neurohypophysis. *Nature (London), 298,* 161–162.

Bolwerk, E. L. M., & Swanson, H. H. (1984). Does oxytocin play a role in the onset of maternal behavior in the rat? *Journal of Endocrinology, 101,* 353–357.

Brinton, R. E., Wamsley, J. K., Gee, K. W., Wan, Y. P., & Yamamura, H. I. (1984). ^{3}H-Oxytocin binding sites demonstrated in the rat brain by quantitative light microscopic autoradiography. *European Journal of Pharmacology, 102,* 365–367.

Brownstein, M. J., Russell, J. T., & Gainer, H. (1980). Synthesis, transport, and release of posterior pituitary hormones. *Science, 207,* 373–378.

Buijs, R. (1978). Intra- and extrahypothalamic vasopressin and oxytocin pathways in the rat: Pathways to the limbic system, medulla oblongata and spinal cord. *Cell and Tissue Research, 252,* 355–365.

Buijs, R. M., & van Heerikhuize, J. J. (1982). Vasopressin and oxytocin release in the brain—A synaptic event. *Brain Research, 252,* 71–76.

Caldwell, J. D., Greer, E. R., Johnson, M. F., Prange, A. J., Jr., & Pedersen, C. A. (1987). Oxytocin and vasopressin immunoreactivity in hypothalamic and extrahypothalamic sites in late pregnant and post-partum rats. *Neuroendocrinology, 46,* 39–47.

Caldwell, J. D., Jirikowski, G. F., Greer, E. R., & Pedersen, C. A. (1989). Medial preoptic area oxytocin and female sexual receptivity. *Behavioral Neuroscience, 103(3),* 655–662.

Caldwell, J. D., Prange, A. J., Jr., & Pedersen, C. A. (1986). Oxytocin facilitates the sexual receptivity of estrogen-treated female rats. *Neuropeptides, 7,* 175–189.

Carmichael, M. S., Humbert, R., Dixen, J., Palmisano, G., Greenleaf, W., & Davidson, J. M. (1987). Plasma oxytocin increases in the human sexual response. *Journal of Clinical Endocrinology and Metabolism, 64(1),* 27–31.

Carter, C. S., Williams, J. R., & Witt, D. M. (1990). The biology of social bonding in a monogamous mammal. In J. Balthazart (Ed.), *Hormones, brain and behavior in vertebrates 2: Behavioral activation in males and females—Social interactions and reproductive endocrinology.* pp. 154–164 Basel: Karger.

Crowley, W. R., O'Donohue, T. L., George, J. M., & Jacobowitz, D. M. (1978). Changes in pituitary oxytocin and vasopressin during the estrous cycle and after ovarian hormones: Evidence for mediation by norepinephrine. *Life Sciences, 23,* 2579–2586.

De Kloet, E. R., Rotteveel, F., Voorhuis, T. A. M., & Terlou, M. (1985). Topography of binding sites for neurohypophyseal hormones in rat brain. *European Journal of Pharmacology, 110,* 113–119.

De Souza, E., Perrin, M., Insel, T. R., Rivier, J., Vale, W., & Kuhar, M. (1984). Corticotropin-releasing factor receptors in rat forebrain: Autoradiographic identification. *Science, 224,* 1449–1451.

Dewsbury, D. A. (1981). An exercise in the prediction of monogamy in the field from laboratory data on 42 species of muroid rodents. *The Biologist, 63(4),* 138–162.

Dewsbury, D. A. (1988). The comparative psychology of monogamy. In D. W. Leger (Ed.), *American zoology Nebraska symposium on motivation* (pp. 1–50). Lincoln: University of Nebraska Press.

Drago, F., Pedersen, C. A., Caldwell, J. D., & Prange, A. J., Jr. (1986). Oxytocin potently enhances novelty-induced grooming behavior in the rat. *Brain Research, 368,* 287–295.

Duffy, K. R., & Pardridge, W. M. (1987). Blood brain barrier transcytosis of insulin in developing rabbits. *Brain Research, 420,* 32–39.

Elands, J., Barberis, C., Jard, S., Tribollet, E., Dreifuss, J., Bankowski, K., Manning, M., & Sawyer, W. (1987). ^{125}I-labelled d(CH$_2$)$_5$[Tyr(Me)^2,Thr^4,Tyr-NH$_2$^{9}] OVT: A selective oxytocin receptor ligand. *European Journal of Pharmacology, 147,* 197–207.

Elands, J., Beetsma, A., Barberis, C., & de Kloet, E. R. (1988). Topography of the oxytocin receptor system in rat brain: An autoradiographical study with a selective radioiodinated oxytocin antagonist. *Journal of Clinical Neuroanatomy, 1,* 293–302.

Fabre-Nys, C., Meller, R. E., & Keverne, E. B. (1982). Opiate antagonists stimulate affiliative behavior in monkeys. *Pharmacology and Biochemistry of Behavior, 16,* 653–660.

Fahrbach, S. E., Morrell, J. J., & Pfaff, D. W. (1984). Oxytocin induction of short-latency maternal behavior in nulliparous, estrogen-primed female rats. *Hormones and Behavior, 18,* 267–286.

Fahrbach, S. E., Morrell, J. I., & Pfaff, D. W. (1985a0. Role of oxytocin in the onset of estrogen-facilitated maternal behavior. In J. A. Amico & A. G. Robinson (Eds.), *Oxytocin: Clinical and laboratory studies* (pp. 372–388). Amsterdam: Elsevier Science Publishers.

Fahrbach, W. E., Morrell, J. I., & Pfaff, D. W. (1985b). Possible role for endogenous oxytocin in estrogen-facilitated maternal behavior in rats. *Neuroendocrinology, 40,* 526–532.

Fox, C. A., & Knaggs, G. S. (1969). Milk-ejection activity (oxytocin) in peripheral venous blood in man during lactation and in association with coitus. *Journal of Endocrinology, 45,* 145–146.

Freund-Mercier, M. J., Stoeckel, M. E., Palacios, J. M., Pazos, A., Reichart, J. M., Porte, A., & Richard, P. (1987). Pharmacological characteristics and anatomical distribution of [^{3}H]oxytocin-binding sites in the Wistar rat brain studied by autoradiography. *Neuroscience, 20(2),* 599–614.

Fuchs, A.-R. (1985). Oxytocin in animal parturition. In J. Amico & A. G. Robinson (Eds.), *Oxytocin: Clinical and laboratory studies* (pp. 207–235). New York: Elsevier.

Fuchs, A. R., Cubile, L., & Dawood, M. Y. (1981). Effects of mating on levels of oxytocin and prolactin in the plasma of male and female rabbits. *Journal of Endocrinology, 90,* 245–253.

Fuchs, A., Periyasamy, S., Alexandrova, M., & Soloff, M. S. (1983). Correlation between oxytocin receptor concentration and responsiveness to oxytocin in pregnant rat myometrium: Effects of ovarian steroids. *Endocrinology, 113(2),* 742–749.

Getz, L. L., Carter, C. S., & Gavish, L. (1981). The mating system of the prairie vole Microtus ochragaster: Field and laboratory evidence for pair bonding. *Behavioral Ecology and Sociobiology, 8,* 189–194.

Getz, L. L., & Hofman, J. E. (1986). Social organization in free living prairie voles, *Microtus ochragaster. Behavioral Ecology and Sociobiology, 18,* 275–282.

Gorzalka, B. B., & Lester, G. L. L. (1987). Oxytocin-induced facilitation of lordosis behavior in rats is progesterone-dependent. *Neuropeptides, 10,* 55–65.

Gray, J. A., & McNaughton, N. (1983). Comparison of the behavioral effects of septal and hippocampal lesions: A review. *Neuroscience and Biobehavioral Reviews, 7,* 119–188.

Greer, E. R., Caldwell, J. D., Johnson, M. F., Prange, A. J., Jr., & Pedersen, C. A. (1986). Variations in the concentrations of oxytocin and vasopressin in the paraventricular nucleus of the hypothalamus during the estrous cycle in rats. *Life Sciences, 38,* 2311–2318.

Grimm, C. T., & Bridges, R. S. (1983). Opioid regulation of maternal behavior in the rat. *Pharmacology and Biochemistry of Behavior, 19,* 609–616.

Hennessy, M. B., & Ritchey, R. L. (1987). Hormonal and behavioral attachment responses in infant guinea pigs. *Developmental Psychobiology, 20,* 613–625.

Herman, B. H., & Panksepp, J. (1980). Ascending endorphin inhibition of distress vocalization. *Science, 221,* 1060–1062.

Hinde, R. A. (1983). *Primate social relationships: An integrated approach.* Sunderland, Massachusetts: Sinauer.

Insel, T. R. (1986). Postpartum increases in brain oxytocin binding. *Neuroendocrinology, 44,* 515–518.

Insel, T. R. (1990). Corticotropin releasing factor in development. In E. B. De Souza & C. Nemeroff (Ed.), *Corticotropin releasing factor: Basic and clinical aspects* (pp. 69–90). Boca Raton, Florida: CRC Press.

Insel, T. R. (1992). Oxytocin: A neuropeptide for affiliation—Evidence from behavioral, receptor autoradiographic, and comparative studies. *Psychoneuroendocrinology, 17,* 3–33.

Insel, T. R., Battaglia, G., Fairbanks, D. W., & De Souza, E. B. (1988). The ontogeny of brain receptors for corticotropin-releasing factor and the development of their functional association with adenylate cyclase. *Journal of Neuroscience, 8(11),* 4151–4158.

Insel, T. R., Gelhard, R. E., & Shapiro, L. E. (1991). The comparative distribution of neurohypophyseal peptide receptors in monogamous and polygamous mice. *Neuroscience, 43,* 623–630.

Insel, T. R., & Harbaugh, C. R. (1989). Lesions of the hypothalamic paraventricular nucleus disrupt the initiation of maternal behavior. *Physiology and Behavior, 45,* 1033–1041.

Insel, T. R., & Shapiro, L. E. (1992). Oxytocin receptor distribution reflects social organization in monogamous and polygamous voles. *Proceedings National Academy Sciences, U.S.A., 89,* 5981–5985.

Insel, T. R., & Winslow, J. T. (1991). Rat pup ultrasonic cells: An ethologically relevant behavior for neurobiological study. In B. Olivier, J. Mos, & J. L. Slangen (Eds.), *Animal models in psychopharmacology* (pp. 15–36). Boston: Birkhauser.

Insel, T. R., & Winslow, J. T. (1991). Central oxytocin administration reduces rat pup isolation calls. *European Journal of Pharmacology, 203,* 149–152.

Jannett, F. J. (1980). Social dynamics in the montane vole *Microtus montanus* as a paradigm. *The Biologist, 62,* 3–19.

Jirikowski, G. F., Caldwell, J. D., Pilgrim, C., Stumpf, W. E., & Pedersen, C. A. (1989). Changes in immunostaining for oxytocin in the forebrain of the female rat during late pregnancy, parturition, and early lactation. *Cell and Tissue Research, 256,* 411–417.

Johnson, A. E., Ball, G. F., Coirini, H., Harbaugh, C. R., McEwen, B. S., & Insel, T. R. (1989). Time course of the estradiol-dependent induction of oxytocin receptor binding in the ventromedial hypothalamic nucleus of the rat. *Endocrinology, 125(3),* 1414–1419.

Jones, P. M., Robinson, I. C. A. F., & Harris, M. C. (1983). Release of oxytocin into blood and cerebrospinal fluid by electrical stimulation of the hypothalamus or neural lobe in the rat. *Neuroendocrinology, 37,* 454–458.

Kalin, N. H., Shelton, S. E., & Barksdale, C. M. (1988). Opioid modulation of separation-induced distress in nonhuman primates. *Brain Research, 440,* 285–292.

Kendrick, K. M., Keverne, E. B., & Baldwin, B. A. (1987). Intracerebroventricular oxytocin stimulates maternal behavior in the sheep. *Neuroendocrinology, 46,* 56–61.

Kendrick, K. M., Keverne, E. B., Baldwin, B. A., & Sharman, D. F. (1986). Cerebrospinal fluid levels of acetylcholinesterase, monoamines and oxytocin during labor, parturition, vaginocervical stimulation, lamb separation and suckling in sheep. *Neuroendocrinology, 44,* 149–156.

Kendrick, K. M., Keverne, E. B., Chapman, C., & Baldwin, B. A. (1988a). Intracranial dialysis measurement of oxytocin, monoamines and uric acid release from the olfactory bulb and substantia nigra of sheep during parturition, suckling, separation from lambs and eating. *Brain Research, 439,* 1–10.

Kendrick, K. M., Keverne, E. B., Chapman, C., & Baldwin, B. A. (1988b). Microdialysis measurement of oxytocin, aspartate, GABA and glutamate release from the olfactory bulb of sheep during vaginocervical stimulation. *Brain Research, 442,* 171–177.

Keverne, E. B., Levy, F., Poindron, P., & Lindsay, D. R. (1983). Vaginal stimulation: An important determinant of maternal bonding in sheep. *Science, 219,* 81–83.

Keverne, E. B., Martensz, N., & Tuite, B. (1989). B-endorphin concentrations in CSF of monkeys are influences by grooming relationships. *Psychoneuroendocrinology, 14,* 155–161.

Kleinman, D. G. (1977). Monogamy in mammals. *Quarterly Review of Biology, 52,* 39–69.

Kling, A. (1972). Effects of amygdalectomy on social-affective behavior in nonhuman primates. In B. E. Eleftheriou (Ed.), *The neurobiology of the amygdala* (pp. 511–536). New York: Plenum.

Kraemer, G. W., Ebert, M. H., Lake, C. R., & McKinney, W. T. (1984). Cerebrospinal fluid measures of neurotransmitter changes associated with pharmacological alteration of the despair response to social separation in rhesus monkeys. *Psychology Research, 11,* 303–315.

Leake, R. D., Weitzman, R. E., & Fisher, D. A. (1981). Oxytocin concentrations during the neonatal period. *Biology of the Neonate, 39,* 127–131.

Levine, S., & Wiener, S. G. (1988). Psychoendocrine aspects of mother-infant relationships in nonhuman primates. *Psychoneuroendocrinology, 13,* 143–154.

Lightman, S. L., & Young, W. S. I. (1987). Vasopressin, oxytocin, enkephalin, dynorphin, corticotrophin releasing factor mRNA stimulation in the rat. *Journal of Physiology, 394,* 23–39.

Lincoln, D. W., & Russell, J. A. (1985). The electrophysiology of magnocellular oxytocin neurons. In J. Amico & A. G. Robinson (Eds.), *Oxytocin: Clinical and laboratory studies* (pp. 53–76). New York: Elsevier.

Maggi, M., Malozowski, S., Kassis, S., Guardabasso, V., & Rodbard, D. (1987). Identification and characterization of two classes of receptors for oxytocin and vasopressin in porcine tunica albuginea, epididymis, and vas deferens. *Endocrinology, 120,* 986–994.

Mansour, A., Khachaturian, H., Lewis, M. E., Akil, H., & Watson, S. J. (1987). Autoradiographic differentiation of mu, delta, and kappa opioid receptors in the rat forebrain and midbrain. *Journal of Neuroscience, 7,* 2445–2464.

Melin, P., & Kihlstrom, J. E. (1963). Influence of oxytocin on sexual behavior in male rabbits. *Endocrinology, 73,* 433–435.

Melis, M. R., Argiolas, A., & Gessa, G. L. (1986). Oxytocin-induced penile erection and yawning: Site of action in the brain. *Brain Research, 398,* 259–265.

Mendoza, S. P., Smotherman, W. P., Miner, M. T., Kaplan, J., & Levine, S. (1978). Pituitary-adrenal response to separation in mother and infant squirrel monkeys. *Developmental Psychobiology, 11,* 169–175.

Morris, B., & Morris, R. (1974). The absorption of ^{125}I-labeled immunoglobulin G by different regions of the gut in young rats. *Journal of Physiology, 241,* 761–770.

Murphy, M. R., Seckl, J. R., Burton, S., Checkley, S. A., & Lightman, S. L. (1987). Changes in oxytocin and vasopressin secretion during sexual activity in men. *Journal of Clinical Endocrinology and Metabolism, 65(4),* 738–741.

Newman, J. D. (1988). *The physiologic control of mammalian vocalization.* New York: Plenum Press.

Newton, N. (1973). Interrelationships between sexual responsiveness, birth, and breast feeding. In J.

Zubin & J. Money (Eds.), *Contemporary sexual behavior: Critical issues in the 1970s* (pp. 77–98). Baltimore: The Johns Hopkins University Press.

Numan, M. (1988). Maternal behavior. In E. Knobil & J. Neill (Eds.), *The physiology of reproduction* (pp. 1569–1645). New York: Raven Press.

Palacios, J. M., Pazos, A., Dietl, M. M., Schlumpf, M., & Lichtensteiger, W. (1988). The ontogeny of brain neurotensin receptors studied by autoradiography. *Neuroscience, 25,* 307–317.

Panksepp, J. B., Herman, B., Conner, R., Bishop, P., & Scott, J. P. (1978). The biology of social attachments: Opiates alleviate separation distress. *Biological Psychiatry, 13,* 607–613.

Pedersen, C. A., Ascher, J. A., Monroe, Y. L., & Prange, A. J., Jr. (1982). Oxytocin induces maternal behavior in virgin female rats. *Science, 216,* 648–649.

Pedersen, C. A., Caldwell, J. D., Fort, S. A., & Prange, A. J., Jr. (1985). Oxytocin antiserum delays onset of ovarian steroid-induced maternal behavior. *Neuropeptides, 6,* 175–182.

Pedersen, C. A., & Prange, A. J., Jr. (1979). Induction of maternal behavior in virgin rats after intracerebroventricular administration of oxytocin. *Proceedings of the National Academy of Science, U.S.A., 76,* 6661–6665.

Perlow, M. J., Reppert, S. M., Artman, H. A., Fisher, D. A., Self, S. M., & Robinson, A. G. (1982). Oxytocin, vasopressin, and estrogen-stimulated neurophysin: Daily patterns of concentration in cerebrospinal fluid. *Science, 216,* 1416–1418.

Quirion, R., & Dam. T.-V. (1988). The ontogeny of substance P binding sites in rat brain. *Journal of Neuroscience, 6,* 2187–2199.

Reubi, J. C., Cortes, R., Maurer, R., Probst, A., & Palacios, J. M. (1986). Distribution of somatostatin receptors in the human brain: An autoradiographic study. *Neuroscience, 18,* 329–346.

Roberts, J. S., & Share, L. (1968). Oxytocin in plasma of pregnant, lactating and cycling ewes during vaginal stimulation. *Endocrinology, 83,* 272.

Rodriguez-Sierra, J. F., Crowley, W. R., & Komisaruk, B. R. (1977). Induction of lordosis responsiveness by vaginal stimulation in rats is independent of anterior or posterior pituitary hormones. *Hormones and Behavior, 8,* 348–355.

Rosenblatt, J. S., & Siegel, H. I. (1981). Factors governing the onset and maintenance of maternal behavior among nonprimate mammals. In D. J. Gubernick & P. H. Klopfer (Eds.), *Parental care in mammals* (pp. 1–76). New York: Plenum Press.

Rubin, B. S., Menniti, F. S., & Bridges, R. S. (1983). Intracerebral administration of oxytocin and maternal behavior in rats after prolonged and acute steroid pretreatment. *Hormones and Behavior, 17,* 45–53.

Sarkar, D. K., & Gibbs, D. M. (1984). Cyclic variation of oxytocin in the blood of pituitary portal vessels of rats. *Neuroendocrinology, 39,* 481–483.

Sawchencko, P. E., & Swanson, L. W. (1982). Immunohistochemical identification of neurons in the paraventricular nucleus of the hypothalamus that project to the medulla or to the spinal cord in the rat. *Journal of Comparative Neurology, 205,* 260–272.

Schumacher, M., Coirini, H., Frankfurt, M., & McEwen, B. S. (1989). Localized actions of progesterone in hypothalamus involve oxytocin. *Proceedings of the National Academy of Sciences, U.S.A., 86,* 6798–6801.

Shapiro, L. E., & Dewsbury, D. A. (1990). Differences in affiliative behavior, pair bonding, and vaginal cytology in two species of vole. *Journal of Comparative Psychology, 104,* 268–274.

Shapiro, L. E., & Insel, T. R. (1989). Ontogeny of oxytocin receptors in rat forebrain: A quantitative study. *Synapse, 4,* 259–266.

Shapiro, L. E., & Insel, T. R. (1990a). Infant's response to social separation reflects adult differences in affiliative behavior: A comparative developmental study in prairie and montane voles. *Developmental Psychobiology, 23,* 375–394.

Singh, P. J., & Hofer, M. A. (1978). Oxytocin reinstates maternal olfactory cues for nipple orientation and attachment in rat pups. *Physiology and Behavior, 20,* 385–389.

Snijdewint, F. G. M., Van Leeuwen, F. W., & Boer, G. J. (1989). Ontogeny of vasopressin and oxytocin binding sites in the brain of Wistar and Brattleboro rats as demonstrted by light microscopical autoradiography. *Journal of Chemistry and Neuroanatomy, 2,* 3–17.

Sofroniew, M. V., & Weindl, A. (1981). Central nervous system distribution of vasopressin, oxytocin, and neurophysin. In J. L. Martinez, R. A. Jensen, R. B. Mesing, H. Rigter, & J. L. McGaugh (Eds.), *Endogenous peptides and learning and memory processes* (pp. 327–369). New York: Academic Press.

Soloff, M. S., Alexandrova, M., & Fernstrom, M. J. (1979). Oxytocin receptors: Triggers for parturition and lactation? *Science, 204,* 1313–1314.

Stanton, M. E., Wallstrom, J., & Levine, S. (1987). Maternal contact inhibits pituitary-adrenal stress responses in preweanling rats. *Developmental Psychobiology, 20,* 131–145.

Stock, S., & Uvnas-Moberg, K. (1988). Increased plasma levels of oxytocin in response to afferent electrical stimulation of the sciatic and vagal nerves and in response to touch and pinch in anesthetized rats. *Acta Physiologica Scandinavica, 132,* 29–34.

Stoneham, M. D., Everitt, B. J., Hansen, S., Lightman, S. L., & Todd, K. (1985). Oxytocin and sexual behavior in the male rat and rabbit. *Journal of Endocrinology, 107,* 97–106.

Swanson, L. W., & Kuypers, H. G. J. M. (1980). The paraventricular nucleus of the hypothalamus: Cytoarchitectonic subdivisions and organization of projections to the pituitary, dorsal vagal complex, and spinal cord as demonstrated by retrograde fluorescence double-labeling methods. *Journal of Comparative Neurology, 194,* 555–570.

Tamborski, A., Lucot, J. B., & Hennessy, M. B. (1990). Central dopamine turnover in guinea pig pups during separation from their mothers in a novel environment. *Behavioral Neuroscience, 104,* 607–611.

Theodosis, D. T., Montagnese, C., Rodriguez, F., Vincent, J., & Poulain, D. A. (1986). Oxytocin induces morphological plasticity in the adult hypothalamo-neurohypophysial system. *Nature (London), 322,* 738–740.

Theodosis, D. T., & Poulain, D. A. (1987). Oxytocin-secreting neurones: A physiological model for structural plasticity in the adult mammalian brain. *Trends in Neuro Science, 10(10),* 426–430.

Tribollet, E., Audigier, S., Dubois-Dauphin, M., & Dreifuss, J. J. (1990). Gonadal steroids regulate oxytocin receptors but not vasopressin receptors in the brain of male and female rats. An autoradiographical study. *Brain Research, 511,* 129–140.

Tribollet, E., Charpak, S., Schmidt, A., Dubois-Dauphin, M., & Dreifuss, J. J. (1989). Appearance and transient expression of oxytocin receptors in fetal, infant, and peripubertal rat brain studied by autoradiogrphy and electrophysiology. *Journal of Neuroscience, 9(5),* 1764–1773.

van Leengoed, E., Kerker, E., & Swanson, H. H. (1987). Inhibition of postpartum maternal behavior in the rat by injecting an oxytocin antagonist into the cerebral ventricles. *Journal of Endocrinology, 112,* 275–282.

Van Leeuwen, F. W., Heerikhuize, J. V., Van der Meulen, G., & Wolters, P. (1985). Light microscopic autoradiographic localization of ^{3}H-oxytocin binding sites in the rat brain, pituitary, and mammary gland. *Brain Research, 359,* 320–325.

van Tol, H. H. M., Bolwerk, E. L. M., Liu, B., & Burbach, J. P. H. (1988). Oxytocin and vasopressin gene expression in the hypothalamo-neurohypophyseal system of the rat during the estrous cycle, pregnancy, and lactation. *Endocrinology, 123(3),* 945–951.

van Wimersma Greidanus, T. B., Kroodsma, J. M., Pot, M. L. H., Stevens, M., & Maigret, C. (1990). Neurohypophyseal hormones and excessive grooming behavior. *European Journal of Pharmacology, 187,* 1–8.

Voorn, P., & Buijs, R. M. (1983). An immuno-electronmicroscopical study comparing vasopressin, oxytocin, substance P and enkephalin containing nerve terminals in the nucleus of the solitary tract of the rat. *Brain Research, 270,* 169–173.

Wamboldt, M. Z., & Insel, T. R. (1987). The ability of oxytocin to induce short latency maternal behavior is dependent on peripheral anosmia. *Behavioral Neuroscience, 101,* 439–441.

Wiener, S. G., Bayart, F., Faull, K. F., & Levine, S. (1990). Behavioral and physiologic responses to maternal separation in squirrel monkeys. *Behavioral Neuroscience, 104,* 108–115.

Winslow, J. T., & Insel, T. R. (1991a). Endogenous opioids: Do they mediate the rat pup's response to social isolation? *Behavioral Neuroscience, 105,* 253–263.

Winslow, J. T., & Insel, T. R. (1991b). Social status in pairs of male squirrel monkeys determines response to central oxytocin administration. *Journal of Neuroscience, 11,* 2032–2038.

Witt, D. M., Carter, C. S., & Insel, T. R. (1991). Oxytocin receptor binding in female prairie voles: Endogenous and exogenous estradiol stimulation. *Journal of Neuroendocrinology, 3,* 155–161.

Witt, D. M., & Insel, T. R. (1990). *Interation of gonadal steroids and oxytocin: Effects on sexual and social behavior in rats.* (Abstr.) Atlanta, Georgia.

Witt, D. M., & Insel, T. R. (1991). A selective oxytocin antagonist attenuates progesterone facilitationof female sexual behavior. *Endocrinology, 128,* 3269–3276.

Yang, Q. Z., & Hatton, G. I. (1988). Direct evidence for electrical coupling among rat supraoptic nucleus neurons. *Brain Research, 463,* 47–56.

Neurohormones in Depression and Anxiety

Margaret Altemus[*] and Philip W. Gold[†]

Laboratory of Clinical Science and
†*Clinical Neuroendocrinology Branch*
Division of Intramural Research Programs
National Institute of Mental Health
Bethesda, Maryland 20892

I. Introduction

Depression and anxiety disorders are accompanied by prominent physiological symptoms including changes in appetite and sleep, loss of libido, augmentation of startle responses, and disruption of circadian rhythms. These observations have focused investigations into the pathophysiology of these disorders on neuroendocrine functions. We present here a model of depression and anxiety as dysregulation of central stress-responsive systems and emphasize the potential role of central nervous system (CNS) neuropeptides and noradrenergic systems, attempting to integrate animal studies with studies of humans suffering from depression and anxiety.

Multiple hypothalamic releasing hormones and peptides also are synthesized in cell bodies outside the hypothalamus, are distributed widely throughout the brain, and act in the CNS to exert specific receptor-mediated biological actions and coordinate complex behavioral and physiological processes. Dissociation of the location of neuropeptides from their receptors is the rule rather than the exception in the CNS (Herkenham, 1987), suggesting that hormonally mediated events constitute a significant mode of information processing in the brain. In support of this model is evidence that small proteins can diffuse readily from the interstitial space along the intraparenchymal vasculature into the cerebroventricular system (Rennel, Gregory, Blaunanis, Fujimoto, & Grady, 1985).

In animals, hormones and peptides and their antagonists can be administered intracerebroventricularly (icv) or administered in specific CNS locations to study behavioral effects. To infer the behavioral activity of these hormones in humans, however, we must rely on cerebrospinal fluid (CSF) measures obtained during episodes of psychiatric illness. Human studies of hypothalamic peptide release into plasma in response to peripheral administration of centrally active drugs are also available, but whether hypothalamic peptide release into plasma reflects the more behaviorally relevant peptide activity in other parts of the brain is often unclear.

We first discuss the role of arousal-producing neuropeptides and noradrenergic systems in adaptive responses to stress; then we describe how dysregulation of these responses may produce the symptomatology of depression and anxiety disorders.

II. Arousal, Anxiety, and Depression

Arousal can occur in response to a wide range of stimuli, pleasant or unpleasant, real or imagined. Arousal in response to threat or novelty is an adaptive response including such features as a subjective feeling of foreboding, autonomic nervous system activation, vigilance, and a narrowed focus of attention. Prolonged or excessive arousal, however, has maladaptive features including anxiety, insomnia, and weight loss. Similarly, sadness is an inevitable response to loss of a valued attachment or expectation but severe, pervasive, and prolonged sadness also impairs functioning. Although no clear demarcation exists between normal reactions and pathological states, the Diagnostic and Statistical Manual of Mental Disorders (DSM-IIIR) published by the American Psychiatric Association (APA, 1987) lists current diagnostic criteria for a number of anxiety and depressive disorders.

Anxiety symptoms were described as distinct from depressive symptoms by Darwin (1872), Freud (1926) and Kraepelin (1946), but most work in the phenomenology and biological basis of anxiety as distinct from depression has occurred in the last 20 years, spurred by the development of benzodiazepine medications which preferentially relieve anxiety rather than depressive symptoms. However, comorbidity of anxiety and depression in individuals and their family members is very high (Leckman, Merikangas, Pauls, Prusoff, & Weissmann, 1983). In fact, approximately 60% of patients with major depression meet criteria for generalized anxiety disorder, and 30–90% of anxiety disorder patients have a history of major depression (Noyes, 1988). This association may be the result of depression and demoralization emerging in response to prolonged anxiety symptoms or may reflect common underlying biological mechanisms. Much comorbidity also exists among individuals and their family members for a number of different anxiety disorders (Boyd et al., 1984; Kendler, Heath, Martin, & Eaves, 1986). Again these familial associations may reflect diagnostic ambiguities as well as common underlying physiological processes.

The success of behavioral and cognitive therapies points to the conditioned nature of depressive and anxiety states, whereas family studies support the involvement of genetic factors in depression and anxiety disorders. These etiological models are most likely interdependent, since conditioning is enhanced by physiological arousal. Thus, maladaptive conditioning may be exacerbated in persons with constitutional vulnerabilities to hyperarousal to stimuli. Prospective studies of abused children (Cole & Putnam, 1992) and children with constitutionally reactive nervous systems (Kagan, Reznick, & Snidman, 1987) will help clarify these interactions.

III. Central Arousal Systems

The two principal central effectors of the generalized stress response are the hypothalamic–pituitary–adrenal (HPA) axis and the locus ceruleus–norepinephrine (LC–NE) system (Figure 1). The hypothesis that adrenal cortical and sympathetic nervous system activity can reflect the degree of emotional arousal or distress of an individual has been studied extensively in the past several decades. Over time, a consensus has developed that novelty and unpredictability, particularly if they overwhelm coping resources (Levine & Whener, 1989), are potent inducers of adrenal cortical and sympathetic nervous system activation. We review some preclinical studies that indicate that these two symptoms, as well as other arousal producing neuropeptides, exert many overlapping functions and seem to participate in a mutually reinforcing, positive feedback loop.

The well-documented rise in plasma cortisol in response to stress is initiated by release of corticotropin releasing hormone (CRH) from the paraventricular nucleus of the hypothalamus. CRH is transported through the median eminence and the portal blood vessels to cause adrenal corticotropic hormone (ACTH) release from the pituitary. ACTH then acts at the adrenals to stimulate release of glucocorticoids. The HPA axis, like all neuroendocrine axes, has feedback loops that maintain a homeostatic set point. Glucocorticoids and ACTH act via hypothalamic and pituitary feedback to limit activation of the system (Dallman et al., 1987).

Hypothalamic CRH is released not only into the portal system to cause ACTH release at the pituitary, but also through projections to numerous areas of the brain including the hippocampus, amygdala, and other areas of the limbic system

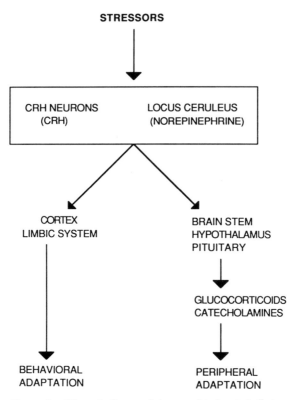

Figure 1. Schematic diagram of stress-mediated central effectors.

(Swanson and Sawchenko, 1980) where the emotional or learned aspects of the stress or fear responses are likely to be mediated (Panksepp, 1990). Although the relationship is far from clear, some evidence suggests that CRH release into the pituitary portal blood vessels and other areas of the brain are regulated in parallel. For example, both hypothalamic and extrahypothalamic CRH levels are increased after exposure to stress (Chappell et al., 1987). Moreover, adrenalectomized and hypophysectomized animals show increased CRH content in the amygdala and the bed nucleus of the stria terminalis, as well as in the paraventricular nuclei of the hypothalamus (Sawchenko, 1987).

The other major central component of the generalized stress response, the locus ceruleus (LC), is a nucleus of noradrenergic neurons located in the mid-pons with terminal fields in the hypothalamus, hippocampus, and amygdala, and throughout the cerebral cortex (Moore & Bloom, 1979). The LC provides much of the brain's supply of norepinephrine, numerous findings have demonstrated the importance of the LC in mediating arousal. Electrical stimulation of the LC in unanesthetized primates produces intense anxiety, hypervigilance, and inhibition of exploratory behavior, and spontaneous firing of the LC increases during threatening situations and diminishes during sleep, grooming, and feeding (Aston-Jones, Foote, & Bloom, 1984). In addition, acute stress in animals causes increased release of norepinephrine in several brain areas, including the hypothalamus and LC (Glavin, 1985) and increased production of tyrosine hydroxylase, the rate-limiting enzyme for norepinephrine synthesis, in the LC (Smith et al., 1991). Reports of the effects of chronic stress on brain norepinephrine levels and the activity of tyrosine hydroxylase are mixed (Weiss, Glazer, Pohorecky, Brick, & Miller, 1975; Smith et al., 1991). Although the LC is related closely anatomically to nuclei of the peripheral sympathetic nervous system and sympathomedullary system, both of which release catecholamines in the periphery during arousal, the degree of functional integration of the central and peripheral noradrenergic systems remains to be defined (Vieth, 1991).

Several lines of evidence indicate that the CRH and LC systems may reinforce each other's functional activity. Anatomic studies demonstrate paraventricular CRH neuron projections to autonomic nuclei and CRH-containing cell bodies and fibers in multiple brainstem autonomic nuclei, including the LC (Olschoska, O'Donohue, Mueller, & Jacobowitz, 1982). CRH-containing neurons in the central nucleus of the amygdala and the bed nucleus of the stria terminalis also have been shown to project to brainstem autonomic nuclei (Sakanaka, Shibasaki, & Lederis, 1986). In physiological studies, central administration of CRH produces stress-like changes in peripheral sympathetic and parasympathetic nervous system activity and adrenomedullary outflow (Brown et al., 1982; Kurosawa, Sato, Swenson, & Takahashi, 1986). In addition, the direct application of CRH onto LC neurons in intact rats markedly increases the LC firing rate (Valentino, Foote, & Aston-Jones, 1986). Propranolol, a beta-adrenergic receptor blocker, has been shown to attenuate the arousal producing effects of centrally administered CRH (Cole & Koob, 1988). Conversely, norepinephrine is a potent stimulator of CRH release from the hypothalamus (Calogero, Galluci, Chrousor, & Gold, 1988); noradrenergic cells in the nucleus tractus solitarius, and other brainstem nuclei of the autonomic nervous system, project to CRH neurons in the paraventricular nucleus (Cunningham, Bone, & Sawchenko, 1990).

The distribution of CRH within and beyond the boundaries of the hypothalamus and its close association with the LC–NE system provide an anatomical context for the observation that these systems simultaneously can activate and coordinate a series of metabolic, cardiovascular, and behavioral responses that are adaptive during threatening or stressful situations. For example, in the rat, the icv administration of CRH leads not only to activation of the pituitary–adrenal axis (producing energy mobilization to meet acute metabolic demands) and of the sympathetic nervous system, but also to several behavioral changes that are characteristic of the stress response. These changes include increased context-dependent changes in motor activity, decreased feeding, hypothalamic hypogonadism, and decreased sexual behavior. In larger doses, CRH produces behavioral effects that can be construed as anxiogenic, including hyperresponsiveness to acoustic startle, improved acquisition of learned responses, and decreased exploration in an open field (Dunn & Berridge, 1990; Koob, 1991).

Electrical brain stimulation studies and lesion studies have identified the central nucleus of the amygdala as important in the generation of the conditioned and affective components of stress and fear responses (Turner, Mishkin, & Knapp, 1980; Hitchcock and Davis, 1986). Anatomical and physiological data suggest that activation of the amygdala by the LC–NE system enhances retrieval of emotional and learned information pertinent to the stressor. Then efferent projections from the amygdala promote further stimulation of the CRH and LC–NE systems (Kapp, Pascoe, & Bixle, 1985; Gray, 1991).

Several brain neurotransmitters and peptides can modulate the activity of the CRH as well as the LC–NE system and may play a role in the development of psychiatric syndromes (Table 1). *In vitro* studies show that acetylcholine (Calogero et al., 1988) and serotonin (Calogero et al., 1990) are potent stimuli to the hypothalamic CRH neuron and the LC–NE system (Guynet & Aghajanian, 1979; Bagdy, Calogero, Murphy, & Szemerdy, 1989), whereas γ-aminobutyric acid (GABA) (Foote, Bloom, & Aston-Jones, 1983; Kalogeras et al., 1991), glucocorticoids (Calogero et al., 1988; Szemeredi et al., 1988), and the opiod peptides (Yajima et al., 1986; Pepper & Henderson, 1990) are inhibitory. *In vivo* studies measuring portal blood levels of CRH or plasma levels of ACTH after stimulation with chemical stimuli generally agree with the *in vitro* studies (Fisher & Brown, 1991). The LC–NE and CRH–glucocorticoid systems do not constitute the only central neuroendocrine and neurotransmitter responses to stress. Vasopressin in particular is known to be stress responsive (Angulo, Ledoux, & McEw-

TABLE I
Central Nervous System Modulation of Arousal and
the Stress Response

Stimulatory	Inhibitory
CRH	GABA
Norepinephrine	Glucocorticoids
Acetylcholine	Opiod peptides
Serotonin	Oxytocin
Vasopressin	Somatostatin
TRH	

en, 1991) and arousal producing, to have synergistic effects with CRH at the pituitary and centrally on behavior (Elkabir, Wyatt, Vellacci, & Herbert, 1990), and to potentiate noradrenergic activity at the hippocampus (Brinton & McEwen, 1989). Like CRH, vasopressin is synthesized in the hypothalamus, bed nucleus of the stria terminalis, amygdala, and LC. From the hypothalamus, projections extend to the forebrain, autonomic nuclei of the brain stem, and limbic system (Bujis, 1990). Conversely, multiple brainstem noradrenergic nuclei send projections to vasopressin-containing cells in the hypothalamus (Cunningham et al., 1990). These anatomical and functional connections suggest a close integration of the vasopressin and CRH/LC–NE systems. Similarly, somatostatin is released in response to stress and produces arousal (Halicek, Rezek, & Friesen, 1976). Somatostatin neurons are localized in the periventricular nucleus of the hypothalamus, the amygdala, the bed nucleus of the stria terminalis, and the LC (Geola, Yamada, Warwick, Tourtelotte, & Hersham, 1981), and descend from the hypothalamus and the amygdala to noradrenergic brain stem nuclei (Swanson & Sawchencko, 1980; Higgins & Schwaber, 1983). Like CRH and the LC–NE system, vasopressin and somatostatin activity seem to be suppressed by glucocorticoids (Davis et al., 1986; Wolkowitz et al., 1990). Thyrotropin releasing hormone (TRH), another arousal producing neuropeptide, also seems to be released in response to acute stress. Oxytocin and opioid peptides also are released centrally in response to stress, and seem to play a role in stress-related analgesia and counterregulation of CRH and NE activation (Plotsky, 1986; Overton & Fisher, 1989). The interplay of these stress-responsive systems allows the maintenance of homeostasis while responding to changes in the environment.

IV. Pathophysiology of Hyperarousal

Although the coping response of an individual will be determined partly by social milieu, experience, and cognitive maturity, a significant genetic component appears to contribute to individual variability in stress responsiveness and vulnerability to anxiety disorders (Last, Hersen, Kazdin, Orvaschel, & Perrin, 1991). Inhibited or fearful reactivity to novel stimuli in childhood is a trait that appears to show significant stability from infancy into childhood and adolescence (Thomas, Chess, & Korn, 1982; Kagan et al., 1987). Twin studies in humans suggest that this trait is inherited (Daniels & Plomin, 1985; Bouchard, Lykken, McGue, Segal, & Tellegen, 1990). Hyperresponsivity to novel stimuli and inhibited traits also have been identified in a number of animal species, appear to be stable across the life-span of individual animals, and also predict vulnerability to maladaptive, anxiety-like, or depression-like responses to stressful circumstances, especially separation from familiar animals (Suomi, 1987). Physiological studies indicate that the degree of activation of the stress-responsive HPA axis and autonomic nervous system is correlated closely to measures of behavioral inhibition in both children (Kagan et al., 1987) and monkeys (Suomi, 1987).

Despite the evidence of inherited sensitivity of stress reactivity, animal work also suggests that hyperresponsiveness to stress may be created by repeated stimulation of central stress systems. In rhesus monkeys, intraindividual differences in reactions to social separations show correlations with the separation responses of genetic relatives, but also with environmental variables, such as the mothering style of natural or foster mothers, and with variations in early rearing experiences (Suomi, 1987). Rats subjected to maternal deprivation during the neonatal period or situations of inescapable stress [the "learned helplessness" model of depression first noted by Curt Richter (Richter, 1957)] show long-lasting hyperactivity of the HPA axis in response to stress in adulthood (Thoman, Levine, & Arnold, 1968). Similarly, children with poor social supports and poor adjustment after loss of a parent have been reported to have higher plasma cortisol concentrations in adulthood (Breier et al., 1988). In the inescapable stress paradigm, central noradrenergic activity is increased also (Tsuda & Tanaka, 1985), and GABA release, which inhibits CRH and LC activation, is attenuated (Petty & Sherman, 1981). In contrast to deprivation, other types of environmental manipulations in the neonatal period can reduce HPA responses to stress later in life (Meany et al., 1991).

These findings are intriguing in light of clinical evidence in humans that repeated trauma predisposes an individual to depression and anxiety disorders. The results raise the possibility that a "kindling mechanism" could be involved in altering individual vulnerability to depression or other maladaptive neuroendocrine and behavioral responses to stress. Kindling refers to the process by which the repeated application of electrical or pharmacological stimulations to the brain that is originally insufficient to produce seizures eventually leads to a long-lasting alteration in seizure threshold, so the previously subthreshold stimuli produce motor seizures (Goddard, MacIntyre, & Leech, 1969). Although when defined narrowly, kindling refers to a process resulting in the production of motor seizures, the model also may be applicable to other long-term alterations in neuronal excitability. Repeated subthreshold stimulation of the amygdala can kindle limbic seizures (Adamec & Stark-Adamec, 1986); icv administration of CRH (Weiss et al., 1986) can sensitize the amygdala to kindling, and depletion of somatostatin inhibits kindling (Nemeroff, Kaliva, Golder, & Prange, 1984). Thus, repeated stress could increase the sensitivity of stress-responsive neuronal systems in the limbic system and contribute to the development of posttraumatic anxiety and depressive disorders. In support of this model, lasting inhibited behavior has been produced in noninhibited cats by repeated electrical stimulation of the amygdala. In addition, inhibited cats show greater propagation of electrical stimuli from the amygdala than noninhibited cats (Adamec & Stark-Adamec, 1986).

V. Psychiatric Syndromes—Hyperarousal

Exaggerated physiological reactivity may be a trait that places individuals at risk for development of depression and other psychiatric disorders. In these individuals, heightened reactivity to novel stimuli could impair development of social skills, self-esteem, and a sense of self-regulation, which in turn would exacerbate the aversive qualities of novel stimuli. We discuss several psychiatric syndromes that can be framed as disorders of hyperarousal. In the following section, we describe syndromes that seem to represent disorders of hyperarousal (Table 2).

A. Melancholic Depression

Melancholic depression, a highly consistent association of symptoms, points to disturbance of arousal systems. The hallmark of melancholic depression is an intensely painful dysphoric hyperarousal, excessive guilt, and obsessional preoccupation with personal inadequacy and the inevitability of loss. Patients with this syndrome display early morning awakening, loss of appetite, and worsening of mood in the morning.

Pathological activation of the HPA axis in melancholic depression is the most consistent and most widely investigated biological abnormality in psychiatry. Several types of measures have been used to demonstrate activation of the HPA axis in melancholic depression. Blood sampling studies over 24 hr have shown elevation of plasma cortisol in patients with melancholic depression, with preservation of a circadian rhythm (Sachar et al., 1973). Measurements of urinary free cortisol are elevated also (Carroll, 1976). The dexamethasone suppression test has been studied widely in these patients, since melancholically depressed patients tend not to suppress cortisol secretion in response to a dose of the long-acting synthetic glucocorticoid dexamethasone (Gwirtsman, Gerner, & Stembach, 1982). Finally, functional and anatomical adrenal hypertrophy occur in patients with melancholic depression (Amsterdam, Winokur, Abelman, Lucki, & Rickels, 1983;

TABLE 2
Disorders Associated with Dysregulation of the Arousal and Stress System

Hyperarousal	Hypoarousal
Melancholic depression	Atypical depression
Obsessive–compulsive disorder	Cushing's syndrome
Panic disorder	Hypothyroidism
Post traumatic stress disorder	Seasonal affective disorder
Anorexia nervosa	Chronic fatigue syndrome
	Bulimia nervosa

Dorovini-Zis & Zis, 1987), consistent with the hypertrophy that occurs in response to chronic administration of exogenous ACTH in humans.

Some evidence suggests that hypercortisolemia in major melancholic depression reflects hypothalamic hypersecretion of CRH rather than autonomous hypersecretion of ACTH or cortisol. First, plasma ACTH responses to synthetic CRH are attenuated in major depression, indicating that the pituitary corticotrope cell is restrained appropriately by the negative feedback effects of elevated glucocorticoids (Gold et al., 1986). Second, normal controls given a continuous infusion of CRH have a pattern and magnitude of hypercortisolemia consistent with that seen in depression (Schulte et al., 1985). Third, CRH in the CSF of depressed patients correlates positively with indices of pituitary–adrenal activation (Roy et al., 1987). In addition, elevated levels of CRH in the CSF of patients with major depression have been reported (Nemeroff et al., 1985).

Several studies in recent years also point to activation of the LC–NE system in melancholic depression. Increased CSF, plasma, and urinary levels of norepinephrine and a major metabolite, 3-methoxy-4-hydroxyphenyl glycol (MHPG), have been described. In addition beta-adrenergic receptor function is down-regulated. Positive correlations between CSF norepinephrine and CSF CRH in depression also have been reported (Vieth, 1991).

In a well-functioning homeostatic system, after CRH stimulates ACTH and cortisol release, cortisol feeds back to contain the activation of the HPA, LC–NE, and sympathetic nervous systems (Munck, Guyre, & Holbrook, 1984; Brown & Fisher, 1986; Szemeredi et al., 1988). In this regard, one could conceptualize the melancholic depressive syndrome and other disorders of hyperarousal as pathological activations of the principal effectors of the generalized adaptational response that have escaped their usual glucocorticoid-mediated counterregulation (Table 3).

Successful responses to antidepressant medications, even those that do not act principally on the norepinephrine system are associated consistently with decreases in CSF and plasma MHPG (Linnoila, Karoun, Calil, Kopin, & Potter, 1982). These clinical data are consistent with preclinical data indicating that monoamine oxidase inhibitors and tricyclic antidepressants decrease the firing rate of the LC (Nyback, Walters, Aghajanian, & Roth, 1975; Svensson & Usdin, 1978) and reduce the level of norepinephrine metabolites in the brain (Nielsen & Braestrup, 1977). Also, tricyclics and electroconvulsive treatment down-regulate the cortical beta-receptors (Sulser, 1983), which are thought to mediate the arousal producing effects of norepinephrine. Also, consistent is the finding that, in rats, tricyclic antidepressant medications decrease expression of messenger RNA for synthesis of both CRH and tyrosine hydroxylase (Brady, Whitfield, Fox, Gold, & Herkenham, 1991). Tricyclics also increase the expression of type 1 mineralocorticoid receptor genes (Brady et al., 1991) and the density of glucocorticoid receptors (Kitayamia et al., 1988) in the hippocampus, both of which are thought to play a principal role in restraining CRH-mediated pituitary–adrenal activation.

TABLE 3

Parallels between the Acute Stress Response and Psychiatric Syndromes
of Hyperarousal

Type of change	Stress	Depression

Redirection of the behavior by the central nervous system	Acute facilitation of adaptive neural pathways
	   Arousal, alertness, increased vigilance, focused attention, aggressiveness when appropriate
		Chronic maladaptive facilitation of neural pathways
		   Dysphoric hyperarousal and anxiety; hypervigilance, constricted focus, obsessionalism; assertiveness inappropriately restrained by anxiety

	Acute inhibition of nonadaptive pathways
	   Decreased eating; decreased libido and sexual behavior; appropriate caution or restraint
		Maladaptive inhibition of neural pathways
		   Decreased eating; decreased libido and sexual behavior; excessive caution, regardless of context

Redirection of energy in the periphery	Oxygen and nutrients to the stressed body site
	   Increased blood pressure, heart, and respiratory rate; increased gluconeogenesis; increased lipolysis; inhibition of programs for growth and reproduction
		Oxygen and nutrients to the central nervous system
		   Increased blood pressure, heart, and respiratory rate; increased gluconeogenesis; increased lipolysis; inhibition of programs for growth and reproduction

	Acute glucocorticoid-mediated counter-regulatory responses (containment)
	   Restraint of the corticotropin-releasing-hormone system and the pituitary–adrenal axis; restraint of the norepinephrine–locus ceruleus system
		Chronic inadequate or maladaptive counterregulatory responses (containment)
		   Inadequate restraint of the corticotropin-releasing-hormone system and the pituitary–adrenal axis; inadequate restraint of the norepinephrine–locus, ceruleus system

Hypersecretion of TRH also may play a role in the symptom complex of melancholic depression. In two of three studies, CSF TRH levels have been reported to be elevated in depressed patients (Kirkegaard, Faber, Jummer, & Rogowski, 1979; Banki, Bisette, Arato, & Nemeroff, 1988). In animals, TRH reverses hibernation and barbiturate-induced sedation, stimulates locomotor activity and peripheral sympathetic nervous system activity, and inhibits food and water intake (Nemeroff et al., 1984). Thyroid stimulating hormone responses to TRH infusion often are blunted in depression (Loosen, 1988), consistent with increased hypothalamic secretion of TRH. In addition, depressed patients have been reported to show attenuation of the normal nocturnal rise of thyroid stimulating hormone release from the pituitary (Kjellman, Beck-Friis, Lunggren, & Wetterber, 1984; Bartalena et al., 1990), a pattern that is seen in normals with chronic TRH administration (Spencer, Greenstadt, Wheeler, Kletzky, & Nicoloff, 1980) and in response to stress and increased levels of glucocorticoids (Bartalena et al., 1990,1991).

Centrally directed vasopressin and somatostatin, as assessed by the measurement of vasopressin in CSF, is decreased in depressed patients compared with controls (Gjerris, Hummer, Vendsborg, Christienson, & Rafaelson, 1985; Gold, Goodwin, Post, & Robinson, 1981), possibly in response to hypercortisolimia. Reduced release of centrally directed vasopressin theoretically could contribute to the memory disturbances that are an intrinsic component of the depressive syndrome. In animals, icv vasopressin delays the extinction of behaviors acquired during aversive conditioning and enhances retrievability of information learned during a stress situation (Koob, 1988). Terminal sites of centrally directed vasopressin neurons include several areas known to be involved in memory processing, for example, the hippocampus and amygdala (Kozlowski & Nilaver, 1986).

In summary, in melancholic depression, activity of LC–NE systems and the arousal producing neuropeptides CRH and TRH seems to be enhanced. These changes may play a role in the symptom complex of melancholic depression and the mechanism of action of antidepressant medications.

B. Obsessive–Compulsive Disorder

Obsessive–compulsive disorder is another illness with symptoms pointing to overactivation of arousal systems. Patients with this illness are plagued by recurrent intrusive thoughts that often involve the fear that some potential danger has been left unchecked or that they are about to perform an act that is harmful to themselves or others. Compulsive, stereotyped, repetitive behaviors or cognitions such as handwashing, hoarding, or counting often are conducted to magically forestall the imagined danger and to relieve anxiety. Behavioral treatment is particularly effective in this illness, suggesting that these behaviors are conditioned responses.

We found that patients with obsessive–compulsive disorder demonstrate enhanced secretion of vasopressin (Figure 2) and CRH (Figure 3) and somatostatin centrally, and enhanced secretion of vasopressin into the plasma in response to osmotic stimulation (Altemus et al., 1991b; 1992). The finding of elevated levels of vasopressin in the CSF and plasma of patients with obsessive–compulsive disorder may contribute to their symptomatology in light of data in experimental animals that central administration of vasopressin (Koob, Lebrun, Bluthe, Dantzer, & LeMoal, 1989), CRH (Koob, 1988) and somatostatin (Vescei, Bollock, Penke, & Telegdy, 1986) delays the extinction of active avoidance behaviors whereas

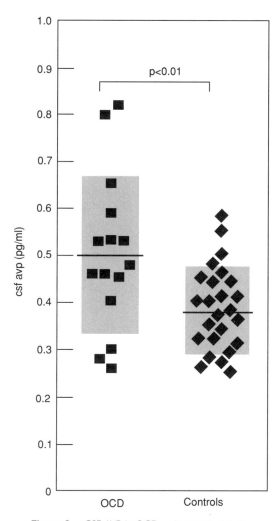

Figure 2. CSF AVP in OCD patients and controls.

central administration of vasopressin antagonists enhances extinction (Koob et al., 1989).

Preliminary data in unimpaired human subjects also show that retention of learned material is enhanced by intranasal administration of vasopressin and vasopressin analogs (Weingartner et al., 1981; Millar, Jeffcoate, & Walder, 1987). The reported effects of vasopressin, CRH, and somatostatin to enhance retention of learned information and behaviors may be analogous to the great difficulty

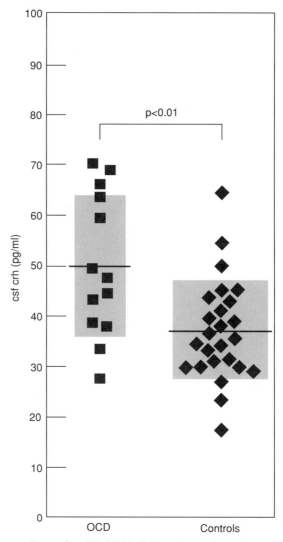

Figure 3. CSF CRH in OCD patients and controls.

patients with obsessive–compulsive disorder have in relinquishing anxiety-related thoughts or familiar rituals. In addition, a number of forms of laboratory- and stimulant-induced arousal also can enhance retention of learned information (Baddeley, 1972; McGaugh, Wang, Bennett, & Sternberg, 1984). Although strengthening memories of information acquired during stressful circumstances and narrowing the focus of attention during stress is adaptive, a constitutional vulnerability toward overactivation of these mechanisms could produce rigid perseverative pathology such as melancholic depression or obsessive–compulsive disorder. Because CSF vasopressin and CRH have not been examined in other anxiety disorders, whether the elevated peptide levels are specific to obsessive–compulsive disorder or whether they are characteristic of anxiety disorders in general remains to be seen.

Patients with obsessive–compulsive disorder often show symptoms of major depression. Unlike patients with melancholic depression, subjects with obsessive–compulsive disorder are not hypercortisolimic (Coryell, Black, Kelly, & Noyes, 1989). This characteristic may account for the differential elevation of vasopressin and somatostatin in obsessive–compulsive disorder since, in melancholic depression, hypercortisolemia seems to suppress release of somatostatin (Rubinow et al., 1983), although the relationship of plasma cortisol and CSF vasopressin has not been explored. As mentioned earlier, reduced activity of vasopressin and somatostatin may contribute to the memory and concentration deficits in depression and stand in contrast to the overconcern with details, consequences, and checking in obsessive–compulsive disorder. Further indirect evidence for effects of these peptides on cognition is that CSF and brain levels of CRH, vasopressin, and somatostatin are reduced in dementing conditions, including Parkinson's disease (Dupont, Christenson, Hanen, Olivarius, & Orskov, 1982) and Alzheimer's disease Sorenson, Gerris, & Hammer, 1985; Nemeroff, Kizer, Reynolds, & Bisette, 1989), which are characterized by decreased ability to attend to and anticipate consequences.

C. Panic Disorder

Panic disorder is characterized by discrete episodes of intense anxiety and somatic symptoms of autonomic arousal, lasting 15–60 min and occurring up to several times per day. Between panic episodes, patients often experience low grade persistent anxiety and tend to become easily aroused in unfamiliar or novel situations. A high incidence of depression is found in patients with panic disorder and in their family members (Brier, Charney, & Heninger, 1984). In addition, panic disorder patients respond well to treatment with a wide range of antidepressant medications. The central anxiogenic effects of CRH and the LC system dovetail with the cardinal symptoms of central activation during panic attacks. However, blood sampling studies and heart rate recordings of patients during panic attacks have shown only inconsistent and moderate activation of the peripheral sympathetic nervous system (Taylor et al., 1986; Vieth, 1991) and little evidence of pituitary adrenal activation (Cameron, Lee, Curtis, & McCann, 1987; Woods et al., 1987). Most studies of panic patients during the basal state between panic attacks have not found consistent evidence of increased peripheral sympathetic nervous system reactivity (Vieth, 1991; Roth et al., 1992) or hypercortisolemia (Curtis & Glitz, 1988; Uhde, Joffe, Jimerson, & Post, 1988).

Patients with panic disorder do seem to show exaggerated responses to alpha-2 adrenergic agonists and antagonists, suggesting increased central receptor sensitivity (Woods et al., 1987b; Nutt, 1986). These data suggest that central or behavioral activation occurs relatively independently of sympathetic nervous system activation. Three substances known to precipitate panic attacks—lactate (Liebowitz et al., 1984); cholecystokinin, a peptide hormone (Woodruff, 1991); and MCPP, a serotonin receptor agonist (Charney, Woods, Goodman, & Heninger, 1987)—are also potent stimuli to hypothalamic CRH release (Bagdy et al., 1989; Kamilaris et al., 1992; A. E. Calogero and P. W. Gold, unpublished observations). Two other known stimulants to panic attacks, caffeine (Boulanger, Uhde, Wolf, & Post, 1984) and yohimbine (Charney, Heninger, & Redmond, 1983), both increase LC firing. To date no studies of CSF neuropeptides or monoamines in panic disorder have been reported. Two studies have shown a blunted ACTH response to CRH infusion, suggesting hypersecretion of CRH and down-regulation of pituitary ACTH receptors (Roy-Byrne, Bierer, & Uhde, 1985; Holsboer, von Bardeleben, Buller, Hauser, & Steiger, 1987).

In light of the proposed common hyperresponsivity of CRH in panic disorder and depression, Why do the two disorders manifest themselves as distinct clinical entities? One speculation is that, in contrast to depression which likely involves chronic activation of the CRH neuron, CRH and LC neuron activation may be intermittent and more intense in panic disorder. In depressive illness, a more chronic central CRH elevation and the greater exposure of the CRH system to high cortisol levels could restrain the system sufficiently so it is incapable of showing the kind of explosive activation postulated for panic disorder.

D. Posttraumatic Stress Disorder

Posttraumatic stress disorder (PTSD) patients constitute another chronically anxious group with symptoms of pronounced startle, nightmares, and sleep disturbance as well as panic attacks. As do panic disorder patients, patients with PTSD improve with antidepressant medication. Also, PTSD patients may have activation of central arousal systems in the context of inappropriately low peripheral cortisol levels, allowing severe explosive exacerbations of anxiety. PTSD patients have been found to have the lowest urinary free cortisols before and after treatment compared with other patient groups, including schizophrenics and depressives (Mason, Giller, Kosten, Ostroff, & Podd, 1986). The CRH stimulation test demonstrates blunted ACTH response in depressed and nondepressed PTSD patients (Smith, Brady, Gold, 1989). As in panic disorder, this reduced ACTH response may reflect a history of episodic acute increases in CRH during episodes of severe anxiety known to occur periodically in these patients. Even episodic

CRH hypersecretion may down-regulate pituitary ACTH receptors for CRH, causing a blunted response. Moreover, some evidence suggests that glucocorticoid receptor feedback sensitivity is even enhanced in PTSD patients, which could account further for lack of hypercortisolemia and blunted ACTH responses to CRH (Yehuda, Lowy, Southwick, Schaffer, & Giller, 1991). Evidence for activation of central noradrenergic systems in PTSD includes exaggerated responses to alpha-2 adrenergic agents (Southwick et al., 1991).

E. Anorexia Nervosa

Women with anorexia nervosa achieve a low body weight by restricting their food intake and increasing their activity. These women have strong personal and family histories of depression and anxiety disorders (Halmi et al., 1991). They are preoccupied obsessively with fears of gaining weight and, like patients with obsessive–compulsive disorder, show significant hypersecretion of CRH (Hotta et al., 1986; Kaye et al., 1987) and vasopressin (Gold, Kaye, Robertson, & Ebert, 1983) into the CSF. Oxytocin, a peptide closely related to vasopressin in structure and anatomical localization, also is released into the plasma from the pituitary and transported to disparate sites in the brain (Kozlowski & Nilaver, 1986). In light of evidence that oxytocin seems to have antagonistic effects to vasopressin on cognition (Bohus, Kovacs, & Deweid, 1965), reduced CSF oxytocin levels in anorexia nervosa during the underweight phase may have functional consequences (Demitrack et al., 1990). Deficiency of this peptide in underweight anorexics may contribute further to their exaggerated sense of the adverse consequences of eating.

Unlike obsessive–compulsive disorder, however, the underweight phase of anorexia nervosa and starvation from other causes is associated with profound hypercortisolemia (Smith, Bledsoe, & Chetri, 1975; Walsh et al., 1978). A high incidence of concurrent melancholic depression occurs in the low weight state. In contrast to melancholic depression, indices of the LC–NE system function in anorexia nervosa generally are diminished (Kaye, Jimerson, Lake, & Ebert, 1985). Down-regulation of noradrenergic systems is necessary for survival in the low-weight state, since sympathetic tone is a major determinant of thermogenesis and metabolic rate (Landsberg & Young, 1984), which must be reduced in situations of low caloric intake.

VI. Psychiatric Syndromes—Hypoarousal

Since CRH biases the central nervous system in the direction of arousal by activating pathways subserving fight-or-flight responses and inhibiting pathways subserving vegetative functions of sleep and appetite, the possibility of a pathological inactivation of the CRH and/or LC–NE system in some psychiatric syndromes characterized by increased sleep and appetite should be considered. In addition, several medical syndromes of diverse etiologies may be associated with a pathological inactivation of the CRH–NE system as a final common pathway. We review data that support inactivation of central arousal systems in atypical depression, premenstrual syndrome, seasonal affective disorder, bulimia nervosa, and the medical syndromes of Cushing's disease, chronic fatigue syndrome, and hypothyroidism. Evidence for hypoactivity of the stress-response systems in these disorders is preliminary and indirect. In some of the disorders discussed, the validity of the diagnoses is not as well established as for the disorders discussed in the previous section.

A. Atypical Depression

The pathological hyperarousal of melancholic depression represents the classic and best described major depressive syndrome. In contrast to this state of intense arousal, however, is the pathological inactivation of the other major depressive syndrome, commonly referred to as atypical depression. in many respects, this syndrome seems to be the antithesis of melancholic depression. The cardinal melancholic symptoms of anorexia and early morning awakening stand in marked contrast to the hyperphagia and hypersomnia of atypical depression. Moreover, in contrast to the intense anxiety about self and the ruminative preoccupation with the inevitability of loss that are characteristic of melancholia, patients with atypical depression seem passive, anergic, and apathetic. Their sensitivity to loss is more reactive in nature. Although hypercortisolemia is the cardinal neuroendocrine manifestation of melancholic depression, this finding is generally absent in patients with atypical depression (Mendelwicz, Charles, & Franckson, 1982). Further evidence for differences in arousal in these two subtypes of depression is found in treatment studies of atypical depression in which monoamine oxidase inhibitors and fluoxetine, drugs that increase the expression of the tyrosine hydroxylase gene in the LC (Brady et al., 1992) are more effective than tricyclic antidepressants (Himmelhoch, Thase, Mallinger, & Houck, 1991), which decrease the expression of the LC tyrosine hydroxylase gene.

B. Cushing's Disease

Depression is often the first symptom of Cushing's disease, often preceding the onset of the physical stigmata of hypercortisolemia by months or years. We initially noted that, in contrast to the hypercortisolemia of melancholia, the hypercortisolemia of Cushing's disease reflected a pituitary rather than a CNS defect

(Gold et al., 1986b). Possibly then, the CRH neuron in Cushing's disease is suppressed by elevated plasma cortisol rather than activated as it is in melancholic depression. Several lines of data suggest that this is the case. First, patients with Cushing's disease show depressions that often are associated with lethargy, fatigue, and hypersomnia rather than with the hyperarousal classically associated with melancholia. Second, postoperative patients with Cushing's disease show pronounced adrenal insufficiency that could reflect either suppression of residual pituitary corticotrophs by long-standing hypercortisolemia or suppression of the hypothalamic CRH neuron. The latter idea is suggested by data that postoperative Cushing's patients show attenuated, although palpable, plasma ACTH responses to exogenous CRH (Gold et al., 1986b). In addition, we also showed that patients with Cushing's disease show a profound decrease in the secretion of CRH into the CSF (Kling et al., 1991). In concert with evidence of hypofunctioning of the CRH neuron in Cushing's disease, we also have preliminary data suggesting decreased functioning of the LC–NE system, as indicated by a significant reduction of CSF MHPG levels in patients with this disorder (M. A. Kling and P. W. Gold, unpublished observations).

C. Hypothyroidism

Patients with hypothyroidism also often complain of depression associated with lethargy, fatigue, and hypersomnia. Although the direct effects of the hypothyroidism per se cannot be discounted, we have advanced several lines of evidence suggesting that experimentally induced hypothyroidism is associated with a central CRH deficiency and associated pituitary–adrenal insufficiency. Hence, hypothyroid rats show diminished basal plasma ACTH and corticosterone secretion, decreased CSF corticosterone levels, a significant reduction in adrenal weight, and diminished plasma cortisol responses either to ACTH itself or to the ACTH released during stimulation with CRH stimulation or during hypoglycemic stress (T. C. Kamilaris, et al., 1991). In addition, hypothyroid rats show decreased expression of CRH messenger RNA in the paraventricular nucleus of the hypothalamus (M. A. Smith and P. W. Gold, unpublished observations), decreased paraventricular CRH content, and decreased CRH release from hypothalamic organ cultures in response to KCL and a variety of secretogogues (T. C. Kamilaris and P. W. Gold, unpublished observations). Also, hypothyroid rats show increased glucocorticoid receptor number in the hippocampus (T. C. Kamilaris, unpublished observations), which could enhance negative feedback on hypothalamic CRH. Preliminary data also suggest that patients with hypothyroidism show a reduction in evening basal plasma cortisol levels and responses to exogenous CRH, suggestive of a possible central CRH deficiency (Kamilaris et al., 1987).

D. Seasonal Affective Disorder

Patients with seasonal affective disorder present with depressions characterized by hyperphagia and hypersomnia. Studies indicate that patients with this disorder show normal to slightly reduced 24-hr basal circadian corticosteroid secretion (Skwerer et al., 1988). These subjects show significant attenuation and delay in the ACTH response to ovine CRH (Vander Pool, Rosenthal, Chrousos, Wehr, & Gold, 1991). The blunted ACTH response to CRH in eucortisolemic or slightly hypocortisolemic patients with seasonal affective disorder is also compatible with a subtle central adrenal insufficiency. In this regard, the blunted response to CRH in these patients is more analogous to that seen in postoperative patients with Cushing's disease than to that seen in hypercortisolemic melancholically depressed patients, in whom the pituitary corticotroph cell is restrained in its response to synthetic CRH by sustained hypercortisolemia.

E. Chronic Fatigue Syndrome

We have advanced several lines of evidence suggesting a subtle central adrenal insufficiency in patients with the chronic fatigue syndrome. First, patients with this disorder show significant reduction in 24-hr urinary free cortisol (UFC) excretion, as well as in evening basal total and free cortisol concentrations. In association with these findings, patients with the chronic fatigue syndrome show an exaggerated plasma cortisol response to low doses of ACTH but a blunted response to high ACTH doses. These data indicate that patients with the chronic fatigue syndrome do not show a primary adrenal insufficiency, but hyperresponsiveness of the adrenal cortex to low doses of ACTH because of chronic understimulation. The blunted plasma cortisol response to high doses of ACTH on the other hand most likely reflects some adrenal atrophy due to an overall decrease in hypothalamic–pituitary stimulation of the adrenals. Despite their hypocortisolemia, patients with the chronic fatigue syndrome show a significant blunting rather than an exaggeration of the plasma ACTH response to synthetic CRH, analogous to our findings in patients with Cushing's disease and the seasonal affective disorder (Demitrack et al., 1991). Hence, although by no means definitively established, the available data suggest that patients with the chronic fatigue syndrome show evidence of a subtle central adrenal insufficiency. In addition to the possible consequences of a deficiency of an arousal producing neurohormone (i.e., CRH), the chronic fatigue syndrome also could reflect basal hypocortisolemia per se, which is well known to be associated with fatigue states. Moreover, symptoms of immune activation in chronic fatigue syndrome, including arthralgias, myalgias, feverishness, and increases in the titer of a variety of viral antigens (Strauss et al., 1985), also could result from a relative adrenal insufficiency. This finding is of interest in light of evidence that the syndrome may be postviral in etiology, and that a pronounced increase is seen in the expression of glucocorticoid receptor genes in Epstein–Barr transformed cells (Tomita et al., 1986), which could enhance negative feedback on the HPA axis.

F. Premenstrual Syndrome

Women with premenstrual syndrome typically complain of increased sleep and appetite as well as depressed mood during the end of the luteal phase. One study of HPA axis activity in these patients found evidence of decreased circulating cortisol in the evening in both the follicular and luteal phases. Perhaps because of decreased cortisol negative feedback at the pituitary, ACTH and cortisol responses to CRH infusion were increased compared with those in women without premenstrual syndrome (Rabin et al., 1990). Although many studies have not found differences in peripheral measures of catecholamine activity, increases in circulating progesterone levels in the luteal phase may decrease central noradrenergic activity, since this hormone is known to increase the activity of the metabolizing enzyme monoamine oxidase (McEwen, 1988). In rats, brain tissue levels of monamine oxidase have been shown to change with the estrous cycle (Youdim, 1972; Belmaker, Murphy, Wyatt, & Loriaux, 1974). In addition, the progesterone metabolite 3α, 5α-dihydroxyprogesterone is a potent agonist at the GABA receptor and also may contribute to reduced arousal (Majewska, 1986). These examples are only two of multiple interactions among gonadal hormones and central neurotransmitters and neuropeptides that could contribute to menstrual mood changes (Majewska, 1987).

G. Bulimia Nervosa

The cardinal feature of bulimia nervosa is binge eating, accompanied by a desire to purge, either by vomiting, diuretics, laxatives, exercise, or fasting. Bulimia can be found in conjunction with anorexia nervosa or independently. The condition is preceded almost invariably by dieting. Moreover, approximately one-third of bulimics have a history of anorexia (Russell, 1979). Sufferers feel guilty about binges, are preoccupied with food, and typically restrict their intake between binges. Thus, even bulimics who maintain a normal body weight show metabolic evidence of starvation (Pirke, Fichter, Schweiger, & Warnhoff, 1985). Half of bulimics have concurrent major depressive episodes. The illness seems to respond to antidepressant medications (Levy, Dixon, & Stern, 1989).

Unlike anorexics, hospitalized normal-weight bulimia nervosa patients show normal indices of HPA function (Gold et al., 1986a). Like anorexics, bulimics exhibit down-regulation of noradrenergic (Heufelder, Warnhoff, & Pirke, 1986; Pirke et al., 1988) and thyroid functions (Pirke et al., 1985), presumably in an attempt to reduce energy expenditure in the face of their restricted caloric intake. These reductions in central and peripheral noradrenergic activity and thyroid hormone may converge to produce a hypoaroused state in these patients. This model is consistent with the fatigue and increased appetite characteristics of the disorder. In anorexic patients with more severe starvation, elevated CSF CRH levels may counteract the hypoarousal expected form low noradrenergic and thyroid tone. Evidence suggests that thyroid hormone activity and central and peripheral NE activity fall further after abstinence from binge eating and vomiting (Kaye et al., 1990; Altemus et al., 1991), suggesting that bingeing behavior counteracts the metabolic down-regulation resulting from starvation. Large releases of catecholamines occur during bingeing (Kaye et al., 1990a) and may be reinforcing, which may explain the tendency of these patients to pursue other arousing behaviors such as drug abuse and theft.

H. Summary

The syndrome of atypical depression seems to occur across the boundaries of a variety of illnesses, including those traditionally classified as medical (e.g., Cushing's disease, hypothyroidism) or psychiatric (e.g., seasonal affective disorder, bulimia nervosa), and those that defy classification (e.g., chronic fatigue syndrome). We have advanced preliminary data suggesting that each of these disorders may be associated with hyposecretion of the arousal producing neurohormone CRH, and that this hyposecretion may be reflected in the lethargy and fatigue that are the cardinal manifestations of the depressive syndromes associated with these disorders.

VII. Developmental Differences

HPA activity is elevated less consistently in studies of children and adolescents with major depression, consistent with the observation that the melancholic syndrome is less common in younger age groups. In studies using the dexamethasone suppression test, abnormal results have been found in several studies in prepubertal and adolescent depressed inpatients, but rates of abnormality are generally lower than in adult depressives. Unlike adults, prepubertal children and adolescents with major depression do not hypersecrete cortisol, when assessed by 24-hr blood sampling studies. In addition, urinary free cortisol measures are not reliably elevated in adolescents with major depression. This relative lack of HPA axis activation in depressed children and adolescents may be explained by multiple reports of a positive correlation of cortisol secretion and age in both normal and depressed adults. An age effect on dexamethasone nonsuppression also has been demonstrated. (Puig-Antich, 1987)

Differences in the pathophysiology of depression and anxiety in children and adults may also reflect maturational changes in neurotransmitter systems and brain pathways. From infancy to adolescence, brain growth and development is complex, including increasing relative thickness of cerebral cortex and progressive regional myelination (Ornitz, 1991). In rats, catecholamine systems are not anatomically and functionally mature until the beginning of adulthood.

VIII. Gender Differences

In community surveys, women report extreme levels of psychiatric distress twice as often as men do. Compared with men, women have two to three times the incidence of major depressive disorder and anxiety disorders, but the same rates of schizophrenia and dementia (Weissman & Klerman, 1977; Regier et al., 1988). Although part of this effect may be social influences for women in our society to invest more in interpersonal relationships and to describe psychological distress more readily, biological differences between men and women also may influence the susceptibility to development of psychiatric syndromes and resilience in the face of sustained stress. Unfortunately, the available data to support such a hypothesis are meager.

Although the studies are small and hard to interpret, men seem to exhibit larger increases in blood pressure and higher levels of catecholamines in response to an acute stressor (Frankenhauser, Dunne, & Lundberg, 1976; Stoney, Davis, & Matthews, 1987). On the other hand, ACTH responses to CRH are elevated in women compared with men, accompanied by a prolonged cortisol response (Gallucci et al., 1992). This finding is consistent with animal data showing increased corticosterone responses to restraint stress and serotonin agonists in female rats (Haleem, Kenneth, Whitton, & Curzon, 1989). Monoamine oxidase, an important enzyme for metabolic inactivation of central monoamine neurotransmitters, is higher in women than in men (Murphy et al., 1976), which theoretically could bias women toward reductions in central noradrenergic tone.

Possibly, also, the large gonadal hormone fluxes that occur with the menstrual cycle and during pregnancy increase the susceptibility of women to anxiety and depression. Childbirth is a well-known precipitant of depressive episodes and often worsens anxiety disorders (George, Landenheim, & Nutt, 1987). Although hormonal factors associated with postpartum depression have not yet been identified, the abrupt and profound drop in circulating cortisol that occurs postpartum may disrupt HPA axis equilibrium. Alternatively, oxytocin, which is released centrally in rats during lactation (Langraf, Neumann, & Pittman, 1991), also may be released centrally in lactating women and contribute to anecdotal reports of decreased anxiety symptoms during the months of breast feeding (Cowley & Roy-Byrne, 1990). In support of this speculation, in rats, lactation has been reported to reduce CRH messenger RNA expression and stress-mediated secretion of CRH and corticosterone (Lightman & Young, 1989).

IX. Conclusion

In summary, dysregulation of central arousal systems seems to play an important role in the pathophysiology of depression and anxiety disorders. The dichotomy of hyperaroused and hypoaroused disorders proposed here is clearly reductionist and is at variance with current efforts to subdivide psychiatric disorders into many discrete but overlapping syndromal entities. However, this classification enables detection of physiological similiarities among syndromes and may have implications for development of more specific treatments that employ agents particularly effective in modulating the functional activity of the stress-responsive, arousal producing neurotransmitter systems. The pathophysiological overlap between depressive, anxiety, and eating disorders points out the descriptive nature of current psychiatric diagnostic categories. As the underlying biology of these syndromes becomes more clear, diagnostic classification will continue to be refined. Independent physiological syndromes may be identified that transcend current psychiatric diagnostic categories.

References

Adamec, R. E., & Stark-Adamec, C. (1986). Limbic hyperfunction, limbic epilepsy and interictal behavior: Models and methods of detection. In *The limbic system: Functional organization and clinical disorders* (pp. 129–145). B. K. Doane, & K. E. Livingston (Eds.), New York: Raven Press.

Altemus, M., Hetherington, M. M., Flood, M., Licinio, J., Nelson, M. L., Bernat, A. S., & Gold, P. W. (1991). Decrease in resting metabolic rate during abstinence form bulimic behavior. *American Journal of Psychiatry, 148,* 1071–1072.

Altemus, M., Pigott, T., L'Heureux, F., Davis, C. L., Rubinow, D. R., Murphy, D. L., & Gold, P. W. (1993) CSF Somalostatin in Obsessive-Compulsive Disorder. *American Journal of Psychiatry, 150,* 460–464.

Altemus, M., Pigott, T., Kalogeras, K. T., Demitrack, M., Dubbert, B., Murphy, D. L., & Gold, P. W. (1992). Abnormalities in the regulation of vasopressin and corticotropin releasing factor secretion in obsessive–compulsive disorder. *Archives of General Psychiatry, 49,* 9–20.

American Psychiatric Association (1987). *Diagnostic and statistical manual of psychiatric disorders,* 3d ed., rev. Washington, D.C.: American Psychiatric Press.

Amsterdam, J., Winokur, A., Abelman, E., Lucki, E., & Rickels, K. (1983). Co-syntropin (ACTH 1–24) test in depressed patients and healthy subjects. *American Journal of Psychiatry, 140,* 907–909.

Angulo, J. A., Ledoux, M., & McEwen, B. S. (1991). Genomic effects of cold and isolation stress on magnocellular vasopressin mRNA-containing cells in the hypothalamus of the rat. *Journal of Neurochemistry, 56,* 2033–2038.

Aston-Jones, G., Foote, S. L., & Bloom, F. E. (1984). Anatomy and physiology of locus ceruleus neurons: Functional implications. In M. G. Ziegler and C. R. Lake (Eds.), *Norepinephrine* (pp. 92–116). Baltimore: Williams & Wilkins.

Baddeley, A. D. (1972). Selective attention and performance in dangerous environments. *British Journal of Psychology, 63*, 537–546.

Bagdy, G., Calogero, A. E., Murphy, D., & Szemerdy, K. (1989). Serotonin agonists cause parallel activation of the sympathoadrenomedullary system and the hypothalamo–pituitary–adrenocortical axis in conscious rats. *Endocrinology, 165*, 2664–2669.

Banki, C. M., Bissette, G., Arato, M., & Nemeroff, C. B. (1988). Elevation of immunoreactive CSF TRH in depressed patients. *American Journal of Psychiatry, 145*, 1526–1531.

Bartalena, L., Martino, E., Brandi, L. S., Falcone, M., Pachhiarotti, A., Ricci, C., Bogazzi, F., Grasso, L., Mammoli, C., & Pinchera, A. (1990a). Lack of nocturnal serum thyrotropin surge after surgery. *Journal of Clinical Endocrinology and Metabolism, 70*, 293–436.

Bartalena, L., Martino, E., Petrini, L., Velluzzi, F., Loviselli, A., Grasso, L., Mammoli, C., & Pinchera, A. (1991). The nocturnal serum thyrotropin surge is abolished in patients with adrenocorticotropin (ACTH)-dependent or ACTH-independent Cushing's syndrome. *Journal of Clinical Endocrinology and Metabolism, 72*, 1195–1199.

Bartalena, L., Placidi, G. F., Martino, E., Falcone, M., Pellegrini, L., Dell'Osso, L. Pacchiarott, A., & Pinchera, A. (1990b). Nocturnal serum thyrotropin (TSH) surge and the TSH response to TSH-releasing hormone: Dissociated behavior in untreated depressives. *Journal of Clinical Endocrinology and Metabolism, 70*, 293–296.

Belmaker, R. H., Murphy, D. L., Wyatt, R. J., & Loriaux, D. L. (1974). Human platelet monoamine oxidase changes during the menstrual cycle. *Archives of General Psychiatry, 31*, 553.

Bohus, B., Kovacs, G. L., & DeWeid, D. (1965). Oxytocin, vasopressin, and memory: Opposite effects on consolidation and retrieval processes. *Brain Research, 157*, 414–417.

Bouchard, T. J., Lykken, D. T., McGue, M., Segal, N. L., & Tellegen, A. (1990). Sources of human psychological differences: The Minnesota study of twins reared apart. *Science, 250*, 223–250.

Boulanger, J. P., Uhde, T. W., Wolff, E. A., & Post, R. M. (1984). Increased sensitivity to caffeine in patients with panic disorder. *Archives of General Psychiatry, 41*, 1067.

Boyd, J. H., Burke, J. D., Gruenberg, E., Holzer, C. E., Rae, D. S., George, L. K., Karno, H., Stolzman, R., McEvoy, L., & Nestadt, G. (1984). Exclusion criteria of DSM-III: A study of co-occurrence of hierarchy-free syndromes. *Archives of General Psychiatry, 41*, 983–989.

Brady, L. S., Gold, P. W., Herkenham, M., Lynn, A. B., Whitfield, H. J. Jr., The antidepressants fluoxetine idozoxan and phenelzine alter corcotropin-releasing hormone and tyrosine hydroxylase mRNA levels in rat brain: therapeutic implications (1992). *Brain Res. 572*, 117–125.

Brady, L. S., Whitfield, H. J., Fox, R. J., Gold, P. W., & Herkenham, M. (1991). Long-term antidepressant administration alters corticotropin-releasing hormone, tyrosine hydroxylase, and mineralocorticoid receptor gene expression in rat brain. *Journal of Clinical Investigations, 87*, 831–837.

Breier, A., Kelsoe, J. R., Kirwin, P. D., Bellar, S. A., Wolkowitz, O. M., & Pickar, D. (1988). Early parental loss and development of adult psychopathology. *Archives of General Psychiatry, 45*, 987–993.

Breier, A., Charney, D. S., & Heninger, G. R. (1984). Major depression in patients with agorophobia and panic disorder. *Archives of General Psychiatry, 41*, 1129–1135.

Brinton, R. E., & McEwen, B. S. (1989). Vasopressin neuromodulation in the hippocampus. *Journal of Neuroscience, 9*, 752–759.

Brown, M. R., & Fisher, L. A. (1986). Glucocorticoid suppression of the sympathetic nervous system and adrenal medulla. *Life Sciences, 39*, 1003–1012.

Brown, M. R., Fisher, L. A., Spiess, J., Rivier, C., Rivier, C., Rivier, J., & Vale, W. (1982). Corticotropin-releasing factor: Actions on the sympathetic nervous system and metabolism. *Endocrinology, 111*, 928–931.

Bujis, R. M. (1990). Vasopressin and oxytocin localization and putative functions in the brain. *Acta Neurochirurgica Suppl. (Wien), 47*, 86–89.

Calogero, A. E., Bagdy, G., Szemerdy, K., Tartaglia, M. E., Gold, P. W., & Chrousos, G. P. (1990). Mechanisms of serotonin agonist-induced activation of the hypothalamic–pituitary–adrenal axis in the rat. *Endocrinology, 126*, 1888–1894.

Calogero, A. E., Gallucci, W. T., Bernadini, R., Saoutis, C., Gold, P. W., & Chrousos, G. P. (1988a). Effect of cholinergic agonists and antagonists on rat hypothalamic corticotropin releasing hormone secretion in vitro. *Neuroendocrinology, 47*, 303–308.

Calogero, A. E., Gallucci, W. T., Chrousos, G. P., & Gold, P. W. (1988b). Catecholamine effects upon rat hypothalamic ccorticotropin releasing hormone secretion in vitro. *Journal of Clinical Investigations, 82*, 839–846.

Calogero, A., Gallucci, W. T., Gold, P. W., & Chrousos, G. P. (1988c). Multiple regulatory feedback loops on hypothalamic corticotropin releasing hormone secretion. *Journal of Clinical Investigations, 82*, 767–774.

Cameron, O. G., Lee, M. A., Curtis, G. C., & McCann, D. S. (1987). Endocrine and physiological changes during "panic attacks. *Psychoneuroendocrinology, 12*, 321–331.

Carroll, B. J. (1976). Limbic-system-adrenal cortex regulation in depression and schizophrenia. *Psychosomatic Medicine, 38*, 106–121.

Chappell, P. B., Smith, M. A., Kilts, C. D., Bissette, G., Ritchie, J., Anderson, C., & Nemeroff, C. B. (1987). Alterations in corticotropin-releasing factor-like immunoreactivity in discrete rat brain regions after acute and chronic stress. *Journal of Neuroscience, 6*, 2908–2914.

Charney, D. S., Heninger, G. R., & Redmond, E. (1983). Yohimbine induced anxiety and increased noradrenergic function in humans: Effects of diazepam and clonidine. *Life Sciences, 33*, 19.

Charney, D. S., Woods, S. W., Goodman, W. K., & Heninger, G. R. (1987). Serotonin function in anxiety. II: Effects of the serotonin agonist mCPP in panic disorder patients and healthy subjects. *Psychopharmacology, 92*, 14–24.

Cole, B. J., & Koob, G. F. (1988). Propranolol antagonizes the enhanced conditioned fear produced by CRF. *Journal of Cell Biochemistry Suppl. 17*, 131–148.

Cole, P. M., & Putnam, F. W. (1992). Effects of incest on self and social function: Developmental psychopathological perspectives. *Journal of Consulting Clinical Psychology, 60*, 174–184.

Coryell, W. H., Black, D. W., Kelly, M. W., & Noyes, R. (1989). HPA axis disturbance in obsessive–compulsive disorder. *Psychiatry Research, 30*, 243–251.

Cowley, D. S., & Roy-Byrne, P. P. (1990). Panic disorder during pregnancy. *Journal of Psychosomatic Obstetrics & Gynecology, 10*, 193–210.

Cunningham, E. T., Bohn, M. C., & Sawchencko, P. E. (1990). The organization of adrenergic inputs to the paraventricular and supraoptic nuclei of the rat hypothalamus. *Journal of Comparative Neurology, 292*, 651–667.

Curtis, G. C., & Glitz, D. A. (1988). Neuroendocrine findings in anxiety disorders. *Endocrinology & Metabolism Clinics of North America, 17*, 131–148.

Dallman, M. F., Akana, S., Casio, C. S., Darlington, D. N., Jacobson, L., & Levin, N. (1987). Regulation of ACTH secretion: Variations on a theme of B. *Recent Progress in Hormone Research, 43*, 113–173.

Daniels, D., & Plomin, R. (1985). Origins of individual differences in shyness. *Developmental Psychology, 21*, 118–121.

Darwin, C. (1872). *The expression of the emotions in man and animals.* London: John Murray.

Davis, L. G., Arentzen, R., Reed, J. M., Manning, R. W., Wofson, B., Lawrence, K. L., & Baldeno, F. (1986). Glucocorticoid sensitivity of vasopressin mRNA levels in the paraventricular nucleus of the rat. *Proceedings of the National Academy of Sciences, U.S.A., 83*, 1145–1149.

Demitrack, M. A., Dale, J. K., Strauss, S. E., Laue, L., Listwak, S. J., Kreusi, M. J. P., Chrousos, G. P., & Gold, P. W. (1991). Evidence for impaired activation of the hypothalamic–pituitary–adrenal axis in patients with chronic fatigue syndrome. *Journal of Clinical Endocrinology and Metabolism, 73*, 1224–1234.

Demitrack, M. A., Lesem, M. D., Listwak, S. J., Brandt, H. A., Jimerson, D. C., & Gold, P. W. (1990). Cerebrospinal fluid oxytocin in anorexia nervosa and bulimia nervosa: Clinical and pathophysiological considerations. *American Journal of Psychiatry, 147*, 882–886.

Dorovini-Zis, K., & Zis, A. P. (1987). Increased adrenal weights in victims of violent suicide. *American Journal of Psychiatry, 144*, 1214–1215.

Dunn, A. J., & Berridge, C. W. (1990). Physiological and behavioral responses to corticotropin-releasing factor administration: Is CRF a mediator of anxiety or stress responses? *Brain Research Review 15*, 71–100.

Dupont, E., Christensen, S. E., Hanen, A. P., Olivarius, B. F., & Orskov, H. (1982). Low cerebrospinal fluid somatostatin in Parkinson disease: An irreversible abnormality. *Neurology, 32*, 312–314.

Elkabir, D. R., Wyatt, M. E., Vellucci, S. V., & Herbert, J. (1990). The effects of separate or combined infusions of corticotropin-releasing hormone and vasopressin either intraventricularly or into the amygdala on aggressive and investigative behavior in the rat. *Regulatory Peptides, 28*, 199–214.

Fisher, L. A., & Brown, M. R. (1991). CRF-41 and stress responses. *Balliere's Clinical Endocrinology & Metabolism, 5*, 35–50.

Foote, S. L., Bloom, F. E., & Aston-Jones, G. (1983). Nucleus locus ceruleus: New evidence for anatomical and physiological specificity. *Physiology Reviews, 63*, 844–914.

Frankenhauser, M., Dunne, E., & Lundberg, U. (1976). Sex differences in sympathetic-adrenal medullary reactions induced by different stressors. *Psychopharmacology, 47*, 1–5.

Freud, S. (1926). Inhibitions, symptoms and anxiety. In J. Strachey (Ed.), *The complete psychological works of Sigmund Freud* (Vol. XX). London: Hogarth Press.

Gallucci, W. T., Baum, A., Laue, L., Rabin, D. S., Chrousos, G. P., Gold, P. W., & Kling, M. A. (1993). Sex differences in sensitivity of the hypothalamic pituitary adrenal axis. *Health Psychology*, in press.

Geola, F. L., Yamada, T., Warwick, R. J., Tourtelotte, W. W., & Hersham, J. M. (1981). Regional distribution of somatostatin-like immunoreactivity in the human brain. *Brain Research, 229*, 35–42.

George, D. T., Landenheim, J. A., & Nutt, D. J. (1987). Effect of pregnancy on panic attacks. *American Journal of Psychiatry, 144*, 1078–1079.

Gjerris, A., Hummer, M., Vendsborg, P., Christienson, P., & Rafaelson, O. J. (1985). Cerebrospinal fluid vasopressin-changes in depression. *British Journal of Psychiatry, 147*, 696–701.

Glavin, G. B. (1985). Stress and brain noradrenaline: A review. *Neurology and Biobehavioral Research, 9*, 233–243.

Goddard, G. V., MacIntyre, D. C., & Leech, C. K. (1969). A permanent change in brain resulting from daily electrical stimulation. *Experimental Neurology, 25*, 295–308.

Gold, P. W., Goodwin, F. K., Post, R. M., & Robertson, G. L. (1981). Vasopressin functions in depression and mania. *Psychopharmacological Bulletin, 17*, 7–9.

Gold, P. W., Gwirtsman, H., Avgerinos, P. C., Nieman, L. K., Galluci, W. T., Kaye, W., Jimerson, D., Ebert, M., Rittmaster, R., Loriaux, D. L., & Chrousos, G. P. (1986a). Abnormal hypothalamic–pituitary-adrenal function in anorexia nervosa. *New England Journal of Medicine, 314*, 1335–1342.

Gold, P. W., Kaye, W., Robertson, G. L., & Ebert, M. (1983). Abnormalities in plasma and cerebrospinal fluid arginine vasopressin in patients with anorexia nervosa. *New England Journal of Medicine, 308*, 1117–1123.

Gold, P. W., Loriaux, D. L., Roy, A., Kling, M. A., Calabrese, J. R., Kellner, C. H., Nieman, L. K., Post, R. M., Pickar, D., & Gallucci, W. (1986b). Responses to corticotropin releasing hormone in the hypercortisolism of depression on Cushing's disease: Pathophysiologic and diagnostic implications. *New England Journal of Medicine, 314*, 1329–1335.

Gray, T. S. (1991). Amygdala: Role in autonomic and neuroendocrine responses to stress. In J. A. McCubbin, P. G. Kaufman, & C. B. Nemeroff (Eds.), *Stress, neuropeptides and systemic disease* (pp. 37–53). New York: Academic Press.

Guynet, P. G., & Aghajanian, G. K. (1979). Acetylcholine, substance P and met-enkephalin in the locus coeruleus: Pharmacological evidence for independent sites of action. *European Journal of Pharmacology, 53*, 319–328.

Gwirtsman, H. E., Gerner, R. H., & Sternbach, H. (1982). The overnight dexamethasone suppression test: Clinical and theoretical review. *Journal of Clinical Psychiatry, 43*, 321–327.

Haleem, D. J., Kennett, G. A., Whitton, P. S., & Curzon, G. (1989). 8-OH-DPAT increases corticosterone but not other 5-HT1a receptor-dependent responses more in females. *European Journal of Pharmacology, 164*, 435–443.

Halicek, V., Rezek, M., & Friesen, H. (1976). Somatostatin and thyrotropin-releasing hormone: Central effects on sleep and motor systems. *Pharmacology and Biochemistry of Behavior, 5*, 73–77.

Halmi, K. A., Eckert, E., Marchi, P., Sampugnaro, V., Apple, R., & Cohen, J. (1991). Comorbidity of psychiatric diagnoses in anorexia nervosa. *Archives of General Psychiatry, 48*, 712–718.

Herkenham, M. (1987). Mismatches between neurotransmitter and receptor localizations in brain: Observations and implications. *Neuroscience, 23*, 1–38.

Heufelder, A., Warnhoff, M., & Pirke, K. M. (1986). Platelet alpha-2 adrenoceptor and adenylate cyclase in patients with anorexia nervosa and bulimia. *Journal of Clinical Endocrinology and Metabolism, 61*, 1053–1060.

Higgins, G. A., & Schwaber, J. S. (1983). Somatostatinergic projections from the central nucleus of the amygdala to the vagal nuclei. *Peptides, 4*, 663–668.

Himmelhoch, J. M., Thase, M. E., Mallinger, A. G., & Houck, P. (1991). Tranylcypromine versus imipramine in anergic bipolar depression. *American Journal of Psychiatry, 148*, 910–916.

Hitchcock, J., & Davis, M. (1986). Lesions of the amygdala but not of the cerebellum or red nucleus block conditioned fear as measured with the potentiated startle paradigm. *Behavioral Neuroscience, 100*, 11–22.

Holsboer, F., von Bardeleben, U., Buller, R., Heuser, I., & Steiger, A. (1987). Stimulation response to corticotropin-releasing hormone (CRH) in patients with depression, alcoholism and panic disorder. *Hormones and Metabolism Research (Suppl.), 16*, 80–88.

Hotta, M., Shibasaki, T., Masuda, A., Imaki, T., Demura, H., Ling, N., & Shizume, K. (1986). The responses of plasma corticotropin and cortisol to corticotropin-releasing hormone (CRH) and cerebrospinal fluid immunoreactive CRH in anorexia nervosa patients. *Journal of Clinical Endocrinology and Metabolism, 62*, 319–324.

Kagan, J., Reznick, J. S., & Snidman, D. (1987). The physiology and psychology of behavioral inhibition in children. *Child Development, 58*, 1459–1473.

Kalogeros, K. T., Calogero, A. E., Kuribayiashi, T., Kahn, I., Gallucci, W. T., Kling, M. A., Chrousos, G. P., & Gold, P. W. (1990). In vitro and in vivo effects of the triazolobenzodiazepine, alprozolam, on hypothalamic–pituitary–adrenal function: Pharmacologic and clinical implications. *Journal of Clinical Endocrinology and Metabolism.* 70:1462–71.

Kamilaris, T. C., DeBold, R. C., Paylou, S. N., Island, D. P., Hoursanidis, A., & Orth, D. N. (1987). Effect of altered thyroid hormone levels on hypothalamic-pituitary-adrenal function. *Journal of Clinical Endocrinology and Metabolism, 65*, 994–999.

Kamilaris, T. C., Johnson, E. O., Calogero, A. E., Kalogerur, K. T., Bernardini, R., Chrousos, G. P., & Gold, P. W. (1992). Cholecystokinin-octapeptide stimulates hypothalamic–pituitary–adrenal function in rats: Role of corticotropin-releasing hormone. *Endocrinology.* 130:1764–74.

Kamilaris, T. C., DeBold, C. R., Johnson, E. O., Mamalaki, E., Listwak, S. J., Calogero, A. E., Kalogeras, K. T., Gold, P. W., Orth, D. N. (1991). Effects of short and long duration hypothyroidism on the plasma adrenocorticotropin and corticosterone responses to ovine corticotropin-releasing hormone in rats. *Endocrinology, 128*; 2567–2576.

Kapp, B. S., Pascoe, J. P., & Bixler, M. A. (1985). The amygdala: A neuroanatomical systems approach to its contribution to aversive conditioning. In N. Butters and L. Squire (Eds.), (pp. 473–488). New York: Guilford Press.

Kaye, W. H., Gwirtsman, H. E., & George, D. T. (1990a). The effect of bingeing and vomiting on hormonal secretion. *Biological Psychiatry, 5*, 768–780.

Kaye, W. H., Gwirtsman, H., George, D. T., Ebert, M. H., Jimerson, D. C., Tomai, T. P., Chrousos, G. P., & Gold, P. W. (1987). Elevated cerebrospinal fluid levels of immunoreactive corticotropin releasing hormone in anorexia nervosa: Relation to state of nutrition, adrenal function, and intensity of depression. *Journal of Clinical Endocrinology and Metabolism, 64*, 203–208.

Kaye, W. H., Gwirtsmann, H. E., George, D. T., Jimerson, D. C., Ebert, M. H., & Lake, R. C. (1990b). Disturbances of noradrenergic systems in normal weight bulimia: Relationship to diet and menses. *Biological Psychiatry, 27*, 4–21.

Kaye, W. H., Jimerson, D. C., Lake, C. R., & Ebert, M. H. (1985). Altered norepinephrine metabolism following long-term weight recovery in patients with anorexia nervosa. *Psychiatry Research, 14*, 333–342.

Kendler, K. S., Heath, A., Martin, M. G., & Eaves, L. J. (1986). Symptoms of anxiety and depression in a volunteer twin population. *Archives of General Psychiatry, 43*, 213–221.

Kirkegaard, C., Faber, J., Jummer, L., & Rogowski, P. (1979). Increased levels of TRH in cerebrospinal fluid from patients with endogenous depression. *Psychoneuroendocrinology, 4*, 227–235.

Kitayamia, I., Janson, A. M., Cintra, A., Fuxe, K., Agnati, L. F., Encroth, P., Aronsson, M., Harfstrand, A., Steinbush, H. W., & Visser, T. J. (1988). Effects of chronic imipramine treatment on glucocorticoid receptor immunoreactivity in various regions of the rat brain. *Journal of Neural Transmission, 73*, 191–203.

Kjellman, B. F., Beck-Friis, J., Lunggren, J.-G., & Wetterber, L. (1984). 24-hr serum levels of TSH in affective disorder. *Acta Psychiatrica Scandinavica, 69*, 491–502.

Kling, M. A., Roy, A., Doran, A. R., Calabrese, J. R., Rubinow, D. R., Whitfield, H. J., May, C., Post, R. M., Chrousos, G. P., & Gold, P. W. (1991). Cerebrospinal fluid immunoreactive corticotropin-releasing hormone and adrenocorticotropin secretion in Cushing's disease and major depression: Potential clinical implications. *Journal of Clinical Endocrinology and Metabolism, 72*, 260–271.

Koob, G. F. (1988). Behavioral actions of corticotropin-releasing factor in the central nervous system. *Journal of Cell Biochemistry, Suppl., 12D*, 299.

Koob, G. F. (1991). Behavioral responses to stress. In M. R. Brown, G. F. Koob, and C. Rivier (Eds.), *Stress: Neurobiology and neuroendocrinology* (pp. 255–271). New York: Marcel Dekker.

Koob, G. F., Lebrun, C., Bluthe, R., Dantzer, R., & Le Moal, M. (1989). Role of neuropeptides in learning versus performance: Focus on vasopressin. *Brain Research Bulletin, 23*, 359–64.

Kozlowski, G. P., & Nilaver, G. (1986). Localization of neurohypophyseal hormones in the mammalian brain. In D. DeWeid, W. H. Gispen, and G. T. B. Van Winersma (Eds.), *Neuropeptides and behavior* (pp. 23–38). Oxford: Pergamon Press.

Kraepelin, E. (1946). *Psychiatrie: Ein Lehrbuch fur Studierende und Ärzte*, 8th ed. Leipzig: J. A. Bart.

Kurosawa, M., Sato, A., Swenson, R. S., & Takahashi, Y. (1986). Sympatho-adrenal medullary functions in response to intracerebroventricularly injected corticotropin-releasing factor in anesthetized rats. *Brain Research, 367*, 250–257.

Landsberg, L., & Young, J. B. (1984). Endocrine changes in anorexia nervosa: an interpretation based on the metabolic adaptation to caloric restriction. In G. M. Brown, S. H. Koslow, and S. Reichlin (Eds.), *Neuroendocrinology and psychiatric disorder* (pp. 349–358). New York: Raven Press.

Langraf, R., Neumann, I., & Pittman, Q. J. (1991). Septal and hippocampal release of vasopressin and oxytocin during late pregnancy and parturition in the rat. *Neuroendocrinology, 54*, 378–383.

Last, C. G., Hersen, M., Kazdin, A., Orvaschel, H., & Perrin, S. (1991). Anxiety disorders in children and their families. *Archives of General Psychiatry, 48,* 928–934.

Leckman, J. F., Merikangas, K. R., Pauls, D. L., Prusoff, B. A., & Weissmann, M. M. (1983). Anxiety disorders and depression: Contradiction between family study data and DSM-III conventions. *American Journal of Psychiatry, 140,* 880–882.

Levine, S., & Wiener, S. G. (1989). Coping with uncertainty: A paradox. In D. S. Palermo (Ed.), *Coping with uncertainty* (pp. 1–16). Hillsdale, New Jersey: Erlbaum Associates.

Levy, A. B., Dixon, K. N., & Stern, S. L. (1989). How are depression and bulimia related? *American Journal of Psychiatry, 146,* 162–169.

Liebowitz, M. R., Fyer, A. J., Gorman, J. M., Dillon, D., Appleby, I. L., Levy, G., Anderson, S., Levitt, M., Palij, M., Davies, S. O., & Klein, D. F. (1984). Lactate provocation of panic attacks II. Biochemical and physiological findings. *Archives of General Psychiatry, 42,* 709.

Lightman, S. L., & Young, W. C. (1989). Lactation inhibits stress-mediated secretion of corticosterone and oxytocin and hypothalamicaccumulation or corticotropin releasing factor and enkephalin messenger ribonucleic acids. *Endocrinology, 124,* 2358–2364.

Linnoila, M., Karoun, F., Calil, H. M., Kopin, I. J., & Potter, W. Z. (1982). Alteration of norepinephrine metabolism with desipramine and zimeldine in depressed patients. *Archives of General Psychiatry, 39,* 1025–1028.

Loosen, P. T. (1988). Thyroid function in affective disorders and alcoholism. *Endocrinology and Metabolism Clinics of North America, 17,* 55–82.

McEwen, B. S. (1988). Basic research perspective: ovarian hormone influences on brain neurochemical functions In L. H. Gise (Ed.), *Contemporary issues in obstetrics and gynaecology* Vol. 2, (pp. 21–33). New York: Churchill Livingstone.

McGaugh, J. L., Liang, K. C., Bennett, C., & Sternberg, D. B. (1984). Adrenergic influences on memory storage: interaction of peripheral and central systems. In G. Lynch, J. L. McGaugh, and N. M. Weinberger (Eds.) *Neurobiology of learning and memory.* (pp. 313–332). New York: Guilford Press.

Majewska, M. D. (1987). Actions of steroids on neurons: role in personality mood, stress and disease. *Integrative Psychiatry, 5,* 258–273.

Mason, J. W., Giller, E. L., Kosten, T. R., Ostroff, R. B., & Podd, L. (1986). Urinary free-cortisol levels in posttraumatic stress disorder patients. *Journal of Nervous and Mental Disorders, 74,* 145–149.

Meany, M. J., Mitchell, J. B., Aitken, D. H., Bhatnagar, S., Bodnoff, S. R., Iny, L. J., & Sarrieau, A. (1991). The effects of neonatal handling on the development of the adrenocortical response to stress: Implications for neuropathology and cognitive deficits in later life. *Psychoneuroendocrinology, 16,* 85–103.

Mendelwicz, J., Charles, G., & Franckson, G. M. (1982). The DST in affective disorder: Relationship to clinical and genetic subgroups. *British Journal of Psychiatry, 141,* 464–470.

Millar, K., Jeffcoate, W. J., & Walder, C. P. (1987). Vasopressin and memory: Improvement in normal short-term recall and reduction of alcohol-induced amnesia. *Psychology and Medicine, 17,* 335–341.

Moore, R. Y., & Bloom, F. E. (1979). Central catecholamine neuron systems: Anatomy and physiology of the norepinephrine and epinephrine systems. *Annual Review of Neuroscience, 2,* 113–168.

Munck, A., Guyre, P. M., & Holbrook, N. J. (1984). Physiological functions of glucocorticoids in stress and their relation to pharmacological actions. *Endocrinology Review, 5,* 25–44.

Murphy, D. L., Wright, C., Buschbaum, M., Nichols, A., Costa, J. L., & Wyatt, R. J. (1976). Platelet and plasma amine oxidase activity in 680 normals: Sex and age differences and stability over time. *Biochemistry and Medicine, 16,* 254–265.

Nemeroff, C. B., Kaliva, P. W., Golder, R. N., & Prange, A. J. (1984). Behavioral effects of hypothalamic hypophysiotrophic hormones, neurotensin, substance P and other neuropeptides. *Pharmacology and Therapeutics, 24,* 1–56.

Nemeroff, C. B., Kizer, J. S., Reynolds, G. P., & Bissette, G. (1989). Neuropeptides in Alzheimer's disease: A postmortem study. *Regulatory Peptides, 25*, 123–130.

Nemeroff, C. B., Widerlov, E., Bissette, G., Walleus, H., Karlsson, I., Eklund, K., Kilts, C. D., Loosen, P. T., & Vale, W. (1985). Elevated concentrations of corticotropin-releasing factor like immunoreactivity in depressed patients. *Science, 226*, 1342–1344.

Nielsen, M., & Braestrup, C. (1977). Chronic treatment with desipramine causes a sustained decrease of 3,4-dihydroxyphenyglycol-sulfate and total 3-methoxy-4-hydroxyphenylglycol-sulfate in the rat brain. *Naunyn Schmiedebergs Archives of Pharmacology, 300*, 87–92.

Noyes, R. (1988). The natural history of anxiety disorders In M. Roth, R. Noyes, and G. D. Burrows (Eds.), *Handbook of anxiety* (Vol. 1, pp. 115–133). New York: Elsevier.

Nutt, D. J. (1986). Increased central alpha2 adrenoreceptor sensitivity in panic disorder. *Psychopharmacology, 90*, 268–269.

Nyback, H. V., Walters, J. R., Aghajanian, G. K., & Roth, R. H. (1975). Tricyclic antidepressants: Effects on the firing rate of brain noradrenergic neurons. *European Journal of Pharmacology, 32*, 302–312.

Olschoska, J. A., O'Donohue, T. L., Mueller, G. P., & Jacobowitz, D. M. (1982). The distribution of corticotropin releasing factor-like immunoreactive neurons in rat brain. *Peptides, 3*, 995–1015.

Ornitz, E. M. (1991). Developmental aspects of neurophysiology. In M. Lewis (Ed.), *Child and adolescent psychiatry: A comprehensive textbook* (pp. 38–51). Baltimore: Williams and Wilkins.

Overton, J. M., & Fisher, L. A. (1989). Modulation of central nervous system actions of corticotropin-releasing factor by dynorphin-related peptides. *Brain Research, 488*, 233–240.

Panksepp, J. (1990). Animal models of fear. In G. D. Burrows, M. Roth, and R. Noyes (Eds.), (Vol. 3, pp. 3–58). Amsterdam: Elsevier.

Pepper, C. M., & Henderson, G. (1980). Opiates and opiod peptides hyperpolarize locus coeruleus neurons in vitro. *Science, 209*, 394–396.

Petty, F., & Sherman, A. D. (1981). GABAergic modulation of learned helplessness. *Pharmacology and Biochemistry of Behavior, 15*, 567–572.

Pirke, K. M., Fichter, M. M., Schweiger, V., & Warnhoff, M. (1985). Metabolic and endocrine indices of starvation in bulimia: a comparison with anorexia nervosa. *Psychiatry Research, 15*, 33–39.

Pirke, K. M., Riedel, W., Tuschl, R., Schweiger, U., Schweiger, B. S., & Spyra, B. (1988). Effect of standardized test meals on plasma norepinephrine in patients with anorexia nervosa and bulimia. *International Journal of Eating Disorders, 7*, 369–373.

Plotsky, P. M. (1986). Opiod inhibition of immunoreactive corticotropin releasing factor secretion into the hypophysial-portal circulation of rats. *Regulatory Peptides, 16*, 235–242.

Puig-Antich, J. (1987). Affective disorders in children and adolescents: Diagnostic validity and psychobiology. In H. Meltzer (Ed.), *Psychopharmacology: Third generation of progress* (pp. 843–859). New York: Raven Press.

Rabin, D. S., Schmidt, P. J., Cambell, G., Gold, P. W., Jensvold, M., Rubinow, D. R., & Chrousos, G. P. (1990). Hypothalamic–pituitary–adrenal function in patients with the premenstrual syndrome. *Journal of Clinical Endocrinology and Metabolism, 71*, 1158–1162.

Regier, D. A., Boyd, J. H., Burke, J. D. J., Rae, D. S., Myers, J. K., Kramer, M., Robins, L. N., George, L. K., Karno, M., & Locke, B. Z. (1988). One-month prevalence of mental disorder in the United States. *Archives of General Psychiatry, 45*, 977–986.

Rennel, M. L., Gregory, T. F., Blaumanis, O. R., Fujimoto, K., & Grady, P. A. (1985). Evidence for a paravascular fluid circulation in the mammalian central nervous system, provided by the rapid distribution of tracer protein throughout the brain from subarachnoid space. *Brain Research, 326*, 47–63.

Richter, C. (1957). On the phenomenon of sudden death in animals and man. *Psychosomatic Medicine, 19*, 319–329.

Roth, W. T., Jurgen, M., Ehlers, A., Haddad, J. M., Maddock, R. J., Agras, W. S., & Taylor, C. B.

(1992). Imipramine and alprazolam: Effects on stress test reactivity in panic disorder. *Biological Psychiatry, 31,* 35–51.

Roy, A., Pickar, D., Paul, S., Doran, A., Chrousos, G. P., & Gold, P. W. (1987). CSF corticotropin releasing hormone in depressed patients and normal control subjects. *American Journal of Psychiatry, 144,* 641–644.

Roy-Byrne, P. P., Bierer, L. M., & Uhde, T. W. (1985). The dexamethasone suppression test in panic disorder: comparison with normal controls. *Biological Psychiatry, 20,* 1234–1237.

Rubinow, D. R., Gold, P. W., Post, R. M., Ballenger, J. C., Cowdry, R., Bollinger, J., & Reichlin, S. (1983). CSF somatostatin in affective illness. *Archives of General Psychiatry, 40,* 409–412.

Russell, G. (1979). Bulimia nervosa: An ominous varient of anorexia nervosa. *Psychology and Medicine, 9,* 429–448.

Sachar, E. J., Hellman, L., Roffwarg, H. P., Halpern, F. S., Fukushima, K. D., & Gallagher, T. F. (1973). Disrupted 24-hr patterns of cortisol secretion in pshychotic depression. *Archives of General Psychiatry, 28,* 19–26.

Sakanaka, M., Shibasaki, T., & Lederis, K. (1986). Distribution and efferent projections of corticotropin-releasing factor-like immunoreactivity in the rat amygdaloid complex. *Brain Research, 382,* 213–238.

Sawchenko, P. E. (1987). Adrenalectomy-induced enhancement of CRF and vasopressin immunoreactivity in parvocellular neurosecretory neurons: anatomic, peptide and steroid specificity. *Journal of Neuroscience,* 7:1093–1106.

Schulte, H. M., Chrousos, G. P., Gold, P. W., Booth, J. D., Oldfield, E. H., Cutler, G. B., & Loriaux, D. L. (1985). Continuous administration of synthetic ovine corticotropin releasing factor in man: Physiological and pathophysiological implications. *Journal of Clinical Investigation, 75,* 1781–1785.

Skwerer, R. G., Duncan, C., Jacobsen, F. M., Sack, D. A., Tamarkin, L., Wehr, T. A., & Rosenthal, N. E. (1988). Neurobiology of seasonal affective disorder and phototherapy. *Journal of Biological Rhythms, 3,* 135–154.

Smith, M. A., Brady, L. S., Glowa, J., Gold, P. W., Herkenham, M. E. (1991). Effects of stress and adrenalectomy on tyrosine hydroxylase mRNA levels in the locus ceruleu by in situ hybridization. *Brain Research, 544,* 26–32.

Smith, S. R., Bledsoe, T., & Chetri, M. K. (1975). Cortisol metabolism and the pituitary-adrenal axis in adults with protein-calorie malnutrition. *Journal of Clinical Endocrinology and Metabolism, 40,* 93–97.

Sorenson, P. S., Gjerris, A., & Hammer, M. (1985). Cerebrospinal fluid vasopressin in neurologic and psychiatric disorders. *Journal of Neurology and Neurosurgical Psychiatry, 48,* 50–57.

Southwick, S., Krystal, J., Morgan, A., Nagy, L., Yehuda, R., & Charney, D. (1991). Noradrenergic dysregulation in PTSD. *30th Annual Meeting, American College of Neuropsychopharmacology,* San Juan.

Spencer, C. A., Greenstadt, M. A., Wheeler, W. S., Kletzky, O. A., & Nicoloff, J. T. (1980). The influence of long term low-dose thyrotropin releasing hormone infusion on serum thyrotropin and prolactin concentrations in man. *Journal of Clinical Endocrinology and Metabolism, 51,* 771–775.

Stoney, C. M., Davis, M. C., & Matthews, K. A. (1987). Sex differences in physiological response to stress and in coronary heart disease: A causal link? *Psychophysiology, 24,* 127–131.

Strauss, S. E., Tosato, G., Armstrong, G., Lawley, T., Preble, O. T., Henle, W., Davey, R., Pearson, G., Epstein, J., & Brus, I. (1985). Persisting illness and fatigue in adults with evidence of Epstein–Barr virus infection. *Annals of Internal Medicine, 102,* 7–16.

Sulser, F. (1983). Mode of action of antidepressant drugs. *Journal of Clinical Psychology* 44:14–20.

Suomi, S. J. (1987). Genetic and maternal contributions to individual differences in Rhesus monkey biobehavioral development. In N. A. Krasnegor, E. M. Blass, M. A. Hofer, and W. P. Smotherman (Eds.), *Perinatal development—A psychobiological perspective* (pp. 397–419). Orlando: Academic Press.

Svensson, T. H., & Usdin, T. (1978). Feedback inhibition of brain noradrenaline neurons by tricyclic antidepressants: alpha receptor mediation. *Science, 202*, 2089.

Swanson, L. W., & Sawchencko, P. E. (1980). Paraventricular nucleus: A site for the integration of neuroendocrine and autonomic mechanisms. *Neuroendocrinology, 31*, 410–417.

Szemeredi, K., Bagdy, G., Stull, R., Calogero, A. E., Kopin, I. J., & Goldstein, D. S. (1988). Sympathoadrenomedullary inhibition by chronic glucocorticoid treatment in conscious rats. *Endocrinology, 123*, 2585–2590.

Taylor, C. B., Sheikh, J., Agras, S., Roth, W. T., Margraf, J., Ehlers, A., Maddock, R. J., & Gossard, D. (1986). Ambulatory heart rate changes in patients with panic attacks. *American Journal of Psychiatry, 143*, 478–482.

Thoman, E. B., Levine, E. S., & Arnold, W. J. (1968). Effects of maternal deprivation and incubation rearing on adrenocortical activity in the adult rat. *Developmental Psychobiology, 1*, 21–23.

Thomas, A., Chess, S., & Korn, S. J. (1982). The reality of difficult temperament. *Merrill–Palmer Quarterly, 28*, 1–20.

Tomita, M., Brandon, D. D., Chrousos, G. P., Vingerhoeds, A. C., Foster, C. M., Fowler, D., Loriaux, D. L., Lipsett, M. B. (1986). Glucocorticoid receptors in Epstein–Barr virus-transformed lymphocytes from patients with glucocorticoid resistence and a glucocorticoid resistant New World primate species. *Journal of Clinical Endocrinology and Metabolism, 62*, 1145–1154.

Tsuda, A., & Tanaka, M. (1985). Differential changes in noradrenaline turnover in specific regions of rat brain produced by controllable and uncontrollable shocks. *Behavioral Neurosciences, 99*, 802–817.

Turner, B. H., Mishkin, M., & Knapp, M. (1980). Organization of the amygdalopetal projections from modality-specific cortical association areas in the monkey. *Journal of Comparative Neurology, 19*, 515–543.

Uhde, T. W., Joffe, R. T., Jimerson, D. C., & Post, R. M. (1988). Normal urinary free cortisol and plasma MHPG in panic disordera: Clinical and theoretical implications. *Biological Psychiatry, 15*, 575–585.

Valentino, R. J., Foote, S. L., & Aston-Jones, G. (1986). Corticotropin-releasing factor activates noradrenergic neurons of the locus ceruleus. *Brain Research, 270*, 363–367.

Vander Pool, J., Rosenthal, N., Chrousos, G. P., Wehr, T., & Gold, P. W. (1991). Evidence for hypothalamic CRH deficiency in patients with seasonal affective disorder. *Journal of Clinical Endocrinology and Metabolism, 72*, 1382–1387.

Vecsei, L., Bollock, L., Penke, B., & Telegdy, G. (1986). Somatostatin and (D-Trp8,D-Cys 14) somatostatin delays extinction and reverses electroconvulsive shock induced amnesia in rats. *Psychoneuroendocrinology, 11*, 111–115.

Vieth, R. C. (1991). Sympathetic nervous system function in depression and panic disorder. In M. R. Brown, G. F. Koob, and C. Rivier (Eds.), *Stress: Neurobiology and neuoendocrinology* (pp. 395–435). New York: Marcel Dekker.

Walsh, B. T., Katz, J. L., Levin, J., Kream, J., Fukushima, D., Hellman, L., Weiner, H., & Zumoff, B. (1978). Adrenal activity in anorexia nervosa. *Psychosomatic Medicine, 40*, 499–510.

Weingartner, H., Gold, P., Ballenger, J. C., Smallber, S. A., Summers, R., Rubinow, D. R., Post, R. M., & Goodwin, F. K. (1981). Effects of vasopressin on human memory functions. *Science, 211*, 601–603.

Weiss, J. M., Glazer, H. I., Pohorecky, L., Brick, J., & Miller, N. E. (1975). Effects of chronic exposure to stressors on avoidance escape behavior and on brain norepinephrine. *Psychosomatic Medicine, 37*, 522–534.

Weiss, S. R., Post, R. M., Gold, P. W., Chrousos, G., Sullivan, T. L., Walker, D., & Pert, A. (1986). Corticotropin releasing factor-induced seizures and behavior: Interaction with amygdala kindling. *Brain Research, 372*, 345–351.

Weissman, M. M., & Klerman, G. (1977). Sex differences and the epidemiology of depression. *Archives of General Psychiatry, 34*, 98–104.

Wolkowitz, O. M., Rubinow, D., Doran, A. R., Breier, A., Berrettini, W. H., Kling, M. A., & Pickar, D. (1990). Prednisone effects on neurochemistry and behavior. *Archives of General Psychiatry, 47*, 963–968.

Woodruff, G. N. (1991). CCK peptides in anxiety. *Biological Psychiatry, 29*, 20S.

Woods, S. W., Charney, D. S., McPherson, C. A., Gradman, A. H., & Heninger, G. R. (1987a). Situational panic attacks. *Archives of General Psychiatry, 44*, 365–375.

Woods, S. W., Goodman, W. K., & Heninger, G. R. (1985). Neurobiological mechanisms of panic anxiety: Biochemical and behavioral correlates of yohimbine-induced panic attacks. *American Journal of Psychiatry, 144*, 1030–1036.

Yajima, F., Suda, T., Tomori, N., Sunitomo, T., Kakagami, Y., Ushiyama, T., Demura, H., & Shizume, K. (1986). Effects of opioid peptides on immunoreactive corticotropin-releasing factor release from the rat hypothalamus in vitro. *Life Sciences, 39*, 181–186.

Yehuda, R., Lowy, M. T., Southwick, S. M., Shaffer, D., & Giller, E. (1991). Lymphocyte glucocorticoid receptor numbers in post-traumatic stress disorder. *American Journal of Psychiatry, 148*, 499–504.

Youdim, M. B. H. (1972). In E. Costa and M. Sandler (Eds.), *Monoamine oxidase—New vistas* pp. 67. New York: Raven Press.

CHAPTER

10

Hormones, Rhythms, and the Blues

Donald L. McEachron

Department of Psychiatry
University of Pennsylvania School of Medicine and
Biomedical Engineering and Science Institute
Drexel University
Philadelphia, Pennsylvania 19104

Jonathan Schull

Department of Psychology
Haverford College
Haverford, Pennsylvania 19041

Animals have evolved diverse behavioral and physiological processes for coping with physical and social challenges. One of the most ubiquitous of these processes is an endogenous or self-generated rhythmicity by which organisms adjust to the various periodic influences of their environment, for example, day–night, lunar, and seasonal cycles. In an important series of early investigations, Curt Richter established that these biological rhythms influence normal and abnormal physiology and behavior in a variety of ways. Functions as diverse as the waxing and waning of attention and activity levels, the swelling of joints, the spiking of fevers, and the onset and remission of depressive and manic episodes all show temporal cycles with recurrent periods ranging from hours to months. By removing various endocrine organs and neural tissue, Richter determined that many near-daily (or *circadian*) rhythms in mammals are driven by a master pacemaker that resides in a region near the anterior hypothalamus. This area subsequently was identified as the suprachiasmatic nuclei. Much of the research that followed focused on the physiological and biochemical properties of these neural oscillators. Although Richter also demonstrated that endocrine manipulations exert significant effects on circadian and other biological cycles, and suggested that abnormalities of hormones and rhythms may play a crucial role in the etiology and treatment of psychopathology, the extent of neuroendocrine influences on biological rhythms has remained relatively unexplored (for a previous review, see Turek & Gwinner, 1982).

In humans, abnormalities in biological rhythms have been associated with disturbances of behavior, physiology, and psychological state. Patients with affective disorders often display an annual periodicity in the incidence of affective episodes, as well as cyclicity of episodes and diurnal changes in symptom severity within each episode (Pflug & Tölle, 1971; Dirlich et al., 1981; Faust, Sarreither, & Wehner, 1974; Sitaram, Gillin, & Bumey, 1978; Maas, 1979; Sachar & Baron, 1979; Templer, Ruff, Ayers, & Beshai, 1982). These individuals also tend to show characteristic alterations in various biological rhythms when compared with the general population (see McEachron, 1984; Halaris, 1987). Strong support for an etiological role involving the circadian system comes from observations that successful treatments aimed specifically at rhythms—such as phase shifts, sleep deprivation, and light therapy—are often effective in relieving depressive and manic–depressive symptoms (Pflug & Tölle, 1971; Wehr, Wirz-Justice, Goodwin, Duncan, & Gillin, 1979; Doust & Christie, 1980; King, 1980; Kripke, Gillin, Mullaney, Risch, & Janousky, 1987; Lewy, Sack, Miller, & Hobar, 1987a; Lewy, Sack, & Singer, 1987b, 1988; Terman, 1988). Moreover, many of the standard treatments for affective disorders, such as lithium or tricyclic antidepressants, simultaneously alter overt circadian rhythms (Johnsson, Engelmann, Pflug, & Klemke, 1980; Kripke & Wyborney, 1980; Naber, Witz-Justice, Kafka, & Wehr, 1980; Wirz-Justice et al., 1980, 1982; McEachron, Kripke, & Wyborney, 1981; McEachron et al., 1982, 1985). These results led several investigators to propose that abnormal rhythms are a major causal factor in affective disease (Halberg, 1968; Jenner, 1968; Papousek, 1976; Pflug, 1976; Kripke, Mullaney, Atkinson, & Wolf, 1978).

Neuroendocrine abnormalities have also been associated with affective disorders. For example, unipolar and bipolar patients are more likely to suffer from thyroid abnormalities than individuals in the general population (Kostin, Ehrensing, Schalch, & Anderson, 1972; Prange, Wilson, Lara, Alltop, & Breese, 1972; Whybrow, Coppen, Prange, Noguera, & Bailes, 1972; Cowdry, Wehr, Zis, & Goodwin, 1983; Bauer & Whybrow, 1986). Interestingly, many standard treatments such as lithium also effect thyroid state. Specific thyroid treatments (T_3 supplements) have proven effective as antidepressants as well (Wilson, Prange, McClane, Rabon, & Lipton, 1970; Coppen, Whybrow, Noguera, Maggs, & Prange, 1972; Wheatley, 1972; Banki, 1975). Other neuroendocrine systems also appear to be involved. Abnormalities in the hypothalamic–pituitary–adrenal axis often have been suggested to be contributing factors in depression (Brown, 1987; Reichlin, 1987; Rose, 1987). Use of the dexamethasone suppression test as a putative biological marker for depression suggests that changes in the ACTH–cortisol system are indeed associated with mood disorders, although only approximately 40–50% of patients with major depressive disorder actually fail to show suppression (Arana & Baldessarini, 1987; Rose, 1987). The predominance of affective disease in women (Sachar & Baron, 1979; Christie, Little, & Gordon, 1980; Glick, Quitkin, & Bennett, 1987) implicates yet another set of neuroendocrine factors—the sex steroids estrogen and progesterone—in the etiology of depression. This latter implication is strengthened by claims that oral contraceptives can induce depression in some women (Lewis & Hoghughi, 1969; Herzberg, Johnson, & Brown, 1970). Finally, pineal melatonin has been recently associated with certain affective disorders (Beck-Friis & Wetterberg, 1987; Kripke, Drennan, & Elliott, 1992). The exact role of neuroendocrine abnormalities in the etiology of affective disease is not yet clear, but the use of hormones in treatment suggests that neuroendocrine changes may be causal factors.

Clear associations have thus been established between circadian abnormalities and mood disorders and between neuroendocrine state and affective disease. In addition, most, if not all, hormones are well documented to be secreted on a circadian schedule, thus providing a link between hormones and rhythms. The question remains, however, about the extent to which this link can operate in the opposite direction—how substantial is the endocrine feedback and modulation of biological cycles? If significant, this factor could be critical to understanding the development of affective disorders. Given that circadian abnormalities contribute to affective symptoms, changes in an individual's endocrine environment might be a cause of these abnormalities and, thus, the ultimate reason for the pathological changes in mood (see Curtis, 1972). Significantly, all four neuroendocrine factors cited earlier as associated with affective disease—thyroid hormones, glucocorticoids, sex steroids, and pineal melatonin—have also been shown to modify circadian rhymicity. Thus, a common mechanism associated with affective disease may be the influence of neuroendocrine factors on individual vulnerability to circadian disruption. To date, however, the ability of hormones to alter or disturb biological rhythms remains, with the exception of melatonin, relatively unstudied. We hope that this chapter provides an incentive to change that situation.

Our objective in this chapter is to provide a biopsychological perspective on the interrelations of biological rhythms, hormones, behavior, and psychopathology. We believe that evolutionary, physiological, and psychological processes must be understood in relation to each other if they are to be understood at all, and that the causal interactions across these domains are at least as important as the more commonly studied causal processes that occur within each of them. Thus, after an introductory description of biological rhythms, we discuss the evolutionary selection pressures that give rise to endogenous oscillators for daily and seasonal rhythms. We then examine the structural properties and physiological mechanisms of these biological clocks. Next, we review the relationships between biological rhythms and several endocrine systems and secretions—pineal melatonin, the adrenal hormone corticosterone, gonadal steroids, and the thyroid hormones (see Figure 1). Several major themes will emerge, which we will preview here.

One theme concerns the evolution of vertebrate biological rhythm systems. Although the pineal gland in many nonmammalian vertebrates tranduces environmental light–dark cycles and serves as a biological clock, the pineal's role in mammals has been almost completely subordinated to circuitry that goes from retina to hypothalamus to suprachiasmatic nucleus (SCN) and thence to the pineal and elsewhere. In mammals and humans, the SCN appears to be a primary circadian pacemaker and the pineal's residual role is to modulate the SCN, and possibly to couple and uncouple secondary circadian rhythms from the SCN-controlled circadian cycles. In addition, the pineal plays a critical role in the regulation of certain seasonal rhythms (and possibly seasonal affective disorders).

The existence of secondary, indeed multiple, oscillatory systems is another important theme. Secondary circadian systems have been identified in most mammals, although the anatomical loci of the controlling oscillators are not known. The association of many of these secondary systems with food-restriction schedules, stress, and with rhythms of temperature and corticosterone will lead us to examine closely the relationship of these non-SCN rhythms with the adrenal

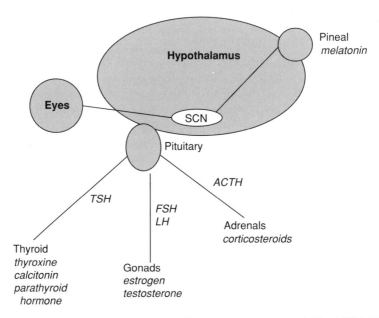

Figure 1. Overview of the endocrine system. Blood-borne hormones are italicized. TSH, Thyroid stimulating hormone; FSH, follicle stimulating hormone; LH, luteinizing hormone; ACTH, adrenalocorticotropic hormone.

glands. A third theme concerns the adaptive function of such multioscillator circadian systems. One possible advantage lies in the ability of secondary pacemakers to adjust to temporary periodic phenomena, a food source or stressful event, while the main circadian clock remains linked to a more permanent source of temporal information, such as the day–night cycle. A fourth theme concerns the relationships and interaction of these circadian rhythm systems with noncircadian cycles; particularly the pineal-based circannual rhythms of seasonal mammals and the gonadally-based estrous and menstrual rhythms. Finally, the last two themes concern the neuroendocrine mechanisms, and the sometimes pathological consequences, of these cross-system interactions. It strikes us as noteworthy that sex steroids and thyroid hormones exert a similar set of effects on circadian rhythms in animals, that both thyroid and steroid imbalances can produce emotional disorders in a sex-linked fashion, that both these hormone classes affect catecholamine systems of the brain (including the catecholaminergic pathway that links the SCN to the pineal gland), and that catecholamine systems are important sites of antidepressant action. A pattern emerges here and, although it is amenable to multiple interpretations, we think we can identify a number of important research questions, clinical as well as basic, that should be addressed in further theoretical and empirical investigation.

I. Endogenous Biological Rhythms

Biological rhythms can be classified by the manner in which their cycles are generated. Exogenous rhythms are imposed on a biological system by factors external to the system whereas endogenous rhythms are produced by activity within the biological system itself. Note that the term endogenous is related to the biological system being examined. An oscillator constructed of several cells may serve as an endogenous rhythm generator for a multicellular organism, but such a cycle is exogenous from the perspective of any single cell within that structure.

Endogenous rhythms can be broadly categorized as *simple* or *circa*. Simple endogenous rhythms, such as the cardiac cycles, are not synchronized to any particular external environmental fluctuation. Circa-rhythms, on the other hand, are adapted to and influenced by geophysical environmental cycles. Four such circa-rhythms have been identified, each associated with a characteristic cycle time or period: circatidal (about 12 hr), circadian (about 24 hr), circalunar (about 28 days), and circannual (about 1 year). The extreme prevalence of circadian rhythms in eukaryotic organisms has led to categorizing other endogenous rhythms in relation to 24-hr cycles: ultradian rhythms are biological cycles with periods $\ll$20 hr whereas infradian rhythms have periods $\gg$30 hr. An example of an ultradian rhythm would be the 90-min REM–nonREM cycle in human sleep; the 4- to 5-day estrous cycles in rats and hamsters would be considered infradian rhythms.

In the presence of appropriate *entraining* stimuli (a light–dark cycle, for example), the frequency and phase relations of circa-rhythms can be synchronized and aligned with the environmental rhythm. However, when such Zeitgebers (literally, in German, time-givers) are eliminated in the laboratory (e.g., under conditions of constant light or constant darkness), circa-rhythms *free run* with periods close to, but not equal to, that of their geophysical cycle. The phase of the endogenous oscillation (often defined by the time at which the biological cycle peaks) gradually drifts away from that of the geophysical oscillation, thus providing the clearest evidence for an endogenous oscillator whose fluctuations are not driven but only synchronized by environmental cues.

Thus, by definition, all endogenous circa-rhythms can free run under constant conditions. Under natural circumstances, however, organismic and environmental rhythms are typically synchronized or entrained. This attunement of organism and environment is thus a fundamental fact of life, and requires two different and complementary explanations. First, we must understand how environmental periodicities create selection pressures favoring the evolution of biological clocks. Second, we must understand the physiological mechanisms by which entrainment occurs.

II. Evolutionary Significance of Biological Rhythms

The cliche "The early bird gets the worm" represents one popular acknowledgment of the importance of proper timing. Although apparently very simple, the implications of this statement are more profound than is evident at first. Why does an early bird get the worm? Assuming that worms are active and readily available to avian predators only at a certain time each day (for example, early morning), the penalty for arriving at feeding areas late is readily apparent. Late arriving individuals may find that the worms are no longer obtainable, either because of a change in worm activity or location or because conspecific competitors already have eaten the available prey. If this is the case, however, why do birds not arrive earlier and earlier until some begin to arrive in the middle of the night? The answer is that penalties for arriving too early exist also. Reaching feeding areas before dawn may expose individuals to metabolic costs in terms of thermoregulation in the predawn cooler temperatures or to danger from nocturnal predators. Under these circumstances, a right time exists to maximize feeding efficiency and minimize risk. Therefore, a strong selection pressure exists to arrive at that time. A consistently periodic environment, such as that provided by the earth's rotation, promotes the evolution of such temporal niches. In this sense, it is not really the *early* bird that gets the worm but rather the *on time* bird. To be on time, however, an organism must always know what time it is. One major, and perhaps obvious, selective advantage to a biological clock is to tell local time.

Circa-rhythms subserve other important biological functions as well. Biological clocks allow animals to estimate the passage of time (Pittendrigh, 1981), a critical function, for example, when using celestrial cues for navigational purposes. A final adaptive function of biological clocks lies less with accommodation to a particular external environment and more with the patterning of physiological and biochemical processes within the internal environment of the organisms themselves. Most biochemical pathways are inherently oscillatory; and all physiological processes fail if their precursors are unavailable when needed. Biological clocks can thus impose an overall temporal order for the coordination of interdependent processes within the organism, as well as coordinating these processes with time-critical demands imposed by the environment.

These functions of biological clocks—estimating local time, measuring the passage of time, and maintaining internal temporal order—often are interrelated both adaptively and mechanistically in ways that vary from species to species. Further, the functions and mechanisms of the various circa-rhythms often are linked together. Consider photoperiodism. Reproductive state in many plants and animals is synchronized with the annual cycle of day length (photoperiod). Male Syrian hamsters, for example, are typically reproductively active in the summer. When males are initially exposed to less than 12 hr light per day, their testes and seminal vesicles become small and they avoid female conspecifics (who can be dangerous if approached at the wrong time). However, when exposed to more than 12.5 hr light per day, the size of the testes can increase 10-fold, and males become more active, more aggressive, and more sexually active. [The relevance of this pattern to winter depression characterized by weight gain and low levels of libido has not escaped notice, although it remains controversial (Mrosovsky, 1988).] Furthermore, these effects of light on behavioral and hormonal state are themselves modulated by circadian rhythms. In hamsters, even a few minutes of light can induce reproductive activity if presented at the optimal time of day; in sheep, the duration of light appears to be more important than its temporal location.

Photoperiodism thus illustrates the interrelationships and interactions among functions and mechanisms of biological rhythms, and shows how biological rhythms can modulate hormonal states. Our major focus, however, is on influences in the other direction—the modulation of biological rhythms, especially circadian rhythms, by hormones.

III. Framework for Investigating Neuroendocrine Influences on Circadian Systems

Understanding at the outset that the notion of a single organismic "circadian system" is a simplification in many if not all species is important. Current research indicates that daily patterns of physiological and behavioral activity are influenced by the combined effects and interactions of multiple circadian oscillators with each other and with various Zeitgebers. Within limits, these multiple rhythmic influences are stably yet dynamically coordinated, but if they become "misaligned" or "abnormally balanced," pathology can result. Thus, it is widely believed that the tendency of affective disorders to show cyclic patterns of onset and remittance may be the consequence of abnormal and shifting phase relations between oscillator systems within the individual and/or between the cycling of the individual and the geophysical or biosocial environment. Support for this notion comes from experiments in which human volunteers have been phase shifted in the laboratory. Even small shifts from 2 to 4 hr lead to significant changes in mood and behavior (Taub & Berger, 1974a,b). Within this conceptual framework, seeing how hormones that differentially affect one circadian system or another could produce, and sometimes be used to ameliorate, emotional and physiological disturbances is easy. However, to analyze the ways in which hormones might affect circadian systems, we must review the organization of the circadian system itself.

IV. Characteristics of Circadian Clocks

Every circa system has several functional components; different hormones affect different components. Circadian systems have 5 basic elements:

(1) An endogenous oscillator, or pacemaker. As mentioned earlier, the SCN appear to serve as master pacemakers in most mammals, whereas the pineal gland serves as a primary clock in many other vertebrate species.

(2) A sensor capable of recognizing the appropriate Zeitgebers. For most vertebrates, light is the most important circadian Zeitgeber, and sensors are located in the eyes and in the pineal of nonmammals (which receives light through the skull). However, restricted feeding schedules can entrain activity rhythms in rats and hamsters, even in individuals whose SCN have been lesioned. The search for the sensor(s) and oscillator(s) for this second circadian system is a matter of current investigation (Mistlberger & Rusak, 1991).

(3) Any circa-system must have a coupling mechanism that links the sensor with the clock. In mammals, one constituent is the retino–hypothalamic tract, which links visual information to the SCN.

(4) Another coupling mechanism must link the clock with the output rhythm. This link is mediated by a variety of neural and hormonal processes.

(5) The last component is the output rhythm itself. One such output rhythm is behavioral activity, which, in turn, may determine when light exposure occurs and thus the pattern of entrainment. Similarly, other neural and hormonal output rhythms may modulate the sensitivity and activity of the

other components. Such influences thus provide potentially critical feed-back loops that link the last component with the others, and reinforce the idea that the causal interactions among the components of circadian systems are multidirectional and involve multiple levels.

To measure local time accurately, an entrained circadian clock must fulfill two basic requirements. First, the cycle length of the rhythm must match that of the entraining cycle or Zeitgeber. A free-running circadian rhythm is analogous to a watch running fast or slow; the entrainment process must correct this free run to match the environment. Second, a stable phase relationship must exist between the circadian and environmental cycles. If an animal's most adaptive behavior is to rise 1 hr before dawn, then some neural or hormonal event must be triggered consistently by the circadian system 1 hr before dawn, which requires that the system be in a similar state each day in the hours before sunrise. These two features, a frequency match and a stable phase angle, are characteristic of entrained circa-rhythms.

Two methods of achieving these results can be suggested. First, the period of the circadian rhythm can be altered to match that of the Zeitgeber by changing the angular velocity of the clock. This change would result in the actual period or frequency of the rhythms becoming the same as that of the Zeitgeber and is called *parametric* entrainment. The second approach is analogous to resetting a watch. Suppose you owned a watch that ran 5 min fast every 24 hr. Rather than spend the money to fix the angular velocity of the watch (the parametric solution), you reset the watch by moving the hands back by 5 min each day (a *phasic* solution). A similar solution is easily envisioned for a slow running watch by moving the hands ahead. By *phase shifting* your timepiece, you changed its *apparent* length or period without changing the actual underlying speed of the mechanism. With many circadian clocks, light pulses can accomplish the same result. By proper application of light pulses at selected times in the circadian cycle, the apparent period of the rhythm is altered to match that of the light cycle, but the underlying period or angular velocity of the clock need not be changed at all. The vast majority of organisms which have been examined seem to use this second or phasic approach to achieve synchronization with their environment although parametric changes may occur as a longer-term adaptation.

By presenting light pulses at different times in the circadian cycle and plotting the phase advances and delays which result, so-called *phase-response curves* (PRCs) can be produced which summarize the essential features of a circadian system entrainment pattern. For a typical PRC, light pulses produce slight phase advances early in the subjective day (the time in the organism's cycle when light would normally be available), slight delays late in the subjective day, and large changes during subjective night; most of the subjective day is insensitive (see Moore-Ede, Sulzman, & Fuller, 1982, for complete discussion).

The PRC also displays another characteristic of biological clocks, that is, a limited range of Zeitgeber cycles to which a given rhythm will entrain. Suppose the free-running period of an organism is 25 hr whereas the maximum phase advance is 3 hr. Thus, the shortest light–dark cycle capable of synchronizing this individual is 22 hr (25 – 3). If the largest delay were 2 hr then the longest cycle to which the organism could adjust would be 27 hr (25 hr + maximum 2-hr delay). This synchronization interval from 22 hr to 27 hr is called the *range of entrainment* and is characteristic for both individuals and species. This range is also adaptable. A PRC can be changed by increasing or decreasing the strength of a Zeitgeber. For example, a light–dark cycle with a light intensity of 20 lux during light is much weaker than one with a light intensity of 1000 lux; a stronger Zeitgeber may result in a larger range of entrainment or other changes in the form of the PRC. Similar results might be achieved without changing the actual light *intensity* if the organism's light *sensitivity* were to be altered. Some bipolar patients have been discovered to be considerably more light sensitive than the normal human population (Lewy et al., 1985), a factor that may contribute to the development of the disease.

The pattern of circadian behavior during entrainment might be altered in yet another way. This change involves the phase angle between Zeitgeber cycle and biological rhythm. The time between particular points in a Zeitgeber cycle, dusk for example, and corresponding points in an animal's circadian cycle such as activity onset, constitute the phase angle. As discussed earlier, this angle is supposed to remain stable during entrainment. However, this is not always the case. For various mathematical reasons (see Aschoff, 1965), phase angles are partly dependent on the ratio between the length of the Zeitgeber period and that of the entrained circadian clock. Practically speaking, if the period of a circadian clock is lengthened a phase delay occurs and if it is shortened a phase advance occurs. This effect can actually be quite useful. Suppose a nocturnal rodent begins activity 8 hr after dawn, that is, the animal has an 8-hr phase angle, in December. If this animal were to retain this phase angle in August, it would begin its activity in bright light, significantly increasing the risk of predation. Instead, however, the animal is phase delayed and may display a 14-hr or longer phase angle. How is this alteration achieved? Increasing the light intensity or duration lengthens the period of the circadian clocks of most nocturnal animals. This change in period results in the observed phase delay while light acting on the PRC prevents loss of entrainment. Humans are also sensitive to light intensity and duration (Wever, 1979, 1986, 1989; Lewy, Wehr, Goodwin, Newsome, & Markey, 1980). The phase angle changes that result may play an important role in a subset of affective diseases called seasonal affective disorders (SAD). Changes in the period of biological clocks associated with Zeitgeber intensities are known as *parametric effects*. Again, it must be emphasized that actual changes in light intensity are not required to observe parametric effects; the only requirement is that the circadian system receive more or less light information, which can be achieved just as easily through alterations in light sensitivity as it can through actual changes in illumination.

The following organizational framework will be our guide in evaluating neuroendocrine influences on circadian rhythms. For each of the neuroendocrine systems we consider—pineal melatonin, thyroid hormones, sex steroids, and adrenal glucocorticoids—the following questions will be addressed. First, is the gland or system a circadian clock, that is, is it an endogenous oscillator that is entrainable by external environmental cycles and responsible for the generation of other rhythms? Second, does the gland or system exert parametric effects on any circadian clocks, increasing or decreasing their endogenous periods? If parametric effects are observed, does the hormone alter the sensitivity of the circadian system to Zeitgeber cycles, either by changing primary sensory reception or the coupling of sensor to clock? Finally, we consider the possibility that the hormone affects the coupling of endogenous oscillators with their output rhythms and/or the coupling of oscillators with each other. Having reviewed the data on rhythmic effects, we briefly discuss the neural mechanisms that might mediate these results.

V. Pineal Melatonin

Although the pineal gland contains numerous peptides (Norris, 1985), the most well studied of its secretory products is melatonin, a hormone released nightly under conditions of darkness. The pineal gland is responsible for driving overt activity rhythms in some nonmammals (Gaston & Menaker, 1968; Garg & Sundaraj, 1986; Panrt & Chandola-Saklani, 1992) and is also involved in the photoperiodic control of reproduction and other seasonal behavior in mammals (Hoffmann, 1981). These facts, along with the recognition of seasonal affective disorders and reports that manic–depressives may be hypersensitive to the suppressive effects of light on melatonin production (Lewy et al., 1985), have led to increased clinical interest in the pineal.

For most nonmammalian vertebrates, the pineal gland serves as a biological clock. Even *in vitro*, the pineals of most birds, reptiles, and fish that have been examined respond directly to light and many show persistent circadian oscillations (Binkley, 1976; Wainwright & Wainright, 1979; Deguchi, 1979; Menaker & Wisner, 1983; Takahashi & Menaker, 1984; Tamotsu & Morita, 1986; van Veen et al., 1986; Falcon & Collin, 1989; Falcon, Marmillian, Claustrat, & Collins, 1989; Kezuka, Aida, & Havnu, 1989; Sato, 1990). In at least some of these species—house sparrow, spotted munia (finch), catfish, and some lizards—pinealectomy also abolishes circadian activity rhythms (Gaston & Menaker, 1968; Garg & Sundararaj, 1986; Pant & Chandola-Saklani, 1992; Underwood, 1977, 1981, 1983). Furthermore, in some species, pineal gland transplants have been shown to restore these rhythms with the phase of the rhythm adopted from the donor, a finding that strongly suggests that the pineal gland of the donor is now driving the system (Zimmerman & Menaker, 1979). In most other species, however, more subtle effects on activity rhythms have been reported (Underwood 1989). Still, the pineal gland in most nonmammalian vertebrate species displays all the characteristics required of a biological clock: endogenous oscillations, entrainment by an external Zeitgeber, and the ability to drive overt circadian rhythms (for an exception, see Janik & Menaker, 1990, a discussion of desert iguanas). For these groups, the endocrine pineal serves as an endogenous oscillator and responds to light with entrainment. Interestingly, although some evidence suggests that the pineal gland may be involved in photoperiodic responses in some reptiles and fish, apparently it is not the critical factor in birds (Hoffman, 1970; de Vlaming & Vodicnik, 1978; Follett, Foster, & Nicholls, 1985; Crews, Hingorani, & Nelson, 1988), although this difference may be species specific (Ohta, Kadota, & Kanishi, 1989).

In mammals, however, the pineal has lost both its photoreceptive ability and its capacity for endogenous oscillation. The gland no longer serves as a secondary oscillator, much less as a master pacemaker. Two features of mammalian evolutionary history may help explain this reduction of the pineal's circadian role. First, mammals began as a predominantly nocturnal group, which reduced the amount of light available to the pineal gland. Second, the skulls of mammals became thicker and more opaque, again reducing the amount of light able to reach the pineal directly. The evolutionary solution to this problem seems to have been that mammalian pineal glands became increasingly dependent on indirect sources of light information relayed from the eyes via the suprachiasmatic nuclei; the mammalian pineal gland now has lost its photoreceptive ability and many of the photosensitive biochemicals characteristic of nonmammalian pineal glands (van Veen et al., 1986; Sato, 1990; Shi, Furr, & Olson, 1991). Thus, mammalian pineals became isolated from direct Zeitgeber contact and came to rely on obtaining temporal information about the external environment by indirect internal means.

Furthermore, an apparent shift in the control of circadian rhythms occurred, with the SCN becoming predominant, while the role of the pineal gland has waned in comparison. This change may have been an independent evolutionary development unrelated to the loss of direct light information. Many birds, which retain a photosensitive and oscillatory pineal gland, also have a circadian clock located within their SCN, although the relative importance of the two clocks is a matter of some conjecture (Cassone & Menaker, 1984; Cassone, Forsyth, & Woodlee, 1990). It should also be noted that the pineal and SCN are not the only structures which might act as dominant biological clocks. In at least one avian species, the Japanese quail, the eyes themselves may be acting in this role (Underwood, Barrett, & Siopes, 1990). Nevertheless, the SCN in mammals, in contrast to most other species, appear to be the master pacemaker(s) in overall command of the circadian system. Lesions of the SCN usually eliminate rhythms in melatonin production, indicating that the mammalian pineal gland no longer oscillates independently *in vivo* (Illnerova, 1991; Klein, 1985; Moore & Klein, 1974; Romero, 1978; Underwood & Groos, 1982). The results of numerous experiments reflect the reduced role of the mammalian pineal gland in circadian clock function. For example, pinealectomy fails to (1) alter the period of free-running activity rhythms, phase angle to a 10:14 light/dark cycle or rates of reentrainment to 6 hour phase shifts in ground squirrels (Martinet & Zucker, 1985); (2) change the rates of reentrainment to photoperiod shifts in male Syrian hamsters (Hastings, Walker, & Herbert, 1987); (3) alter the free-running activity rhythms in constant light or dark in female rats (Cheung & McCormack, 1982); (4) change the circadian rhythm of core body temperature in female rats exposed to either light–dark cycles or constant darkness despite reducing the mean temperature (Spencer, Shirer, & Yochim, 1976); (5) significantly modify the diurnal rhythm in plasma prolactin in male rats under a 1:23 LD cycle (Niles, Brown & Grota, 1977); (6) influence circadian rhythms of taste preference, drinking, feeding or body weight in rats (Scalera, Banassi, & Porro, 1990). Additionally, continuous melatonin treatment via silastic capsule failed to alter activity rhythms in female rats (Cheung & McCormack, 1982) and neutralizing circulating plasma melatonin failed to change circadian cycles in plasma prolactin in male rats (Niles, Brown, & Grota, 1977). Finally, timed melatonin infusions fail to entrain either maternal uterine activity or plasma steroid rhythms in pregnant rhesus macaques (Matsumoto, Hess, Kaushal, Valenzuela, Yellon, and Ducsay, 1991). There is at least one exception to this general pattern. Bobbert and Riethoven (1991) have reported that pinealectomy lengthened, and melatonin administration shortened, rhythms in feeding behavior in adult rabbits. Further experimentation will be needed to determine if rabbits are simply a single exception to the general pattern in mammals or whether parametric effects with melatonin are more widespread than we at present believe.

Comparative neuroanatomy further supports this hypothesis about the evolutionary shift of pineal function. Although neural connections exist that link the pineal gland with the brains of lower vertebrates, this *tractus pinealis* is absent in mammals. In addition, the pineal glands of most vertebrates receive only a few autonomic fibers. In contrast, the pineal glands of most species of birds and mammals display well-developed sympathetic and parasympathetic innervation (Norris, 1985; Ueck & Wake, 1977). These changes reflect the increasing dominance of the SCN as a master pacemaker and, in mammals, the dependence of the pineal gland on the neural circuits for lighting information. However, although the mammalian pineal gland may have lost its photoreceptive and oscillatory functions, it has retained the ability to act as internal Zeitgeber or coupling agent. In addition, the pineal gland evolved or retained a critical role as part of the system controlling mammalian photoperiodism.

The role of the pineal gland in coupling or synchronizing circadian rhythms has been clarified by the finding that injections of melatonin are able to entrain the free-running activity rhythms in male rats, provided the time of injection coincides with the onset of activity (Redman, Armstrong, & Ng, 1983). A similar finding was reported in female rats, although entrainment could only be demonstrated in constant dark (DD) but not in constant light (LL) (Thomas & Armstrong, 1988). Melatonin injection also appears to promote internal synchronization of multiple components of rat activity rhythms when these components are disrupted in LL (Armstrong, Cassone, Chesworth, Redman, & Short, 1986; Cassone, 1992) and may affect the direction, and possibly the speed, of reentrainment of activity rhythms under selected circumstances (Armstrong & Redman, 1985). In rats, melatonin also may play a role in controlling its own secretions. Melatonin injected subcutaneously entrains the circadian rhythm of pineal N-acetyl transferase (NAT) activity (Humlova & Illnerova, 1990) and accelerates the rate of reentrainment of pineal melatonin rhythms to an 8-hr phase advance of the light–dark cycle by one full cycle (Illnerova, Trentini, & Maslova, 1989). In contrast, Stetson and Hamilton (1981) reported that melatonin injections were ineffective in altering plasma melatonin rhythms in male golden hamsters.

One important clue to the mechanism by which artificial and natural melatonin pulses exert their effects comes from the finding that melatonin injections fail to synchronize activity cycles in SCN-lesioned rats (Armstrong et al., 1986). In contrast, large bilateral lesions of the anterior paraventricular nuclei of thalamus (where significant melatonin binding sites have been found) failed to alter activity rhythms in hamsters (Ebling, Maywood, Humby, & Hastings, 1992). These results strongly imply that a prime target of pineal melatonin is the SCN.

Melatonin receptors have been localized in numerous brain regions, including the hypothalamus, hippocampus, striatum, midbrain structures, and pituitary (Cardinali, Vacos, & Boyer, 1979; Niles, Wong, Mishra, & Brown, 1979; Morgan & Williams, 1989). Melatonin injected into blood or cerebrospinal fluid (CSF) is concentrated in the hypothalamus and midbrain (Anton-Tay & Wurtman, 1969; Cardinali, Hyyppä, & Wurtman, 1973). Most interestingly, however, high levels of melatonin binding have been reported in the SCN of rats (Vanecek, Pavlik, & Illnerova, 1987; Laitinen, Castran, Vakkuri, & Saavedra, 1989; Vanecek, 1988), hamsters (Weaver, Rivkees, & Reppert, 1988; Weaver, Namboodri, & Reppert, 1986), and humans (Reppert, Weaver, Rivkees, & Stopa, 1988). Melatonin has been reported to alter SCN neuron firing rates, further supporting the hypothesis that melatonin effects are mediated via the SCN. These neurons, which normally show decreased electrical and metabolic activity in darkness (Inouye & Kawamura, 1979; Gillette, 1991; Schwartz, 1991) when melatonin levels are highest, also show reduced activity when exposed to melatonin *in vitro* (Mason & Brooks, 1988; Stehle, Vanacek, & Vollrath, 1989) and *in vivo* (Cassone, Roberts, & Moore, 1988). In all studies, the effects of melatonin were dependent on time of day. Stehle and co-workers (1989) reported, for example, that 80–100% of SCN neurons tested were sensitive to melatonin from 3 hr before to 3 hr after lights off, compared with only 30% at other times. Experiments conducted by McArthur and colleagues (1991) demonstrated that the electrical activity of the SCN maintained in an *in vitro* brain slice preparation could be phase shifted by melatonin only near the time of the day-night transition (McArthur, Gillette, & Prosser, 1991). These data agree nicely with the previously described observation that melatonin injections can entrain activity rhythms only if they are delivered at the onset of activity (Redman et al., 1983).

The somewhat ironic implication of these findings is that the one pineal circadian function that has not been usurped by the SCN (that of internal coupling agent) is, nonetheless, mediated at least in part by the SCN. In this regard, pineal melatonin may act as a kind of edge detector, reinforcing the actions of light and darkness on the SCN. As darkness acts to reduce SCN neural activity, the increase in melatonin synthesis (which also is induced by darkness) feeds back on the SCN to reduce neural activity still further. Thus, melatonin enhances the original effect of darkness on SCN neurons.

Despite the reduced importance of the pineal gland in controlling circadian rhythms, a fundamental role exists for the gland in mammalian photoperiodism (Bartness & Goldman, 1989). Both short-day and long-day responsive species exist that seasonally change their reproductive status. Short-day breeders, such as sheep and deer, are stimulated by periods of reduced daylength whereas long-day breeders, such as hamsters and ferrets, respond positively when daylength exceeds a certain duration (Hoffman, 1981). In a short-day species, the ewe, the duration of high melatonin secretion appears to be the critical variable (Arendt, Symons, English, Poulton, & Tobler, 1988; English, Arendt, Symons, Poulton, & Tobler, 1988). Melatonin administration that mimics the secretion found during long or short days results in the same pattern of photoperiodic responses observed in the analogous artificial photoperiods (Karsch, 1986; Arendt et al., 1988). Pinealectomized ewes do not respond to altered photoperiods (Bittman, 1985); melatonin administered to mimic short days is fully capable of inducing the reproductive system in pinealectomized animals (Karsch, 1986; Arendt et al., 1988), demonstrating that the effect of melatonin is not on the pineal gland itself.

Duration of elevated melatonin concentrations may be the crucial factor in long-day breeders as well. Early experiments with light pulses suggested that the circadian phase at which light is presented was critical (Elliott, 1976; Hoffman, 1979). However, other experiments indicate that male hamsters, like the ewe, respond to the duration of melatonin infusions (Darrow & Goldman, 1985; Bartness & Goldman, 1989). Bartness and Goldman (1989) suggested that earlier reports in pineal-intact hamsters in which timing of melatonin injection appeared to play a critical role could be explained by two factors: (1) a need for a continuously elevated melatonin level for gonadal regression and (2) the fact that melatonin injections could provide such a continuous elevation by extending the natural increase resulting from pineal gland activity. Since melatonin injections could only accomplish this extention by overlapping with the normal rise in endogenous melatonin, phase appeared to be more critical than it really was. Duration by itself is not sufficient either, however, since equivalent length infusions given at different frequencies do not have the same effects (Maywood, Buttery, Vance, Herbert, & Hastings, 1990). Pinealectomy and SCN lesions are equally effective in blocking these photoperiodic responses, demonstrating again that mammalian pineal glands obtain their light information from the SCN (Rusak & Morin, 1976; Morin, Fitzgerald, & Zucker, 1977b; Miernicki, Karp, & Powers, 1990; Bittman, Bartness, Goldman, & DeVries, 1991). Although the pineal gland itself is not required for a response to melatonin to develop (Bartness & Goldman, 1988), the data are more equivocal concerning the role of the SCN mediating melatonin effects. In Syrian hamsters, SCN ablation does not interfere with male responses to melatonin infusions (Maywood et al., 1990) whereas SCN lesions do block Siberian hamster responses to similar infusions (Bartness, Goldman, & Bittman, 1991). Obviously, the system is more complex than was imagined at first and more research is needed. Two conditions seem fairly certain, however. First, the pineal gland secretion that transduces the effects of photoperiod to the reproductive axis is melatonin rather than some other pineal substance. Second, melatonin can be stimulatory or inhibitory to reproduction, depending on the species and photoperiod. Melatonin is not simply an antigonadal hormone.

Results in female hamsters are similar in some respects. Injections of melatonin are capable of inducing acyclicity seen with exposure to short photoperiods when given at certain times of day and not at others in golden hamsters (Bridges, Tamarkin, & Goldman, 1976; Badura & Nunez, 1989). Interestingly, and in contrast to males, the pineal gland appears to be required, since melatonin injections are ineffective in pinealectomized females (Bridges et al., 1976). In females, then, melatonin may act through other pineal secretions, in contrast to our hypothesized direct action in males. However, despite the ability of melatonin to disrupt estrous cycles, photoperiodic control of estrogen-stimulated sexual behaviors appears to be independent of the pineal gland. Pinealectomized female golden hamsters exposed to long days (16:8) display the same behavioral response to hormone treatment as intact controls exposed to the same photoperiod (Badura & Nunez, 1989).

Some other long-day breeders show similar behavior to that observed in hamsters, although considerably less research has been done with other species. Male voles, for example, appear to have a particular light-sensitive phase in their circadian system for transducing day length (Grocock & Clarke, 1974). Female ferrets, which become acyclic on exposure to 8:16 light–dark cycles, will maintain estrous cycles in short photoperiods if pinealectomized (Thorpe & Herbert, 1976). Melatonin injections are capable of promoting the change from cycling to acyclicity, therefore mimicking some of the effects of short days (Thorpe & Herbert,

1976). Although periodic breeding behavior and estrous cycles are maintained in pinealectomized ferrets, pinealectomy abolishes the link between breeding season and season of the year. Groups of pinealectomized females become increasing asynchronous (Herbert, 1972; Herbert, Stacey, & Thorpe, 1978).

Many of the findings we have reviewed can be summarized by suggesting that melatonin is a "second messenger" that reinforces the circadian rhythm of the SCN and conveys information about the timing of darkness back to the SCN and to other tissues. This loop, particularly the catecholamine connection from SCN to pineal gland, will prove particularly relevant when we examine the effects on circadian rhythms and depression of steroid hormones, thyroid hormones, and antidepressant medications.

The adaptive or evolutionary reason that the SCN–pineal–SCN circuit continues to play any role at all in nonseasonal mammals such as rats and humans is not completely clear. The pineal gland might be vestigially "reflecting" photic information from the SCN back to the SCN. Alternatively, seasonality and nightlength information may, in fact, be more functionally important in nonseasonal mammals than we know, so the ability of the pineal gland to monitor nightlength remains adaptive. It is also possible, however, that the circuit may serve to reinforce and modulate the SCN's own endogenous oscillations. A positive feedback circuit of this sort would be buffered against slight perturbations. By evolving such a system, mammals would have the advantage of a strong local time clock unaffected by small environmental variations and resistant to disruption if secondary clocks were to become uncoupled from the primary environmental light–dark cycle. As we shall see, secondary clocks do seem to exist in many species. The selective uncoupling of secondary oscillators may be a way of adapting to periodic events, such as food availability, which occur at arbitrary or changeable times of day. The pineal gland's buffering of the SCN could keep the primary oscillator "on schedule." When desynchronization was no longer necessary, pineal melatonin might help bring the secondary oscillators back in line with the primary one. Consistent with this view, Armstrong (1989) has speculated that pineal melatonin serves as an overall internal synchronizer for the "total circadian structure of mammals."

In rodents, and especially rats, two circadian systems have been identified— the photic and nonphotic clock systems. The light-sensitive system uses light–dark cycles as the primary Zeitgeber and the master clock is the SCN. The primary nonphotic system responds differentially to restricted feeding (RF) cycles as the primary Zeitgeber, but the location of the RF-sensitive clock(s) is not known (although in the next section of this chapter, we examine the possibility that at least one set is located in the adrenal glands). In intact rodents, RF cycles are powerful Zeitgebers, affecting rhythms of activity, temperature, and corticosterone but, interestingly, not influencing the SCN or pineal rhythms (Inouye, 1982; Yamazaki, Tsujimoto, Inome, & Nakagawa, 1984; Ho, Burns, Grota, & Brown, 1985).

In SCN lesioned animals, circadian RF schedules can entrain rhythms that had been eliminated by the SCN lesion, thus demonstrating that an RF-sensitive oscillator exists (or can exist) outside the SCN (Stephan, Swann, & Sisk, 1979).

Based on such data, one of us proposed a model of rat circadian structure in which the SCN serves as a primary clock, receiving light–dark information and entraining numerous secondary clocks that, in turn, drive overt rhythms (Figure 2; McEachron, 1987). The hypothesis was that the SCN normally couples the secondary clocks (which are linked directly to overt rhythms) by entraining them to its own cycle. When the SCN is destroyed, the clocks become free to express their own inherent periods, which differ slightly from each other. As each pacemaker drifts out of sychrony with the others, the overt rhythms affected by these clocks become more unstable and incoherent. The arrhythmicity observed in SCN-lesioned animals is, therefore, the result of a population of uncoupled secondary clocks. In SCN-lesioned animals, then, RF cycles can serve as substitute Zeit-gebers, synchronizing secondary clocks to the RF cycle and, thus, to each other.

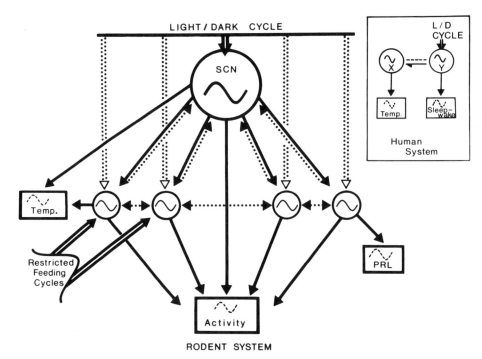

Figure 2. Model of rat circadian organization. Solid lines indicate strong effects; the arrow shows the direction of entrainment. Dotted lines indicate weaker effects. Single lines represent internal effects; double lines show external influences. Reprinted with permission from McEachron (1987), *Progress in Clinical and Biological Research*, Vol. 227B: Advances in Chronobiology, Part B, J.D. Pauly and L.E. Scheving, eds. Copyright © 1987. Reprinted by permission of Wiley-Liss, a Division of John Wiley and Sons, Inc.

In the context of our discussion of the pineal gland, melatonin rhythms may be supposed to buffer the SCN rhythm and its most closely allied secondary oscillators against RF cycles to aid in reentraining rhythms after phase or period shifts induced by RF cycles. In either case, the major function of the pineal gland in rodents appears to involve internal coupling.

As in rats, the human circadian system also has been characterized as being composed of at least two circadian systems. In humans, however, the systems cannot be characterized as photic and nonphotic. Instead, the two master clocks are identified by the fundamental rhythms they control, one clock primarily linked with the sleep–wake cycle and associated rhythms and the other influencing body temperature and its associated cycles, such as cortisol (Moore-Ede & Sulzman, 1977; Moore-Ede et al., 1982). The exact identity of human circadian pacemakers is not known, although the SCN has been suggested as the location of the sleep–wake oscillator (Moore-Ede et al., 1982). As in rats, melatonin does have subtle effects on some circadian rhythms, possibly those rhythms that are related more closely to the SCN and less closely to the "other" system. Thus, melatonin has been tested successfully as a treatment for jet lag (Petrie, Conaglen, Thompson, & Chamberlain, 1989; Samuel et al., 1991), speeding resynchronization and alleviating most symptoms, and as a treatment for delayed sleep phase syndrome, significantly phase advancing components of the sleep–wake cycle without altering the peak of alertness rhythms (Dahlitz et al., 1991). Melatonin also has been reported to entrain rhythms in blind people (Sack, Lewy, Blood, Stevenson, & Keith, 1991) and, similar to studies with rats, melatonin may entrain its own rhythms by influencing the pineal (Lewy, Ahmed, Jackson, & Sack, 1992). Similarly, Arendt and colleagues (1985; Arendt, 1986) reported that timed administration of melatonin was able to affect fatigue and some sleep variables under uncontrolled normal environmental conditions. However, melatonin did not affect the 24-hr patterns of growth hormone, leutinizing hormone, testosterone, or thyroxine in these individuals. Arendt and co-workers also examined the notion that melatonin could alter entrainment characteristics of various human circadian rhythms differentially using a procedure called "fractional desynchronization" (Wever, 1983). This procedure involves exposing organisms to Zeitgeber cycles that continually expand or contract by small amounts; if multiple clocks exist within a single organism, each with its own range of entrainment, the clocks should fall out of entrainment independently and, thus, be individually identifiable. Using this approach, Arendt and colleagues demonstrated that melatonin extended the range of entrainment for behavioral variables including the rhythms of self-rated fatigue, but had no consistent effect on rhythms of core body temperature, cortisol, performance tasks, verbal reasoning, urine volume, or electrolyte balance (Arendt et al., 1985). Similar findings were reported by Wever (1986), who was unable to synchronize body temperature rhythms using melatonin in isolation, despite clear entrainment of sleep–wake cycles. In other studies, periodic melatonin treatment has failed to: (1) alter the phase of adrenal and pituitary rhythms in men (Paccotti et al., 1989), (2) entrain either temperature or cortisol rhythms in a blind man (Folkard, Arendt, Aldhous, & Kennett, 1990), (3) alter the pattern of prolactin or thyroid hormones or significantly phase advance the cortisol rhythm in adult men, despite a significant phase advance of their testosterone rhythms (Tarzolo et al., 1990), or (4) alter the nocturnal pattern of cortisol or thyrotropin (TSH) in normal sleep or after sleep deprivation (Strassman, Peake, Quails, & Lisansky, 1988). Most studies show little effect of melatonin on temperature or related rhythms, although there are contradictory reports (Cagnacci, Elliott, & Yen, 1992; Cassone, 1990). Despite these reports, a reasonable preliminary conclusion is that melatonin in humans has a similar effect to that found in rats—reinforcing an SCN–pineal gland circuit and promoting synchronization of selected rhythms.

Although human beings are not classified as directly photoperiodic, the role of pineal melatonin in modulating internal coupling among rhythms still may provide important clues to understanding the pathophysiology of human affective disorders. Pineal gland modulation of biological rhythms in general, and of the SCN in particular, is clearly part of our phylogenetic heritage. Many of the relevant physiological mechanisms can still be seen in humans. Subtle influences of melatonin on the SCN, which are normally undetectable, may play a significant role in humans whose SCN-based biological rhythms are impaired. Thus, in clinical patients with weak or abnormal SCN cycles, it could be that pineal or other rhythms that are normally dominated by the endogenous oscillations of the SCN become pathologically influenced by melatonin's still-robust sensitivity to light cycles. Alternatively, impairment of the pineal gland's normal ability to reinforce the SCN rhythm may increase the SCN's vulnerability to temporal perturbations. In such cases, environmental stimuli that are ineffective in disturbing circadian cycles under normal circumstances could lead to pathological changes in clock function. Finally, low or high melatonin levels might alter the state of coupling among various components in the human ciradian system, leading either to internal desynchronization or to rhythms that are linked together so tightly that adaptive flexibility to temporal changes in the external environment is seriously impaired.

Collectively, the literature on pineal gland involvement in biological rhythms might be summarized as follows. In birds and lower vertebrates, the pineal gland is a light-sensitive endogenous oscillator that acts as an important biological clock. In mammals, suprachiasmatic nuclei have come to dominate the pineal gland, which has lost its endogenous oscillators and its photoreceptive apparatus. However, the pineal still produces melatonin nightly (driven by the endogenous rhythm of the SCN and inhibited by photic information relayed through the SCN–pineal catecholaminergic connection), and produces profound effects on reproductive and motivational states by way of the SCN and other tissues. In humans (and other nonphotoperiodic mammals such as rats), the pineal gland has been shown to produce subtle effects on circadian rhythms that have come to light only recently. The existence of annual rhythms of motivation, libido, and affect are also only just being recognized in humans, primarily through the recognition of seasonal affective disorders. We suspect that, as seasonal fluctuations of the human spirit become more widely understood, the role of the pineal gland will emerge more clearly.

VI. Adrenal Hormones

Under control of an SCN-based rhythm, the pituitary releases adrenocorticotrophic hormone (ACTH) into the blood on a circadian basis. As its name implies, ACTH causes the adrenal cortex to secrete glucocorticoid hormones, such as corticosterone and cortisol, which are involved in a wide array of functions including maintenance of serum glucose levels, feeding, modulation of inflammatory responses, and immune function. The release of ACTH and glucocorticoids is also triggered by physical and emotional stress, hypoglycemia, hypoxemia, and hemorrhage. Abnormal levels and rhythms of glucocorticoids have been linked to emotional and affective disorders. Under chronic stress, glucocorticoid-producing tissues tend to hypertrophy, presumably to provide ample quantities of these hormones for resource mobilization and redistribution.

This situation poses an interesting puzzle. Despite its coupling to the SCN, the circadian rhythm of glucocorticoid release of the adrenal gland is entrained more easily to RF schedules than is the SCN rhythm that (at least sometimes) drives the adrenal rhythm. For this and other reasons, a second circadian oscillator sensitive to RF schedules is believed to exist in many species, although the anatomical location(s) and physiological mechanisms have not been identified. One intriguing possibility is that the adrenal glands themselves are RF-sensitive circadian oscillators. Glucocorticoid functions are, after all, involved in many aspects of resource mobilization; glucocorticoids do influence and entrain at least some other biological rhythms; and endogenous rhythmicity has been reported in isolated adrenal glands. Thus, the possibility cannot be ruled out that the adrenal glands are secondary circadian pacemakers that uncouple from the SCN rhythm when food-getting and other metabolically demanding activities are required at nontypical times. In this section, we review the available literature on adrenal gland hormones and biological rhythms, and suggest that the matter deserves considerably more scientific attention than it has received to date.

Corticosteroid levels show a circadian rhythm in all mammalian species examined except dogs (Dallman et al., 1987). In nocturnal as well as diurnal species, cortisol levels rise gradually and peak around the time when daily behavioral activity begins—at dusk for rats (Krieger, 1979; McEachron et al., 1982) and at dawn for humans (Aschoff & Wever, 1981). This rhythm is reportedly blunted and/or phase advanced in depressed humans, although whether this condition is cause or effect is far from clear (Sachar et al., 1973; Halbreich, 1987). As mentioned earlier, little doubt exists that light–dark cycles can entrain corticosteroid rhythms. Although the circadian rhythm of ACTH is clearly under control of the SCN (Szafarczyk et al., 1979, 1981), the light–dark cycle is not the strongest Zeitgeber for the adrenal glands, at least in rats. RF cycles exert a far stronger effect, and the pattern of corticosterone secretion is determined primarily by the time of feeding in these animals (Morimoto, Arisue, & Yamamura, 1977; Wilkinson, Shinsako, & Dallman, 1979). This result is especially intriguing since the SCN is virtually insensitive to RF schedules (Inouye, 1982) and that RF schedules can call forth and entrain rhythms in both activity and temperature in SCN-lesioned rats (Krieger, Hauser, & Krey, 1977; Stephan et al., 1979).

Although the adrenal glands have received considerably less attention than the pineal gland, evidence in support of a possible clock function for these glands is surprisingly strong. Two separate research groups have demonstrated endogenous oscillatory capacity in hamster adrenals. Andrews and Folk (1964) reported that circadian patterns of steroid secretion and respiration were maintained by adrenal glands in culture. The glands displayed Q10 values of 0.96 between 15° and 25°C and 1.11 between 25° and 37.5°C. (A Q10 value refers to the rate of change of a biochemical or physiological process with each 10°C change in temperature.) This kind of temperature independence or compensation is quite characteristic of circadian processes and is considered a fundamental feature of any biological clock (Bünning, 1973). Andrews and Folk (1964) also demonstrated that the isolated glands showed some responses to light, although complete phase shifts could not be demonstrated. These experiments were repeated several times (Andrews, 1968a,b, 1971). Shiotsuka and co-workers (Shiotsuka, Jovonovich, & Jovonovich, 1974) also reported persistence of circadian rhythms in isolated hamster adrenals, as well as the existence of ultradian components in adrenal function.

Also consistent with the possibility of an endogenous adrenal oscillator are reports of a persistent *in vitro* adrenal rhythm in ACTH sensitivity in mice (Ungar & Halberg, 1963) and hamsters (Andrews, 1969) and *in vivo* rhythms in rats (Kaneko, Hiroshige, Shinsako, & Dallman, 1980, 1981). Kaneko and colleagues (1981) reported that, in rats, the timing of ACTH sensitivity mimicked the pattern of plasma corticosterone levels, with a peak at the light-to-dark transition, whereas in mice the *in vitro* rhythm was out of phase with previously measured serum corticosterone (Ungar & Halberg, 1963). However, this discrepancy may indicate simply that the usual coincidence of ACTH levels, ACTH sensitivity, and corticosteroid levels is facultative rather than obligatory, and that adjustment of ACTH sensitivity may be a mechanism for uncoupling adrenal and SCN rhythms. Changes in the ACTH sensitivity rhythm could, for example, help explain reports of significant circadian variation in corticosterone levels in the absence of ACTH rhythms (Wilkinson et al., 1979, 1981) and of dissociation between adrenal and pituitary rhythms under conditions of constant light (Fischman, Kastin, Graf, & Moldow, 1988). Dallman and co-workers (1987) have suggested that alpha-MSH is the controlling factor for ACTH sensitivity rhythms. As stated earlier, however, the *in vitro* rhythm of mouse adrenal glands was out of phase with previously measured serum corticosterone, making it unlikely that a pituitary factor was responsible. A recent report by West and Bassett (1992) concluded that this apparent ACTH sensitivity rhythm reflects circadian variations in actual size of the adrenal glands and thus is exogenously, rather than endogenously, generated. In the case of hamster adrenal glands, however, the sensitivity rhythm was blocked by *in vitro* addition of puromycin, a protein synthesis inhibitor, strongly suggesting that an endogenous adrenal cycle of protein synthesis is responsible for the changes in ACTH responsiveness. Perhaps the role of alpha-MSH is to entrain endogenous cycles of ACTH sensitivity. ACTH itself is capable of phase shifting adrenal corticosterone levels *in vitro* (Andrews, 1969), so the two peptides may represent a redundant entrainment mechanism.

Another mechanism that could account for dissociations of adrenal and ACTH rhythms *in vivo* consists of the adrenal gland inputs from the autonomic nervous system (Ottenweller & Meier, 1982; Holzwarth, Cunningham, & Kleitman, 1987). However, this hypothesis is weakened by the finding that 24-hr rhythms of corticosterone are also seen in rats with neurally isolated or auto-transplanted adrenal glands (Gibson & Krieger, 1981; Wilkinson et al., 1981). These results have been assumed to represent another example of the changes in adrenal sensitivity to ACTH (Wilkinson et al., 1981) but an observation made by Gibson and Krieger (1981) suggests otherwise. These researchers reported that transplanted adrenal glands showed no significant response to ACTH injections of 0.5 mU or 2.0 mU, compared with sham-operated controls that did respond significantly when injected at the same time. Insofar as the rhythm of plasma corticosterone showed the exact same phase relationship in adrenalectomized autotransplants and sham-operated animals, the lack of sensitivity to ACTH suggests that another factor is regulating corticosterone rhymicity in those animals with autotransplants. This result leaves the possibility of endogenous oscillation. Interestingly, Gibson and Krieger (1981) reported that, in 50% of the cases, only one functional adrenal gland could be identified in the rats with autotransplants. This result would eliminate the problem of generating a coherent circadian corticosterone rhythm with two independently oscillating adrenal glands, a situation that likely would result in arrhythmic overt cycles.

The data are mixed about whether corticosteroid rhythms can persist in animals whose ACTH and other rhythms are abolished by SCN lesions. Szafarcyzk and colleagues (1979, 1980) and Assenmacher (1982) report that corticosteroid rhythms do persist, whereas Abe and co-workers (Abe, Kroning, Greer, Critchlow, 1979) and Moore and Eichler (1972) report that corticosteroid rhythms are eliminated. Note, however, that the former result may be more telling than the latter. SCN lesions simply may leave adrenal oscillators uncoupled to each other and, thus, free to drift out of phase. Experimental designs that sample the aggregate blood corticosterone rhythm (Abe et al., 1979) or designs that assess adrenal state by destructive sampling from a population of SCN-lesioned animals (Moore & Eichler, 1972) might fail to detect these ongoing but asynchronous oscillations. Multiple out-of-phase circadian oscillators also might produce an ultradian-like pattern of fluctuation in corticosteroid levels, which has been reported in animals with lateral fornix lesions (Fischette, Edinger, & Siegel, 1981) as well SCN lesions (Assenmacher, 1982).

The data are thus consistent with the hypothesis that the adrenal glands are endogenous oscillators that are capable of entrainment to RF cycles, but the question remains as to whether these glands might serve as biological clocks. To be so considered, the glands must be shown to control other biological rhythms. Adrenalectomy in rats does eliminate circadian rhythms of corticosterone binding protein (Hsu & Kuhn, 1988), opiate-induced prolactin release (Kiem, Kanyicksa, Stark, & Fekete, 1987), hypothalamic multiple unit activity in females (Terkel, Johnson, Whitmayer, & Sawyer, 1974), paradoxical sleep (Johnson & Sawyer, 1971), certain liver enzymatic activities (Edwards, 1973), and midbrain norepinephrine levels (Freidman & Walker, 1968) and aggressive behavior rhythms in hamsters (Landau, 1975). Cortisol injections under constant light are capable of entraining rhythms of urinary potassium secretion in squirrel monkeys (Sulzman, Fuller, & Moore-Ede, 1978) and a PRC to dexamethasone injections involving temperature rhythms in rats has been reported (Horseman & Ehret, 1982). These and other data led Angeli (1983) to suggest that glucocorticoid secretion is an important synchronizer of human circadian rhythms. However, not all rhythms in a given species are altered in the same manner by manipulation of adrenal glucocorticoids. For example, temperature and feeding rhythms are not entrained by cortisol in the squirrel monkey (Sulzman et al., 1978) and prolactin patterns are unaffected by dexamethasone in humans (Banovac, Tolis, McKenzie, Guyda, & Sekso, 1977). Albers and co-workers (Albers, Gerall, & Axelson, 1985) also demonstrated that hamster activity cycles were not entrained by timed infusion of cortisol. In addition, body temperature rhythms and rhythms in the neurotransmitters histamine and serotonin found in midbrain areas of the rat remain intact in adrenalectomized rats, albeit with different wave-forms (Friedman & Walker, 1968). Similar findings of pattern changes have been reported for hypothalamic dopamine-β-hydroxylase activity in rats (Banerji, Quay, & Kachi, 1978) and a 3.5-hr phase advance was reported in hypothalamic corticotropin releasing hormone (CRH) rhythms associated with adrenalectomy (Takebe, Sakakura, & Mashimo, 1972). Thus, although glucocorticoids exhibit clear coupling or entraining function, the effects of the steroids depend on the rhythm and species being studied.

A complex scenario such as the one presented here is consistent with the idea that adrenal rhythms usually are redundantly coupled to SCN rhythms. In an excellent report that has not received sufficient attention, Hiroshige (1984) analyzed a number of different RF patterns on plasma corticosterone rhythms in rats. This investigator concluded that the rhythms were driven by three factors: a circadian oscillator not entrainable to RF cycles, a weaker oscillator that is entrainable to RF cycles, and an interval timer with a length of approximately 20 hr. If this analysis is correct, and if corticosterone usually is only a secondary pacemaker for other rhythms, then the effects of adrenalectomy on those rhythms in SCN-intact animals should indeed be difficult to interpret. Stephan and associates (1979) conducted a study of RF entrainment in animals that had been given both SCN lesions and adrenalectomies. The ability to entrain to RF survives this dual surgical insult as well as other hypothalamic lesions (Mistlberger & Rechtschaffen, 1984; Honma, Honma, Nagasaka, & Hiroshige, 1987), leading Mistlberger and Rusak (1991) to suggest that RF entrainment is controlled by multiple redundant systems, both within the hypothalamus and without. However, reexamination of Stephan and colleagues' (1979) published activity data from combined adrenalectomized and SCN-lesioned rats does reveal some suggestive patterns. RF entrainment was clearly present, but took longer in the dually damaged animals than in the animals with SCN lesions alone. Further, whereas a stable phase angle between activity and the time of food availability was evident before surgery, this stability was clearly impaired or absent in the animals lacking both adrenal glands and SCN. Stephan was appropriately circumspect in interpreting these results, but note that the ability of adrenalectomized-plus-SCN-lesioned animals to synchronize to food presentation does not establish entrainment of a circadian clock. Adrenalectomized animals may be left with an intact interval timer but have lost their weak oscillators. Further experimentation is needed to determine the true situation.

Just as homing pigeons could be shown to use magnetic cues for orientation only when primary sun-based cues were unavailable (Keeton, 1979), experiments that seek to localize secondary pacemakers are likely to yield negative results until these oscillators are studied in animals for whom secondary pacemakers have been made primary. One design of this sort would explore the effect of adrenalectomy on rhythms in animals that previously had been food-entrained after SCN lesion. Another informative experiment would investigate whether glucocorticoid injections become capable of entraining activity rhythms in SCN-lesioned animals (although they are not able to do so in SCN-intact animals). Our point is not that the adrenals are the (or a) food-entrainable pacemaker for locomotor activity, but that the critical experiments to answer such questions have yet to be done.

To summarize, adrenal corticoid rhythms can be entrained by both light–dark and RF cycles, although the RF cycle appears to be the more potent in rats (Morimoto et al., 1977). Control of corticoid secretion is complex and involves autonomic innervation, ACTH stimulation, and probably endogenous oscillations (of the adrenal glands) in synthesis of glucocorticoids and in ACTH sensitivity. Timed glucocorticoid injections or infusions are capable of entraining oscillations in squirrel monkeys (Sulzman et al., 1978) and rats (Horseman & Ehret, 1982) but have not been shown to be effective in hamsters (Albers, Yoger, Todd, & Goldman, 1985). Evidence indicates that not all rhythms react similarly to corticoids (Sulzman et al., 1978). Thus, glucocorticoid secretion probably does not act as a master synchronizer. Corticoids do act as coupling agents for a number of biological rhythms, however (Friedman & Walker, 1968; Johnson & Sawyer, 1971; Edwards, 1973; Terkel et al., 1974; Landau, 1975; Kiem et al., 1987; Hsu & Kuhn, 1988).

Adrenalectomy in rats does not eliminate circadian activity rhythms (Richter, 1965), although removing the glands does reduce the levels of general locomotor activity (Richter, 1936; Moberg & Clark, 1976). This effect can be reversed by glucocorticoid treatment (Pedersen-Bjergaard & Tonnesen, 1954), so these steroid do appear to be able to alter activity levels directly. Micco and co-workers (Micco, Myer, & McEwen, 1980) reported that corticosterone replacement (constant release via implanted pellet) in adrenalectomized rats increased activity levels just prior to the dark-to-light transition, effectively sharpening the rhythm and increasing the amplitude. In contrast, Iuvone and van Hartesveldt (1977) reported that adrenalectomy had little effect on open field activity in rats. Few studies have measured long-term activity under conditions that would allow assessment of adrenalectomy effects on free-running period; those that have (Stephan et al., 1979; Albers et al., 1985) did not report any results on this factor. Effects of adrenalectomy on free-running period would not be surprising, but would be of interest given the issues discussed earlier: the fact that adrenalectomy reduces overall activity levels (Richter, 1936; Moberg & Clark 1976) and the recognition that rhythm and phase changes may be secondary to activity or arousal changes induced by a variety of drugs (Turek, 1989), hormonal alterations (Schull et al., 1988,1989), or situational manipulations (Mrosovsky, 1988).

Recent experiments suggest a rather intriguing connection between exercise, glucocorticoids, and melatonin secretion which may help to explain the effects of activity on circadian behavioral rhythms. Reebs and Mrosovsky (1989) demonstrated that 2 hr of forced exercise led to phase shifts of activity rhythms in a subset of Syrian hamsters maintained in constant darkness. Not all animals responded and there was considerable variation in the extent of the shifts which were observed in those animals that did react. Nevertheless, a PRC of a sort was obtainable indicating a real effect of exercise on rhythm parameters. One reported effect of physical exercise is an increase in serum glucocorticoid concentrations, described for both hamsters (Borer, Bestervelt, Mannheim, Brosamer, Thompson, Swamy & Piper, 1992) and humans (Monteleone, Fuschino, Nolfe & Maj, 1992). Interestingly, when humans are forced to exercise at night, a decrease in serum melatonin follows the increase in serum cortisol levels. In addition, the circadian pattern of melatonin shows a slight decrease in amplitude and a phase delay (Monteleone, Fuschino, Nolfe & Maj, 1992). This provides two potential mechanisms for translating the effects of exercise into changes in circadian patterns: (1) a direct effect of glucocorticoids; or (2) an indirect effect of glucocorticoids mediated via melatonin suppression. The latter possibility might help to explain the variability of Reebs and Mrosovsky's (1989) results insofar as the rodent circadian system appears sensitive to changes in melatonin only at certain times in the circadian cycle. This association between glucocorticoids and melatonin may be only one-way in human beings. Suppression of melatonin secretion by bright light has been reported to be without effect on cortisol levels (McIntrye, Norman, Burrows & Armstrong, 1992).

The associations between alterations in the hypothalamic–pituitary–adrenal axis and affective disorders and the clear evidence of rhythm abnormalities in those same disorders increase the probability that the data discussed here that link glucocorticoids with circadian cycles are clinically relevant. A closer and more encompassing examination of the interactions between glucocorticoids and rhythms appears warranted. Further research on glucocorticoid effects on other motivationally relevant rhythms also is warranted because of the effects of glucocorticoids on catecholaminergic systems. These hormones have a variety of site-specific effects on alpha- and beta-receptors in the brain and the periphery (Jhanwar-Uniyal & Leibowitz, 1986; Stone, 1987; Stone, McEwen, Herrera, & Carr, 1987; Malbon & Hadcock, 1988; Stone, Mitsuo, & Colbjorensen, 1989; & Kendall, 1990) and, as we shall see, catecholaminergic influence is a hallmark of estrogen and thyroid hormones, which also alter circadian rhythms and which also have been implicated in affective disorders.

VII. Sex Steroids

Reproductive behavior in mammals involves a multiplicity of rhythmic phenomena. Females can show ultradian rhythms in gonadotropic hormone secretion, circadian rhythms in behavior, infradian rhythms in sexual responsiveness, and annual breeding cycles. Many of these patterns are interrelated, as are the circadian cycles and seasonal photoperiodism. Rodents and primates appear to have fundamentally different mechanisms controlling their sexual cycles. For rats and hamsters, circadian rhythms seem to control the timing of estrous cycles; disruption of these rhythms can eliminate female reproductive cycling. In primates, on the other hand, a self-contained ultradian oscillator in the medial basal hypothalamus (MBH) is linked with the ovaries to produce a hypothalamic–gonadal clock that is apparently independent of circadian and photoperiodic influences. Neither model seems able to explain all the experimental data, so the two models may represent the extremes of a continuum.

Male and female sex steroids affect the expression of circadian behavioral rhythms. However, in this chapter we concentrate almost exclusively on female sex steroids. In rodents, where most attention has focused, estrogens (1) shorten free-running activity periods, (2) phase advance entrained rhythms, (3) increase activity levels, and (4) consolidate rhythm patterns. Progesterone anatogonizes these effects, but only in conjunction with estrogens; progesterone by itself apparently has little influence on circadian organization. Testosterone has been reported to have similar effects in males of some rodent species but not others. The consensus is that estradiol strengthens the coupling between multiple oscillators.

Whereas reflex ovulators such as rabbits and cats ovulate in response to sexual behavior, ovulation in many mammalian species is controlled by a multiday estrous and/or menstrual cycle mediated by a causal loop involving gonadotropic releasing hormone (GnRH), pituitary luteinizing hormone (LH), and sex steroids of ovarian origin that feed back on hypothalamus and pituitary. The females of many of these species, including rodents and most primates, undergo a regular cyclic alteration in sexual receptivity as well as in fertility. Thus, estrous cycles normally involve both physiological and behavioral changes. When attempting to interpret experimental findings and link various estrous stages with hormonal patterns, it is important to note that behavioral and some physiological measures do not correlate exactly. For example, female rats show a regular alteration in the cells found with vaginal wipes (vaginal estrous cycles) as well as changes in sexual receptivity and responsiveness to males (behavioral estrous cycles); however, behavioral estrus and vaginal estrus are offset from each other by 12 hr or more (Feder, 1981). Thus, caution is required when associating a level of hormone or neurotransmitter with estrus, to be certain which definition of estrus is being used. In species in which female sexual cycles include bleeding caused by loss of the endometrial lining, sexual rhythms are called menstrual cycles. Menstrual and estrous cycles are not mutually exclusive; primates with well-defined menstrual bleeding also may display behavioral estrus.

The regulations of estrous or menstrual cycles in the two best studied nonhuman mammals, rats and rhesus monkeys, have a number of common features. Both species are spontaneous ovulators, show some form of behavioral estrus, and require a large increase in LH to achieve ovulation. The LH surge, in turn, requires prior increases in serum estrogens from developing follicles and leads to a rise in progesterone, which exerts a negative influence on further LH secretion. Both species also rely on ultradian rhythms of GnRH release from the hypothalamus as part of the timing mechanism for these sexual cycles (Fox & Smith, 1985; Kalra & Kalra, 1991; Yen, 1991). Despite these similarities, however, a significant difference exists in the control of sexual cycles in the two species. In rodents, ultimate control of the GnRH ultradian rhythms involves the light-sensitive circadian system, whereas a self-contained endogenous ultradian oscillator independent of photic cycles and partially controlled by ovarian steroids has been proposed as the mechanism for primates (Alleva, Waleski, & Alleva, 1971; Stetson, 1978; Knobil, 1980,1989; Ferin, 1983; Kalra & Kalra, 1991; Yen, 1991).

The obligatory role for the circadian system in rodent sexual cycles is supported by lesioning and deafferentation experiments. Lesioning the SCN, sectioning the connection between the SCN and MBH, or frontal deafferentation of the MBH all lead to a constant vaginal estrus syndrome (Blake, Weiner, Gorski, & Sawyer, 1972; Blake & Sawyer, 1974; Koves, Gottschall, & Arimura, 1989). Constant light, which is known to cause arrhythmia at select intensities, simultaneously disrupts locomotor and estrous rhythms in hamsters; changes in free-running periods of activity also are reflected in changes in estrous cycle periods in this species (Stetson, 1978). A more precise account of the role of the circadian system role was elucidated in an experiment on a mutant strain of hamsters with endogenous circadian periods of about 20, rather than 24 hr, (Ralph & Menaker, 1988). Whereas the hamster estrous period is normally 4 circadian cycles (and, thus, approximately 96 hr), in *tau* mutants the estrous period is typically 5 circadian cycles or 100 hr (Refinetti, Rissman, & Menaker, 1991). Thus, the estrous cycle is slightly longer than in the wild type, although the circadian period is shorter, suggesting that the primary role of the circadian system is to *gate* behavioral estrus and ovulation, limiting these activities to particular phases of the circadian cycle. By this hypothesis, the ovarian–pituitary axis determines the basic period of the estrous cycle within the constraints set by the circadian gate. If this is so, the major difference between rodents and primates may be the dependence of rodent cycles on this circadian gate.

Estrous cycles, in turn, exert marked influences on circadian rhythms. Period, phase angle, and activity time all vary consistently over the estrous cycle in both rats and hamsters (Carter, 1972; Albers et al., 1981). Figure 3 displays an example of such a pattern from a blinded (a) and a sighted (b) female rat. Albers and co-workers (1981) reported that shortest estimated period (from onset to onset) earliest activity onset (phase angle from lights off to activity onset), and longest duration of activity all occurred on the day of estrus in both 4- and 5-day cycle rats, whereas the longest period, latest activity onset, and shortest duration of activity were observed during diestrus days. While Albers and co-workers defined the stages of estrous based on behavioral criteria only, Wollnik and Turek (1988) reported the same results and were able to confirm the days of the estrous cycle in a subset of their animals by vaginal smearing. In addition, numerous experiments have demonstrated that estradiol can shorten circadian activity rhythms in rodents. Estradiol implants shorten the free-running period of activity in blinded ovariectomized hamsters compared with pretreatment cycles and with controls implanted with blank pellets (Morin, Fitzgerald, Rusak, & Zucker, 1977a; Takahashi & Menaker, 1980). Albers (1981) reported similar data for sighted ovariectomized rats maintained in constant dim red light. As expected from the effect of changing period on phase angle during entrainment (see previous discussion),

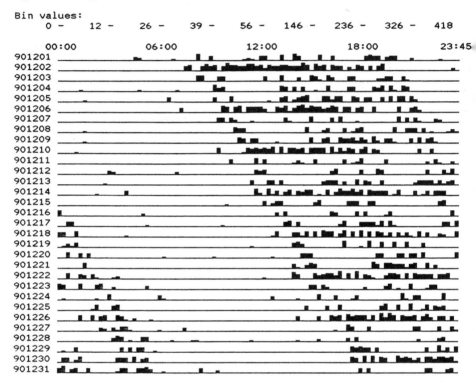

Figure 3.   Circadian and estrous rhythms in a female rat. (A) Blinded animal, modulo 24 hr, single plot. Each data bin represents 15 min of running wheel activity. (B) Animal free-running in constant darkness. (*Left*) Actogram of wheel running activity, modulo 24 hr. To preserve visual continuity from one day to the next, the actogram is double plotted so each 24-hr period of activity is represented twice: first on the right side, then on the left side, one row down. The major band of activity drifting down and to the right reflects circadian activity. The heavier activity every fifth day is indicative of an estrous rhythm. Each data bin represents 10 min (*Right*) Actogram of wheel running activity, modulo 97.24 hr. To illustrate the estrous rhythm, the actogram is double plotted at the free-running period the estrous cycle (97.24 hr). The average curves at the bottom of the panel show how the circadian waveform of activity varies in shape and in amplitude over the course of the estrous cycle. Figures produced using TAU circadian software (Schull, 1991). (*Figure continues*)

B

97.24 hours		97.24 hours

```
870314 0100
870318 0300
870322 0400
870326 0500
870330 0600
870403 0800
870407 0900
870411 1000
870415 1100
870419 1300
870423 1400
870427 1500
```

349.0

mean
77.20

min
0.00

0.00

24 hours		24 hours

```
S:allfm039
From 870310
To   870425
Cell*bin=min
  1 * 10=10
Periodogram
Lo:23.0 Inc
Hi:25.0 0.10
Depict:24.00
   Waveform
   Actogram
Guide \ Rule

Next Subject
Modify...
Output...
Erase, ^Quit
data clips
----- | -----
wav/act lims
    0 ----- |
```

```
870311 0000
870313 0000
870315 0000
870317 0000
870319 0000
870321 0000
870323 0000
870325 0000
870327 0000
870329 0000
870331 0000
870402 0000
870404 0000
870406 0000
870408 0000
870410 0000
870412 0000
870414 0000
870416 0000
870418 0000
870420 0000
870422 0000
870424 0000
```

estradiol treatment phase advanced the onset of activity on ovariectomized hamsters exposed to either 6:18 (Widmaier & Campbell, 1980) or 12:12 light–dark cycles (Morin et al., 1977a), although not during entrainment to a 16:8 photoperiod (Widmaier & Campell, 1980). Presumably the short duration of the active cycle under 16:8 light–dark precluded any possible phase shift.

Thus, estrogen effects on circadian rhythms have been amply demonstrated in both intact and ovariectomized rodents. The data from estradiol treatment of ovariectomized rats, both entrained and free-running, fit well with observations from intact cycling females and support the hypothesis that estrogens shorten activity rhythms. Several indications suggest that this effect is mediated by strengthening the coupling between pacemakers in a multipacemaker system. Morin and co-workers (1977a) reported that estradiol treatment consolidated activity bouts (a few long sessions of activity rather than many short sessions) in 10 of 16 treated ovariectomized hamsters; Takahashi and Menaker (1980) determined that estradiol treatment significantly reduced the variability of activity onsets. More importantly, estradiol reduces the number of ultradian components in activity rhythms. When investigating sexual dimorphism in circadian activity patterns, Wollnik and Dohler (1986) discovered that male LEW/Ztm rats displayed significant ultradian patterns (4- and 4.8-hr) in activity that were absent in females. Ovariectomy led to the appearance of these ultradian components, which were then eliminated by estradiol treatment. The authors interpreted the ultradian patterns to reflect the output of loosely coupled circadian pacemakers, the coupling of which was strengthened by estradiol, leading to a more coherent and consolidated pattern. Consistent with this position, Wollnik and Turek (1988) later found that intact female rats displayed ultradian rhythms in activity during diestrus days but that these components disappeared during proestrus and estrus. Also supportive is a study by Morin and Cummings (1982) on constant light-induced splitting in hamsters. These rodents often will display two components in running wheel activity when subjected to intense constant light (60 lux or more), a result that has been interpreted to mean that two pacemakers have become temporarily uncoupled. Estradiol implants significantly reduce the incidence of splitting in ovariectomized animals, again suggesting that estradiol strengthens the linkage between pacemakers. Finally, Albers (1981) found that the effect of estradiol on the frequency of activity rhythms often was related to the pretreatment period observed in the animals. Albers was able to model this effect mathematically using coupled oscillators and assuming that the main effect of estradiol was to strengthen coupling. Although the effects of estrogen on rhythms are more clear, robust, and reliable in females, manipulations of testosterone and androgens in males produce similar but less consistent effects when detected (Daan, Damassa, Pittendrigh, & Smith, 1975; Ellis & Turek, 1979; Morin & Cummings, 1981,1982; Rowsemitt, 1989).

Estradiol has been shown to affect rhythms other than activity. Cohen and Wise (1988) reported that estradiol treatment of ovariectomized rats altered numerous rhythms in serotonin turnover in various hypothalamic areas such as the median eminence (ME), SCN, medial preoptic nuclei (MPO), and arcuate nuclei (ARC). In fact, estradiol-treated animals displayed a rhythm in the ME that was not found in ovariectomized animals and lost the rhythm displayed in the ARC through a large increase in 5-hydroxytryptamine (5-HT) turnover at all times. Estradiol treatment reversed rhythms in SCN and MPO. Rhythms in hypothalamic β-endorphin concentration also apparently are under estrogen control, since ovariectomized rats show no rhythm whereas intact and estradiol-treated ovariectomized rats display a prominent diurnal variation (Genazzani et al., 1990).

Finally, the ovaries play a necessary role in initiating and maintaining circadian rhythms in LH and follicle stimulating hormone (FSH) secretion. In golden hamsters, a prominent daily rhythm in LH with a peak at 1700 hours appears at 17 days of age and is maintained for the following 3 weeks until the 4-day estrous cycle is begun. Ovariectomy before day 17 prevents the appearance of any diurnal rhythm in gonadotropins; ovariectomy between days 25 and 30 eliminates the rhythm that already exists (Donham, DePinto, & Stetson, 1985). Clearly some signal is being provided by the ovary that is necessary for periodic gonadotropin secretion. Adult ovariectomized rodents also lack LH rhythms unless provided with estradiol. Experiments have shown that a rise in estradiol is required for the LH surge and that the LH surge, in turn, is necessary for the subsequent increase in progesterone levels in a normal estrous cycle (Freeman, Dupke, & Croteau, 1976; Feder, 1981). Although estradiol is sufficient for restoring vaginal cornification, the LH surge, and sexual behavior to ovariectomized animals, sexual behavior is facilitated greatly by pretreatment with estradiol, followed by progesterone some 12–24 hr later (Feder, 1981). Interestingly, injections of estradiol into ovariectomized rats result in timed LH surges that occur at 1600–1700 hours beginning 30–34 hr after injection and continuing for up to 3 days at approximately the same time (Ramirez & Sawyer, 1974; Freeman et al., 1976). These recurring LH surges can be blocked effectively by progesterone injections given at 1600, at or about the time of the first surge (Freeman et al., 1976). Thus, the basic pattern appears to be an increase in estradiol leading to an LH surge approximately 24 hr later that generates an increase in progesterone about 2 hr afterwards. Progesterone facilitates the expression of sexual behavior about 12 hr prior to ovulation (also caused by the LH surge), while blocking subsequent LH surges. Given all these results, one might suspect that the initiating signal in developing the diurnal pattern in prepubertal golden hamsters is an estrogen. The switch from a daily LH surge to the normal 4-day cycle most likely is caused by the appearance of sufficient quantities of progesterone or progesterone sensitivity in the system. These data reinforce the notion of an ovarian clock exerting control over reproductive cycles.

In summary, at least three different oscillatory systems appear to be involved in the control of female sexual and reproductive cycles. One is the light-sensitive circadian system that sets a specific temporal gate in rodents but is relegated to a minor role in primates. The second is the ultradian oscillator controlling GnRH secretion. The third is the ovarian clock, which sets a lower limit on the timing of ovulation presumably by developmental constraints. These three temporal systems interact to create the rhythms of estrus and menstruation. As we have seen, these actions are bidirectional—the circadian system may set a specific time for ovulation in rodents even as ovarian steroids alter the expression of circadian rhythms. Estrogens increase motor activity levels, consolidate activity bouts, and shorten free-running activity periods, resulting in an earlier onset of motor behavior. Overt reproductive and sexual cycles thus represent the final output of a series of mutually interacting temporal loops. Discussion of the mechanisms or all possible interactions is beyond the scope of this chapter, so we continue to focus on the effects of estrogens on circadian rhythms.

Estradiol affects several brain regions and neurotransmitter systems relevant to circadian organization. A circadian rhythm is seen in the cytosolic estrogen receptors (CERs) with reductions during darkness in the hypothalamus, preoptic area, and amygdala of ovariectomized hamsters (O'Connor, Morin, & Feder, 1985) and rats (Roy & Wilson, 1981; Wilson, Clark, Clyde, & Roy, 1983). Catecholaminergic systems also may be involved in CER rhythms, since reductions of norepinephrine and epinephrine and higher levels of dopamine increase specific uptake of tritiated estradiol in the pituitary and several brain regions of ovariectomized female rats (Thompson, Woolley, Gietzen, & Conway, 1983).

Several studies have associated sex steroids with adrenergic activity, and generally indicate that estrogens decrease activity associated with beta-receptor function and increase alpha-1 receptor-mediated responses. These effects appear to be quite dependent on the neural region being analyzed. Weiland and Wise (1989), for example, found that treating ovariectomized rats with estradiol led to significant decreases in beta-1 receptor densities in the SCN and pineal gland but increased such receptors in the medial preoptic nuclei of the same animals. Carlberg and Fregly (1986) treated intact female rats with estradiol in Silastic capsules and measured reactions of those animals to isoproterenol (IP), a beta-1 agonist, compared with those of untreated controls. A significant reduction in responsiveness to IP was detected, supporting the hypothesis of a decrease in the beta-1 adrenergic system, as well as an increase in urinary excretion of catecholamines, which suggested that the observed loss of responsiveness was due to down-regulation secondary to an increase in release of catecholamines caused by estradiol. Additional support for a decrease in beta receptor function comes from Petitti and Etgen (1990), who examined the *in vitro* response of brain slices, including hypothalamic and preoptic areas, to IP. These individuals found that previous *in vivo* exposure to estradiol significantly reduced the synthesis of cAMP in response to IP stimulation. These effects of estradiol on beta-1 receptors in the SCN are especially noteworthy given the similar finding for hypothyroid rats (discussed subsequently) and the fact that these animals show the same behavioral effects, that is, shorter and more consolidated activity patterns.

Estradiol treatment also has been shown to affect catecholamine turnover and concentration in certain neural regions. Norepinephrine levels were found to decrease in association with estradiol treatment in medial preoptic and paraventricular nuclei, while increasing in the lateral septum (Crowley, O'Donohoe, Wachslicht, & Jacobowitz, 1978b). At the same time, turnover was decreased by estradiol in lateral septum and central gray and increased in anterior hypothalamus (Crowley, O'Donohoe, & Jacobowitz, 1978a).

In addition to affecting catecholaminergic systems, estrogens apparently increase pineal melatonin (Johnson, Vaughn, Richardson, Petterborg, & Reiter, 1982); change the pattern and turnover rates of serotonin in the SCN, median eminence, medial preoptic, and arcuate nucleus without affecting 5-HT activity in the globus pallidus (Cohen & Wise, 1988); and alter β-endorphin levels and binding (LaBella, 1985; Genazzini et al., 1990). Ovariectomy reduces cholinergic activation of the rat SCN and abolishes alpha-bungerotoxin binding to those nuclei (Mistlberger & Rusak, 1986).

Since many of the studies cited here used blinded animals, the observed changes in free-running periods cannot be attributed to changes in the visual sensitivity of the circadian system. Nevertheless, a few studies bear mentioning that indicate changes in sensory perception. Freidman and Meares (1978), for example, reported that visual sensitivity is enhanced in women during the late follicular phase of the cycle, as ovulation approaches. Other investigators also have reported fluctuations in visual sensitivity throughout the menstrual cycle, with nadirs in the premenstrual period (DeMarchi & Tong, 1972; Ward, Stone, & Sandman, 1978). Although the exact nature of the hormonal control of these changes has not been established, changes in light sensitivity certainly could alter circadian parameters. Speculating that the nadir in light sensitivity associated with the period just prior to menstruation might be involved with premenstrual syndrome is tempting. Changes in light sensitivity have been associated with other affective disorders, such as manic depression (Lewy et al., 1985). These data seem worth further investigation. In other species, testosterone has been linked with developmental increases in visual sensitivity in chicks (Meyer, Parker, & Salzen, 1976), and estradiol with enhanced auditory evoked potentials in frogs (Yovanof & Feng, 1983), so modulation of sensory perception by sex steroids appears to be a widespread phenomenon in vertebrates.

VIII. Thyroid and Parathyroid Hormones

Richter reported that thyroid impairment in animals could produce marked cycles of activity and inactivity that he likened to manic depression; he suggested that thyroid hormones regulated biological rhythms (Richter, 1933,1965). A substantial body of evidence now exists that associates various affective disorders in humans with thyroid abnormalities. Thyroid axis aberrations frequently are seen in depressed patients (Kostin et al., 1972; Prange et al., 1972; Whybrow et al., 1972; Kirkegaard et al., 1975a,b), although these abnormalities have been subjected to a variety of interpretations (e.g., Whybrow & Prange, 1982; Joffe, Roy-Byrne, Uhde, & Post, 1984). Also, studies in unipolar and bipolar patients often show a blunted TSH response to thyrotropin releasing hormone (TRH) infusion, even when levels of thyroid hormones in the blood are normal (Amsterdam et al., 1983). Furthermore, most studies have revealed that depressives show increased T_4 or free T_4 levels during the acute phase of their disease that fall to near normal during remission (Bauer & Whybrow, 1986). In addition, alterations in thyroid function appear to be able to precipitate behavioral pathology resembling depression and mania. Symptoms of depression can be produced by hypothyroidism (Whitlock, 1982), hyperthyroidism, or hyperparathyroidism (Mandel, 1960; Carmen &Wyatt, 1979), whereas symptoms of mania and hyperactivity can be produced by hyperthyroidism and thyroid hormones can precipitate manic episodes in individuals with family histories of affective disease (Carmen & Wyatt, 1979; Whybrow, 1986). Interestingly, some depressed patients only respond to antidepressant medications, or respond more rapidly, when these drugs are augmented by exogenous thyroid hormones (Prange et al., 1969,1972; Earle, 1970; Wilson et al., 1970; Coppen et al., 1972; Wheatley, 1972; Goodwin, Prange, Post, Muscettola, & Lipton, 1970; Banki, 1975,1977; Schwarcz et al., 1984). Moreover, certain drugs used in the treatment of affective disorders (e.g., lithium) also have effects on thyroid status (Cooper & Simpson, 1969; Gershon & Shopsin, 1973). Thus, little doubt exists that thyroid function is involved at some stage in the progress of affective disease, although the extent to which this involvement contributes to the development of the illnesses is not clear.

The first indication that thyroid state and circadian function might interact in the etiology of any affective illness is found in studies of a particularly virulent form of manic depression known as rapid-cycling bipolar disorder. Patients with rapid cycling (RC) disorder have, by definition, at least four episodes per year (Dunner & Fieve, 1974). However, most patients exhibit a much higher rate of recurrence with a fairly fixed pattern of episodes. This group also has a high prevalence of thyroid abnormalities, particularly hypothyroidism, either *de novo* (Cowdry et al., 1983; Bauer, Droba, & Whybrow, 1987) or during the course of treatment with lithium carbonate (Cho, Bone, Dunner, Colt, & Fieve, 1979). These thyroid abnormalities are distinct from those found in unselected bipolar patients (M. Bauer, personal communication). Further, RC symptoms reportedly remit when high doses of thyroxine are added to their treatment (Stancer & Persad, 1981; Bauer & Whybrow, 1986; Bauer et al., 1987). Interestingly, 80–90% of RC bipolars are women, compared with a fairly even distribution of males and females in non-RC manic depression (Dunner & Fieve, 1974; P. Whybrow, personal communication). No other demographic characteristics distinguish RC and non-RC bipolars, suggesting that RC bipolars represent the extreme end of a continuum of episodic cyclicity (P. Whybrow, personal communication). The striking periodicity in those symptoms—coupled with a high prevalence of hypothyroidism, and the ability of thyroid hormone to induce remission in otherwise intractable symptoms, and the much greater incidence of thyroid disorders in women than in men—reinforces the link among the thyroid axis, affective illness, and episodic behavior.

Work in our laboratories supports such a link. Thyroidectomy and thyroparathyroidectomy in rats can produce similar abnormalities of circadian periods and activity levels in a sex-specific manner. In this section we review the literature linking thyroid hormones and biological rhythms, and explore the mechanisms and implications of these effects.

The thyroid and the parathyroid glands typically are fused to each other in the pharyngeal region. Parathyroid hormone (PTH) is a major regulator of calcium homeostasis: it elevates blood calcium by increasing bone resorption, enhances calcium uptake along the GI tract, and reduces calcium excretion. Calcitonin, manufactured by C cells scattered throughout the thyroid gland proper, counteracts all these effects, although the importance of this hormone for calcium regulation in humans is thought to be minor. The major products of the thyroid are triiodothyronine (T_3) and tetraiodothyronine or thyroxine (T_4). Conversion of T_4 to T_3 typically takes place in the periphery and is required for most, if not all, of T_4's physiological effects. In mammals, the primary function of the thyroid is considered to be control of basal metabolism and thermogenesis. In response to acute cold exposure, for example, the pituitary of many mammals secretes TSH; the T_4 release that results causes many tissues all over the body to increase their rates of metabolism (Edelman, 1974). TSH secretion is, in turn, regulated in part by TRH from the hypothalamus. The thyroid is directly and indirectly involved in many other functions as well—among them are regulation of cardiac output, reproductive and neural function, sympathetic autonomic activity, and early development, including metamorphosis in amphibians (Eayers, 1964a,b; Ford, 1968; Fisher, Dussault, Sack, & Chopra, 1977; Slotkin & Sleptis, 1984; Hadley, 1992).

Control of thyroid hormone secretion is complex and not completely understood. As for the pineal and adrenal glands, several levels of regulation appear to exist, including autoregulation at the tissue level of iodide uptake, stimulation by TSH from the anterior pituitary, sympathetic innervation, and negative feedback from T_3 and T_4 on TSH release from the pituitary and possibly on TRH release from the hypothalamus. The main positive regulator of thyroid hormone secretion is TSH whereas the major negative regulators are the thyroid hormones themselves (Sterling & Lazarus, 1977; Norman & Litwack, 1987). Chemically induced hypothyroidism in rats leads to increased levels of TSH (Rondeel et al., 1988; Taylor,

Wondisford, Blaine, & Weintraub, 1990) whereas hyperthyroidism caused by T_4 injections reduces TSH concentrations (Rondeel et al., 1988) and, thus, T_3 and T_4 secretion. The major effect of thyroid hormones in this case appears to be to inhibit TSH release through a reduction in the sensitivity of the pituitary cells to TRH (Sterling & Lazarus, 1977). This result indicates that hypothalamic release of TRH from the dorsomedial and paraventricular nuclei into the portal blood is probably the primary stimulus for TSH release from the anterior pituitary. Further evidence has strengthened this hypothesis. Reducing the level of TRH has been shown to result in hypothyroidism in humans (Sterling & Lazarus, 1977) and other animals (Faglia & Persani, 1991), whereas T_3 or T_4 injections reduce TRH concentrations in various hypothalamic areas such as the paraventricular nuclei, posterior hypothalamus, and median eminence (Rondeel et al., 1988; Yamada, Rogers, & Wilber, 1989). The primary regulatory pathway is thus hypothalamic TRH stimulating release of TSH from the anterior pituitary, which stimulates thyroid hormone synthesis and release. Thyroid hormones then act to inhibit the release of TRH and TSH. TRH release is itself under complex control, being stimulated by histamine, norepinephrine, and dopamine and inhibited by serotonin, somatostatin, and thyroid hormones (Norman & Litwack, 1987; Segerson et al., 1987).

The last regulatory factor influencing thyroid hormones is sympathetic innervation from the superior cervical ganglion via the external carotid nerve (Melander, Ericson, & Sundler, 1974a; Melander, Ericson, Sundler, & Ingbar, 1974b; Romeo et al., 1986). Both norepinephrine and dopamine are thought to increase T_3 and T_4 release via alpha receptors (Melander et al., 1974a,b) and superior cervical ganglionectomy reduces the magnitude of TSH-stimulated T_4 release (Cardinali et al., 1982; Cardinali, Romeo, Boado, & Deza, 1986). Measurement of the norepinephrine content of rat thyroids across 24 hr led Barontini and co-workers (Barontini, Romeo, Armando, & Cardinali, 1988) to suggest that sympathetic stimulation is responsible for the nocturnal rise in serum T_4 when TSH levels are consistently low. One report, however, indicated that alpha-1 stimulation actually decreases T_4 and T_3 release from mouse thyroid tissue (Muraki, Nakaki, & Kato, 1987), so the exact nature of the sympathetic nervous system's role remains to be determined. One interesting note is that least three of the four hormones mentioned as probable modulators of mammalian circadian rhythms all share some level of sympathetic control and even ovarian function may be modulated by the actions of the sympathetic nervous system (Cardinali, Romeo, Ochatt, & Moguilevsky, 1989). This common characteristic suggests that circadian modifications may be related to the general adaptation syndrome put into effect when animals are faced with stressful circumstances.

A pronounced circadian rhythm of thyroid hormone is detectable, but the significance of this rhythm is not clear, since the conversion of T_4 to T_3 determines functional thyroxine activity and the regulation of this conversion is not well understood (Nicolau & Haus, 1992). Whether the thyroid can be considered a biological clock in its own right is also unclear. No data are available whether thyroid tissue shows endogenous oscillations in isolation. On the other hand, the thyroid clearly has significant effects on biological rhythms *in vivo*.

Under conditions of entrainment to 12:12 light–dark schedules, thyroidectomy attenuates or alters rhythms of feeding (Bellinger, Williams, & Bernardis, 1979) and of plasma corticosterone and prolactin (Ottenweller & Hedge, 1981). Under free-running conditions, the period of behavioral activity rhythms has been shortened by thyroidectomy in canaries (Wahlstrom, 1967,1969) and by thyroparathyroidectomy in rats (Schull et al., 1988,1989; McEachron, Levine, & Adler, 1989; Lauchan & McEachron, 1991; McEachron, 1991). Further evidence that thyroidectomy shortens the endogenous period of activity rhythms comes from McEachron and associates (1989) finding that thyroparathyroidectomized rats are impaired in their ability to entrain to long (e.g., 26-hr) light–dark cycles. In hamsters, however, the thyroid inhibitors propylthiouracil and thiourea reportedly lengthen free-running rhythms (Beasley & Nelson, 1982; Morin, Gavin, & Ottenweiler, 1986). This discrepancy probably is not caused by a species difference, but by independent but unidentified effect of thyroid inhibitors (Morin, 1988). We have shown that the ability of another thyroid inhibitor—lithium—to lengthen circadian periods and reduce activity levels is unimpaired in thyroparathyroidectomized animals (Schull et al., 1988). This result demonstrates that the effects of lithium (on these variables at least and also, we suspect, on manic depression) are not mediated by way of the thyroid or parathyroid.

Thyroidectomy and parathyroidectomy impair functions relevant to both calcium regulation and thyroxine action. However, thyroxine appears to be the critical factor in shortening rat free-running periods, because comparable effects were obtained in both thyroparathyroidectomized and thyroidectomized rats (Schull et al., 1988), and because this effect is reversed in animals given T_4 replacement therapy (Lauchan & McEachron, 1991). Further evidence against a role for calcium deficiency in the shortening of free-running periods was obtained in an experiment in which this deficiency was induced in thyroparathyroidectomized animals given calcium deficient diets. No consistent effects on free-running periods were observed despite the successful induction of calcium deficiency. However, calcium deficiency did modify the activity rhythm in a manner that suggests a role in the coupling of the oscillator to the output rhythm. As Figure 4 shows, calcium deficiency produced an apparent phase advance of activity onset. When normal calcium levels were restored, however, the underlying clock was seen to have been running at the same speed all along, since the timing of the postdeficiency wheel-running was predictable by extrapolating from predeficiency activity. The activity records that resulted from this experiment were strikingly similar to records of manic–depressive humans, a population in whom disturbances of calcium and calcitonin physiology also have been reported (Carman, Wyatt, Smith, Post, & Ballenger, 1984).

Thus, the thyroid–parathyroid system may have two distinct effects on circadian clocks. Thyroxine deficiency shortens the period of circadian activity rhythms and calcium deficiency (regulated by parathyroid hormone and calcitonin) affects the coupling of the clock to activity (and possibly other) output rhythms. Either or both of these effects may be relevant to the physiology of affective disorders.

Thyroidectomy also has important effects on activity levels in a sex-specific manner. In humans, lethargy is a common symptom of hypothyroidism, and the vast majority of patients presenting with these symptoms are female. Richter (1933), Lee and van Buskirk (1928), and Stern (1970) similarly report that thyroid impairment lowers activity in rats. A close examination of these early reports, however, indicates that most of their subjects were females and that the effects on males were less consistent and tended to be in the opposite direction. In our studies, thyroidectomy in male rats consistently produced a large and reliable increase in activity levels, whereas activity reduction in females was nonsignificant (Schull et al., 1989). This sex-specific difference may be clinically relevant as well, since mania is more common in male bipolar patients whereas depression is more common in female manic–depressives. The results also have led us to speculate that hypothyroidism in human males may produce a male hyperactivity syndrome (Schull et al., 1989) that simply has escaped attention because moderately excessive energy is rarely a cause for complaint. The effects of thyroidectomy on activity level may be linked to the effects on period, since activity rhythms have been recognized to feed back on the oscillators that drive them (Turek, 1989). We have conducted a preliminary investigation of the possibility that activity level differences in males were in some way causing the differences in free-running period by manipulating running wheel access in four sham-operated and four thyroparathyroidectomized male rats (J. Schull & Lazaroff, unpublished observations). Despite other reports that access to running wheels could affect free-running periods in mice (Edgar, Martin, & Dement, 1991), we found no indication that this was true for sham-operated or thyroparathyroidectomized rats. Our sample size was too small, however, for firm conclusions. When examining the effects of T_4 treatment on thyroparathyroidectomized male rats, a -0.46 correlation was found between changes in period and activity levels (McEachron, Lauchlan, & Midgley, in press) indicating that greater reductions in activity were associated with longer periods in the treated animals. The correlation was not statistically significant, however, and the question of activity levels influencing circadian periods in these animals remains open.

Another hypothesis to explain the effects of thyroparathyroidectomy on the period of the activity rhythm is that the resultant hypothyroidism alters the light sensitivity of the circadian system. In the studies described here, our free-running conditions consisted of constant dim red illumination at a luminance level that should normally be virtually invisible to rats. However, since increasing light intensity under constant conditions reliably lowers activity and lengthens circadian periods in rats and other nocturnal animals, the difference in period between thyroparathyroidectomized and sham-operated rats might be explained by a thyroid-induced decrease in light sensitivity. This possibility was tested in two experiments. First, we manipulated the luminance levels of red light to which both thyroparathyroidectomized and sham-operated rats were exposed. Increased levels of light did, indeed, lead to longer activity rhythms but the difference in period between the thyroparathyroidectomized sham-operated rats remained intact (D. McEachron, unpublished observations). In a second experiment, we examined blinded (orbitally enucleated) male and female rats. Thyroparathyroidectomized animals still had significantly shorter free-running periods, indicating that this effect is not a result of thyroparathyroidectomy-induced reduction of tonic photic input to the SCN (McEachron, 1991). Note that thyroparathyroidectomy may increase light sensitivity to tonic white illumination. In constant bright light, the effects of thyroparathyroidectomy on activity levels and periods may be reversed relative to controls: activity levels become nonsignificantly lower and periods become nonsignificantly longer (J. Schull, unpublished observations). We have seen other indications of increased light sensitivity in thyroparathyroidectomized rats as well (Schull et al., 1988, 1989), and speculate that the physiology involved is relevant to the report by Lewy and co-workers (1985) that manic–depressive humans are more sensitive than normal individuals to the melatonin-suppressing effects of light. Because of the different results for red and white illumination, wavelength might be an important factor, although the explanation probably lies in the greater intensities used in the white light experiments. Although light sensitivity changes thus do not account for the effects of thyroidectomy on the pacemaker in constant darkness, other tonic inputs may play a role since hypothyroidism and thyroxine affect a variety of sensory systems in humans and animals.

Various studies in the literature associate decrements in sensory systems with hypothyroidism but not hyperthyroidism. Hearing loss has been associated with hypothyroidism for many years (Meyerhoff, 1976) but visual and somatosensory systems also may be affected. Himelfarb and co-workers (Himelfarb, Lakretz, Gold, & Shanon, 1981) found only a mild decrease in brain stem conduction velocity (BSCT) in hyperthyroid patients, but marked increases in BSCT with decreased amplitude and poor synchronization in hypothyroid patients. In addition, 50% of the hypothyroid patients suffered from some level of hearing loss. Ben-Tovim and colleagues (1985) found that propylthiouracel-induced hypothyroidism in adult rats was associated with changes in the third wave of auditory evoked potentials, again showing that thyroid hormones can change sensory

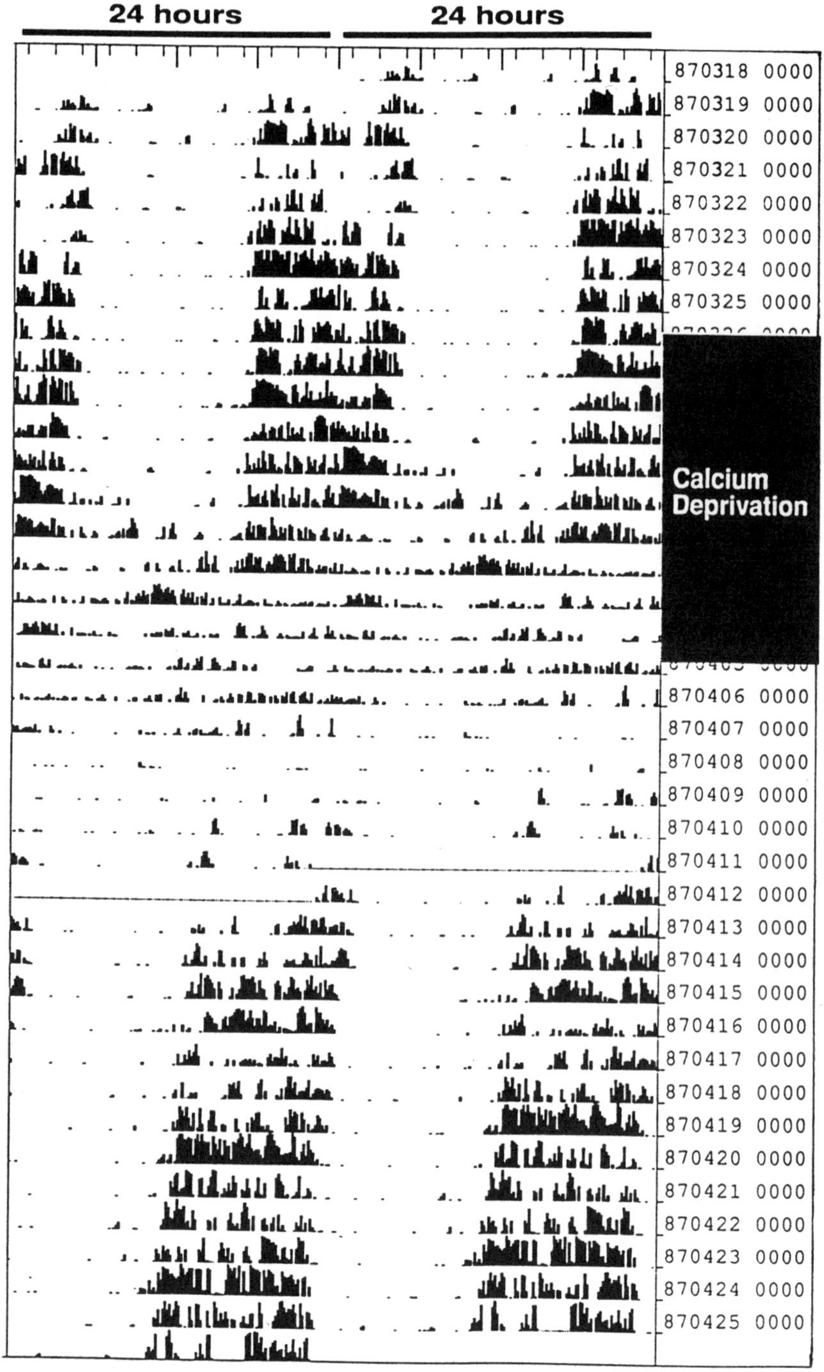

24 hours   **24 hours**

| 870318 0000 |
| 870319 0000 |
| 870320 0000 |
| 870321 0000 |
| 870322 0000 |
| 870323 0000 |
| 870324 0000 |
| 870325 0000 |

Calcium Deprivation

| 870406 0000 |
| 870407 0000 |
| 870408 0000 |
| 870409 0000 |
| 870410 0000 |
| 870411 0000 |
| 870412 0000 |
| 870413 0000 |
| 870414 0000 |
| 870415 0000 |
| 870416 0000 |
| 870417 0000 |
| 870418 0000 |
| 870419 0000 |
| 870420 0000 |
| 870421 0000 |
| 870422 0000 |
| 870423 0000 |
| 870424 0000 |
| 870425 0000 |

Figure 4. Effects of dietary calcium deficiency on wheel running in three representative thyroparathyroidectomized (TPX) rats. Double plotted actograms. During the time indicated, animals were fed a calcium deficient diet. The onset of calcium deficiency was accompanied by an apparent phase advance of wheel running activity, culminating in arrhythmia. When normal levels of calcium were restored by dietary

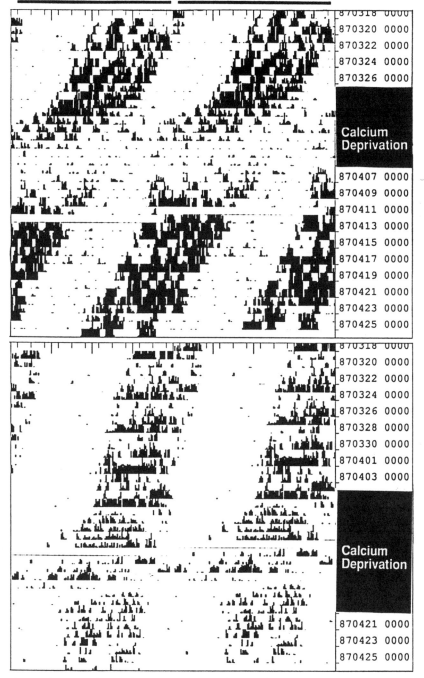

24 hours	**24 hours**

870318 0000
870320 0000
870322 0000
870324 0000
870326 0000

Calcium Deprivation

870407 0000
870409 0000
870411 0000
870413 0000
870415 0000
870417 0000
870419 0000
870421 0000
870423 0000
870425 0000

870318 0000
870320 0000
870322 0000
870324 0000
870326 0000
870328 0000
870330 0000
870401 0000
870403 0000

Calcium Deprivation

870421 0000
870423 0000
870425 0000

supplementation, the activity rhythms reemerged with phases and periods consistent with predeprivation values. This observation suggests that calcium deprivation had relatively little effect on phase or periodicity of the circadian itself, but rather interfered with the coupling of the oscillator with the locomotor output rhythm.

parameters in adult mammals. Several authors have associated increased latencies in human visual evoked potentials with hypothyroidism (Ladenson, Stakes, & Ridgeway, 1984; Huang, Chang, Lee, Chen, & Chopra, 1989; Timiras & Nzekwe, 1989), changes that could be reversed with T_4 treatment (Ladenson et al., 1984; Timiras & Nzekwe, 1989). Even aspects of evoked potentials from human somatosensory systems can be altered by hypothyroidism (Huang et al., 1989). Again, hyperthyroidism does not seem to exert any major effects (Huang et al., 1989; Timiras & Nzekwe, 1989). As stated earlier, the fact that thyroparathyroidectomized rats showed a shortening of their rhythms even after bilateral orbital enucleation rules out visual sensitivity changes as the causal factor for the rhythm alterations. This evidence does not mean that such changes do not occur or do not contribute in some fashion. Additionally, experiments with blinded animals do not rule out changes in other sensory systems as causal factors. Much care must be taken to ascertain how sensory systems are changed in thyroparathyroidectomized animals, as well as what these changes may mean to the expression of biological rhythms.

Similarly, although thyroid elevation of activity does not seem to be the cause of shorter periods, an inverse correlation of activity level and period length crops up suspiciously often—in response to thyroidectomy, lithium, and (arguably) sex. (Female rats in the Schull et al., 1988, study were more active than males and also had shorter periods; female thyroparathyroidectomies were not more active than sham-operated controls, nor were their periods significantly shorter.) As mentioned earlier, drugs and other manipulations that increase spontaneous locomotor activity often produce phase shifts or other chronobiological alterations. Most of these studies have failed to show that the elevated activity causes shifts, but the correlation is probably not coincidental.

We suspect that a more central aminergic arousal state is influenced by thyroid manipulations as well as by antidepressants and other manipulations that affect activity levels, biological rhythms, or both. One indication of this possibility is the finding that the monoamine oxidase inhibitor clorgyline has larger and longer lasting effects on levels and wave-forms of wheel-running activity in thyroparathyroidectomized rats than in controls (Duncan & Schull, 1993). Another indication is that thyroid hormones are known to influence alpha- and beta-receptor densities in both central and peripheral nervous systems (Atterwill, Bunn, Atkinson, Smith, & Heal, 1984; Swann, 1988; Viticchi, Grinta, & Piantanneli, 1990), and hypothyroidism has been correlated with decreases of up to 30% in hypothalamic beta receptor concentrations as well as significant reductions of cortical beta-receptor (Gross, Brodde, & Schumann, 1980; Swann, 1988), alpha-2 receptor concentrations (Atterwill et al., 1984), and binding of dopamine in the striatum (Kalaria & Prince, 1986). Dopamine and norepinephrine concentrations in hypothalamus and median eminence also change with surgical thyroidectomy, but can be reversed with T_3 or T_4 treatment (Anderson & Eneroth, 1987). We examined all four major catecholamine receptor subtypes (alpha 1 and 2; beta 1 and 2) in rats obtained from one of our circadian experiments. The most consistent result was a significant loss of beta-receptors at the preoptic and suprachiasmatic nuclei. (Vessotskie, 1992). When thyroparathyroidectomized animals were treated with T_4, beta-1 receptor densities were significantly increased in both SCN and ventromedial hypothalamic nuclei (Vessotskie, McGonigle, Molthen, & McEachron, in press). Thus, thyroxine may act fairly directly on the pacemaker or its inputs, and on some of the same pathways involved in psychopathology and psychotherapy. Finally, another indication that a central aminergic arousal state mediates thyroid effects on the pacemaker or the SCN is that sex steroids produce similar effects and influence some of the same neurotransmitter systems in the same parts of the brain.

A final piece of this puzzle should be mentioned. In the experiments in which T_4 was replaced in thyroparathyroidectomized animals, we also implanted thyroxine pellets into sham-operated animals. Although T_4 treatment significantly lengthened the activity rhythms of the thyroparathyroidectomies, it had no effect on the sham-operated rats (Lauchlan & McEachron, 1991). One possible interpretation of this finding is that an excess of TRH or TSH is responsible for the behavioral results. The reasoning is that these two peptide hormones presumably would be at low concentration in intact sham-operated animals, and excess thyroid hormone would not be able to reduce levels to any great extent. Clearly, the situation is much different for the thyroparathyroidectomized animals. In their case, high levels of TRH and TSH are going to be reduced a great deal. Some support for this hypothesis comes from the observation that intraperitoneal injections of TRH enhance locomotor activity in rats (Agarwal, Rastogi, & Singhal, 1977). In preliminary experiments, intraventricular infusion of TRH has resulted in activity patterns in intact rats that are similar to those observed with thyroparathyroidectomy (Figure 5). The results require confirmation but suggest that TRH concentrations may play an important role in rodent circadian organization.

IX. Conclusion

The orchestration of organismic activities is the function of both biological rhythms and neuroendocrine systems. Therefore, the idea that biological rhythms and hormones are intimately related should not be surprising. Biological clocks clearly are affected by all the hormones we have examined. The activities of the hormones are controlled critically by biological rhythms over periods ranging from a few minutes to a full year. To some extent, the idea that one can separate

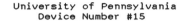

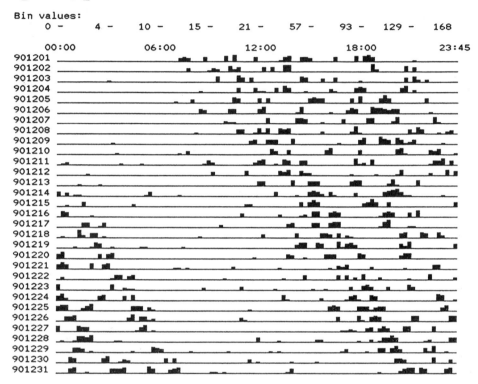

Figure 5. Comparison of activity patterns from thyroparathyroidectomized, sham-operated, TRH-in-fused, and saline-infused male rats. All actograms represent single plots of running wheel activity under constant dim red light (RR). Each line or bin displays 15 min of data. (A) A sham-operated male rat. (B) A TPX male. Note the shorter period and more consolidated pattern (see page 333). (C) A male rat with constant intraventricular (icv) infusion of saline (see page 334). (D) A male with constant icv infusion of TRH. Note the similarity of the pattern to that seen in the TPX animal. *(figure continues.)*

biological clocks from their hormonal mechanisms (and vice versa) is fictional. However, the concept is useful since clocks and hormones often can be manipulated independently. The effort to do so provides valuable insights into causal relationships.

Another source of insight, one that is underused by basic researchers, is the clinical relationship between rhythmic, hormonal, and psychological disturbances. The clinical literature usually is concerned with identifying "diagnostic markers," evaluating treatments, and pinpointing this neurotransmitter or that hormone as a

```
                         University of Pennsylvania
 B                          Device Number #22

Lowest=1,Average=223,Highest=601
L_width=55,H_width=94

Bin values:
     0 -      54 -    110 -    165 -    223 -    317 -    411 -    505 -    601

        00:00              06:00           12:00           18:00           23:45
901201
901202
901203
901204
901205
901206
901207
901208
901209
901210
901211
901212
901213
901214
901215
901216
901217
901218
901219
901220
901221
901222
901223
901224
901225
901226
901227
901228
901229
901230
901231
```

Figure 5 (B).

"cause" of a given disorder. However, this same literature can provide the basic researcher with valuable clues. We originally were led to explore the relationship between thyroid hormones and biological rhythms by the clinical observation that abnormalities of both are unusually common in manic–depressives. Similarly, the differential incidence of mania and depression in men and women, and the clinically observed hypersensitivity of manic–depressives to antidepressant medications led us to conjecture that hypothyroidism might have different effects in males and females, and might make animals hypersensitive to antidepressants. Both conjectures have been supported. The convergence of clinical and basic research observations on catecholamine systems in general, and on the catecholaminergic actions at the levels of the SCN and the pineal gland in particular, seems a lead worth pursuing in the future.

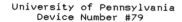

University of Pennsylvania
Device Number #79

C

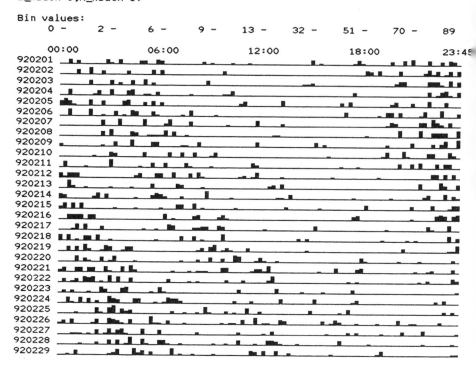

Lowest=1,Average=13,Highest=89
L_width=3,H_width=19

Bin values:
     0 -   2 -   6 -   9 -   13 -   32 -   51 -   70 -   89

Figure 5 (C).

Note also that such investigations can be fruitful without adopting over-simplified models of physiological cause and effect. Willner (1985) argued with extraordinary clarity that the notion of "a catecholamine theory of depression" is incoherent and an incomplete notion unless it is accompanied by an account of how neurotransmitter imbalances are translated into emotional states. Similarly, the circular nature of hormonal feedback loops, chronobiological mechanisms, and organism–environment interactions puts real constraints on the kinds of "explanations" that should be considered acceptable. Nonetheless, explanatory frameworks are indispensable. Our own attempt at a preliminary synthesis of the biological and clinical material we have reviewed follows.

Internal and external environments are temporally complex, and animals must adapt to temporal niches as well as spatial niches. This adaptation requires that the temporal environment be monitored systematically and that appropriate organis-

```
                        University of Pennsylvania
D                          Device Number #80

Lowest=1,Average=111,Highest=395
L_width=27,H_width=71

Bin values:
     0 -      26 -     54 -     81 -    111 -    182 -    253 -    324 -    395

        00:00           06:00           12:00           18:00           23:45
920201
920202
920203
920204
920205
920206
920207
920208
920209
920210
920211
920212
920213
920214
920215
920216
920217
920218
920219
920220
920221
920222
920223
920224
920225
920226
920227
920228
920229
```

Figure 5 (D).

mic activities occur at the right place and the right time. In some mammals, two specialized systems seem to exist for temporal monitoring of the circadian environment. Light–dark cycles are monitored by a circuit that proceeds from the eyes to the hypothalamus, to the SCN, and to the pineal gland. A secondary system (possibly involving the adrenal glands) seems linked more closely to periodic food availability at locally variable times of day, and possibly to the periodic occurrence of stressors. Each of these systems has the ability to entrain to the stimuli it monitors and, thus, to provide neural and endocrine signals that drive diverse output rhythms in tissues all over the body. Each of these systems also is modulated by other hormones that function to coordinate them, both with each other and with other physiological and environmental demands. Thus, steroid and thyroid hormones both affect the period and phase of SCN-based rhythms, possibly by direct effects on catecholamine receptor function. Other neuroendocrine factors, possibly involving the pineal gland and the adrenal glands, adjust the coupling and uncoupling of various rhythms to each other and to the environment. In some cases the function of these modulations is obvious, for example, when female rodents are made to become active earlier on nights when reproductive activity is called for; in other cases the adaptive function of these modulations is less clear.

Some of the neuro-chronobiological interactions we have reviewed may, in fact, be maladaptive side effects of systems that have broken down or gone awry. The induction of manic episodes by sleep deprivation is certainly amenable to this kind of interpretation, as are the blunted rhythms and neuroendocrine imbalances associated with depressive and manic–depressive illnesses. Even in understanding the often bizarre syndromes of human psychopathologies, however, an evolutionary or physiological perspective may be useful. The human bio-psycho-cultural niche is, in the grand scheme of vertebrate evolution, a truly bizarre and very recent occurrence. Only since the advent of fire have circadian systems been exposed to bright light at any, and sometimes all, hours of the day and night. In addition, temporal adaptation to the social environment may have become much more difficult as human societies became bound to external rather than internal clocks. In other species, low-ranking animals often can shift their time schedules to avoid dominant individuals or other periodic stressors (Calhoun, 1962; Bovet, 1972). In the human workplace, the sanctions against temporo-spatial escape may preclude such a behavioral strategy, increase stress levels, exhaust coping mechanisms, and force us to be active behaviorally at times when our endocrine systems are most urgently demanding retreat. Shiftwork, the fact that indoor work environments are significantly darker than the sunlit outdoors, and the widespread use of chronobiologically active drugs such as caffeine, and so on also affect our rhythms. In modern society, human beings constantly are being exposed to conflicting and often contradictory temporal demands. As a result, human life has become a chronobiologically "heady brew" to which our neuroendocrine systems have not had time to adapt evolutionarily. Perhaps understandably, then, affective disorders are not uncommon and chronobiological treatments (such as lithium, bright lights, and imposed sleep schedules) are among the most effective therapies.

How can we apply these ideas to the clinical situation? We propose, as a working hypothesis, that all species have a range of rhythmic flexibility which depends upon the evolutionary history and adaptations of that species. The circadian system of rats, for example, is clearly adapted to altering some of its components to take advantage of temporarily periodic food sources while remaining anchored to the light–dark cycle. Further, we suggest that pathological physiological and/or behavioral changes may result if an individual adopts a temporal organization which is beyond this species-typical range. This could be in either direction, i.e., a rhythm system which is too flexible and thus prone to internal desynchronization or the exact opposite, a rhythmic organization which is too structured and thus incapable of adjusting to social and/or environmental demands. How might an individual's neuroendocrine environment contribute to this abnormal temporal order?

As an example, consider the possibilities of unusual melatonin levels or patterns of secretion. If we accept that a major functional role of melatonin is to synchronize internal rhythms, three different pathologies can easily be envisioned. First, if melatonin secretion is inadequate, an individual may be prone to internal desynchronization and abnormal phase relationships. Second, if melatonin levels are unusually high, then the individual may become over-synchronized and temporally inflexible. Third, if an anomalous pattern of melatonin secretion develops, for example, if a person is unusually light-sensitive, aberrant temporal signals may be sent to the rest of the circadian system, potentially leading again to internal desynchronization and abnormal phase relationships. Notice that the behavioral pathology may seem superficially similar, but that the underlying circadian and neuroendocrine abnormalities will be quite different in these three instances. If one were to group individuals from all three categories together in a single study, the conclusion might well be that no consistent circadian or melatonin can be associated with behavioral pathology in this population. Further complicating the matter is the fact that changes in the *effectiveness* of melatonin might be achieved without changes in either *levels* or *patterns*. All that is necessary is that the effected systems display enhanced or diminished responsiveness to the hormone.

There is a great need to determine if subgroups exist within the present categories of affective disease which can be associated with specific rhythmic and neuroendocrine abnormalities. To do this properly, we need working hypotheses to guide the search. This chapter was written with that goal in mind. To review briefly, melatonin seems to act as an internal Zeitgeber, synchronizing various internal rhythms to create a coherent temporal order. Adrenal glucocorticoids also appear to act in this role, but insofar as the adrenal glands may themselves be pacemakers, the glucocorticoids may also exert considerable independent control on selected rhythms. Thyroid hormones and estrogens seem to influence coupling strength between multiple oscillators, albeit in opposite directions. Estrogen and hypothyroidism both result in activity patterns in rodents which have been interpreted as indicating a more tightly linked circadian system. Thus, if we accept the original proposition that circadian systems which are either too loosely or too tightly organized can result in behavioral pathology, the actions of these neuroendocrine systems on rhythmic phenomena are perfectly situated to predispose individuals to developing that pathology. Only more research will be able to either confirm or deny these hypotheses.

References

Abe, K., Kroning, J., Greer, M. A., & Critchlow, V. (1979). Effects of destruction of the suprachiasmatic nuclei on the circadian rhythms in plasma corticosterone, body temperature, feeding and plasma thyrotropin. *Neuroendocrionology, 29,* 119–131.

Agarwal, R. A., Rastogi, R. B., & Singhal, R. L. (1977). Enhancement of locomotor activity and catecholamine and 5-hydroxytryptamine metabolism by thyrotropin releasing hormone. *Neuroendocrinology, 23,* 236–247.

Albers, H. E. (1981). Gonadal hormones organize and modulate the circadian system of the rat. *American Journal of Physiology, 241,* R62–R66.

Albers, H. E., Gerall, A. A., & Axelson, J. F. (1981). Effect of reproductive state on circadian periodicity in the rat. *Physiology & Behavior, 26,* 21–25.

Albers, H. E., Yogev, L., Todd, R. B., & Goldman, B. D. (1985). Adrenal corticoids in hamsters: Role in circadian timing. *American Journal of Physiology, 248,* R434–R438.

Alleva, J. J., Waleski, M. V., & Alleva, F. R. (1971). A biological clock controlling the estrous cycle of the hamster. *Endocrinology, 88,* 1368–1379.

Amsterdam, J. D., Winokur, A., Liucki, I., Caroff, S., Snyder, P., & Rickels, K. (1983). A neuroendocrine test battery in bipolar patients and healthy subjects. *Archives of General Psychiatry, 40,* 515–521.

Anderson, K., & Eneroth, P. (1987). Thyroidectomy and central catecholamine neurons of the male rat. *Neuroendocrinology, 45,* 14–27.

Andrews, R. V. (1968a). Daily variation in membrane flux of cultured hamster adrenals. *Comparative Biochemistry and Physiology, 26,* 479–488.

Andrews, R. V. (1968b). Temporal secretory responses of cultured hamster adrenals. *Comparative Biochemistry and Physiology, 26,* 179–193.

Andrews, R. V. (1969). Temporal relationships of puromycin and ACTH effects on cultured hamster adrenal glands. *Comparative Biochemistry and Physiology, 30,* 123–128.

Andrews, R. V. (1971). Circadian rhythms in adrenal organ cultures. *Gegenbauers Morphologie Jahrbuch, 117,* 89–98.

Andrews, R. V., & Folk, G. E., Jr. (1964). Circadian metabolic patterns in cultured hamster adrenal glands. *Comparative Biochemistry and Physiology, 11,* 393–409.

Angeli, A. (1983). Glucocorticoid secretion: A circadian synchronizer of the human temporal structure. *Journal of Steroid Biochemistry, 19(1),* 545–554.

Anton-Tay, F., & Wurtman, R. J. (1969). Regional uptake of ^{3}H-melatonin from blood or cerebrospinal fluid by rat brain. *Nature (London), 221,* 474–475.

Arana, G. W., & Baldessarini, R. J. (1987). Development and clinical application of the dexamethasone suppression test in psychiatry. In U. Halbreich (Ed.), *Hormones and depression.* New York: Raven Press.

Arendt, J. (1986). Assay of melatonin and its metabolites: Results in normal and unusual environments. *Journal of Neural Transmission [Supplement], 21,* 11–33.

Arendt, J., Bojkowski, C., Folkard, S., Franey, C., Marks, V., Minors, D., Waterhouse, J., Wever, R. A., Wildgruber, C., & Wright, J. (1985). Some effects of melatonin and the control of its secretion in humans. *Ciba Foundation Symposium, 117,* 266–283.

Arendt, J., Symons, A. M., English, J., Poulton, A. L., & Tobler, I. (1988). How does melatonin control seasonal reproductive cycles? *Reproductive and Nutritional Development, 28,* 387–397.

Armstrong, S. M. (1989). Melatonin and circadian control in mammals. *Experientia, 45,* 932–938.

Armstrong, S. M., Cassone, V. M., Chesworth, M. J., Redman, J., & Short, R. V. (1986). Synchronization of mammalian circadian rhythms by melatonin. *Journal of Neural Transmission [Supplement], 21,* 375–394.

Armstrong, S. M., & Redman, J. (1985). Melatonin administration: Effects on rodent circadian rhythms. *Ciba Foundation Symposium, 117*, 188–207.

Aschoff, J. (1965). The phase angle difference in circadian periodicity. In J. Aschoff (Ed.), *Circadian clocks* (pp. 264–276). Amsterdam: North Holland.

Aschoff, J., & Wever, R. (1981). The circadian system of man. In J. Aschoff (Ed.), *Handbook of behavioral neurobiology* (pp. 311–331). New York: Plenum Press.

Assenmacher, I. (1982). CNS structures controlling circadian neuroendocrine and activity rhythms in rats. In J. Aschoff, S. Daan, & G. Groos (Eds.), *Vertebrate circadian systems*. 87–95 Berlin: Springer-Verlag.

Atterwill, C. K., Bunn, S. J., Atkinson, D. J., Smith, S. L., & Heal, D. J. (1984). Effects of thyroid status on presynaptic a2-adrenoreceptor function and b-adrenoreceptor binding in the rat brain. *Journal of Neural Transmission, 59*, 43–55.

Badura, L. L., & Nunez, A. A. (1989). Photoperiodic modulation of sexual and aggressive behavior in female golden hamsters (*Mesocricetus auratus*): Role of the pineal gland. *Hormones and Behavior, 23*, 27–42.

Banerji, T. K., Quay, W. B., & Kachi, T. (1978). Hypothalamic dopamine-β-hydroxylase activity: Fluctuations with time of day and their modifications by intracranial surgery, adrenalectomy and pinealectomy. *Neurochemical Research, 3*, 281–293.

Banki, C. (1975). Triiodothyronine in the treatment of depression. *Orv. Hetil., 116*, 2543–2547.

Banki, C. (1977). Cerebrospinal fluid amine metabolites after combined amitriptyline-triiodothyronine treatment of depressed women. *European Journal of Clinical Pharmacology, 11*, 311–315.

Banovac, K., Tolis, G., McKenzie, J. M., Guyda, H., & Sekso, M. (1977). Dissociation of thyrotropin and prolactin circadian periodicity induced by glucocorticoids. *Clinical Research, 25*, 678A.

Barontini, M., Romeo, H. E., Armando, I., & Cardinali, D. P. (1988). 24-Hour changes in catecholamine content of rat thyroid and submaxillary glands. *Journal of Neural Transmission, 71*, 189–194.

Bartness, T. J., & Goldman, B. D. (1988). Effects of melatonin on long-day responses in short-day housed adult Siberian hamsters. *American Journal of Physiology, 255(5)*, R823–R830.

Bartness, T. J., & Goldman, B. D. (1989). Mammalian pineal melatonin: A clock for all seasons. *Experientia, 45*, 939–945.

Bartness, T. J., Goldman, B. D., & Bittman, E. L. (1991). SCN lesions block responses to systematic melatonin infusions in Siberian hamsters. *American Journal of Physiology, 260(1)*, R102–R112.

Bauer, M., Droba, M., & Whybrow, P. (1987). Disorders of thyroid and parathyroid. In P. Loosen & C. Nemeroff (Eds.), *Handbook of clinical psychoneuroendocrinology* (pp. 41–70). New York: Guilford.

Bauer, M., & Whybrow, P. (1986). The effect of changing thyroid function on cyclic affective illness in a human subject. *American Journal of Psychiatry, 143*, 633–636.

Beasley, L. J., & Nelson, R. J. (1982). Thyroid gland influences the period of hamster circadian oscillations. *Experientia, 38*, 870–871.

Beck-Friis, J., & Wetterberg, L. (1987). Melatonin and the pineal gland in depressive disorders. In U. Halbreich (Ed.), *Hormones and depression* (pp. 195–206). New York: Raven Press.

Bellinger, L. L., Williams, F. E., & Bernardis, L. L. (1979). Effect of hypophysectomy, thyroidectomy, castration and adrenalectomy on diurnal food and water intake in rats. *Proceedings of the Society for Experimental Biology and Medicine, 16*, 12–16.

Ben-Tovim, R., Zohar, Y., Zohar, S., Laurian, N., & Laurian, L. (1985). Auditory brain stem response in experimentally induced hypothyroidism in albino rats. *Laryngoscope, 95*, 982–986.

Binkley, S. (1976). Pineal gland biorhythms: *N*-acetyltransferase in chickens and rats. *Federation Proceedings, 35*, 2347–2352.

Bittman, E. L. (1985). The role of rhythms in the response to melatonin. *Ciba Foundation Symposium, 117*, 149–169.

Bittman, E. L., Bartness, T. J., Goldman, B. D., & De Vries, G. J. (1991). Suprachiasmatic and paraventricular control of photoperiodisim in Siberian hamsters. *American Journal of Physiology, 260,* R90–R101.

Blake, C. A. (1976). Effects of pinealectomy on the rat oestrous cycle and pituitary gonadotropin release. *Journal of Endocrinology, 69,* 67–75.

Blake, C. A., & Sawner, C. H. (1974). Effects of hypothalamic deafferentation on the pulsatile rhythm in plasma concentrations of luteinizing hormone in ovariectomized rats. *Endocrinology, 94(3),* 730–736.

Blake, C. A., Weiner, R. I., Gorski, R. A., & Sawyer, C. H. (1972). Secretion of pituitary luteinizing hormone and follicle stimulating hormone in female rats made persistently estrous or diestrous by hypothalamic deafferentation. *Endocrinology, 90(4),* 855–861.

Bobbert, A. C., & Riethoven, J. J. (1991). Feedback in the rabbit's central circadian system, revealed by the changes in its free-running food intake pattern induced by blinding, cervical sympathectomy, pinealectomy, and melatonin administration. *Journal of Biological Rhythms 6(3),* 263–278.

Borer, K. T., Bestervelt, L. L., Mannheim, M., Brosamer, M. B., Thompson, M., Swamy, U., & Piper, W. N. (1992). Stimulation by voluntary exercise of adrenal glucocorticoid secretion in mature female hamsters. *Physiology & Behavior, 51,* 713–718.

Bovet, J. (1972). On the social behavior in a stable group and long-tailed mice (*Apodemus sylvanticus*). II. Its relations with distribution of daily activity. *Behavior, 41,* 55–67.

Bridges, R., Tamarkin, L., & Goldman, B. D. (1976). Effects of photoperiod and melatonin on reproduction in the Syrian hamster. *Annales de Biologie Animale, Biochimie, Biophysique, 16,* 399–408.

Brown, W. A. (1987). Clinical correlates of pituitary-adrenocortical activity in depression. In U. Halbreich (Ed.), *Hormones and depression* (pp. 135–149). New York: Raven Press.

Bunning, E. (1973). *The physiological clock: Circadian rhythms and biological chronometry,* Rev. 3d London: The English Universities Press.

Cagnacci, A., Elliott, J. A., & Yen, S. S. C. (1992). Melatonin: A major regulator of the circadian rhythm of core temperature in humans. *Journal of Clinical Endocrinology and Metabolism, 75,* 447–452.

Calhoun, J. B. (1962). *The ecology and sociology of the Norway rat. Public Health Document No. 1008.* Washington, D.C.: United States Department of Housing, Education, and Welfare.

Cardinali, D. P., Hyyppä, M. T., & Wurtman, R. J. (1973). Fate of intracisternally injected melatonin in the rat brain. *Neuroendocrinology, 12,* 30–40.

Cardinali, D. P., Pisarev, M. A., Barontini, M., Juvenal, G. J., Boado, R. J., & Vacas, M. I. (1982). Efferent neuroendocrine pathways in sympathetic superior cervical ganglia. Early depression of the pituitary-thyroid axis after ganglionectomy. *Neuroendocrinology, 35,* 248–254.

Cardinali, D. P., Romeo, H. E., Boado, R. J., & Deza, S. (1986). Early inhibition and changes in diurnal rhythmicity of the pituitary-thyroid axis after superior cervical ganglionectomy. *Journal of the Autonomic Nervous System, 16,* 13–21.

Cardinali, D. P., Romeo, H. E., Ochatt, C. M., & Moguilevsky, J. A. (1989). Estrous cycle delay and inhibition of gonadotropin and prolactin release during sympathetic nerve degeneration after superior cervical ganglionectomy of rats. *Neuroendocrinology, 50,* 59–65.

Cardinali, D. P., Vacas, M. I., & Boyer, E. E. (1979). Specific binding of melatonin in bovine brain. *Endocrinology, 105(2),* 437–441.

Carlberg, K. A., & Fregly, M. J. (1986). Catecholamine excretion and beta-adrenergic responsiveness in estrogen-treated rats. *Pharmacology, 32,* 147–156.

Carmen, J. S., & Wyatt, R. J. (1979). Calcium: Bivalent cation in bivalent psychosis. *Biological Psychiatry, 14,* 295–336.

Carmen, J. S., Wyatt, E. S., Smith, W., Post, R. M., & Ballenger, J. C. (1984). Calcium and cacitonin in bipolar affective disorder. In R. M. Post & J. C. Ballenger (Eds.), *Frontiers of clinical neuroscience, Vol. 1. Neurobiology of mood disorders.* Baltimore: Williams and Wilkins.

Carter, S. B. (1972). The relationship of the time of ovulation to the activity pattern in the rat. *Journal of Endocrinology, 53*, 333–334.

Cassone, V. M. (1990). Effects of melatonin on vertebrate circadian systems. *Trends in Neuroscience, 13(11)*, 457–464.

Cassone, V. M., Forsyth, A. M., & Woodlee, G. L. (1990). Hypothalamic regulation of circadian noradrenergic input to the chick pineal gland. *Journal of Comparative Physiology A, 167*, 187–192.

Cassone, V. M., & Menaker, M. (1984). Is the avian circadian system a neuroendocrine loop? *Journal of Experimental Zoology, 232*, 539–550.

Cassone, V. M., Roberts, M. H., & Moore, R. Y. (1988). Effects of melatonin on 2-deoxy-[1-^{14}C]glucose uptake within the rat suprachiasmatic nucleus. *American Journal of Physiology, 255(2)*, R332–337.

Cheung, P. W., & McCormack, C. E. (1982). Failure of pinealectomy or melatonin to alter circadian rhythm of the rat. *American Journal of Physiology, 242*, R261–R264.

Cho, J. T., Bone, S., Dunner, D. L., Colt, E., & Fieve, R. R. (1979). The effect of lithium treatment on thyroid function in patients with primary affective disorder. *American Journal of Psychiatry, 136*, 115–116.

Christie, M. J., Little, B. C., & Gordon, A. M. (1980). Peripheral indices of depressive states. In H. M. Van Praag, M. H. Lader, O. J. Rafaelson, & E. J. Sachar (Eds.), *Handbook of biological psychiatry. Part II. Brain mechanisms and abnormal behavioral psychophysiology.* New York: Marcel Dekker.

Cohen, I. R., & Wise, P. M. (1988). Effects of estradiol on the diurnal rhythm of serotonin activity in microdissected brain areas of ovariectomized rats. *Endocrinology, 122(6)*, 2619–26925.

Cooper, T. B., and Simpson, G. M. (1969). Preliminary report of a longitudinal study on the effects of lithium on iodine metabolism. *Current Therapeutic Research, 11*, 603.

Coppen, A., Whybrow, P., Noguera, R., Maggs, R., & Prange, A. (1972). The comparative antidepressant value of L-tryptophan and imipramine with and without attempted potentiation by liothyronine. *Archives of General Psychiatry, 26*, 234–241.

Cowdry, R., Wehr, T., Zis, A., & Goodwin, F. (1983). Thyroid abnormalities associated with rapid-cycling bipolar illness. *Archives of General Psychiatry, 46*, 414–20.

Crews, D., Hingorani, V., & Nelson, R. J. (1988). Role of the pineal gland in the control of annual reproductive behavioral and physiological cycles in the red-sided garter snake (*Thamnophis sirtalis parietalis*). *Journal of Biological Rhythms, 3*, 293–302.

Crowley, W. R., O'Donohoe, T. L., & Jacobowitz, D. M. (1978a). Changes in catecholamine content in discrete brain nuclei during the estrous cycle of the rat. *Brain Research, 147*, 315–326.

Crowley, W. R., O'Donohoe, T. L., Wachslicht, H., & Jacobowitz, D. M. (1978b). Effects of estrogen and progesterone on plasma gonadotropins and on catecholamine levels and turnover in discrete brain regions of ovariectomized rats. *Brain Research, 154*, 345–357.

Curtis, G. C. (1972). Psychosomatics and chronobiology: Possible implications of neuroendocrine rhythms. *Psychosomatic Medicine, 34(3)*, 235–256.

Daan, S., Damassa, D., Pittendrigh, C. S., & Smith, E. R. (1975). An effect of castration and testosterone replacement on a circadian pacemaker in mice (*Mus musculus*). *Proceedings of the National Academy of Science, U.S.A., 72*, 3744–3747.

Dahlitz, M., Alvarez, B., Vignau, J., English, J., Arendt, J., & Parkes, J. D. (1991). Delayed sleep phase syndrome response to melatonin. *The Lancet, 337*, 1121–1124.

Dallman, M. F., Akana, S. F., Casico, C. S., Darlington, D. N., Jacobsen, L., & Levin, N. (1987). Regulation of ACTH secretion: Variations on theme of B. In *Recent Progress in Hormone Research* (Vol. 43) (pp. 113–167). Orlando, Florida: Academic Press.

Darrow, J. M., & Goldman, B. D. (1985). Circadian regulation of pineal melatonin and reproduction in the Djungarian hamster. *Journal of Biological Rhythms, 1*, 39–54.

Deguchi, T. (1979). A circadian oscillator in cultured cells of chicken pineal gland. *Nature (London), 282*, 94–96.

DeMarchi, G. W., & Tong, J. E. (1972). Menstrual, diurnal, and activation effects on the resolution of temporally paired flashes. *Psychophysiology, 9*, 362–367.

de Vlaming, V. L., & Vodicnik, M. J. (1978). Seasonal effects of pinealectomy on gonadal activity in the goldfish, *Carassius auratus. Biology of Reproduction, 19*, 57–63.

Dirlich, G., Kammerloher, A., Schultz, H., Lund, R., Doerr, P., & Von Zerssen, D. (1981). Temporal coordination of rest-activity cycle, body temperature, urinary free cortisol, and mood in patient with 48-hour unipolar-depressive cycles in clinical and time-cue free environment. *Biological Psychiatry, 16*, 163–179.

Donham, R. S., Di Pinto, M. N., & Stetson, M. H. (1985). Effects of ovariectomy on clock-timed daily gonadotropin rhythms in prepubertal golden hamsters. *Biology of Reproduction, 32*, 284–289.

Doust, J. W. L., & Christie, H. (1980). Repeated sleep deprivation as a therapeutic zeitgeber for circular type manic depressive disturbance. *Chronobiologia, 7*, 505–511.

Dunner, D., and Fieve, R. (1974). Clinical factors in lithium carbonate prophylaxis failure. *Archives of General Psychiatry, 30*, 229–233.

Earle, B. (1970). Thyroid hormone and tricyclic antidepressants in resistant depressions. *American Journal of Psychiatry, 126*, 1667–1669.

Eayrs, J. T. (1964a). Effect of neonatal hyperthyroidism on maturation and learning in the rat. *Animal Behaviour, 12*, 195–199.

Eayrs, J. T. (1964b). Effects of thyroid hormones on brain differentiation. In *Brain–Thyroid Relationships* (pp. 60–74). Boston: Little, Brown.

Ebling, F. J. P., Maywood, E. S., Humbly, T., & Hastings, M. H. (1992). Circadian and photoperiodic time measurement in male Syrian hamsters following lesions of the melatonin-binding sites of the paraventricular thalamus. *Journal of Biological Rhythms, 7*, 241–254.

Edelman, I. S. (1974). Thyroid thermogenesis. *New England Journal of Medicine, 290*, 1303–1308.

Edgar, D. M., Martin, C. E., & Dement, W. C. (1991). Activity feedback to the mammalian circadian pacemaker: Influence on observed measures of rhythm period length. *Journal of Biological Rhythms, 6(3)*, 185–199.

Edwards, P. A. (1973). Effect of adrenalectomy and hypophysectomy on the circadian rhythm of β-hydroxy-β-methylglutaryl coenzyme a reductase activity in rat liver. *The Journal of Biological Chemistry, 248(8)*, 2912–2917.

Elliott, J. A. (1976). Circadian rhythms and photoperiodic time measurement in mammals. *Federation of the American Society of Experimental Biology, 35*, 2339–2346.

Ellis, G. B., & Turek, F. W. (1979). Changes in locomotor activity associated with photoperiodic response of the testes in male golden hamsters. *Journal of Comparative Physiology, 132*, 277–284.

English, J., Arendt, J., Symons, A. M., Poulton, A. L., & Tobler, I. (1988). Pineal and ovarian response to 22- and 24-hr days in the ewe. *Biology of Reproduction, 39*, 9–18.

Faglia, G., & Persani, L. (1991). Thyrotropin-releasing hormone: Basic and clinical aspects. In M. Motta (Ed.), *Brain endocrinology*, 2d ed. (pp. 315–350). New York: Raven Press.

Falcón, J., & Collin, J.-P. (1989). Photoreceptors in the pineal of lower vertebrates: Functional aspects. *Experientia, 45*, 909–913.

Falcón, J., Marmillion, J. B., Claustrat, B., & Collin, J. P. (1989). Regulation of melatonin secretion in a photoreceptive pineal organ: An *in vitro* study in the pike. *Journal of Neuroscience, 9(6)*, 1943–1950.

Faust, V., Sarreither, P., & Wehner, W. F. (1974). The frequency of hospitalization of mentally ill subjects in relation to the time of the year *Journal of Interdisciplinary Cycle Research, 5*, 313–319.

Feder, H. H. (1981). Estrous cyclicity in mammals. In N. T. Adler (Ed.), *Neruoendocrinology of reproduction: Physiology and behavior* (pp. 279–348). New York: Plenum Press.

Ferin, M. (1983). Neuroendocrine control of ovarian function in the primate. *Journals of Reproduction & Fertility, 69*, 369–381.

Fischette, C. T., Edinger, H. M., & Siegel, A. (1981). Temporary desynchronization among circadian rhythms with lateral fornix ablation. *Brain Research, 229*, 85–101.

Fischmann, A. J., Kastin, A. J., Graf, M. V., & Moldow, R. L. (1988). Constant light and dark affect the circadian rhythm of the hypothalamic–pituitary–adrenal axis. *Neuroendocrinology, 47*, 309–316.

Fisher, D. A., Dussault, J. H., Sack, J., & Chopra, I. J. (1977). Ontogenesis by hypothalamic–pituitary–thyroid function and metabolism in man, sheep, and rat. *Recent Progress in Hormone Research, 33*, 59–116.

Folkard, S., Arendt, J., Aldhous, M., & Kennett, H. (1990). Melatonin stabilizes sleep onset time in a blind man without entrainment of cortisol or temperature rhythms. *Neuroscience Letters, 113(2)*, 193–198.

Follett, B. K., Foster, R. G., & Nicholls, T. J. (1985). Photoperiodism in birds. *Ciba Foundation Symposium, 117*, 93–105.

Ford, D. H. (1968). Central nervous system-thyroid interrelationships. *Brain Research, 7*, 329–349.

Fox, S. R., & Smith, M. S. (1985). Changes in the pulsitile pattern of luteinizing hormone secretion during the rat estrous cycle. *Endocrinology, 116*, 1485–1492.

Freeman, M. C., Dupke, K. C., & Croteau, C. M. (1976). Extinction of the estrogen-induced daily signal for LH release in the rat: A role for the proestrous surge of progesterone. *Endocrinology, 99(1)*, 223–229.

Freidman, J., & Meares, R. A. (1978). Comparison of spontaneous and contraceptive menstrual cycles on visual discrimination task. *Australian New Zealand Journal of Psychiatry, 12(4)*, 233–239.

Friedman, A. H., & Walker, C. A. (1968). Circadian rhythms in rat mid-brain and caudate nucleus biogenic amine levels. *Journal of Physiology, 197*, 77–85.

Garg, S. K., & Sudararaj, B. I. (1986). Role of pineal in the regulation of some aspects of circadian rhythmicity in the catfish, *Herteropneustes fossilis. Chronobiologia, 13*, 1–11.

Gaston, S., & Menaker, M. (1968). Pineal function: The biological clock in the sparrow? *Science, 160*, 1125–1127.

Genazzani, A. R., Trentini, G. P., Petraglia, F., De Gaetani, C. F., Criscuolo, M., Ficarra, G., De Ramundo, B. M., & Cleva, M. (1990). Estrogens modulate the circadian rhythm of hypothalamic beta-endorphin contents in female rats. *Neuroendocrinology, 52*, 221–224.

Gershon, S., & Shopsin, B. (1973). Pharmacology–toxicology of the lithium ion. In S. Gershon & B. Shopsin (Eds.), *Lithium: Its role in psychiatric research and treatment* (pp. 118–127). New York: Plenum Press.

Gibson, M. J., & Krieger, D. T. (1981). Circadian corticosterone rhythm and stress response in rats with adrenal autotransplants. *American Journal of Physiology, 240*, E363–E366.

Gillette, M. U. (1991). SCN electrophysiology *in vitro*: Rhythmic activity and endogenous clock properties. In D. C. Klein, R. Y. Moore, & S. M. Reppert (Eds.), *Suprachiasmatic nucleus: The mind's clock*. Oxford: Oxford University Press.

Glick, I. D., Quitkin, F. M., & Bennett, S. E. (1987). The influence of estrogens, progestins, and oral contraceptives on depression. In U. Halbreich (Ed.), *Hormones and depression* (pp. 339–356). New York: Raven Press.

Goodwin, F. K., Prange, A. J., Post, R. M., Muscettola, G., & Lipton, M. A. (1982). Potentiation of antidepressant effects by F-triiodothyronine in tricyclic nonresponders. *American Journal of Psychiatry, 139*, 34–38.

Grocock, C. A., & Clarke, J. R. (1974). Photoperiodic control of testis activity in the vole, *Microtus agrestis*. *Journal of Reproduction and Fertility, 39*, 337–347.

Gross, G., Brodde, O-E., & Schumann, H-J. (1980). Decreased numbers of β-adrenoreceptors in cerebral cortex of hypothyroid rats. *European Journal of Pharmacology, 61*, 191–194.

Hadley, M. E. (1992). *Endocrinology*, 3d ed. Englewood Cliffs, New Jersey: Prentice-Hall.

Halaris, A. (Ed.) (1987). *Chronobiology and psychiatric disorders*. Amsterdam: Elsevier.

Halberg, F. (1968). Physiologic considerations underlying rhythmometry with special reference to emotional illness. In J. de Ajuriaguerra (Ed.), *Symposium Bel-Air III. Cycles biologiques et psychiatrie* (pp. 73–126). Mason & Cie: Paris.

Halbreich, U. (1987). The circadian rhythm of cortisol and MHPG in depressives and normals. In A. Halaris (Ed.), *Chronobiology and psychiatric disorders*. Amsterdam: Elsevier.

Hastings, M. H., Walker, A. P., & Herbert, J. (1987). Effect of assymetrical reductions of photoperiod on pineal melatonin, locomotor activity and gonadal condition of male Syrian hamsters. *Journal of Endocrinology, 114(2)*, 221–229.

Herbert, J. (1972). Initial observations on pinealectomized ferrets kept for long periods in either daylight or artificial illumination. *Journal of Endocrinology, 55*, 591–597.

Herbert, J., Stacey, P. M., & Thorpe, D. H. (1978). Recurrent breeding seasons in pinealectomized or optic-nerve-sectioned ferrets. *Journal of Endocrinology, 78*, 389–397.

Herbert, R., & Dassault, J. H. (1984). Permanent peripheral hearing system alteration following transient neonatal hyperthyroidism in rats. *Brain Research, 316(2)*, 159–164.

Herzberg, A. N. Johnson, A. L., & Brown, S. (1970). Depressive symptoms and oral contraceptives. *British Medical Journal, 4*, 142–145.

Himelfarb, M. Z., Lakretz, T., Gold, S., & Shanon, E. (1981). Auditory brain stem responses in thyroid disfunction. *Journal of Laryngology and Otology, 95*, 679–686.

Hiroshige, T. (1984). Hormonal rhythm and feeding behavior. *Journal of the Autonomic Nervous System, 10*, 337–346.

Ho, A. K., Burns, T. G., Grota, L. J., & Brown, G. M. (1985). Scheduled feeding and 24-hour rhythms of *N*-acetylserotonin and melatonin in rats. *Endocrinology, 116(5)*, 1858–1862.

Hoffman, R. A. (1970). The epiphysial complex in fish and reptile. *American Zoologist, 10*, 191–199.

Hoffmann, K. (1979). Photoperiodic effects in the Djungarian hamster: One minute of light during darktime mimics influence of long photoperiods on testicular recrudescence, body weight and pelage colour. *Experientia, 35*, 1529–1530.

Hoffmann, K. (1981). Photoperiodism in vertebrates. In J. Aschoff (Ed.), *Handbook of Behavioral Neurobiology 4: Biological Rhythms* (pp. 449–473). New York: Plenum Press.

Holzwarth, M. A., Cunningham, L. A., & Kleitman, N. (1987). The role of adrenal nerves in the regulation of adrenocortical functions. *Annals of the New York Academy of Sciences, 512*, 449–464.

Honma, S., Honma, K-I., Nagasaka, T., & Hiroshige, T. (1987). The ventromedial hypothalamic nucleus is not essential for the prefeeding corticosterone peak in rats under restricted daily feeding. *Physiology & Behavior, 39*, 211–215.

Hopper, A. F., & Yatvin, M. B. (1965). Protein metabolism in the liver of thiouracil-treated goldfish. *Growth, 29*, 355–360.

Horseman, N. D., & Ehret, C. F. (1982). Glucocorticosteroid injection is a circadian zeitgeber in the laboratory rat. *American Journal of Physiology, 243*, R373–R378.

Hsu, B. R., & Kuhn, R. W. (1988). The role of the adrenal in generating the diurnal variation in circulating levels of corticosteroid-binding globulin in the rat. *Endocrinology, 122(2)*, 421–426.

Huang, T. S., Chang, Y. C., Lee, S. H., Chen, F. W., & Chopra, I. J. (1989). Visual, brainstem auditory and somatosensory evoked potential abnormalities in thyroid disease. *Thyroidology: Clinical Experiments, 1*, 137–142.

Humlova, M., & Illnerova, H. (1990). Melatonin entrains the circadian rhythm in the rat pineal N-acetyltransferase activity. *Neuroendocrinology, 52(2)*, 196–199.

Illnerova, H. (1991). The suprachiasmatic nucleus and rhythmic pineal melatonin production. In D. C. Klein, R. Y. Moore, & S. M. Reppert (Eds.), *Suprachiasmatic nucleus: The mind's clock* (pp. 197–216). Oxford: Oxford University Press.

Illnerova, H., Trentini, G. P., & Maslova, L. (1989). Melatonin accelerates reentrainment of the circadian rhythm of its own production after an 8-hr advance of the light-dark cycle. *Journal of Comparative Physiology A, 166(1)*, 97–102.

Inouye, S-I. T. (1982). Restricted daily feeding does not entrain circadian rhythms of the suprachiasmatic nucleus in the rat. *Brain Research, 232*, 194–199.

Inouye, S. T., & Kawamura, H. (1979). Persistence of circadian rhythmicity in a mammalian hypothalamic "island" containing the suprachiasmatic nucleus. *Proceedings of the National Academy of Sciences U.S.A., 76*, 5962–5966.

Iuvone, P. M., & Van Hartesveldt, C. (1977). Diurnal locomotor activity in rats: Effects of hippocampal ablation and adrenalectomy. *Behavioral Biology, 19*, 228–237.

Janik, D. S., & Menaker, M. (1990). Circadian locomotor rhythms in the desert iguana. I. The role of the eyes and the pineal. *Journal of Comparative Physiology A, 166(6)*, 803–810.

Jenner, F. A. (1968). Periodic psychoses in the light of biological rhythm research. *International Review in Neurobiology, 11*, 129–169.

Jhanwar-Uniyal, M., & Leibowitz, S. F. (1986). Impact of circulating corticosterone on a_1- and a_2-noradrenergic receptors in discrete brain areas. *Brain Research, 368*, 404–408.

Joffe, R., Roy-Byrne, P., Uhde, T., & Post, R. (1984). Thyroid function and affective illness: A reappraisal. *Biological Psychiatry, 19*, 1685–1691.

Johnson, J. H., & Sawyer, C. H. (1971). Adrenal steroids and the maintenance of a circadian distribution of paradoxical sleep in rats. *Endocrinology, 89*, 507–512.

Johnson, L. Y., Vaughn, B. A., Richardson, L. J., Petterborg, L. J., & Reiter, R. J. (1982). Variation in pineal melatonin content during the estrous cycle of the rat. *Proceedings of the Society of Experimental Biology and Medicine, 169*, 416–419.

Johnsson, A., Engelmann, W., Pflug, B., & Klemke, W. (1980). Influence of lithium ions on human circadian rhythms. *Zeitschrift der Naturforschung, 35c*, 503–507.

Kalaria, R. N., & Prince, A. K. (1986). Decreased neurotransmitter receptor binding in striatum and cortex from adult hypothyroid rats. *Brain Research, 364*, 268–274.

Kalra, S. P., & Kalra, P. S. (1991). Steroid-peptide interactions in the endocrine brain. In M. Motta (Ed.), *Brain endocrinology*, 2d ed. (pp. 177–216). New York: Raven Press.

Kaneko, M., Hiroshige, T., Shinsako, J., & Dallman, M. F. (1980). Diurnal changes in amplification of hormone rhythms in the adrenocortical system. *American Journal of Physiology, 239*, R309–R316.

Kaneko, M., Kaneko, K., Shinsako, J., & Dallman, M. F. (1981). Adrenal sensitivity to adrenocorticotropin varies diurnally. *Endocrinology, 109(1)*, 70–75.

Karsch, F. J. (1986). A role for melatonin as a timekeeping hormone in the ewe. *Journal of Neural Transmission (Suppl.), 21*, 109–124.

Keeton, W. T. (1979). Avian orientation and navigation. *Annual Review of Physiology, 41*, 353–366.

Kezuka, H., Aida, K., & Havnu, I. (1989). Melatonin secretion from goldfish pineal gland in organ culture. *General Comparative Endocrinology, 75(2)*, 217–221.

Kiem, D. T., Kanyicksa, B., Stark, E., & Fekete, M. I. (1987). Diurnal variation in prolactin, adrenocorticotropin and corticosterone release induced by opiate agonists in intact and adrenalectomized rats. *Neuroendocrinology, 46(6)*, 475–480.

King, D. (1980). Sleep deprivation therapy in depressive syndromes. *Psychosomatics, 21*, 404–407.

Kirkegaard, C., Norlem, N., Lauridsen, U., & Bjorum, N. (1975a). Prognostic value of thyrotropin-releasing hormone stimulation test in endogenous depression. *Acta Psychiatrica Scandanavia, 52,* 170–177.

Kirkegaard, C., Norlem, N., Lauridsen, U., Bjorum, N., & Christiansen, C. (1975b). Protirelin stimulation test and thyroid function during treatment of depression. *Archives of General Psychiatry, 32,* 1115–1118.

Klein, D. C. (1985). Photoneural regulation of the mammalian pineal gland. *Ciba Foundation Symposium, 117,* 38–56.

Knobil, E. (1980). The neuroendocrine control of the menstrual cycle. In *Recent Progress in Hormone Research* (Vol. 36) (pp. 53–88). New York: Academic Press.

Knobil, E. (1989). The electrophysiology of the GnRH pulse generator in the rhesus monkey. *Journal of Steroid Biochemistry, 33(4B),* 669–671.

Köves, K., Gottschall, P. E., & Arimura, A. (1989). Gonadotropin-releasing hormone binding sites in ovaries of normal cycling and persistent-estrus rats. *Biology of Reproduction, 41,* 505–511.

Kostin, A., Ehrensing, R., Schalch, D., & Anderson, M. (1972). Improvement in mental depression with decreased thyrotropin response after administration of thyrotropin-releasing hormone. *Lancet, 2,* 704–742.

Krieger, D. T. (1979). Rhythms in CRF, ACTH, and corticosteroids. In D. T. Krieger (Ed.), *Endocrine Rhythms* (pp. 123–142). New York: Raven Press.

Krieger, D. T., Hauser, H., & Krey, L. C. (1977). Suprachiasmatic nuclear lesions do not abolish food-shifted circadian adrenal temperature rhythmicity. *Science, 197,* 398–399.

Kripke, D. F., Drennan, M. D., & Elliott, J. A. (1992). The complex circadian pacemaker in affective disorders. In Y. Touitou & E. Haus (Eds.) *Biologic rhythms in clinical and laboratory medicine.* Berlin: Springer-Verlag, 265–276.

Kripke, D. F., Gillin, J. C., Mullaney, D. J., Risch, S. C., & Janowsky, D. S. (1987). Treatment of major depressive disorders by bright light for 5 days. In A. Halaris (Ed.), *Chronobiology and psychiatric disorders* (pp. 207–218). Amsterdam: Elsevier.

Kripke, D. F., Mullaney, D. J., Atkinson, M., & Wolf, S. (1978). Circadian rhythm disorders in manic-depressives. *Biological Psychiatry, 13,* 335–351.

Kripke, D. F., & Wyborney, V. G. (1980). Lithium slows rat circadian activity rhythms. *Life Sciences, 26,* 1319–1321.

LaBella, F. S. (1985). Opiate receptor activity on 17-α-estrodial and related steroids. In H. Lal, F. LaBella & J. Lane (Eds.), *Endocoids* (pp. pp. 323–328). New York: Liss.

Ladenson, P. W., Stakes, J. W., & Ridgeway, E. C. (1984). Reversible alteration of the visual evoked potential in hypothyroidism. *The American Journal of Medicine, 77,* 1010–1014.

Laitinen, J. T., Castran, E., Vakkuri, O., & Saavedra, J. M. (1989). Diurnal rhythm of melatonin binding in the rat suprachiasmatic nucleus. *Endocrinology, 124(3),* 1585–1587.

Landau, I. T. (1975). Effects of adrenalectomy on rhythmic and non-rhythmic aggressive behavior in the male golden hamster. *Physiology and Behavior, 14,* 775–780.

Lauchlan, C. L., & McEachron, D. L. (1991). The effect of thyroxine (T4) replacement on the free-running circadian activity periods of thyroparathyroidectomized (TPX) and sham-operated (Sham) male rats. *Third World Congress of Neuroscience, Abstracts P38.9,* 254.

Lee, M. O., & Van Buskirk, E. F. (1928). Studies on vigor. XV: The effect of thyroidectomy on spontaneous activity in the rat, with a consideration of the relation of basal metabolism to the spontaneous activity. *American Journal of Physiology, 84,* 321.

Lewis, A., & Hoghughi, M. (1969). An evaluation of depression as a side effect of oral contraceptives. *British Medical Journal, 115,* 687–701.

Lewy, A. J., Nurnberger, J. I., Wehr, T. A., Pack, D., Becker, L., Powell, R., & Newsome, D. (1985). Supersensitivity to light: Possible trait marker for manic-depressive illness. *American Journal of Psychiatry, 142,* 725–727.

Lewy, A. J., Sack, R. L., Miller, S., & Hoban, T. M. (1987a). Antidepressant and circadian phase-shifting effects of light. *Science, 235*, 352–354.

Lewy, A. J., Sack, R. L., & Singer, C. M. (1987b). The phase shift hypothesis for bright light's therapeutic mechanism of action: Theoretical considerations and experimental evidence. *Psychopharmacology Bulletin, 23*, 349–354.

Lewy, A. J., Sack, R. L., & Singer, C. M. (1988). Winter depression and the phase-shift hypothesis for bright light's therapeutic effects: History, theory, and experimental evidence. *Journal of Biological Rhythms, 3*, 121–134.

Lewy, A. J., Wehr, T. A., Goodwin, F. K., Newsome, D. A., & Markey, S. P. (1980). Light suppressed melatonin secretion in humans. *Science, 210*, 1267–1269.

Lewy, A. J., Saeeduddin, A., Jackson, J. M. L., & Sack, R. L. (1992). Melatonin shifts human circadian rhythms according to a phase-responsive curve. *Chronobiology International, 9*, 380–392.

Maas, J. W. (1979). Biochemistry of affective disorders. *Hospital Practices, 14*, 113–120.

McArthur, A. J., Gillette, M. U., & Prosser, R. A. (1991). Melatonin directly resets the rat suprachiasmatic circadian clock in vitro. *Brain Research, 565*, 158–161.

McEachron, D., Kripke, D. F., Sharp, F. R., Lewy, A. J., & McClellan, D. E. (1985). Lithium effects on selected circadian rhythms in rats. *Brain Research Bulletin, 15*, 347–390.

McEachron, D. L. (1984). *Testing the circadian hypothesis of affective disorders: Lithium's effect on selected circadian rhythms in rats.* Ph.D. Thesis, University of California at San Diego.

McEachron, D. L. (1987). Antidepressants and food restriction cycles: Evidence for multiple pacemakers in rats. In J. D. Pauly & L. E. Scheving (Eds.), *Progress in clinical and biological research* (Vol. 227) (pp. 491–506). New York: Alan R. Liss.

McEachron, D. L. (1991). Thyroparathyroidectomy (TPX) shortens the free-running circadian activity periods of blinded male and female rats. *21st Annual Meeting of the Society for Neuroscience, Abstracts, Part I: 292.10*, 730.

McEachron, D., Kripke, D. F., Hawkins, R., Haus, E., Pavlinac, D., & Deftos, L. (1982). Lithium delays biochemical rhythms in rats. *Neuropsychobiology, 8*, 12–29.

McEachron, D. L., Kripke, D. F., & Wyborney, V. G. (1981). Lithium promotes entrainment by rats to long circadian light-dark cycles. *Psychiatry Research, 5*, 1–9.

McEachron, D. L., Levine, L., & Adler, N. T. (1989). The relationship of a circadian rhythm's free-running period to phase during entrainment: A need to reevaluate current theory. *Photochemistry and Photobiology, (Suppl.) 49*, 26S.

McEachron, D. L., Lauchlan, C. L., & Midgley, D. E. (1993). Effects of thyroxine and thyroparathyroidectomy on circadian wheelrunning in rats. *Pharmacology, Biochemistry, & Behavior*, in press.

Malbon, C. C., & Hadcock, J. R. (1988). Evidence that glucocorticoid response elements in the 5'-noncoding region of the hamster β_2-adrenergic receptor gene are obligate for glucocorticoid regulation of receptor mRNA levels. *Biochemical and Biophysical Research Communications, 154(2)*, 676–681.

Mandel, M. (1960). Recurrent psychotic depression associated with hypocalcemia and parathyroid adenoma. *American Journal of Psychiatry, 117*, 234–235.

Martinet, L., & Zucker, I. (1985). Role of pineal gland in circadian organization of diurnal ground squirrels. *Physiology & Behavior, 34*, 799–803.

Mason, R., & Brooks, A. (1988). The electrophysiological effects of melatonin and a putative melatonin antagonist (*N*-acetyltryptamine) on rat suprachiasmatic neurons *in vitro*. *Neuroscience Letters, 95*, 296–301.

Matsumoto, T., Hess, D. L., Kaushal, K. M., Valenzuela, G., Yellon, S. M., & Ducsay, C. A. (1991). Circadian myometrial and endocrine rhythms in the pregnant rhesus macaque: Effects of constant light and timed melatonin infusion. *American Journal of Obstetrics and Gynecology, 165*, 1777–1784.

Maywood, E. S., Buttery, R. C., Vance, G. H. S., Herbert, J., & Hastings, M. H. (1990). Gonadal response of the male Syrian hamster to programmed infusions of melatonin are sensitive to signal duration and frequency but not to signal phase nor to lesions of the suprachiasmatic nuclei. *Biology of Reproduction, 43*, 174–182.

McIntyre, I. M., Norman, T. R., Burrows, G. D., & Armstrong, S. M. (1992). Melatonin, cortisol, and prolactin responses to acute nocturnal light exposure in healthy volunteers. *Psychoneuroendocrinology, 17*, 243–248.

Melander, A., Ericson, L. E., & Sundler, F. (1974a). Sympathetic regulation of thyroid hormone secretion. *Life Sciences, 14*, 237–246.

Melander, A., Ericson, L. E., Sundler, F., & Ingbar, S. H. (1974b). Sympathetic innervation of the mouse thyroid and its significance in thyroid hormone secretion. *Endocrinology, 94*, 959–966.

Menaker, M., & Wisner, S. (1983). Temperature-compensated circadian clock in the pineal of *Anolis*. *Proceedings of the National Academy of Science, U.S.A., 80*, 6119–6121.

Meyer, C. C., Parker, D. M., & Salzen, E. A. (1976). Androgen-sensitive midbrain sites and visual attention in chicks. *Nature (London), 259*, 689–690.

Meyerhoff, W. L. (1976). The thyroid and audition. *Laryngoscope, 86(4)*, 483–489.

Micco, D. J., Meyer, J. S., & McEwen, B. S. (1980). Effects of corticosterone replacement on the temporal patterning of activity and sleep in adrenalectomized rats. *Brain Research, 200*, 206–212.

Miernicki, M., Karp, J. D., & Powers, J. B. (1990). Pinealectomy prevents short photoperiod inhibition of male hamster sexual behavior. *Physiology & Behavior, 47*, 293–299.

Mistlberger, R. E., & Rechtschaffen, A. (1984). Recovery of anticipatory activity to restricted feeding in rats with ventromedial hypothalamic lesions. *Physiology & Behavior, 33*, 227–235.

Mistlberger, R. E., & Rusak, B. (1986). Carbachol phase shifts circadian activity rhythms in ovariectomized rats. *Neuroscience Letters, 72*, 357–362.

Mistlberger, R. E., & Rusak, B. (1991). Food-anticipatory circadian rhythms in rats with paraventricular and lateral hypothalamic ablations. *Journal of Biological Rhythms, 6(3)*, 277–291.

Moberg, G. P., & Clark, C. R. (1976). Effect of adrenalectomy and dexamethasone treatment on circadian running in the rat. *Pharmacology, Biochemistry, & Behavior, 4*, 617–619.

Monteleone, P., Fuschino, A., Nolfe, G., & Maj, M. (1992). Temporal relationship between melatonin and cortisol responses to nighttime physical stress in humans. *Psychoneuroendocrinology, 17*, 81–86.

Moore, R. Y., & Eichler, V. B. (1972). Loss of circadian adrenal corticosterone rhythm following suprachiasmatic lesions in the rat. *Brain Research, 42*, 201–206.

Moore, R. Y., & Klein, D. C. (1974). Visual pathways and the central neural control of a circadian rhythm in pineal serotonin *N*-acetyltransferase activity. *Brain Research, 71*, 17–33.

Moore-Ede, M. C., & Sulzman, F. M. (1977). The physiological basis of circadian timekeeping in primates. *Physiologist, 20(3)*, 17–25.

Moore-Ede, M. C., Sulzman, F. M., & Fuller, C. A. (1982). *The clocks that time us*. Cambridge, Massachusetts: Harvard University Press.

Morgan, P. J., & Williams, L. M. (1989). Central melatonin receptors: Implications for a mode of action. *Experientia, 45*, 955–964.

Morimoto, Y., Arisue, K., & Yamamura, Y. (1977). Relationship between circadian rhythm of food intake and that of plasma corticosterone and effect of food restriction on circadian adrenocortical rhythm in the rat. *Neuroendocrinology, 23*, 212–222.

Morin, L. P. (1988). Propylthiouracil, but not other antithyroid treatments, lengthens hamster circadian period. *American Journal of Physiology, 255*, R1–R5.

Morin, L. P., & Cummings, L. A. (1981). Effect of surgical or photoperiodic castration, testosterone replacement, or pinealectomy on male hamster running rhythmicity. *Physiology & Behavior, 26*, 825–838.

Morin, L. P., & Cummings, L. A. (1982). Splitting of wheel-running rhythms by castrated or steroid treated male and female hamsters. *Physiology & Behavior, 29,* 665–675.

Morin, L. P., Fitzgerald, K. M., Rusak, B., & Zucker, I. (1977a). Circadian organization and neural mediation of hamster reproductive rhythms. *Psychoneuroendocrinology, 2,* 73–98.

Morin, L. P., Fitzgerald, K. M., & Zucker, I. (1977b). Estradiol shortens the period of hamster circadian rhythms. *Science, 196,* 305–307.

Morin, L. P., Gavin, M., & Ottenweller, J. E. (1986). Propylthiouracil causes phase delays and circadian period lengthening in male and female hamsters. *American Journal of Physiology, 250,* R151–R160.

Mrosovsky, N. (1988). Seasonal affect disorder, hibernation, and annual cycles in animals: Chipmunks in the sky. *Journal of Biological Rhythms, 3,* 189–207.

Muraki, T., Nakaki, T., & Kato, R. (1987). Alpha$_1$-adrenoreceptor production of inositol phosphates mediates the inhibition of thyroxine release from the mouse thyroid. *Journal of Endocrinology, 115,* 289–293.

Naber, D., Wirz-Justice, A., Kafka, M. S., & Wehr, T. A. (1980). Dopamine binding in rat striatum: Ultradian rhythm and its modification by chronic imipramine. *Psychopharmacology, 68,* 1–5.

Nagai, K., Yamakazi, K., Tsujimoto, H., Inoue, S., & Nakagawa, H. (1984). Meal feeding schedule and ventromedial hypothalamic lesions do not affect rhythm of pineal serotonin *N*-acetyltransferase activity in rats. *Life Sciences, 35(7),* 769–774.

Nicolau, G. Y., & Haus, E. (1992). Chronobiology of the hypothalamic-pituitary-thyroid axis. In Y. Touitou & E. Haus (Eds.) *Biologic rhythms in clinical and laboratory medicine.* Berlin: Springer-Verlag, 330–347.

Niles, L. P., Brown, G. M., & Grota, L. J. (1977). Effects of neutralization of circulating melatonin and *N*-acetylserotonin on plasma prolactin levels. *Neuroendocrinology, 23,* 14–22.

Niles, L. P., Wong, Y., Mishra, R. K., & Brown, G. M. (1979). Melatonin receptors in brain. *European Journal of Pharmacology, 55,* 219–220.

Norman, A. W., & Litwack, G. (1987). *Hormones* (pp. 221–262). Orlando, Florida: Academic Press.

Norris, D. O. (1985). *Vertebrate Endocrinology.* 2d ed. (pp. 394–403). Philadelphia: Lea & Febiger.

O'Connor, L. H., Morin, L. P., & Feder, H. H. (1985). A diurnal fluctuation in medial basal hypothalamic–preoptic area cytosol estrogen receptors in ovariectomized hamsters. *Brain Research, 347,* 376–380.

Ohta, M., Kadota, C., & Konishi, H. (1989). A role for melatonin in the initial stage of photoperiodism in the Japanese quail. *Biology of Reproduction, 40,* 935–941.

Ottenweller, J. E., & Hedge, G. A. (1981). Thyroid hormones are required for daily rhythms of plasma corticosterone and prolactin concentration. *Life Sciences, 28,* 1033–1040.

Ottenweller, J. E., & Meier, A. H. (1982). Adrenal innervation may be an extrapituitary mechanism able to regulate adrenocortical rhythmicity in rats. *Endocrinology, 111(4),* 1334–1337.

Paccotti, P., Terzolo, M., Piovesan, A., Torta, M., Vignani, A., & Angeli, A. (1988). Effects of exogenous melatonin on human pituitary and adrenal secretions: Hormonal responses to specific stimuli after acute administration of different doses at two opposite circadian stages in man. *Chronobiologia, 15(4),* 279–287.

Pant, K., & Chandola-Saklani, A. (1992). Pinealectomy and LL abolish circadian perching rhythms but did not alter circannual reproductive or fattening rhythms in finches. *Chronobiology International, 9,* 413–420.

Papousek, V. M. (1976). Temporal structures of sleep in endogenous depression. *Arzneimittel Forschung, 26,* 1062–1964.

Pedersen-Bjergaard, K., & Tonnesen, M. (1954). The effects of steroid hormones on muscular activity in rats. *Acta Endocrinology, 17,* 329–337.

Petiti, N., & Etgen, A. M. (1990). Alpha 1-adrenoreceptor augmentation of beta-stimulated cAMP formation is enhanced by estrogen and reduced by progesterone in rat hypothalamic slices. *Journal of Neuroscience, 10(8)*, 2842–2849.

Petrie, K., Conaglen, J. V., Thompson, L., & Chamberlain, K. (1989). Effect of melatonin on jet lag after long haul flights. *British Medical Journal, 289*, 705–707.

Pflug, B. (1976). Methodological problems of the clinical research on rhythms in depressive patients. *Arzneimittel Forschung, 26*, 1065–1968.

Pflug, B., & Tölle, R. (1971). Disturbance of the 24-hour rhythm in endogenous depression and the treatment of endogenous depression by sleep deprivation. *International Pharmacopsychiatry, 6*, 187–196.

Pittendrigh, C. S. (1981). Circadian systems: General perspective. In J. Aschoff (Ed), *Handbook of Behavioral Neurobiology 4: Biological Rhythms* (pp. 57–80). New York: Plenum Press.

Prange, A., Wilson, I., Lara, P., Alltop, L., & Breese, G. (1972). Effects of thyrotropin-releasing hormone in depression. *Lancet, 2*, 999–1002.

Prange, A., Wilson, I., Rabon, A., & Lipton, M. (1969). Enhancement of imipramine antidepressant activity by thyroid hormone. *American Journal of Psychiatry, 126*, 457–469.

Ralph, M. R., & Menaker, M. (1988). A mutation of the circadian system in golden hamster. *Science, 241*, 1225–1227.

Ramirez, V. D., & Sawyer, C. H. (1974). Differential dynamic responses of plasma LH and FSH to ovariectomy and to single injection of estrogen in the rat. *Endocrinology, 94(4)*, 987–993.

Redman, J., Armstrong, S. M., & Ng, K. T. (1983). Free-running activity rhythms in the rat: Entrainment by melatonin. *Science, 219*, 1089–1091.

Reebs, S. G., & Mrosovsky, N. (1989). Effects of induced wheel running on the circadian activity rhythms of Syrian hamsters: Entrainment and phase response curve. *Journal of Biological Rhythms, 4*, 39–48.

Refinetti, R., Rissman, E., & Menaker, M. (1991). Circadian organization of sex behavior in pairs of tau-mutant hamsters. *The FASEB Journal, 5*, Abstract 4378.

Reichlin, S. (1987). Basic research of hypothalamic-pituitary-adrenal neuroendocrinology: An overview. The physiological function of the stress response. In U. Halbreich (Ed.), *Hormones and depression* (pp. 21–30). New York: Raven Press.

Reppert, S. M., Weaver, D. R., Rivkees, S. A., & Stopa, E. G. (1988). Putative melatonin receptors in a human biological clock. *Science, 242*, 78–80.

Richter, C. P. (1933). The role played by the thyroid gland in the production of gross body activity. *Endocrinology, 17*, 73–87.

Richter, C. P. (1936). The spontaneous activity of adrenalectomized rats treated with replacement and other therapy. *Endocrinology, 20*, 657–666.

Richter, C. P. (1965). *Biological clocks in medicine and psychiatry*. Springfield, Illinois: Charles C. Thomas.

Robinson, J. P., & Kendall, D. A. (1990). The influence of adrenalectomy on α-adrenoreceptor responses in rat cerebral cortex slices. *Molecular Brain Research, 7*, 69–74.

Romeo, H. E., González Solveyra, C., Vacas, M. I., Rosenstein, R. E., Barontini, M. B., & Cardinali, D. P. (1986). Origins of sympathetic projections to the rat thyroid and parathyroid glands. *Journal of the Autonomic Nervous System, 17*, 63–70.

Romero, J. A. (1978). Biologic rhythms and sympathetic control of pineal metabolism. In H. V. Samis, Jr. & S. Capobianco (Eds.), *Aging and biological rhythms* (pp. 235–249). New York: Plenum Press.

Rondeel, J. M. M., De Greef, W. J., Van Der Schoot, P., Karels, B., Klootwijk, W., & Visser, T. J. (1988). Effect of thyroid status and paraventricular area lesions on the release of thyrotropin-releasing hormone and catecholamines into hypophysial portal blood. *Endocrinology, 123*, 523–527.

Rose, R. M. (1987). Endocrine abnormalities in depression and stress—An overview. In U. Halbreich (Ed.), *Hormones and depression* (pp. 31–47). New York: Raven Press.

Rosenwasser, A. M. (1989a). Effects of chronic clonidine administration and withdrawal on free-running circadian activity rhythms. *Pharmacology, Biochemistry, & Behavior, 33(2)*, 291–297.

Rosenwasser, A. M. (1989b). Free-running circadian activity rhythms during long-term clonidine administration in rats. *Pharmacology, Biochemistry, & Behavior, 35(1)*, 35–39.

Rowsemitt, C. N. (1989). Activity of castrated male voles: Rhythms of responses to testosterone replacement. *Physiology & Behavior, 45*, 7–13.

Roy, E. J., & Wilson, M. A. (1981). Diurnal rhythm of cytoplasmic estrogen receptors in the rat brain in the absence of circulating estrogens. *Science, 213*, 1525–1527.

Rusak, B., & Morin, L. P. (1976). Testicular responses to photoperiod are blocked by lesions of the suprachiasmatic nuclei in golden hamsters. *Biology of Reproduction, 15*, 366–374.

Sachar, E. J., & Baron, M. (1979). The biology of affective disorders. In W. M. Cowan, Z. W. Hall, & E. R. Kandel (Eds.), *Annual review of neuroscience* (Vol. 2) (pp. 505–518). Palo Alto: Annual Reviews.

Sachar, E. J., Hellman, L., Roffwarg, H. P., Halpern, F. S., Fukushima, D. K., & Gallagher, T. F. (1973). Disrupted 24-hour patterns of cortisol secretion in psychotic depression. *Archives of General Psychiatry, 28*, 19–24.

Sack, R. L., Lewy, A. J., Blood, M. L., Stevenson, J., & Keith, L. D. (1991). Melatonin administration to blind people: Phase advances and entrainment. *Journal of Biological Rhythms, 6(3)*, 249–261.

Samel, A., Wegmann, H. M., Vejvoda, M., Maab, H., Gundel, A., & Schütz, M. (1991). Influence of melatonin treatment on human circadian rhythmicity before and after a simulated 9-hr time shift. *Journal of Biological Rhythms, 6(3)*, 235–248.

Sato, T. (1990). Histochemical demonstration of NADPH-diaphorase activity in the pineal organ of the frog (*Rana esculenta*) but not in the pineal organ of the rat. *Archives of Histology and Cytology, 53*, 141–146.

Scalera, G., Banassi, C., & Porro, C. A. (1990). Pineal involvement in the alimentary behavior and taste preferences in the rat. *Physiology & Behavior, 48(1)*, 97–101.

Schull, J., McEachron, D. L., Adler, N. T., Fiedler, L., Horvitz, J., Noyes, A., Olson, M., & Shack, J. (1988). Effects of thyroidectomy, parathyroidectomy and lithium on circadian wheel-running in rats. *Physiology & Behavior, 42*, 33–39.

Schull, J., Walker, J., Fitzgerald, K., Hilivirta, L., Ruckdeschel, J., Schumacher, D., Stanger, D., & McEachron, D. L. (1989). Effects of sex, thyro-parathyroidectomy, and light regimes on levels and circadian rhythms of wheel-running in rats. *Physiology & Behavior, 46*, 341–346.

Schwarcz, G., Halaris, A., Baxter, L., Escobar, J., Thompson, M., & Young, M. (1984). Normal thyroid function in desipramine nonresponders converted to responders by the addition of L-triiodo-thyronine. *American Journal of Psychiatry, 141*, 1614–1616.

Schwartz, W. J. (1991). SCN metabolic activity *in vivo*. In D. C. Klein, R. Y. Moore, & S. M. Reppert (Eds.), *Suprachiasmatic nucleus: The mind's clock* (pp. 144–156). Oxford: Oxford University Press.

Segerson, T. P., Kauer, J., Wolfe, H. C., Mobtaker, H., Wu, P., Jackson, I. M., & Lechan, R. M. (1987). Thyroid hormone regulates TRH biosynthesis in the paraventricular nucleus of the rat hypothalamus. *Science, 238*, 78–80.

Semm, P., & Demaine, C. (1983). Electrical responses to direct and indirect photic stimulation of the pineal gland in the pigeon. *Journal of Neural Transmission, 58*, 281–289.

Shi, H., Furr, H. C., & Olson, J. A. (1991). Retinoids and carotenoids in bovine pineal gland. *Brain Research Bulletin, 26*, 235–239.

Shiotsuka, R., Jovonovich, J., & Jovonovich, J. A. (1974). *In vitro* data on drug sensitivity: Circadian and ultradian corticosterone rhythms in adrenal organ cultures. In *Chronobiological aspects of endocrinology.* J. Aschoff, F. Ceresa, & F. Halberg (Eds.). F. K. Schattauer Verlag, Stuttgart 255–267.

Shiraishi, I., Homma, K.-I., Honma, S., & Hiroshige, T. (1984). Ethosecretogram: Relation of behavior to plasma corticosterone in freely moving rats. *American Journal of Physiology, 247,* R40–R45.

Sitaram, N., Gillin, J. C., & Bunney, W. E., Jr. (1978). Circadian variation in the time of "switch" of a patient with 48-hour manic–depressive cycles. *Biological Psychiatry, 13,* 567–574.

Slotkin, T. A., & Sleptis, R. J. (1984). Obligatory role of thyroid hormone in the development of peripheral sympathetic and central catecholaminergic neurons: Effect of propylthiouracil-induced hypothyroidism on transmitter levels, turnover, and release. *Journal of Pharmacology and Experimental Therapeutics, 230,* 53–61.

Spencer, F., Shirer, H. W., & Yochim, J. M. (1976). Core temperature in the female rat: Effect of pinealectomy or altered lighting. *American Journal of Physiology, 231(2),* 355–360.

Stancer, H., & Persad, E. (1982). Treatment of intractable rapid-cycling manic-depressive disorder with levothyroxine. *Archives of General Psychiatry, 39,* 311–312.

Stehle, J., Vanecek, J., & Vollrath, L. (1989). Effects of melatonin on spontaneous electrical activity of neurons in rat suprachiasmatic nuclei: An *in vitro* ionophoretic study. *Journal of Neural Transmission, 78(2),* 173–177.

Stephan, F. K., Swann, J. M., & Sisk, C. L. (1979). Anticipation of 24-hr feeding schedules in rats with lesions of the suprachiasmatic nucleus. *Behavior and Neural Biology, 25,* 346–363.

Sterling, K., & Lazarus, J. H. (1977). The thyroid and its control. *Annual Review of Physiology, 39,* 349–371.

Stern, J. J. (1970). The effects of thyroidectomy on the wheel-running activity of female rats. *Physiology and Behavior, 5,* 1277–1279.

Stetson, M. H. (1978). Circadian organization and female reproductive cyclicity. In H. V. Samis, Jr. (Ed.), *Aging and biological rhythms* (pp. 251–274). New York: Plenum Press.

Stetson, M. H., & Hamilton, B. (1981). A comparison of the effects of short photoperiod and daily melatonin injection on the induction and termination of ovarian acyclicity. *Journal of Experimental Zoology, 215,* 173–178.

Stone, E. A. (1987). Central cyclic-AMP-linked noradrenergic receptors: New findings on properties as related to the actions of stress. *Neuroscience Biobehavioral Reviews, 11(4),* 391–398.

Stone, E. A., McEwen, B. S., Herrera, A. S., & Carr, K. D. (1987). Regulation of α and β components of noradrenergic cyclic AMP response in cortical slices. *European Journal of Physiology, 141,* 347–356.

Stone, E. A., Mitsuo, & Colbjornsen, C. M. (1989). Catecholamine-induced desensitization of brain beta adrenoreceptors *in vivo* and reversal by corticosterone. *Life Sciences, 44,* 209–213.

Strassman, R. J., Peake, G. T., Quails, C. R., & Lisansky, E. J. (1988). Lack of an acute modulatory effect of melatonin on human nocturnal thyrotropin and cortisol secretion. *Neuroendocrinology, 48(4),* 387–393.

Sulzman, F. M., Fuller, C. A., & Moore-Ede, M. C. (1978). Extent of circadian synchronization by cortisol in the squirrel monkey. *Comparative Biochemistry and Physiology, 59A,* 279–283.

Swann, A. C. (1988). Thyroid hormone and norepinephrine: Effects on alpha-2, beta, and reuptake sites in cerebral cortex and heart. *Journal of Neural Transmission, 71,* 195–205.

Szafarczyk, A., Alonso, G., Ixart, G., Malaval, F., Nouguier-Soule, J., & Assenmacher, I. (1980). Serotonergic system and circadian rhythms in plasma ACTH, plasma corticosterone, and motor activity. *American Journal of Physiology, 239,* E482–E489.

Szafarczyk, A., Ixart, G., Alonso, G., Malaval, F., Nouguier-Soule, J., & Assenmacher, I. (1981). Neural control of circadian rhythms in plasma ACTH, plasma corticosterone, and motor activity. *Journal of Physiology (Paris), 77,* 969–976.

Szafarczyk, A., Ixart, G., Malaval, F., Nouguier-Soule, J., & Assenmacher, I. (1979). Effects of lesions of the suprachiasmatic nuclei and of p-chlorophenylanine on the circadian rhythms of adrenocorticotropic hormone and corticosterone in the plasma and on the locomotor activity of rats. *Journal of Endocrinology, 83*, 1–6.

Takahashi, J. S., & Menaker, M. (1980). Interaction of estradiol and progesterone: Effects on circadian locomotor rhythm of female golden hamsters. *American Journal of Physiology, 239*, R497–R504.

Takahashi, J. S., & Menaker, M. (1984). Multiple redundant circadian oscillators within the isolated avian pineal gland. *Journal of Comparative Physiology A, 154*, 435–440.

Takebe, K., Sakakura, M., & Mashimo, K. (1972). Continuance of diurnal rhythmicity of CRF activity in hypophysectomized rats. *Endocrinology, 90(6)*, 1515–1520.

Tamotsu, S., & Morita, Y. (1986). Photoreception in pineal organs of larval and adult lampreys, *Lampetra japonica*. *Journal of Comparative Physiology A, 159(1)*, 1–5.

Tarzolo, M., Plovesan, Puligheddu, B., Torta, M., Osella, G., Paccotti, P., & Angeli, A. (1990). Effects of long-term, low-dose, time-specified melatonin administration on endocrine and cardiovascular variables in adult men. *Journal of Pineal Research, 9*, 113–124.

Taub, J. M., & Berger, J. F. (1974a). Acute shifts in the sleep-wakefulness cycle: Effects on performance and mood. *Psychosomatic Medicine, 36*, 164–173.

Taub, J. M., & Berger, J. F. (1974b). Effects of acute shifts in circadian rhythms of sleep and wakefulness on performance and mood. In L. Scheving, F. Halberg, & J. Pauley (Eds.), *Chronobiology*. Tokyo, Japan, Elgaku Shoin. 571–575.

Taylor, T., Wondisford, F. E., Blaine, T., & Weintraub, B. D. (1990). The paraventricular nucleus of the hypothalamus has a major role in thyroid hormone feedback regulation of thyrotropin synthesis and secretion. *Endocrinology, 126*, 317–324.

Templer, D. I., Ruff, C. F., Ayers, J. L., & Beshai, J. A. (1982). Diurnal mood fluctuation and age. *International Journal of Aging and Human Development, 14*, 189–193.

Terkel, J., Johnson, J. H., Whitmoyer, D. I., & Sawyer, C. H. (1974). Effect of adrenalectomy on a diurnal (circadian) rhythm in hypothalamic multiple unit activity in the female rat. *Neuroendocrinology, 14*, 103–113.

Terman, M. (1988). On the question of mechanism in phototherapy for seasonal affective disorder: Considerations of clinical efficacy and epidemiology. *Journal of Biological Rhythms, 3*, 155–172.

Thomas, E. M. V., & Armstrong, S. M. (1988). Melatonin administration entrains female rat activity rhythms in constant darkness but not in constant light. *American Journal of Physiology, 255*, R237–R242.

Thompson, M. A., Woolley, D. E., Gietzen, D. W., & Conway, S. (1983). Catecholamine synthesis inhibitors acutely modulate [^{2}H]estradiol binding by specific brain areas and pituitary in ovariectomized rats. *Endocrinology, 113*, 855–865.

Thorpe, P., & Herbert, J. (1976). Studies on the duration of the breeding season and photorefractoriness in female ferrets pinealectomized or treated with melatonin. *Journal of Endocrinology, 70*, 255–262.

Timiras, P. S., & Nzekwe, E. U. (1989). Thyroid hormone and nervous system development. *Biology of Neonates, 55*, 376–385.

Turek, F. W. (1989). Effects of stimulated physical activity on the circadian pacemaker of vertebrates. *Journal of Biological Rhythms, 4*, 135–147.

Turek, F., & Gwinner, E. (1982). Role of hormones in the circadian organization of vertebrates. In J. Aschoff, S. Daan, & G. Groos (Eds.), *Vertebrate circadian systems*. Berlin: Springer-Verlag.

Ueck, M., & Wake, K. (1977). The pinealocyte—A paraneuron? A review. *Archives of Histology (Japan) (Suppl.), 40*, 261–278.

Underwood, H. (1977). Circadian organization in lizards: The role of the pineal organ. *Science, 195*, 587–589.

Underwood, H. (1981). Circadian organization in the lizard *Sceloporus occidentalis*: The effects of pinealectomy, blinding and melatonin. *Journal of Comparative Physiology, 141*, 537–547.

Underwood, H. (1983). Circadian organization in the lizard *Anolis carolinesis*: A multioscillator system. *Journal of Comparative Physiology A, 152*, 265–274.

Underwood, H. (1989). The pineal and melatonin: Regulators of circadian function in lower vertebrates. *Experientia, 45*, 914–921.

Underwood, H., & Groos, G. (1982). Vertebrate circadian rhythms: Retinal and extraretinal photoreception. *Experientia, 38*, 1013–1021.

Underwood, H., Barrett, R. K., & Siopes, T. (1990). Melatonin does not link the eyes to the rest of the circadian system in Quail: A neural pathway is involved. *Journal of Biological Rhythms, 5*, 349–361.

Ungar, F., & Halberg, F. (1963). *In vitro* exploration of a circadian rhythm in adrenocorticotropic activity of C mouse hypophysis. *Experientia, 19*, 158–160.

Van Veen, T., Östholm, T., Gierschik, P., Spiegel, A., Somers, R., Korf, H. W., & Klein, D. C. (1986). α-Transducin immunoreactivity in retinae and sensory pineal organs of adult vertebrates. *Proceedings of the National Academy of Science, U.S.A., 83*, 912–916.

Vanecek, J. (1988). Melatonin binding sites. *Journal of Neurochemistry, 51*, 1436–1440.

Vanecek, J., Pavlik, A., & Illnerova, H. (1987). Hypothalamic melatonin receptor sites revealed by autoradiography. *Brain Research, 435*, 359–363.

Vessotskie, J. M. (1992). The effect of hypothyroidism on adrenergic receptor density in male rats. Master's Thesis, Drexel University.

Vessotskie, J. M., McGonigle, P., Molthen, R. C. & McEachron, D. L. Thyroid and thyroxine effects on adrenoreceptors in relation to circadian activity. *Pharmacology, Biochemistry, & Behavior*, in press.

Viticchi, C., Grinta, R., & Piantanneli, L. (1990). Influence of age on the thyroid hormone-induced up-regulation of β-adrenoreceptors in mouse brain cortex. *Gerontology, 36*, 286–292.

Wahlstrom, G. (1967). Changes induced in the self-selected circadian rhythm of the canary by triiodothyronine and [131]I. *Acta Pharmacologica Toxicologica (Suppl. 4), 25*, 72.

Wahlstrom, G. (1969). Effects of radiothyroidectomy by iodine on the self-selected circadian rhythms of rest and activity in the canary. *Acta Societa Medica Upsaliensis, 74*, 161–178.

Wainwright, S. D., & Wainwright, L. K. (1979). Chick pineal serotonin acetyltransferase: A diurnal cycle maintained *in vitro* and its regulations by light. *Canadian Journal of Biochemistry, 57*, 700–709.

Ward, M. M., Stone, S. C., & Sandman, C. A. (1978). Visual perception in women during the menstrual cycle. *Physiology & Behavior, 20(3)*, 239–243.

Weaver, D. R., Namboodiri, M. A. A., & Reppert, S. M. (1986). Iodinated melatonin mimics melatonin action and reveals discrete binding sites in fetal brain. *FEBS Letters, 228*, 123–127.

Weaver, D. R., Rivkees, S. A., & Reppert, S. M. (1988). Autoradiographic localization of iodomelatonin binding sites in the hypothalamus of three rodent species. *Endocrinology Society Abstracts, 177*, 265.

Wehr, T. A., Wirz-Justice, A., Goodwin, F. K., Duncan, W., & Gillin, J. C. (1979). Phase advance of circadian sleep-wake cycle as an antidepressant. *Science, 206*, 710–713.

Weiland, N. G., & Wise, P. M. (1989). Diurnal rhythmicity of beta-1- and beta-2-adrenergic receptors in ovariectomized, ovariectomized estradiol-treated and proestrous rats. *Neuroendocrinology, 50*, 655–662.

West, S. H., & Bassett, J. R. (1992). A circadian rhythm in adrenal responsiveness to ACTH is not confirmed by in vitro studies. *Acta Endocrinologica, 126*, 363–368.

Wever, R. A. (1979). *The circadian system of man: Results of experiments under temporal isolation.* Berlin: Springer-Verlag.

Wever, R. A. (1983). Fractional dysynchronization of human circadian rhythms: A method for evaluating entrainment limits and functional interdependencies. *Pflugers Archives, 396*, 128–137.

Wever, R. A. (1986). Characteristics of circadian rhythms in human function. *Journal of Neural Transmission [Supplement], 21*, 323–373.

Wever, R. A. (1989). Light effects on human circadian rhythms: A review of recent andechs experiments. *Journal of Biological Rhythms, 4(2)*, 161–185.

Wheatley, D. (1972). Potentiation of amitriptyline by thyroid hormone. *Archives of General Psychiatry, 26*, 229–233.

Whitlock, F. A. (1982). *Symptomatic affective disorders.* New York: Academic Press.

Whybrow, P. (1986). Hypothyroidism: Behavioral and psychiatric aspects. In S. Ingbar & L. Braveman (Eds.), *Werner's The thyroid*, 5th ed. (pp. 967–973). Philadelphia: Lippincott.

Whybrow, P., Coppen, A., Prange, A., Noguera, A., & Bailey, J. (1972). Thyroid function and the response to liothyronine in depression. *Archives of General Psychiatry, 26*, 242–245.

Whybrow, P., & Prange, A. (1981). A hypothesis of thyroid–catecholamine–receptor interactions: Its relevance to affective illness. *Archives of General Psychiatry, 38*, 106–113.

Widmaier, E. P., & Campell, C. S. (1980). Interaction of estradiol and photoperiod on activity patterns in the female hamster. *Physiology & Behavior, 24*, 923–930.

Wilkinson, C. W., Shinsako, J., & Dallman, M. F. (1979). Daily rhythms in adrenal responsiveness to adrenocorticotropin are determined primarily by the time of feeding in the rat. *Endocrinology, 104(2)*, 350–359.

Wilkinson, C. W., Shinsako, J., & Dallman, M. F. (1981). Return of pituitary–adrenal function after adrenal enucleation or transplantation: Diurnal rhythms and responses to ether. *Endocrinology, 109(1)*, 162–169.

Willner, P. (1985). *Depression: A psychobiological synthesis.* New York: John Wiley & Sons.

Wilson, I., Prange, A., McClane, T., Rabon, A., & Lipton, M. (1970). Thyroid-hormone enhancement of imipramine in nonretarded depression. *New England Journal of Medicine, 282*, 1063–1067.

Wilson, M. A., Clark, A. S., Clyde, V., & Roy, E. J. (1983). Characterization of a pineal-independent diurnal rhythm in neural estrogen receptors and its possible behavioral consequences. *Neuroendocrinology, 37*, 14 22.

Wirz-Justice, A., & Campbell, I. C. (1982). Antidepressant drugs can slow or dissociate circadian rhythms. *Experientia, 38*, 1301–1309.

Wirz-Justice, A., Kafka, M. S., Naber, D., Campbell, I. C., Marangos, P. J., Tamarkin, L., & Wehr, T. (1982). Clorgyline delays the phase-position of circadian neurotransmitter receptor rhythms. *Brain Research, 241*, 115–122.

Wirz-Justice, A., Wehr, T. A., Goodwin, F. K., Kafka, M. S., Naber, D., Marangos, P. J., & Campbell, I. C. (1980). Antidepressant drugs slow circadian rhythms in behavior and brain neurotransmitter receptors. *Psychopharmacology Bulletin, 16*, 45–47.

Wollnik, F., & Dohler, K-D. (1986). Effects of adult or perinatal hormonal environments on ultradian rhythms in locomotor activity of laboratory LEW/Ztm rats. *Physiology & Behavior, 38*, 229–240.

Wollnik, F., & Turek, F. W. (1988). Estrous correlated modulations of circadian and ultradian wheel-running activity rhythms in LEW/Ztm rats. *Physiology & Behavior, 43*, 389–396.

Yamada, M., Rogers, D., & Wilber, J. F. (1989). Exogenous triiodothyrine lowers thyrotropin-releasing hormone concentrations in the specific hypothalamic nucleus (paraventricular) involved in thyrotropin regulation and also in posterior nucleus. *Neuroendocrinology, 50*, 560–563.

Yen, S. S. C. (1991). Hypothalamic gonadotropin-releasing hormone: Basic and clinical aspects. In M. Motta (Ed.), *Brain endocrinology*, 2d ed. (pp. 245–280). New York: Raven Press.

Yovanof, S., & Feng, A. S. (1983). Effects of estradiol on auditory evoked responses from the frog's auditory midbrain. *Neuroscience Letters, 36*, 291–297.

Zimmerman, N., and Menaker, M. (1979). The pineal gland: A pacemaker within the circadian system of the house sparrow. *Proceedings of the National Academy of Sciences U.S.A., 76*, 999–1003.

CHAPTER 11

Bright Light Therapy and Circadian Neuroendocrine Function in Seasonal Affective Disorder

David Avery and Kitty Dahl

Harborview Medical Center
Department of Psychiatry and Behavioral Sciences
University of Washington School of Medicine
Seattle, Washington 98104

Curt Richter, to whom this book is dedicated, had a long-standing interest in biological rhythms as evidenced by his book, *Biological Clocks in Medicine and Psychiatry,* in which he presented his "shock phase" hypothesis (Richter, 1965). Richter proposed that "under certain circumstances the synchronized units of an organ or center may be desynchronized by various forms of interferences— or spontaneously . . . " and that "the units may temporarily or permanently all be thrown into phase by a severe shock, trauma, or other factors." Although the hypothesis was not expressed in contemporary chronobiological language, Richter's general idea that some psychiatric and medical problems are caused by desynchronized rhythms has gained considerable support. This chapter explores the possibility that a desynchrony of the neuroendocrine 24-hr clock and sleep can cause seasonal affective disorder (SAD) and that bright light effectively treats SAD by resynchronizing these processes.

Both neuroendocrine function and behavior are fundamentally circadian. Hormone levels and behavior seem to be programmed genetically to anticipate encountering a light–dark cycle of the 24-hr world that is timed by the spinning of the earth in relation to the sun. Further, many animals appear to be programmed to anticipate the changing light–dark cycle and temperature changes across the seasons of the year. Is SAD associated with circadian or seasonal abnormalities of neuroendocrine function? This chapter reviews clinical aspects of SAD and bright light therapy, circadian rhythms, circadian neuroendocrine function, and circadian neuroendocrine function in SAD. We hypothesize that abnormalities of the timing of neuroendocrine function may account for abnormalities of behavior seen in SAD.

I. Seasonal Affective Disorder

Seasonal affective disorder (winter depression) is a type of depression that recurs every fall or winter and remits in the spring or summer. In psychiatry, depression is considered more than simple sadness; depression is characterized not only by low mood but also by alteration of sleep, appetite, energy, concentration, or self-esteem and interferes with the person's life. Neuroanatomically, depression has been associated with diencephalic and, more specifically, hypothalamic dysfunction (Akiskal & McKinney, 1975).

Although SAD has been noted by various clinicians for over 2000 years, Rosenthal and colleagues (Rosenthal et al., 1984; Rosenthal & Wehr, 1987) were the first to study the disorder systematically. Typically SAD patients have the onset of their symptoms in the fall and remit in the spring. Unlike the classic endogenous melancholic patients who have insomnia with early morning awakening, decreased appetite, and weight loss, the SAD patients usually have hypersomnia, increased eating, and weight gain. Some patients may sleep as long as 16 hr per day, but more typically the sleep duration is 1–3 hr longer than in the summer. Other patients do not have schedules that allow them to oversleep in the morning and report a normal sleep duration with marked difficulty waking up in the morning during the winter, compared with waking up easily in the summer. Excessive daytime drowsiness, particularly in the late afternoon, and napping are common. Even in the hypersomnic patients, the quality of sleep is described as poor; some may have intermittent awakening. Some individuals have difficulty getting to sleep that, combined with the hypersomnia, will rotate their sleep pattern in a "clockwise" fashion. Rare patients will have insomnia with clear-cut early morning awakening.

Increased eating usually is expressed as increased appetite or hunger, but some patients describe eating more to feel less depressed, less anxious, or warmer. The seasonal weight gain may be minimal or sometimes as great as 40 or 50 pounds and usually takes place in the fall. The gained weight may be lost in the subsequent spring and summer, but some individuals may lose only part of the weight, resulting in a more chronic weight problem. Other less specific symptoms that are seen in nonseasonal type of depression are also seen in SAD, for example: low mood, low energy, poor concentration, irritability, anxiety, hopelessness, worthlessness, guilt, feeling too cool or too cold, and suicidal ideation. SAD is usually a less severe depression than the melancholic type.

Some emerging data suggest that the classic endogenous melancholic depression also may have a significant seasonal component with a spring–summer peak in incidence (Wehr, Sack, & Rosenthal, 1987; Wehr & Rosenthal, 1989). Several studies show a peak suicide rate in the northern hemisphere in May or June. The peak rates of admission for endogenous depression and electroconvulsive therapy use occur in the spring (Eastwood & Stiasny, 1978). In this chapter, the focus is on the better-described fall–winter type of SAD.

Patients describe that their symptoms usually begin in their early twenties or late teens, but about 9% report their symptoms to have begun before age 11 (Rosenthal et al., 1986). Women appear to be affected more frequently than men. Early studies, using recruitment through newspaper advertisements, found over 80% of the sample were women (Rosenthal & Wehr, 1987). Subsequent epidemiologic studies of populations using random telephone screening estimates that the ratio of women to men was about 3.5:1 (Kasper, Wehr, Bartko, Gaist, & Rosenthal, 1989). Of depressed outpatients with recurrent depression, about 16% may have a seasonal fall–winter pattern (Thase, 1989). Among menstruating women with SAD, premenstrual syndrome is common and often minimal in the summer.

Although studies clearly have documented strong genetic factors for depression in general (Gershon, 1990), no genetic studies have been done specifically concerning SAD. However, SAD patients have an increased incidence of first degree blood relatives with depression and alcoholism. SAD subjects report a greater incidence of alcoholism among their first degree blood relatives than do individuals in nondepressed control groups (Norden, Avery, & Dahl, 1990). Comparing seasonal with nonseasonal depressed patients, Allen and co-workers (Allen, Lam, Remick, & Sadovnick, 1993) found a greater family history of alcoholism among the SAD patients. However, in a similar comparison, Garvey and colleagues (Garvey, Wesner, & Godes, 1988) found no difference in the family history of alcoholism.

SAD patients often seem to self-medicate with drugs. Many use caffeine to try to stay alert during the day and increase their dose to the point of caffeinism and its attendant symptoms of anxiety, tremor, and poor quality of sleep. Some individuals use cigarettes and alcohol to lift their mood in the winter. Even seasonal cocaine abuse has been reported. The drugs not only aggravate the problems of the patients but also cloud the differential diagnosis.

Epidemiology studies have documented that the incidence of SAD increases with latitude (Potkin, Zetin, Stamenkovic, Kripke, & Bunney, 1986; Rosen et al., 1990). In New Hampshire and Alaska the incidence of SAD is about 10%; another 10–20% of the population may have more mild symptoms that do not fulfill criteria for a major depression. In Florida, SAD is uncommon (about 1%). SAD patients frequently report feeling better when they travel closer to the equator. In the Southern hemisphere, SAD symptoms peak in June and July (Boyce & Parker, 1988).

II. Light Treatment of Seasonal Affective Disorder

A long history exists of scattered reports of clinicians who felt that light might play a therapeutic role in treating mood problems. For example, about 200 years ago, the French psychiatrist Esquirol recommended a trip to Italy for one of his patients with winter depression (Wehr, 1989b). Lewy and colleagues (Lewy, Kern, Rosenthal, & Wehr, 1982) treated a patient with winter depression with bright light (2000 lux) therapy. However, Rosenthal and co-workers (Rosenthal et al., 1984) published the first controlled study of bright light therapy. In any study of depression, a control condition is important because of the high placebo response rates seen in studies of depression (Eastman, 1990). If improvement is associated with administration of a treatment, the improvement may be related to expectation or other factors. For example, the natural history of depression is episodic: most depressed patients eventually will recover without treatment. Rosenthal found that 1 week of bright light (2500 lux) administered 3 hr before dawn and 3 hr after dusk was clearly more effective in relieving winter depression than dim light (100 lux) administered for the same time and duration (Rosenthal et al., 1984). Subsequent studies have documented clearly that bright light therapy is superior to a dim light control condition (Terman et al., 1989a). Recently, a 10,000-lux light treatment for 0.5 hr was found superior to 2500 lux for 0.5 hr and similar to 2500 lux for 2 hr (J. Terman et al., 1990). Thus, both light illuminance and duration are important.

The illuminance required for effective treatment should be viewed relative to indoor and outdoor illuminance. A bright sunny day will have an illuminance of about 50,000–100,000 lux; a cloudy day will show 1000–10,000 lux. However, even bright, well-lit offices omit only about 600 lux; the presence of computers has reduced that illuminance to about 300 lux. The average home emits about 100 lux. To be indoors is to be in darkness. Even bright light therapy is barely comparable to the illuminance of a cloudy day. Staying indoors during the spring and summer may create an "eternal winter." Some depressed patients with hypersomnia, increased appetite, and weight gain and a chronic, nonseasonal pattern have histories of staying indoors and may benefit from either bright light therapy or going outside, particularly in the spring and summer.

Some SAD patients report being symptomatic during successive cloudy days in the summer, despite the long photoperiod. Some SAD patients report relief of their symptoms during successive bright sunny days in the winter, despite the short photoperiod. Whether these behaviors are learned responses to the cloudiness and sunshine associated with the seasons or whether some individuals clearly have a better biological response to 100,000 lux than to 10,000 lux is not known. Aschoff found that the seasonal variation of suicide had a better correlation with the seasonal variation of light intensity than with the seasonal variation of photoperiod (Aschoff, 1981).

The timing of the bright light therapy is important. Several investigators have found that 2 hr of bright light in the morning soon after awakening is superior to 2 hr of bright light in the evening (Lewy, Sack, Miller, & Toban, 1987; Avery et al., 1990a,b, 1991; Sack et al., 1990). The superiority of morning light also is confirmed by the meta-analysis of previous light therapy studies by Terman and co-workers (Terman et al., 1989a) which showed 51% ($N = 136$) responding to morning bright light compared with 38% ($N = 143$) responding to evening bright light. Dawn simulation, a lower intensity light (250 lux) that comes on gradually in the bedroom while the subject is asleep, has been found to be superior to a shorter dimmer control dawn (Avery et al., 1993). However, other studies have shown that the timing of light is not critical for a therapeutic response. A short 8-hr skeleton photoperiod was found equal to a long 15.5-hr skeleton photoperiod (Wehr et al., 1986). In addition, 2 hr of midday bright light were found to be as effective as 2 hr of bright light in the morning (Jacobsen, Wehr, Skwerer, Sack, & Rosenthal, 1987).

The wavelength of the light treatment also may be important. Oren and colleagues found green light superior to red light treatment (Oren et al., 1991a). Although some studies suggest that ultraviolet (UV) light which is present in "full spectrum" lights, might be important to the therapeutic response, good therapeutic responses have been achieved without UV light. In addition, UV light is absorbed by the cornea and lens in adults and generally does not reach the retina (M. Terman, Remé, Rafferty, Gallin, & Terman, 1990). Routine bright light therapy probably should minimize exposure to UV light since UV light exposure has been associated with skin cancer and cataracts.

Bright light therapy given only to the eyes is more effective than therapy given to the skin (with light to the eyes reduced by strong sunglasses) (Wehr, Skwerer, Jacobsen, Sack, & Rosenthal, 1987). Thus, the therapeutic effect of the light probably is mediated through the eyes. The paradox of the efficacy of dawn simulation, a treatment that occurs while the eyes are closed, may be explained by the translucence of the eyelids (Moseley, 1988) and the increased sensitivity of the retinas in the early morning hours (Bassi & Powers, 1986; O'Keefe & Baker, 1987). Among the possible mechanisms for the therapeutic effect of light is its effect on circadian rhythms.

III. Circadian Rhythms

Nearly all physiological functions have a 24-hr or circadian (circa = about, dian = day) rhythmicity. The circadian symphony includes not only hormones, but variables such as neurotransmitter levels, temperature, rapid eye movement (REM) sleep, enzyme activity, blood pressure, and DNA synthesis. Traditionally, the timing or "phase" of the rhythm has been assessed by mathematically finding the best fitting cosine function and noting the time at which the peak value of that cosine function occurs (the "acrophase"). The amplitude of the cosine function is half of the peak-to-trough value. The "mesor" denotes the average value of the fitted cosine function. Initially these rhythms were thought to be simply a direct response to the day–night cycle or the sleep–wake cycle; however, studies eliminating environmental influences and sleep show the existence of endogenous rhythms.

One procedure used to reveal endogenous circadian rhythms is a "constant routine" (Minors & Waterhouse, 1984). During a constant routine, the subject is kept awake, at bed rest, given small frequent feedings, and kept in constant dim light for at least 24 hr. Under these circumstances, the circadian rhythms of temperature, thyroid stimulating hormone (TSH), cortisol, and melatonin persist. Certainly activity, sleep, diet, and light may influence these variables. Activity will "mask" the underlining endogenous temperature rhythm by raising the temperature; sleep will mask the underlying endogenous rhythm by lowering the temperature below a bed rest baseline. Sleep will mask the underlying TSH and cortisol circadian rhythms by partially suppressing their secretions. Figure 1 shows the circadian rhythms for temperature and cortisol in healthy controls in a constant routine. One problem in assessing the structure and timing of the underlying

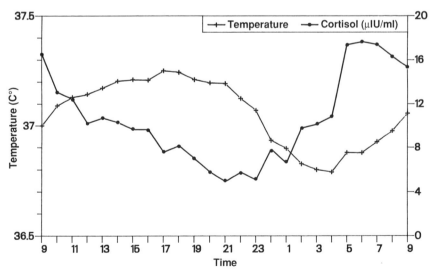

Figure 1. Cortisol and temperature rhythms in control subjects during a constant routine.

endogenous circadian rhythm in a subject outside of the laboratory is the influence of these masking effects; for example, a bout of activity may raise the temperature, causing a "peak" in the temperature that is unrelated to the endogenous rhythm. Thus, investigators have used the constant routine to unmask the structure of the endogenous rhythm and have used the time of the temperature minimum as an estimate of circadian phase (Czeisler et al., 1989).

Studies in time-cue free environments demonstrate that people will tend to "free run," going to bed and waking up about an hour later each day. Thus, each "day" in temporal isolation is about 25 hr (Czeisler, 1987). A plot of the temperature rhythm shows a period, or tau (the time from the beginning of one cycle to the beginning of the next), of about 25 hr. Initially, the sleep–wake cycle and the temperature rhythm are synchronized, the temperature minimum occurs during the middle of sleep. However, after 1–2 weeks, most subjects begin to experience "internal desynchronization," a state in which the period of the temperature continues to cycle with a period of about 25 hr while the sleep–wake cycle period is clearly different, for example, 35 hr or 19 hr. At least two processes are represented by the temperature rhythm on the one hand and the rest–activity process on the other. These processes have been referred to as the circadian (or C) process and the sleep (or S) process; the strong oscillator and the weak oscillator; the X pacemaker and the Y pacemaker; or simply the "clock" and sleep. Variables that are associated temporally with the circadian process include not only temperature but also cortisol, melatonin, TSH, REM sleep, and urinary K^+ and Na^+. The sleep process is associated with delta sleep and growth hormone secretion. The temporal association of variables may reflect a close functional interrelationship. For example, REM sleep is coupled closely with the temperature minimum and the rising phase of the temperature rhythms; heat conservation and thermogenesis are associated with individual REM periods (Avery, 1987).

The circadian and sleep process are both influenced by synchronizing influences or *zeitgebers* (in German, literally, time-givers). Zeitgebers for the sleep process include work schedules, social activity, and the late-night movie. Light appears to be the best-documented zeitgeber for the circadian process, although exercise may be able to shift the clock (Piercy & Lack, 1988). Numerous animal studies have documented that light administered at the end of subjective night will shift the circadian rhythm counterclockwise (a phase advance) whereas light at the beginning of subjective night will shift the rhythm clockwise (a phase delay). The phase response curve of humans has been documented to show similar characteristics (Czeisler et al., 1989). The phase response curve of humans shows that bright light occurring after the endogenous temperature minimum will phase-advance the rhythm, bright light occurring before the temperature minimum will delay the rhythm, and bright light occurring at midday will have little effect on the timing of the rhythm. Bright light administered at the temperature minimum initially will flatten the amplitude and, after a few days, will cause the minimum to emerge 180° from the original minimum. (see Figure 2). In animal studies, a low level of light early in the morning may have a phase-advancing effect that is comparable to a higher intensity light later in the morning (Pittendrigh, 1988). In rats, the action spectrum for entrainment of circadian rhythms is the same as the action spectrum for rhodopsin, suggesting that the rods of the retina, which are sensitive to dim light, can mediate entrainment (McGuire, Rand, & Wurtman, 1973).

The terms phase advance and phase delay not only refer to the direction of a rhythm shift but also can characterize the relationship of a rhythm to its normal phase. For example, normally, for a person sleeping from 10 P.M. to 6 A.M., the temperature minimum, which may be thought of as a "hand of the clock," usually occurs at about 3 A.M. or 4 A.M. If the temperature minimum occurs at 6 A.M. in a person with that sleep pattern, the temperature rhythm is said to be phase delayed relative to its normal timing with that sleep schedule. If, on the other hand, the temperature minimum occurs at 10 P.M., the temperature rhythm is phase advanced. Figure 3 is a schematic representation of a phase advance and a phase delay of the time of melatonin onset under dim light conditions (DLMO) relative to the normal DLMO phase positive in a person sleeping from 10 P.M. to 6 A.M.

Such types of desynchrony may have relevance to circadian disturbances seen in sleep disorders and depression. During internal desynchronization in the temporal isolation studies cited earlier, sleep onset will occur at different times relative to the temperature rhythm. If sleep onset occurs 4 or 5 hr prior to the temperature minimum (a normal phase relationship), the sleep duration will be 7–9 hr (Zulley, Wever, & Aschoff, 1981). If sleep onset occurs at or after the temperature minimum (the temperature rhythm is phase advanced relative to sleep onset), the sleep

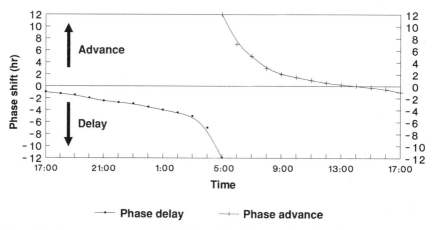

Figure 2. Phase response curve to 10,000 lux × 5 hr in humans. Data from Czeisler et al. (1989).

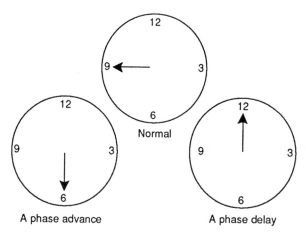

Figure 3.   Schematic representation of a normal phase of the dim light melatonin onset (DLMO) relative to sleep compared to a phase advance and a phase delay.

duration will be short, often 5 hr of less. If sleep onset occurs 8 hr before the temperature minimum (the temperature rhythm is phase delayed relative to sleep onset), the sleep duration will be long, often 9–13 hr. The rising phase of the circadian temperature rhythm is associated with a natural awakening. Natural awakenings rarely occur during the falling phase or at the temperature minimum.

   The attempts at subtyping depression usually have included subtypes that differ in the type of sleep disturbance present: insomnia with early morning awakening on the one hand or hypersomnia or lack of insomnia on the other. Endogenous, psychotic, melancholic, unipolar, familial depressive disorder, and summer depression subtypes all have been associated with insomnia with early morning awakening. Nonendogenous, reactive–neurotic, atypical, nonmelancholic, bipolar, depressive spectrum disorder, and winter depression all have been associated either with a lack of insomnia or hypersomnia. The similarity between the types of sleep disturbances seen in internal desynchronization and the sleep disturbances seen in subtypes of depression raises the hypotheses that the depressions associated with insomnia and early morning awakening are a result of a phase advance of the clock relative to sleep, whereas depressions associated with hypersomnia are a result of a phase delay.

IV. Circadian Neuroendocrine Function

Neuroanatomically, the suprachiasmatic nuclei of the hypothalamus appear to play an important role in the generation of circadian rhythms (Moore, 1990). If the suprachiasmatic nuclei of a rat with a short free-running tau are implanted into a rat with a long tau, the long-tau rat will assume a short free-running tau (Ralph, Foster, Davis, & Menaker, 1990). The tau of the circadian clock is, in part, under genetic control. A mutation in hamster is associated with a short free-running tau and has a pattern of inheritance that suggests a single autosomal locus (Ralph & Menaker, 1988). The retinas have direct connections with both the suprachiasmatic nuclei and the lateral geniculate nuclei. In addition, pulses of light that can shift rhythms will induce c-*fos,* a protooncogene, in the suprachiasmatic nuclei (Rusak, Robertson, Wisden, & Hunt, 1990). Protooncogenes activate synthesis of transcription factor proteins, setting in motion a cascade of events involving multiple gene loci. The hypothalamus plays a crucial role in neuroendocrine function, including the adrenal axis, the thyroid axis, melatonin production, and temperature regulation (see Figure 4). The suprachiasmatic nuclei have neural input to the paraventricular nuclei of the hypothalamus (Palkovits, 1987), where corticotropin releasing factor (CRF) and, possibly, thyroid releasing hormone (TRH) are synthesized (Swanson, Sawchenko, Lind, & Rho, 1987). This neuronal circuitry involving the suprachiasmatic nuclei probably plays an important functional role in the daily and seasonal survival of an animal.

A. Predictive Homeostasis

Moore-Ede (1986) has distinguished between reactive and predictive homeostasis. In classic reactive homeostasis, an organism preserves equilibrium by reacting to an inconstant world. For example, blood glucose is maintained homeostatically. Hypoglycemia triggers mechanisms that increase the blood glucose; hyperglycemia triggers mechanisms that lower glucose.

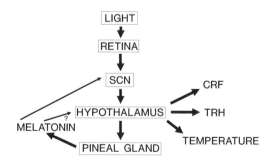

Figure 4. Schematic representation of the effect of light on the suprachiasmatic nucleus (SCN) and the subsequent effects on the hypothalamus, the pineal gland, the temperature rhythm, corticotropin releasing factor (CRF), and thyroid releasing hormone (TRH).

In contrast, in predictive homeostasis, the organism physiologically anticipates predictable environmental changes. From an evolutionary and functional perspective, the circadian organization can be seen as an example of predictive homeostasis. Through evolution, the organism has developed timing mechanisms that allow it to anticipate the predictable day–night changes of the environment and the associated changes in illumination, food availability, temperature, and so on (Moore-Ede, 1986). For example, a diurnal animal seems to anticipate daylight by increasing its core body temperature well before sunrise; a higher body and brain temperature may be optimal for daytime activity. Conversely, the temperature starts dropping before sunset; a lower core temperature may be optimal for sleep and associated energy conservation. The circadian neuroendocrine rhythms of cortisol, TSH, and melatonin are coupled closely to the circadian temperature rhythm and may, themselves, be examples of predictive homeostasis.

B. Adrenocortical Axis

Cortisol release from the adrenal glands is controlled primarily by the pituitary hormone adrenal corticotropic hormone (ACTH). ACTH in turn is controlled not only by feedback from cortisol but also by a hypothalamic factor, CRF. Cortisol, ACTH, and CRF all have documented circadian rhythms that appear to be influenced by output from the suprachiasmatic nucleus (Krieger, 1988). Sleep (Weitzman, Czeisler, & Moore-Ede, 1979) and melatonin (Rimijn, 1978) have suppressing effects on cortisol secretion. Under normal circumstances, cortisol levels start rising at about 3 A.M., reach a peak at 6 A.M., and gradually fall during the day, reaching a nadir in the first half of sleep. Cortisol has a clear circadian rhythm that appears to be coupled closely to the circadian rhythm of temperature; the major cortisol surge begins close to the temperature minimum. Figure 1 shows the cortisol rhythm in healthy controls in a constant routine. Cortisol secretion and temperature appear to be associated closely, even in subjects who are in temporal isolation and are internally desynchronized (Czeisler, 1978). Cortisol is important for energy balance and the regulation of body fat. Cortisol is crucial for gluconeogenesis (Bray, 1984). Patients with a deficiency of cortisol (Addison's disease) are plagued by marked fatigue and weight loss. The early morning surge of cortisol could be seen as an example of predictive homeostasis, an anticipation of the energy requirements of the waking state.

C. Thyroid Axis

TSH, a pituitary hormone, is stimulated by TRH secreted in the hypothalamus. TSH stimulates the synthesis and release of thyroxine (T_4) and triiodothyronine (T_3) from the thyroid gland. T_3 is thought to be the primary agent of thermogenesis within the thyroid axis. T_3, in turn, can exhibit feedback inhibition on TSH. TSH has a well-characterized circadian rhythm, the acrophase usually occurs at about 3 A.M. during sleep deprivation. Because sleep has a suppressing effect on TSH secretion, the peak in a sleeping person usually occurs about midnight and is followed by a lower plateau. The initial rise in TSH typically takes place in the early evening.

The hypothalamic–pituitary–thyroid axis plays a major role in thermogenesis. TSH increases acutely in response to a cold challenge in most mammalian studies; in humans, some studies show an increase whereas others show no change (Fregly, 1989). In humans chronically exposed to cold, the TSH response to exogenous TRH is increased (Reed et al., 1990). In addition, the production and clearance of T_3 are increased, suggesting increased T_3 receptor binding (Reed et al., 1990). In the United Kingdom, normal subjects had higher T_4 and T_3 serum levels during the winter months (Harrop, Ashwell, & Hopton, 1985). High temperatures during long photoperiods caused a marked reduction in serum T_3 in hamsters (Li et al., 1987).

The functional significance of the timing of the TSH circadian rhythm is unclear. Why would a hormone of the thermogenic thyroid system increase at night when the circadian temperature rhythm "thermostat" has a low setting? One possibility is that TSH reacts to changing heat loss over the 24-hr period. The TSH circadian rhythm with its nocturnal peak is related inversely to temperature. Under a heat stress, the 24-hr mesor of the TSH values decreases and the overall amplitude of the TSH rhythm heat decreases as associated with an increase in the average 24-hr rectal temperature rhythm and a decrease in the temperature amplitude (O'Malley, Richardson, Cook, Swart, & Rosenthal, 1984). The heat loss associated with the evening core body temperature drop may be a stimulus for the increase in TSH levels in the evening. With sleep deprivation, an increase in the amplitude of TSH is seen; the night-time peak increases (Sack, James, Rosenthal, & Wehr, 1988). Wehr (1990) has noted that sleep is analogous to a heat stimulus because sleep lowers the temperature set point. Removing this heat stimulus through sleep deprivation may be the stimulus for increased TSH secretion.

Another possibility is that a consistent time lag occurs between TSH secretion and the actual thermogenesis that results from the action of T_3. Assuming an efficient system in which the beginning of thermogenesis occurs at the beginning of the temperature rise, the time lag would be 6–7 hr (the time from the TSH rise at about 9 P.M. to the temperature minimum at 3 or 4 A.M.). Another related possibility is that the TSH rhythm reflects the rhythm of available T_3 at the receptors and that the negative feedback of high T_3 levels during the day suppresses TSH while low T_3 levels during the night allow TSH to increase. However, in humans, the T_3 and T_4 ratio in the serum is elevated at night compared with daytime values (Nimalasuriya, Spencer, Lin, Tse, & Nicoloff, 1986).

Although assuming a circadian thermogenic function for the thyroid axis and that thyroid-related thermogenesis occurs in preparation for awakening is tempting, the data available to date do not permit such a simple conclusion. Although the timing of the TSH rhythm may be an example of predictive homeostasis, this conclusion must remain speculative.

D. Catecholamines

Catecholamines play a major role in the response to cold stimulus (Landsberg & Young, 1984). Catecholamines stimulate increased thermogenesis by a beta receptor mechanism. Norepinephrine will facilitate conversion of T_4 to the metabolically active T_3. In addition, sympathetic outflow increases heat conservation by vasoconstriction (an alpha-adrenergic mechanism) and piloerection. Fasting suppresses sympathetic nervous system activity, whereas overeating stimulates the system.

Plasma norepinephrine has a 24-hr rhythm with a peak that usually occurs at about noon (Souetre et al., 1986). However, the sleep–wake cycle and posture have very significant effects on norepinephrine (Akerstedt & Gillberg, 1983). Thus, although norepinephrine appears to be an activating hormone and has a circadian rhythm, to what extent the norepinephrine rhythm is endogenous is unclear.

E. Melatonin

The secretion of melatonin by the pineal gland is controlled by an elaborate circuitry involving the retino–hypothalamic tracts, the suprachiasmatic nuclei, the lateral hypothalamus, the medial forebrain bundle, the tegmentum, the thoracic interomediolateral nuclei, and the superior cervical ganglion (Lewy, 1983). Beta-adrenergic stimulation of the pineal gland causes melatonin release. Light has an immediate suppressing effect on melatonin production and a sustained effect on the timing of melatonin secretion. Even under dim light conditions, melatonin will have an endogenous circadian rhythm, rising in the evening and falling in the morning. DLMO has been developed as a phase marker for the circadian clock by Lewy and co-workers (Lewy, Sack, & Singer, 1990).

Many functions have been attributed to the pineal gland, including Descartes' hypothesis that the pineal gland is the seat of the soul (Bhatnagar, 1990). In general, the pineal gland can be conceptualized as a tranquilizing organ (Romijn, 1978). Administration of melatonin induces sleep and synchronizes electroencephalogram (EEG) activity, whereas pinealectomy results in disruption of the normal circadian rhythm of REM sleep distribution. Subjectively, normal subjects report feeling more relaxed and increasingly drowsy with melatonin administration (Anton-Tay & Fernandez-Guardiola, 1971; Carman, Post, Buswell, & Goodwin, 1976; Lieberman, Waldhauser, Garfield, Lynch, & Wurtman, 1984). Thus, the timing of the melatonin rhythm may be an example of predictive homeostasis; the nocturnal melatonin secretion facilitates sleep.

Further, the effects of melatonin on temperature and cortisol are consistent with a sleep-inducing role for melatonin. The inverse relationship between the melatonin rhythm and the temperature rhythm (Shanahan & Czeisler, 1991) may not be coincidental. Oral administration of melatonin during the day lowers core body temperature in humans (Cagnacci & Elliott, 1991). Atenolol given at 6 P.M., which completely suppressed the evening rise of melatonin, also attenuated the evening decrement in core body temperature, blunting the decline by 0.35°C, 40% of the circadian temperature rhythm amplitude (Cagnacci & Elliott, 1991). Whether melatonin has a direct effect on the temperature setpoint or an indirect effect by acting through the suprachiasmatic nucleus is not known. Melatonin receptors have been found on the suprachiasmatic nuclei (Reppert, Weaver, Rivkees, & Stopa, 1988). Melatonin itself can shift the circadian clock (Lewy, Ahmed, Jackson, & Sack, 1991).

Several studies indicate that melatonin administration suppresses adrenocortical function and that pinealectomy results in increased levels of corticosteroid hormones (Romijn, 1978). The nocturnal elevation of melatonin is associated temporally with the nadir of the cortisol rhythm. The inverse relationship between melatonin and cortisol also is seen in patients with major depression. Patients with major depression who have dexamethasone nonsuppression (which is indicative of cortisol hypersecretion) are more likely to have low nocturnal melatonin levels than are subjects with normal dexamethasone suppression (Beck-Friis et al., 1985). Thus, the elevated melatonin levels at night may have a functional significance in allowing sleep by lowering temperature and suppressing adrenocortical function.

F. Melatonin and the Two Components of the Complex Circadian Oscillator

DLMO may be only a marker for part of a more complex circadian oscillator. Studies of other animals suggest that two coupled but separable oscillators constitute the "clock" (Pittendrigh & Daan, 1976a[b]). The activity patterns of several mammals and birds will "split" under certain experimental circumstances; rather than one bout of activity per day, the animals will have two distinct bouts of activity per day. Although the existence of two oscillators has not been proven to exist in humans, Illnerova and colleagues (Illnerova, Zvolsky, & Vanacet, 1985) have used assessment of N-acetyl transferase levels in the pineal glands of rats to describe the two oscillator system. N-Acetyl transferase is an enzyme for the rate-limiting step in the production of melatonin. The rise in N-acetyl transferase in the evening (the evening oscillator) can be separated from the decline in N-acetyl transferase in the morning (the morning oscillator). The changing photoperiod causes the phase difference in the two oscillators to change, and therefore controls total melatonin production. The short photoperiod in the winter produces an early rise and late fall in *N*-acetyl transferase, causing greater melatonin production. In the hamster, the long summer photoperiod causes a late rise and early fall, causing less melatonin production (see Figure 5). Melatonin suppresses testosterone production; thus, the lower melatonin levels in the summer allow greater testosterone production (Goldman & Elliott, 1987). The timing of the light pulses on this complex oscillator may have profound effects on sexual maturation (Elliot, Stetson, & Menaker, 1972). Thus, the modulation of a circadian rhythm produces a seasonal rhythm.

Some natural studies of humans have found a lengthening of melatonin duration during the winter compared with the summer (Beck-Friis, Von Rosen, Kjellman, Ljunggren, & Wetterberg, 1984; Kauppila, Kivela, Pakarinen, & Vakkuri, 1987), but others have found no change with the seasons (Illnerova et al., 1985; Broadway, Arendt, & Folkard, 1987). However, some data suggest that the two-component oscillator exists in humans (Wehr, 1991; Wehr et al., 1992). Normal controls exposed to an artificial short photoperiod have long durations of melatonin secretion, low nocturnal temperature, and cortisol suppression. After an artificial long photoperiod, they have short durations of melatonin secretion, low nocturnal temperature, and cortisol suppression. The "compression" of the melatonin secretion during the long photoperiod and an "expansion" during the short photoperiod in humans raises some questions. Does "splitting" of the circadian system occur in humans? Is extreme compression, extreme expansion, or splitting present in affective disorders? Extreme contraction of the circadian system could account for the low melatonin levels often but not always (Thompson, Franey, Arendt, & Checkley, 1988) seen in melancholic endogenous depression (Wetterberg et al., 1979; Wirz-Justice & Arendt, 1979; Nair, Hariharasubraman, & Pilapil, 1984; Beck-Friis et al., 1985; Brown et al., 1985). Is the tendency for endogenous depression to occur in the spring (Eastwood & Stiasny, 1978) related to the rapidly increasing photoperiod? Parry and colleagues (Parry et al., 1990) found that the duration of melatonin secretion was short in women with premenstrual depression. Mania, which may peak in the spring and summer (Carney, Fitzgeral, & Monaghan, 1989), might be associated with a compression. On the other hand, winter depression might be hypothesized to be associated with expansion of the circadian system, as will be discussed subsequently.

Predictive homeostasis appears to be at work in normal circadian neuroendocrine function. The increase in melatonin secretion and drop in core body temperature prior to sleep appear to be adaptive for sleep and calorie conservation during the night. The rise of cortisol and the increasing core body temperature prior to awakening can be thought of as facilitating daytime activity and energy expenditure. Because of limited data, the TSH and norepinephrine rhythms are more difficult to interpret. The complex circadian oscillator may have two components, a morning and an evening oscillator.

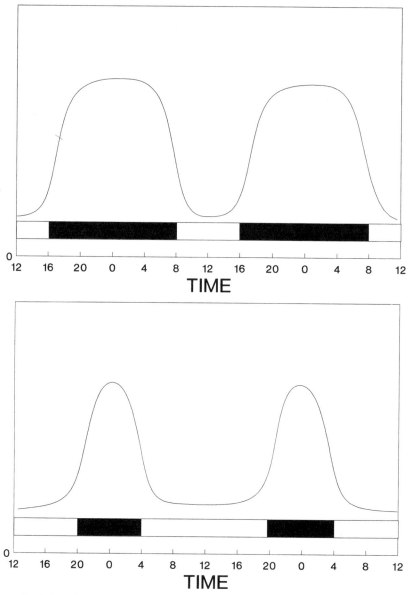

Figure 5. Schematic representation of the effect of an 8-hr photoperiod and a 16-hr photoperiod on melatonin secretion according to animal studies. Adapted from Arendt et al. (1989).

V. Circadian Neuroendocrine Rhythms in Seasonal Affective Disorder

Neuroendocrine abnormalities in depression have been the subject of intense interest over the past 20 years. Hypercortisolemia, dexamethasone nonsuppression, elevated levels of CRF and TRH in the cerebrospinal fluid (CSF), decreased responsiveness of TSH to TRH stimulation, elevated CSF levels of TRH, decreased circadian amplitude of the TSH circadian rhythm, and a blunted melatonin rhythm all have been found in depressed patients (Brown et al., 1984; Souetre et al., 1988; Nemeroff, 1989). However, in these studies, the depressed patients were not characterized according to seasonality, or the abnormalities were found to be more common in patients with an endogenous melancholic type of depression.

Pioneering melatonin studies by Lewy sparked interest in the neuroendocrine function of seasonal affective disorders. Prior to Lewy's finding that bright light suppresses melatonin production in humans (Lewy, Wehr, Goodwin, Newsome, & Markey, 1980), the human melatonin production system was thought to be insensitive to light, since earlier attempts using lower light intensities had failed (Czeisler, Richardson, Zimmerman, Moore-Ede, & Wertzman, 1981). Lewy then found that the melatonin rhythm as measured by DLMO is phase delayed in patients with winter depression compared with nondepressed controls (Lewy et al., 1987; Sack et al., 1990). The DLMO for the SAD group was 1–2 hr later than that of the controls; with 2 hr of bright light from 6 A.M. to 8 A.M., not only did the symptoms improve but the DLMO normalized its timing. In the evening, 2 hr of bright light did not cause a significant clinical response and did not significantly shift the DLMO.

Checkley and co-workers (Checkley, Fancey, Winton, Corn, & Arendt, 1989) found that the melatonin rhythm was not changed significantly after a therapeutic dose of 3 hr of bright light from 7 A.M. to 10 A.M. and 3 hr from 8 P.M. to 11 P.M. However, the melatonin rhythm was not measured under dim light conditions, so a phase advance of the endogenous melatonin rhythm could have been obscured by the acute suppressing effect of light.

Other groups have examined noncircadian measures of endocrine activity. Joffe (1991) found normal levels of thyroid function (T_4, T_3, T_3RU, TSH, free T_4 index) in SAD patients and no significant changes after 2 weeks of successful phototherapy. Dexamethasone suppression tests and responsiveness of TSH to TRH stimulation are normal (James et al., 1986; Skwerer et al., 1989). Norepinephrine levels are normal (Rudorfer, Skwerer, Potter, & Rosenthal, 1990). Depue and colleagues found low serum prolactin in SAD patients in the afternoon (Depue et al., 1990).

We did a study using a constant routine to "unmask" the endogenous rhythms of cortisol, TSH, melatonin, and temperature in controls (N = 7) and winter depressives before (N = 9) and after (N = 8) bright light therapy (2500 lux from 6 A.M. to 8 A.M.; Avery et al., 1990c; Dahl et al., 1990). All subjects were drug-free women, except one in each group who was taking oral contraceptives. For 6 days prior to the constant routine, subjects were allowed to sleep only between the hours of 9 P.M. and 6 A.M. Following an adaptation night, subjects were sleep-deprived at bed rest, with light (60 lux) and temperature held constant for 27 hr. Isocaloric meals were administered every 2 hr. Rectal temperature was monitored continuously (1/min). Blood samples for cortisol and TSH were drawn every hour. Melatonin was drawn every half hour from 6 P.M. to midnight to determine the DLMO.

Preliminary analysis of the data indicates that the temperature acrophase of the winter depressives was significantly ($p < .05$) phase delayed relative to controls (5:56 P.M. vs 4.43 P.M.). With treatment with bright light from 6 A.M. to 8 A.M., the acrophase phase advanced ($p = .07$) from 5:56 P.M. to 4:32 P.M. The cortisol acrophase of the winter depressives was significantly ($p < .05$) phase delayed relative to controls (9:38 A.M. vs 8:08 A.M.). With the light treatment, the cortisol rhythm significantly ($p < .05$) phase advanced (10:03 A.M. vs 8:24 A.M.). The TSH acrophase was significantly ($p = .05$) phase delayed in the winter depressives compared with controls (4:08 A.M. vs 3:19 A.M.). With light treatment, a nonsignificant trend ($p = .09$) for the TSH acrophase to phase advance (4:16 A.M. vs 3:19 A.M.) was seen. A nonsignificant trend ($p = .08$) for the DLMO of the winter depressives to be phase delayed relative to controls (10:38 P.M. vs 9:38 P.M.) was seen, as was a nonsignificant trend ($p < .07$) for morning bright light to phase advance the DLMO (11:10 P.M. vs 9:35 P.M.). No significant differences were seen between the amplitudes of the rhythms or the mesors among the three groups. Significant correlations existed among the acrophases for the four variables, that is, those subjects with an early temperature peak were also likely to have an early peak for cortisol and TSH and an early DLMO.

Our data are consistent with the hypothesis that the neuroendocrine rhythms are phase delayed relative to their normal phase relationship to sleep, and that bright light in the morning is able to phase advance that rhythm—a change that is associated with clinical improvement. Additionally, the data suggest that temperature, cortisol, TSH, and melatonin are all part of the same circadian system.

In general, the data suggest that winter depression is physiologically different than endogenous melancholic depression. Characteristic hypercortisolemia, dexamethasone nonsuppression, and decreased responsiveness of TSH to TRH infusion usually are lacking in SAD patients. Further, the low amplitudes of circadian rhythms (Souetre et al., 1988) and reports of phase-advanced rhythms (Jarrett, Coble, & Kupfer, 1983; Wehr & Goodwin, 1983) seen in depressed patients in general do not appear to be common among SAD patients.

VI. The Phase-Delay Hypothesis

The phase-delay hypothesis states that the circadian clock *relative to sleep* is phase delayed (Lewy, Sack, Singer, White, & Hoban, 1989). The desynchronization of the clock and sleep through a phase delay of the clock is a plausible explanation for some of the symptoms of SAD. Although the phase delay is "only" 1–2 hr relative to normal, the desynchrony may occur for several weeks or months. The difficulty awakening in the morning that is reported by SAD patients may be analogous to a nondepressed person being forced to wake up 1-2 hr before their natural wake-up time for a sustained period of time. Such awakenings would occur before the body temperature has started to rise and before the REM, cortisol, TSH, and melatonin rhythms have reached their usual levels for a natural awakening. The thermostat of the body would still be at a low level. REM sleep would not have been able to exert fully its heat conservation and thermogenic effects. If the cortisol rhythm is phase delayed, the cortisol levels would be relatively low when the person awakens in the morning. The person would have difficulty generating energy needs through gluconeogenesis (see Figure 6). Melatonin levels would still be high and sleepiness and low energy would be predicted. The lethargy and hypersomnia seen in SAD patients is consistent with these altered neuroendocrine rhythms. The normal predictive homeostasis of the circadian neuroendocrine function is disrupted. The SAD patient is trying to awaken when, physiologically, it is "the middle of the night." Effective bright light therapy may work by phase advancing the neuroendocrine clock to its normal relationship with sleep so the

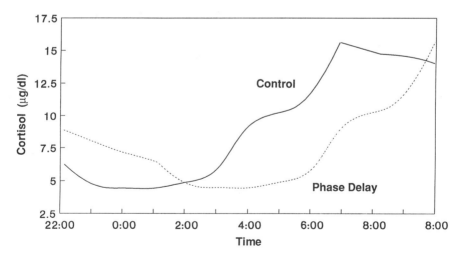

Figure 6. Schematic representation of a phase delay of the cortisol rhythm relative to a normal phase position. Note the low cortisol level that would be present at a 6 A.M. awakening.

person can wake up with normal high cortisol levels, rising temperature, and low melatonin levels.

Anecdotally, some winter depressives experience a reduction of symptoms if they are allowed to sleep until their natural awakening time. In two pilot studies, Lewy found that keeping the timing of the bright light exposure constant and phase delaying the wake-up time of winter depressives resulted in a remission of symptoms (Lewy et al., 1990). These data not only are consistent with the phase delay as a causal factor of winter depression, but also suggest that part of the problem in winter depression may be the early awakening in darkness that is demanded by the "8-to-5" world.

Why does SAD exhibit a phase delay of the circadian neuroendocrine system? The simplest explanation is the lack of morning light in the winter. The lack of morning light may allow the phase-delaying tendency of the endogenous 25-hr rhythm to express itself. In diurnal animals, the free-running tau in constant light conditions varies inversely with the light intensity; that is, constant darkness will be associated with a longer tau (Aschoff's Rule) (Pittendrigh & Daan, 1976b). Decreased sensitivity to light logically might be considered a risk factor for winter depression; the low light levels of the winter might be predicted to cause a light "deficiency" in susceptible people. Some investigators have hypothesized decreased retinal photoreceptor renewal mechanism (Remé, Terman, & Wirz, 1990). Another possibility is that SAD patients have supersensitivity to light, and the low intensity of indoor evening light in the winter is perceived as a high intensity phase-shifting signal, so a phase delay results (Beersma, 1990). Some researchers have shown increased light sensitivity in SAD patients (Oren, Joseph-Vanderpool, & Rosenthal, 1991b; Seggie et al., 1991). Others have found no difference (Murphy et al., 1989). Thompson and colleagues (Thompson, Stinson, & Smith, 1990) found that, relative to controls, SAD patients were supersensitive to the melatonin-suppressing effects of light in the winter but subsensitive in the summer.

Why do only some people develop a phase delay? Some individuals may be constitutionally or genetically vulnerable to SAD. Altered photosensitivity or an inherently long circadian tau, for example, could be risk factors for SAD. Alternatively, individuals vulnerable to winter depression may be unable to synchronize their circadian rhythms with nonlight stimuli whereas others may be able to use other stimuli such as exercise. However, ambient light and sleep schedules may influence whether the vulnerability manifests itself. For example, a person with SAD who wakes up at 5 A.M. and works in a dark office with no windows may have the same vulnerability as another person in the same city who wakes up at 8 A.M., works by a large window, and is asymptomatic. SAD probably results from the interaction of constitutional and environmental variables.

Some data are difficult to reconcile with the phase-delay hypothesis. Why is evening light reported to be effective in several studies? Terman and colleagues (1989a), in their meta-analysis, found that 38% of those receiving evening light (*N*

= 143) met strict criteria for response, compared with only 11% of the dim light controls (N = 77). The phase-delaying effects of evening light would be expected to worsen the winter depression. One possible explanation is the heterogeneity of SAD and the existence of some phase-advanced patients.

The phase-delay hypothesis could be tested further by using a nonlight stimulus to phase advance the clock to a normal phase. For example, using total darkness in the evening would, theoretically, phase advance the clock and would be predicted to improve the mood of SAD patients. In a similar way, an evening dose of melatonin might be predicted to phase advance the clock and cause improvement. If, in either case, the timing of the clock were normalized without improvement of symptoms, the phase-delay hypothesis could be falsified.

Is the phase delay of the circadian clock relative to sleep the cause of winter depression or merely an epiphenomenon? Association is not causation. Bright light may work by a mechanism other than shifting the clock so that the clock is more synchronous with sleep. The phase delay seen in winter depression and the phase advance with bright light may be only incidental findings.

VII. The "Desynchrony within the Clock" Hypothesis

Whereas the phase-delay hypothesis proposes a desynchrony between circadian rhythms and sleep, the "desynchrony-within-the-clock" hypothesis postulates desynchrony *within* the circadian system itself. The phase relationships between, for example, TSH and melatonin would be altered or unstable in SAD during the darkness of winter relative to the phase relationships under the clear light signals of summer. Light therapies then might work by "resetting" or synchronizing the circadian variables. Pittendrigh (1981) has postulated that, with a changing photoperiod, the phase relationships of circadian variables may change. This model was given some empirical support in humans by Wehr (1991, 1992), who studied circadian rhythms in normal controls under experimentally imposed long and short photoperiods. In our constant routine study, we found that the correlations among the temperature, melatonin, cortisol, and TSH phases in SAD subjects became greater after bright light therapy (Avery et al., 1990c). Although this hypothesis is very difficult to test, it deserves further study.

VIII. Other Theories of the Mechanism for Seasonal Affective Disorder

The short photoperiod hypothesis suggests that both the late dawn and the early dusk can create SAD symptoms. Not only would the morning oscillator need to be phase advanced, but the evening oscillator would need to be phase delayed for light therapy to be optimally effective. The phase difference between the melatonin onset and the melatonin decline would have to be decreased. This hypothesis is compatible with studies in animals that show that a shortened photoperiod can have a profound effect on behavior. Wehr and colleagues (1986) compared a short 8-hr skeleton photoperiod of bright light with a long 15.5-hr skeleton photoperiod and found no difference in clinical response. However, under the long photoperiod condition, the urinary 6-hydroxy-melatonin levels were decreased relative to the short photoperiod condition. Isaacs and co-workers (Isaacs, Stainer, Sensky, Moor, & Thompson, 1988) found midday bright light as effective as a long photoperiod. Checkley and colleagues (Checkley et al., 1989) found that a 16-hr photoperiod created by 3 hr of bright light in the morning and 3 hr in the evening was more effective than a 16-hr photoperiod created by 1 hr in the morning and 1 hr in the evening, although the changes in the melatonin rhythms were similar. In a small sample, Lewy and associates (Lewy et al., 1987) found that the DLMO (a marker for the evening oscillator) was phase delayed relative to controls and that morning bright light phase advanced the DLMO to a normal time. If the lack of evening light and a shift of the phase difference had been important for the development of SAD, the DLMO would have occurred earlier. However, Lewy and colleagues did not measure the melatonin decline at the end of the night, so the difference between the morning oscillator and evening oscillator may be increased in SAD, and morning light decreases that differences. Murphy and co-workers (Murphy et al., 1989) found no obvious differences in the melatonin profiles of winter depressives and controls. However, because of the data concerning the two-component model of the circadian system (Wehr, 1991), further studies of the short photoperiod hypothesis are necessary.

Czeisler and colleagues (Czeisler et al., 1987) proposed that bright light may work in winter depression by increasing the circadian amplitude. However, in our constant routine study, we found no differences in amplitudes of temperature, cortisol, or TSH among the controls, SAD patients, and recovered SAD patients. However, Rosenthal and co-workers (Rosenthal et al., 1990) found that, with light treatment, the amplitude of the temperature rhythm of ambulatory SAD patients increased. However, the baseline amplitude was no different from that of controls.

The studies showing no difference in efficacy with different photoperiods have given rise to a more general "photon-counting" hypothesis (Wehr et al., 1986). The photon-counting hypothesis postulates that the critical variable is the total amount of light received during the day. Light may have any number of direct effects on our physiology that are unrelated to circadian phase or amplitude. Lack of light may cause physiological changes that create SAD symptoms; bright light reverses those changes and relieves the symptoms. However, the photon-counting hypothesis must explain why the timing of light appears to be important, that is, why morning light is superior to evening light for hypersomnic winter depressives. would have to postulate that "morning photons" have a greater effect than "evening photons." Some studies have shown a circadian rhythm for the sensitivity to light (Bassi & Powers, 1986; O'Keefe & Baker, 1987).

Wurtman and co-workers (Wurtman & Wurtman, 1989) postulated that SAD may be a result of decreased serotonergic activity. Low serotonin has been linked with low mood and altered sleep. One of the common symptoms of SAD, carbohydrate craving, may have a close connection with low serotonin levels in the brain. Fernstrom and colleagues (Fernstrom & Wurtman, 1971) showed that carbohydrate ingestion increases brain serotonin content. The insulin release from the carbohydrates preferentially will increase the uptake of most amino acids but not tryptophan, a precursor of serotonin. Since serotonin must compete with other amino acids to cross the blood–brain barrier, the lower levels of other amino acids associated with insulin release favors the transport of tryptophan into the brain, where it can be converted to serotonin. Some drugs that increase serotonin release, such as *d*-fenfluramine, will suppress carbohydrate craving. In addition, *d*-fenfluramine (O'Rourke, Wurtman, Wurtman, Chebli, & Gleason, 1989) and tryptophan (McGrath, Buckwald, & Resnick, 1990) have been shown to be effective in treating SAD in double-blind placebo-controlled studies. Buspirone (Buspar®), a 5HT1A agonist, was found to be effective in treating SAD in an open study (Jacobsen, 1990). Anecdotally, many clinicians have seen SAD patients respond to fluoxetine (Prozac®). These findings raise the possibility that the carbohydrate craving seen in SAD is an attempt by the patient to increase low brain serotonin levels. Since serotonin is also a precursor for melatonin, this hypothesis overlaps with a low melatonin hypothesis. Serotonin increases prolactin secretion as well (Coccaro & Murphy, 1990) and prolactin levels are low in SAD (Depue et al., 1990).

The low serotonin hypothesis is also compatible with the phase-delay hypothesis. The serotonergic dorsal and median raphes provide a major input for the suprachiasmatic nucleus. Lesions involving this serotonergic innervation will disrupt circadian rhythms (Levine, Rosenwasser, & Yanovski, 1986; Banky, Molnar, Csernus, & Halasz, 1988). Serotonin agonists can shift circadian rhythms in rats (Miller, Dean, Edgar, Prosser, & Heller, 1991). In addition, clomipramine, a strongly serotonergic antidepressant, may facilitate entrainment of rats to a new light–dark schedule in comparison with desipramine, a strong noradrenergic antidepressant (Brown et al., 1988). If SAD patients have low brain serotonin levels, they might have increased difficulty entraining to the minimal light available in the fall and winter.

Depue and associates (Depue et al., 1990) hypothesized altered dopaminergic functioning in SAD patients, based on low serum prolactin levels and high spontaneous blink rates in SAD patients; these alterations are thought to be independent of season. Dopaminergic activity decreases prolactin secretion. Dopaminergic activity will increase blinking. No simple model emerges to explain the way in which the dopaminergic system might be altered in SAD.

Joseph-Vanderpool and co-workers (Joseph-Vanderpool et al., 1991) found that the CRF-induced increase in ACTH and cortisol was blunted in SAD patients compared with controls and that, after bright light treatment, the responses normalized. These researchers note that CRF increases arousal, locomotor activity, and anorexia in animal studies and that a deficiency of CRF would be compatible with the lethargy, increased sleep, and increased eating seen in SAD.

Explorers of the biological mechanisms of SAD encounter problems such as the heterogeneity of SAD and the dynamism of the symptoms. Although SAD is a much more homogeneous illness than depression in general, some differences in symptoms exist among SAD patients. Given the high prevalence of SAD in high latitudes and the heterogeneity of symptoms, SAD may not result from a single cause. For example, some individuals may have the disorder because of retinal supersensitivity whereas others may have the disorder because of high melatonin production. However, the fall and winter seasons create a seasonal pattern in both groups, and light therapy works for both groups. In addition, winter depression may change physiologically across the fall and winter. For example, for some patients, the carbohydrate craving and weight gain may occur predominantly in the fall and be absent in the winter season itself.

Investigators and clinicians also should recognize that some of these theories are not mutually exclusive. For example, decreased serotonergic function might coexist with a phase delay of the circadian rhythms relative to sleep. If this were true, a serotonergic agent such as fluoxetine rationally could be combined with a morning light treatment.

IX. Do Seasonal Affective Disorder Symptoms Have Evolutionary Significance?

The notion that winter depression may be analogous to hibernation has been mentioned by both patients and scientific investigators. The patients will report spontaneously that they are like bears in the winter. Investigators have noted that the increased caloric intake, weight gain, inactivity, and hypersomnia seen in winter depressives may be analogous to hibernation behaviors (Rosenthal et al., 1984; Wehr et al., 1986). Other scientists have criticized this idea (Mrosovsky, 1988; Zucker, 1988). Mrosovsky (1988) notes that "hibernation is absent or minimal in large mammals" and that the marked temperature decline seen in hibernating animals is not seen in humans. Even the bear only drops its temperature by a few degrees during its winter inactive period. Mrosovsky also lists a number of dissimilarities between SAD symptoms and hibernation behavior, including the unequal sex ratio in SAD patients. Interestingly, he notes that "there is little to suggest that photoperiod is an important zeitgeber for endogenous circannual rhythms in hibernators."

Although clearly humans do not hibernate, one could argue that humans have inherent adaptive physiological responses to the changing seasons that developed through evolution. The alternative null hypothesis is that the increased food intake, weight gain, and hypersomnia seen each fall and winter in as much as 30% of people at high latitudes have no relationship to evolution. However, one could speculate that even a large animal that did not increase its caloric consumption and gain weight in the fall and did not decrease its metabolic rate by increasing its sleep duration might be more likely to leave the gene pool during a harsh calorie-deficient winter.

Genotypes that were adaptive in human evolution may be detrimental in our modern world. The environment created by modern Western culture differs from the environments in which our genotypes evolved in several ways (see Table 1). Food availability, ambient temperature, exercise, light, and sleep patterns were probably quite different. All these variables may have significant impact on neuroendocrine function. Wehr (1989b) emphasized that environmental influences on affective disorder have been noted by clinicians over the past 2000 years, and could be very significant, but that the study of environmental factors has been neglected in general by psychiatrists in recent years. One example of a genotype that was once adaptive but now is detrimental is the "thrifty" genotype postulated by Neel (1962). This investigator hypothesized that changes in food availability and exercise pattern interact with certain genotypes to give rise to diabetes mellitus and obesity. A "thrifty" genotype would have been adaptive to feast-or-famine conditions and would have allowed greater insulin release and greater calorie storage. However, with progress and the feast-and-feast opportunities, such a genotype might be maladaptive and result in overproduction of insulin early in life, leading to obesity, increased insulin resistance, and symptoms of diabetes (Neel, 1962). Research on nondiabetic Pima Indians has shown that the resting metabolic rate is a familial trait and that a low resting metabolic rate is a risk factor for subsequent weight gain (Ravussin & Bogardus, 1990). Not only a thrifty

TABLE I

Summary of the Ways in Which Our Modern Environment Differs from the Environments in Which Humans Evolved

Evolution	Modern World
Feast–Famine	Feast–Feast
Temperature extremes	60–75°F
Exercise ++++	Exercise +
Changing photoperiod	Indoor light (dark)
Long sleep in winter	10 P.M. to 6 A.M. (or shift work)
Short sleep in summer	

assimilation of calories but also thrifty expenditure of calories may play a role in the development of obesity. A thrifty genotype may have made some individuals vulnerable to diabetes mellitus and obesity because of the changes of diet and exercise in modern world. Analogously, a "seasonal" genotype that may have used the changing photoperiod or changing light intensity to aid in calorie conservation may now be more vulnerable to SAD because of life-style changes of the modern world.

Relative to the conditions of evolution, our modern indoor world is dark. Using ambulatory light meters, studies in San Diego documented that modern humans receive an average of 90 min per day of light exposure above 2000 lux during a typical day (Savides, Messin, Senger, & Kripke, 1986). The light signals for the circadian clocks in more northern latitudes during the fall and winter are probably dim for most individuals. For some persons, modern culture allows the opportunity to create an "eternal winter" with respect to light exposure. The impact of the low light level of the modern world on our physiology and other disorders, such as obesity, should be explored further. Conversely, some individuals who are very sensitive to light may experience a 16 hour photoperiod across the entire year through exposure to artificial light. None of our more distant ancestors experienced such a photoperiod schedule.

As humans began to regiment their lives with clocks and develop regular work schedules, they inadvertently changes their sleep schedules. Even in the modern world, people tend to sleep longer in the winter than in the summer (Whybrow & Bahr, 1988; Terman et al., 1989b). As noted earlier, allowing winter depressives to "sleep in" in the morning often is associated with improvement in symptoms. The development of the electric light allowed workers to work at any time of the 24-hr period. Shift work has created an additional opportunity for desynchronization of sleep and the clock.

In our evolution, adaptation to temperature extremes was probably necessary; in the modern world, thermostatically controlled rooms create an artificially neutral thermal environment. One could speculate that this change has had a detrimental effect on thermoregulation. Further, the changes in food, exercise, sleep, and light exposure all may have thermoregulatory importance. The excessive caloric intake of feasting without fasting and the diminished caloric expenditure of a "couch potato" adds a thermal burden to modern humans. Thus, "progress" may have a profound effect on our thermoregulatory environment. Klerman (1988) has found evidence for an increasing prevalence of depression among those at risk for depression in the 20th century. Environmental factors should be among those variables studied in the search for an explanation of this trend.

The study of the physiological and psychological effects of light in SAD will be part of that search. A phase delay of the circadian neuroendocrine clock in relation to sleep appears to be common in SAD, and may explain several behaviors of SAD patients. The normal predictive homeostasis of core temperature, cortisol, and melatonin circadian rhythms may be disrupted. The SAD patients may be waking up at a time when, physiologically, they are not prepared for the metabolic demands of the waking state; the core temperature and cortisol level would be low and the melatonin level high. The phase delay of the neuroendocrine rhythms may be only one piece of the puzzle. However, the placement of the other pieces will depend on a better understanding of biological rhythms, both seasonal and circadian. The search will use Richter's views that behavior is grounded in biology, and that both biology and behavior are inherently rhythmic.

Acknowledgments

This research was supported in part by Research Scientist Development Award MHK5 K01 MH00493 and the Graduate School Research Fund of the University of Washington for David Avery and by an International Fellowship of the American Association of University Women for Kitty Dahl.

References

Akerstedt, T., & Gillberg, M. (1983). Circadian variation of catecholamine excretion and sleep. *European Journal of Applied Physiology, 51*, 203–210.

Akiskal, H., & McKinney, W. (1975). Overview of recent research in depression. *Archives of General Psychiatry, 32*, 285–305.

Allen, J. M., Lam, R. W., Remick, R. A., & Sadovnick, A. D. (1993). Depressive symptoms and family history in seasonal and non-seasonal mood disorder. *American Journal of Psychiatry, 150(3)*, 443–448.

Anton-Tay, F. D. J. L., & Fernandez-Guardiola, G. (1971). On the effect of melatonin upon human brains: Its possible therapeutic implications. *Life Sciences, 10*, 841–850.

Arendt, J., Broadway, J., Folkard, S., & Marks, M. (1989). The effects of light on mood and melatonin in normal subjects. In C. Thompson & T. Silverstone (Eds.), *Seasonal affective disorder* (pp. 133–144). London: Clinical Neuroscience Publishers.

Aschoff, J. (1981). Annual rhythms in man. In J. Aschoff (Ed.), *Handbook of behavioral neurobiology* (Vol. 4, pp. 475–487). New York: Plenum Press.

Avery, D. H. (1987). REM sleep and temperature regulation in affective disorder. In A. Halaris (Ed.), *Chronobiology and psychiatric disorders* (pp. 75–101). New York: Elsevier.

Avery, D. H., Bolte, M. A., Dager, S. R., Wilson, L. G., Weyer, M., Cox, G. B., & Dunner, D. L. (1993). Dawn simulation treatment of winter depression: A controlled study. *American Journal of Psychiatry, 150*, 113–117.

Avery, D. H., Dahl, K., Savage, M., Brengelmann, G., Larson, L., Vitiello, M., & Prinz, P. (1990a). The temperature rhythm is phase-delayed in winter depression. *Biological Psychiatry, 27*, 99A. (Abst.)

Avery, D. H., Kahn, A., Dager, D. R., Cox, G. B., & Dunner, D. L. (1990b). Bright light treatment of winter depression: AM compared to PM light. *Acta Psychiatrica Scandinavica, 82*, 335–338.

Avery, D. H., Khan, A., Dager, S. R., Cohen, S., Cox, G. B., & Dunner, D. L. (1990c). Is morning light superior to evening light in treating seasonal affective disorder? *Psychopharmacology Bulletin, 26*, 521–524.

Avery, D. H., Khan, A. D., Stephen, R., Cohen, S., Cox, G. B., & Dunner, D. L. (1991). Morning or evening light treatment of winter depression? The significance of hypersomnia. *Biological Psychiatry, 29,* 117–126.

Banky, Z., Molnar, J., Csernus, V., & Halasz, B. (1988). Further studies on circadian hormone rhythms after local pharmacological destruction of the serotoninergic innervation of the rat suprachiasmatic region before the onset of the corticosterone rhythm. *Brain Research, 445,* 222–227.

Bassi, C. J., & Powers, M. K. (1986). Daily fluctuations in the detectability of dim lights by humans. *Physiology and Behavior, 38,* 871–877.

Beck-Friis, J., Kjellman, B. F., Aperia, B., Unden, F., von Rosen, D., Ljunggren, J.-G., & Wetterberg, L. (1985). Serum melatonin in relation to clinical variables in patients with major depressive disorder and a hypothesis of a low melatonin syndrome. *Acta Psychiatrica Scandinavica, 71,* 319–330.

Beck-Friis, J., von Rosen, D., Kjellman, B. F., Ljunggren, J.-G., & Wetterberg, L. (1984). Melatonin in relation to body measures, sex, age, season and the use of drugs in patients with major affective disorders and healthy subjects. *Psychoneuroendocrinology, 9(3),* 261–277.

Beersma, D. G. (1990). Do winter depressives experience summer nights in winter? *Archives of General Psychiatry, 47,* 879–880.

Boyce, P., & Parker, G. (1988). Seasonal affective disorder in the southern hemisphere. *American Journal of Psychiatry, 145(1),* 96–99.

Bray, G. (1984). Integration of energy intake and expenditure in animals and man: The autonomic and adrenal hypothesis. *Clinical Endocrinology and Metabolism, 13(3),* 521–545.

Broadway, J., Arendt, J., & Folkard, S. (1987). Bright light phase shifts the human melatonin rhythm during the antarctic winter. *Neuroscience Letters, 79,* 185–189.

Brown, R., Kocsis, J. H., Caroff, S., Amsterdam, J., Winokur, A., Stokes, P. E., & Frazer, A. (1985). Differences in nocturnal melatonin secretion between melancholic depressed patients and control subjects. *American Journal of Psychiatry, 142,* 811–816.

Cagnacci, A., & Elliott, J. (1991). Melatonin a major regulator of the human circadian body temperature rhythm. *Meeting of The Endocrine Society,* (Abst.) 220.

Carman, J., Post, R., Buswell, R., & Goodwin, F. (1976). Negative effects of melatonin on depression. *American Journal of Psychiatry, 133(10),* 1181–1186.

Carney, P. A., Fitzgeral, C. T., & Monaghan, C. (1989). Seasonal variations in mania. In C. Thompson & T. Silverstone (Eds.), *Seasonal affective disorder* (pp. 19–27). London: Clinical Neuroscience Publishers.

Checkley, S. A., Fancey, C., Winton, F., Corn, T., & Arendt, J. (1989). Neuroendocrine study of the mechanism of action of phototherapy in seasonal affective disorder. In C. Thompson & T. Silverstone (Eds.), *Seasonal affective disorder* (pp. 223–232). London: Clinical Neuroscience Publishers.

Coccaro, E. F., & Murphy, D. L. (1990). Clinical significance of central serotonergic system dysfunction in major psychiatric disorders. In E. Coccaro & D. L. Murphy (Eds.), *Serotonin in major psychiatric disorders* (pp. 253–257). Washington, D.C.: American Psychiatric Press.

Czeisler, C. A., Kronauer, R. E., Mooney, J. J., Anderson, J. L., & Allan, J. S. (1987). Biologic rhythm disorders, depression, and phototherapy. *Psychiatric Clinics of North America, 10,* 687–709.

Czeisler, C. A., Kronauer, R. E., Allan, J. S., Duffy, J. F., Jewett, M. E., Brown, E. N., & Ronda, J. M. (1989). Bright light induction of strong (type O) resetting of the human circadian pacemaker. *Science, 244,* 1328–1333.

Dahl, K., Avery, D., Savage, M., Brengelmann, G., Kenny, M., Lewy, A., Larsen, L., Vitiello, M., & Prinz, P. (1990). Temperature, melatonin and TSH in seasonal affective disorder during a constant routine. *Meeting of the Society for Light Treatment and Biological Rhythms,* New York.

Depue, R. A., Arbisi, P., Krauss, S., Iacono, W. G., Leon, A., Muir, R., & Allen, J. (1990). Seasonal independence of low prolactin concentration and high spontaneous eye blink rates in unipolar and bipolar II seasonal affective disorder. *Archives of General Psychiatry, 47,* 356–364.

Eastman, C. I. (1990). What the placebo literature can tell us about light therapy for SAD. *Psychopharmacology Bulletin, 26,* 495–504.

Eastwood, M., & Stiasny, S. (1978). Psychiatric disorder, hospital admission, and season. *Archives of General Psychiatry, 35,* 769–777.

Elliott, J., Stetson, M., & Menaker, M. (1972). Regulation of testis function in golden hamsters: A circadian clock measures photoperiodic time. *Science, 178,* 771–773.

Fernstrom, J., & Wurtman, R. J. (1971). Brain serontonin content: Increase following digestion of carbohydrate diet. *Science, 174,* 1023–1025.

Fregly, M. J. (1989). Activity of the hypothalamic-pituitary-thyroid axis during exposure to cold. *Pharmacological Therapeutics, 41,* 85–142.

Garvey, M., Wesner, R., & Godes, M. (1988). Comparison of seasonal and nonseasonal affective disorders. *American Journal of Psychiatry, 145(1),* 100–102.

Gershon, E. S. (1990). Genetics. In F. K. Goodwin & K. R. Jamison (Eds.), *Manic-depressive illness* (pp. 373–401). New York: Oxford University Press.

Goldman, B., & Elliott, J. (1987). Photoperiodism and seasonality in hamsters: Role of the pineal gland. In M. H. Stetson (Ed.), *Processing of environmental information in vertebrates* (pp. 203–218). New York: Springer-Verlag.

Harrop, J., Ashwell, K., & Hopton, M. (1985). Circannual and within-individual variation of thyroid function tests in normal subjects. *Annals of Clinical Biochemistry, 22,* 371–375.

Isaacs, G., Stainer, D., Sensky, T., Moor, S., & Thompson, C. (1988). Phototherapy and its mechanisms of action in seasonal affective disorder. *Journal of Affective Disorders, 14,* 13–19.

Illnerova, H., Zvolsky, P., & Vanecek, J. (1985). The circadian rhythm in plasma melatonin concentration of the urbanized man: The effect of summer and winter time. *Brain Research, 328,* 186–189.

Jacobsen, F. M. (1990). Buspirone reverses winter worsening and potentiates antidepressant response. *Meeting of the Society for Light Treatment and Biological Rhythms,* New York.

Jacobsen, F. M., Wehr, T. A., Skwerer, R. A., Sack, D. A., & Rosenthal, N. E. (1987). Morning versus midday phototherapy of seasonal affective disorder. *American Journal of Psychiatry, 144,* 1301–1305.

James, S., Wehr, T., Sack, D., Parry, B., Rogers, S., & Rosenthal, N. (1986). The dexamethasone suppression test in seasonal affective disorder. *Comprehensive Psychiatry, 27(3),* 224–226.

Jarrett, D. B., Coble, P. A., & Kupfer, D. J. (1983). Reduced cortisol latency in depressive illness. *Archives of General Psychiatry, 40,* 506–511.

Joffe, R. (1991). Thyroid function and phototherapy in seasonal affective disorder. *American Journal of Psychiatry, 148(3),* 393.

Joseph-Vanderpool, J. R., Rosenthal, N. E., Chrousos, G. P., Wehr, T. A., Skwerer, R., Kasper, S., & Gold, P. W. (1991). Abnormal pituitary–adrenal responses to corticotropin-releasing hormone in patients with seasonal affective disorder: Clinical and pathophysiological implications. *Journal of Clinical Endocrinology and Metabolism, 72(6),* 1382–1387.

Kasper, S., Wehr, T., Bartko, J., Gaist, P., & Rosenthal, N. (1989). Epidemiological findings of seasonal changes in mood and behavior. A telephone survey of Montgomery County, Maryland. *Archives of General Psychiatry, 46(9),* 823–833.

Kauppila, A., Kivela, A., Pakarinen, A., & Vakkuri, O. (1987). Inverse seasonal relationship between melatonin and ovarian activity in humans in a region with a strong seasonal contrast in luminosity. *Journal of Clinical Endocrinology and Metabolism, 65(5),* 823–827.

Klerman, G. L. (1988). The current age of youthful melancholia: Evidence for increase in depression among adolescents and young adults. *British Journal of Psychiatry, 152,* 4–14.

Krieger, D. T. (1988). Abnormalities in circadian periodicity in depression. In D. J. Kupfer, T. H. Monk, & J. D. Barchas (Eds.), *Biological rhythms and mental disorders* (pp. 177–195). London: Guilford Press.

Landsberg, L., & Young, J. (1984). The role of the sympathoadrenal system in modulating energy expenditure. *Journal of Clinical Endocrinology and Metabolism, 13(3)*, 475–499.

Levine, J. D., Rosenwasser, A. M., & Yanovski, J. A. (1986). Circadian activity rhythms in rats with midbrain raphe lesions. *Brain Research, 384*, 240–249.

Lewy, A. J. (1983). Effects of light on melatonin secretion and the circadian system of man. In T. A. Wehr & F. K. Goodwin (Eds.), *Circadian rhythms and psychiatry* (pp. 203–220). Pacific Grove, California: Boxwood Press.

Lewy, A. J., Ahmed, S., Jackson, J. M., & Sack, R. L. (1991). A complete PRC for melatonin administration in humans. *Meeting of the Society for Light Treatment and Biological Rhythms*, Toronto, Ontario, Canada.

Lewy, A., Kern, H., Rosenthal, N., & Wehr, T. (1982). Bright artificial light treatment of a manic-depressive patient with a seasonal mood cycle. *American Journal of Psychiatry, 139*, 1496–1498.

Lewy, A. J., Sack, R. L., Miller, L. S., & Toban, T. M. (1987). Antidepressant and circadian phase-shifting effects of light. *Science, 235*, 352–354.

Lewy, A. J., Sack, R. L., & Singer, C. M. (1990). Bright light, melatonin, and biological rhythms in humans. In J. Montplaisir & R. Godbout (Eds.), *Sleep and biological rhythms: Basic mechanisms and applications to psychiatry* (pp. 99–112). New York: Oxford University Press.

Lewy, A., Wehr, T., Goodwin, F., Newsome, D., & Markey, S. P. (1980). Light suppresses melatonin secretion in humans. *Science, 210*, 1267–1269.

Lewy, A. J., Sack, R. L., Singer, C. M., White, D. M., & Hoban, T. M. (1989). Winter depression: The phase angle between sleep and other circadian rhythms may be critical. In C. Thompson & T. Silverstone (Eds.), *Seasonal affective disorder* (pp. 205–222). London: CNS (Clinical Neuro-science) Publishers.

Li, K., Reiter, R. J., Vaughan, M. K., Oaknin, S., Troiani, M. E., & Esquifino, A. I. (1987). Elevated ambient temperature retards the atrophic response of the neuroendocrine–reproductive axis of male syrian hamsters to either daily afternoon melatonin injections or to short photoperiod exposure. *Neuroendocrinology, 45*, 356–362.

Lieberman, H., Waldhauser, F., Garfield, G., Lynch, H., & Wurtman, R. (1984). Effects of melatonin on human mood and performance. *Brain Research, 323*, 201–207.

McGrath, R., Buckwald, B., & Resnick, E. V. (1990). The effect of L-tryptophan on seasonal affective disorder. *Journal of Clinical Psychiatry, 51(4)*, 162–163.

McGuire, R. A., Rand, W. M., & Wurtman, R. J. (1973). Entrainment of the body temperature rhythm in rats: Effect of color and intensity of environmental light. *Science, 181*, 956–957.

Miller, J., Dean, R., Edgar, D., Prosser, R., & Heller, C. (1991). Serotonin agonists reset the biological clock. *Sleep Research, 20A*, 549 (Abst.).

Minors, D. S., & Waterhouse, J. M. (1984). The use of constant routines in unmasking the endogenous component of human circadian rhythms. *Chronobiology International, 1(3)*, 205–216.

Moore, R. Y. (1990). The circadian system and sleep—Wake behavior. In J. Montplaisir & R. Godbout (Eds.), *Sleep and biological rhythms: Basic mechanisms and applications to psychiatry* (pp. 3–10). New York: Oxford University Press.

Moore-Ede, M. (1986). Physiology of the circadian timing system: Predictive versus reactive homeo-stasis. *American Journal of Physiology, 250*, R737–R752.

Moseley, M. J. (1988). Light transmission through the human eyelid: *In vivo* measurement. *Ophthalmic and Physiological Optics, 8(2)*, 229–230.

Mrosovsky, N. (1989). Seasonal affective disorder, hibernation and annual cycles in animals: Chip-munks in the sky. In N. Rosenthal & M. C. Blehar (Eds.), *Seasonal affective disorders and phototherapy* (pp. 127–148). New York: Guilford Press.

Murphy, D. G. M., Abas, M., Winton, F., Palazidou, L., Franey, C., Arendt, J., Binnie, C., & Checkley, S. A. (1989). Seasonal affective disorder: A neurophysiological approach. In C. Thompson & T.

Silverstone (Eds.), *Seasonal affective disorder* (pp. 233–242). London: Clinical Neuroscience Publishers.

Nair, N. P. V., Hariharasubraman, N., & Pilapil, C. (1984). Circadian rhythm of plasma melatonin in endogenous depression. *Progress in Neuro-Psychopharmacological and Biological Psychiatry, 8,* 715–718.

Neel, J. V. (1962). Diabetes mellitus: A "thrifty" genotype rendered detrimental by "progress"? *American Journal of Human Genetics, 14,* 353–362.

Nemeroff, C. (1989). Clinical significance of psychoneuroendocrinology in psychiatry: Focus on the thyroid and adrenal. *Journal of Clinical Psychiatry, 50(5),* 13–20.

Nimalasuriya, A., Spencer, C., Lin, S., Tse, J., & Nicoloff, J. (1986). Studies on the diurnal pattern of serum 3,5,3'-triiodothyronine. *Journal of Clinical Endocrinology and Metabolism, 62(1),* 153–158.

Norden, M. J., Avery, D., & Dahl, K. (1990). Serotonin and thermoregulation in seasonal affective disorder. *Meeting of the Society for Light Treatment and Biological Rhythms,* New York.

O'Keefe, L. P., & Baker, H. D. (1987). Diurnal changes in human psychophysical luminance sensitivity. *Physiology and Behavior, 41,* 193–200.

O'Malley, B., Richardson, A., Cook, N., Swart, S., & Rosenthal, F. (1984). Circadian rhythms of serum thyrotrophin and body temperature in euthyroid individuals and their responses to warming. *Clinical Science, 67,* 433–437.

Oren, D. A., Brainard, G. C., Johnston, S. H., Joseph-Vanderpool, J. R., Sorek, E., & Rosenthal, N. E. (1991a). Treatment of seasonal affective disorder with green light and red light. *American Journal of Psychiatry, 148(4),* 509–511.

Oren, D., Joseph-Vanderpool, J., & Rosenthal, N. (1991b). Adaptation to dim light in depressed patients with seasonal affective disorder. *Psychiatry Research, 36,* 187–193.

O'Rourke, D., Wurtman, J. J., Wurtman, R. J., Chebli, R., & Gleason, R. (1989). Treatment of seasonal depression with *d*-fenfluramine. *Journal of Clinical Psychiatry, 50,* 343–347.

Parry, B. L., Berga, S. L., Kripke, D. F., Klanben, M. R., Laughlin, G. A., Yen, S. S., & Girin, J. C. (1990). Altered waveform of plamsa nocturnal melatonin secretion in premenstrual depression. *Archives of General Psychiatry, 47,* 1139–1146.

Palkovits, M. (1987). Anatomy of neural pathways affecting CRH secretion. *Annals of the New York Academy of Sciences, 512,* 139–148.

Piercy, J., & Lack, L. (1988). Daily exercise can shift the endogenous circadian phase. *Sleep Research, 17,* 393.

Pittendrigh, C. S. (1981). Circadian systems: General perspective. In J. Aschoff (Ed.), *Handbook of behavioral neurobiology* (Vol. 4, pp. 57–80). New York: Plenum Press.

Pittendrigh, C. S. (1988). The photoperiodic phenomena: Seasonal modulation of the "day within". *Journal of Biological Rhythms, 3,* 173–188.

Pittendrigh, C. S., & Daan, S. (1976a). A functional analysis of circadian pacemakers in nocturnal rodents. IV. Entrainment: Pacemaker as clock. *Journal of Comparative Physiology, 106,* 291–331.

Pittendrigh, C. S., & Daan, S. (1976b). A functional analysis of circadian pacemakers in nocturnal rodents. V. Structural complexity. *Journal of Comparative Physiology, 106,* 333–355.

Potkin, S., Zetin, M., Stamenkovic, V., Kripke, D., & Bunney, W. (1986). Seasonal affective disorder: Prevalence varies with latitude and climate. *Clinical Neuropharmacology, (Suppl.), 9,* 181–183.

Ralph, M., Foster, R., Davis, F., & Menaker, M. (1990). Transplanted suprachiasmatic nucleus determines circadian period. *Science, 247,* 975–978.

Ralph, M., & Menaker, M. (1988). A mutation of the circadian system in golden hamsters. *Science, 241,* 1225.

Ravussin, E., & Bogardus, C. (1990). Energy expenditure in the obese: Is there a thrifty gene? *Infusionstherapie, 17,* 108–112.

Reed, H., Silverman, E., Shakir, K., Dons, R., Burman, K., & O'Brien, J. (1990). Changes in serum triiodothyronine (T3) kinetics after prolonged antarctic residence: The polar T3 syndrome. *Journal of Clinical Endocrinology and Metabolism, 70(4)*, 965–974.

Remé, C., Terman, M., & Wirz, J. A. (1990). Are deficient retinal photoreceptor renewal mechanisms involved in the pathogenesis of winter depression? *Archives of General Psychiatry, 47(9)*, 878–879.

Reppert, S., Weaver, D., Rivkees, S., & Stopa, E. (1988). Putative melatonin receptors in a human biological clock. *Science, 242*, 78–81.

Richter, C. (1965). *Biological clocks in medicine and psychiatry.* Springfield, Illinois: Charles C. Thomas.

Romijn, H. (1978). The pineal, a tranquillizing organ? *Life Sciences, 23*, 2257–2274.

Rosen, L., Targum, S., Terman, M., Bryant, M., Hoffman, H., Kasper, S., Hamovit, J., Docherty, J., Welch, B., & Rosenthal, N. (1990). Prevalence of seasonal affective disorder at four latitudes. *Psychiatry Research, 31(2)*, 131–144.

Rosenthal, N., Carpenter, C., James, S., Parry, B., Rogers, S., & Wehr, T. (1986). Seasonal affective disorder in children and adolescents. *American Journal of Psychiatry, 143(3)*, 356–358.

Rosenthal, N., Levendosky, A., Skwerer, R., Joseph, V. J., Kelly, K., Hardin, T., Kasper, S., Della-Bella, P., & Wehr, T. (1990). Effects of light treatment on core body temperature in seasonal affective disorder. *Biological Psychiatry, 27(1)*, 39–50.

Rosenthal, N. E., Sack, D. A., Gillin, J. C., Lewy, A., Goodwin, F. K., Davenport, Y., Mueller, P., Newsome, D., & Wehr, T. (1984). Seasonal affective disorder: A description of the syndrome and preliminary findings with light therapy. *Archives of General Psychiatry, 41*, 72–80.

Rosenthal, N. E., & Wehr, T. A. (1987). Seasonal affective disorders. *Psychiatric Annals, 17*, 670–674.

Rudorfer, M., Skwerer, R., Potter, W., & Rosenthal, N. (1990). Biochemical actions of phototherapy in SAD: Relationship to response. *Biological Psychiatry, 27(9A)*, 158A.

Rusak, B., Robertson, H., Wisden, W., & Hunt, S. (1990). Light pulses that shift rhythms induce gene expression in the suprachiasmatic nucleus. *Science, 248*, 1237–1240.

Sack, D., James, S., Rosenthal, N., & Wehr, T. (1988). Deficient nocturnal surge of TSH secretion during sleep and sleep deprivation in rapid-cycling bipolar illness. *Psychiatry Research, 23(2)*, 179–191.

Sack, R. L., Lewy, A. J., White, D. M., Singer, C. M., Fireman, M. V., & Van Diver, R. (1990). Morning versus evening light treatment for winter depression: Evidence that the therapeutic effects of light are mediated by circadian phase shifting. *Archives of General Psychiatry, 47*, 343–351.

Savides, T., Messin, S., Senger, C., & Kripke, D. (1986). Natural light exposure of young adults. *Physiology and Behavior, 38*, 571–574.

Seggie, J., MacMillan, H., Griffith, L., Shannon, H. S., Martin, J., Simpson, J., & Steiner, M. (1991). Retinal pigment epithelium response and the use of the EOG and Arden ratio in depression. *Psychiatry Research, 36*, 175–185.

Shanahan, T., & Czeisler, C. (1991). Light exposure induces equivalent phase shifts of the endogenous circadian rhythms of circulating plasma melatonin and core body temperture. *Journal of Clinical Endocrinology and Metabolism, 73(2)*, 227–235.

Skwerer, R. G., Jacobsen, F. M., Duncan, C. C., Kelly, K. A., Sack, D. A., Tamarkin, L., Gaist, P. A., Kasper, S., & Rosenthal, N. E. (1989). Neurobiology of seasonal affective disorder and photo-therapy. In N. Rosenthal & M. C. Blehar (Eds.), *Seasonal affective disorders and phototherapy* (pp. 311–332). New York: Guilford Press.

Souetre, E., Candito, M., Salvati, E., Pringuey, D., Chambon, P., & Darcourt, G. (1986). 24-Hour profile of plasma norepinephrine in affective disorders. *Neuropsychobiology, 16*, 1–9.

Souetre, E., Salvati, E., Wehr, T., Sack, D., Krebs, B., & Darcourt, G. (1988). Twenty-four-hour profiles of body temperature and plasma TSH in bipolar patients during depression and during remission and in normal control subjects. *American Journal of Psychiatry, 145(9),* 1133–1137.

Swanson, L. W., Sawchenko, P. E., Lind, R. W., & Rho, J. H. (1987). The CRH motoneuron: Differential peptide regulation in neurons with possible synaptic, paracrine, and endocrine outputs. *Annals of the New York Academy of Sciences, 512,* 12–23.

Terman, J., Terman, M., Schlager, D., Rafferty, B., Rosofsky, M., Link, M., Gallin, P., & Quitkin, F. (1990). Efficacy of brief, intense light exposure for treatment of winter depression. *Psychopharmacological Bulletin, 26(1),* 3–11.

Terman, M., Remé, C., Rafferty, B., Gallin, P., & Terman, J. (1990). Bright light therapy for winter depression: Potential ocular effects and theoretical implications. *Photochemistry and Photobiology, 51(6),* 781–792.

Terman, M., Botticelli, S. R., Link, B. G., Link, M. J., Quitkin, F. M., Hardin, T. E., & Rosenthal, N. E. (1989a). Seasonal symptom patterns in New York: Patients and population. In C. Thompson & T. Silverstone (Eds.), *Seasonal affective disorder* (pp. 77–96). London: Clinical Neuroscience Publishers.

Terman, M., Terman, J., Quitkin, F., McGrath, P., Stewart, J., & Rafferty, B. (1989b). Light therapy for seasonal affective disorder. A review of efficacy. *Neuropsychopharmacology, 2(1),* 1–22.

Thase, M. E. (1989). Comparison between seasonal affective disorder and other forms of recurrent depression. In N. Rosenthal & M. C. Blehar (Eds.), *Seasonal affective disorders and phototherapy* (pp. 64–78). New York: Guilford Press.

Thompson, C., Franey, C., Arendt, J., & Checkley, S. A. (1988). A comparison of melatonin secretion in depressed patients and normal subjects. *British Journal of Psychiatry, 152,* 260–265.

Thompson, C., & Isaacs, G. (1989). Phototherapy in SAD: Is photoperiod extension necessary? In C. Thompson & T. Silverstone (Eds.), *Seasonal affective disorder* (pp. 159–168). London: Clinical Neuroscience Publishers.

Thompson, C., Stinson, D., & Smith, A. (1990). Seasonal affective disorder and season-dependent abnormalities of melatonin suppression by light. *The Lancet, 336,* 703–706.

Wehr, T. A. (1989). Seasonal affective disorders: A historical overview. In N. Rosenthal & M. C. Blehar (Eds.), *Seasonal affective disorders and phototherapy* (pp. 11–32). New York: Guilford Press.

Wehr, T. A. (1990). Effects of wakefulness and sleep on depression and mania. In J. Montplaisir & R. Godbout (Eds.), *Sleep and biological rhythms* (pp. 42–86). New York: Oxford University Press.

Wehr, T. A. (1991). The durations of human melatonin secretion and sleep respond to changes in daylength (photoperiod). *Journal of Clinical Endocrinology and Metabolism, 73(6),* 1276–1280.

Wehr, T. A., & Goodwin, F. K. (1983). Biological rhythms in manic-depressive illness. In T. A. Wehr & F. K. Goodwin (Eds.), *Circadian rhythms and psychiatry* (pp. 129–184). Pacific Grove, California: Boxwood Press.

Weht, T. A., Jacobsen, F. M., Sack, D. A., Arendt, J., Tamarkin, L., & Rosenthal, N. E. (1986). Phototherapy of seasonal affective disorder: Time of day and suppression of melatonin are not critical for antidepressant effects. *Archives of General Psychiatry, 43,* 870–875.

Wehr, T. A., Moul, D. E., Barbato, G., Giesen, H., Seidel, J., Bender, C., & Barker, C. (1992). Daily patterns of human hormones, temperature and sleep respond to changes in photoperiod. *Meeting of the Society for Light Treatment and Biological Rhythms,* Bethesda, Maryland.

Wehr, T., & Rosenthal, N. (1989). Seasonality and affective illness. *American Journal of Psychiatry, 146(7),* 829–839.

Wehr, T., Sack, D., & Rosenthal, N. (1987). Seasonal affective disorder with summer depression and winter hypomania. *American Journal of Psychiatry, 144(12),* 1602–1603.

Wehr, T. A., Skwerer, R. G., Jacobsen, F. M., Sack, D. A., & Rosenthal, N. E. (1987). Eye versus skin phototherapy of seasonal affective disorder. *American Journal of Psychiatry, 144,* 753–757.

Weitzman, E., Czeisler, C., & Moore-Ede, M. (1979). Sleep–wake, neuroendocrine, and body temperature circadian rhythms under entrained and non-entrained (free-running) conditions in man. In M. Suda, O. Hayaishi, & H. Nakagawa (Eds.), *Biological rhythms and their central mechanisms* (pp. 199–227). Amsterdam: Elsevier North-Holland Biomedical Press.

Wetterberg, L., Beck-Friis, J., Aperia, B., & Peterson, U. (1979). Melatonin/cortisol ratio in depression. *Lancet, 2,* 1361.

Wirz-Justice, A., & Arendt, J. (1979). Diurnal, menstrual cycle and seasonal indole rhythms in man and their modification in affective disorders. In J. Obiolos, C. Ballus & M. Gonzales (Eds.), *Biological psychiatry today* (pp. 294–302). Amsterdam: Elsevier North-Holland.

Whybrow, P., & Bahr, R. (1988). *The hibernation response.* New York: Avon Books.

Wurtman, R. J., & Wurtman, J. J. (1989). Carbohydrates and depression. *Scientific American, 260(1),* 68–75.

Zucker, I. (1988). Seasonal affective disorders: Animal models non fingo. *Journal of Biological Rhythms, 3,* 209–223.

Zulley, J., Wever, R., & Aschoff, J. (1981). The dependence of onset and duration of sleep on the circadian rhythm of rectal temperature. *Pflugers Archive, 391,* 314–318.

Index